DIGITAL SIGNAL PROCESSING
THEORY, APPLICATIONS, AND HARDWARE

Richard A. Haddad
Polytechnic University

and

Thomas W. Parsons
Hofstra University

Computer Science Press
An imprint of W. H. Freeman and Company New York

Library of Congress Cataloging-in-Publication Data

Haddad, Richard A.
    Digital signal processing : theory, applications, and hardware /
by Richard A. Haddad and Thomas W. Parsons.
        p.   cm. — (Electrical engineering communications and signal
processing series)
    Includes bibliographical references and index.
    ISBN 0-7167-8206-5
    1. Signal processing—Digital techniques.   I. Parsons, Thomas W.
II. Title.   III. Series.
TK5102.5.H23   1991
621.382'2—dc20   90-48064
                                                              CIP

Printed in the United States of America

Computer Science Press

An imprint of W. H. Freeman and Company
41 Madison Avenue, New York, NY 10010
20 Beaumont Street, Oxford OX1 2NQ, England

1 2 3 4 5 6 7 8 9 0 RRD 9 9 8 7 6 5 4 3 2 1

To our wives
Marion and Patricia

CONTENTS

■ ■

PREFACE

■ ■

There is now an abundance of textbooks on elementary digital signal processing (DSP), but relatively few going beyond the elements, and such books as there are tend to address relatively specialized areas. We feel there is a need for a text for a follow-up course in modern signal processing at the senior or graduate level. This book is designed to meet that need.

Such a text should go far beyond the traditional focus on the analysis and design of digital filters. Our view of DSP encompasses a broad range of signal conditioning and manipulation for the purposes of filtering, enhancement, storage, and transmission. Coverage of some of these topics has traditionally been relegated to texts on digital communication, and on estimation and control.

We have tried to give a representative coverage of advanced topics, or, in some cases, advanced aspects of more familiar topics. Examples of the former are orthogonal expansions for transform signal coding, two-dimensional DSP, and least-squares adaptive filters. Examples of the latter are non-uniform sampling and quantization, and fast transforms beyond the FFT. Any advanced book must also include a good deal about estimation of various kinds, particularly spectral estimation and parameter estimation, and we have included these as well. We have tried, where possible, to cover topics of interest whose main sources are still in the technical literature, and the reader will notice that in many cases the coverage is an attempt to render classic papers accessible to the graduate student.

Our approach has been, in general, to build from the fundamentals. We have avoided the theorem/proof approach, however, preferring to give informal explanations emphasizing clarity rather than rigor. The practicing engineer must be equipped with an intuitive feel for the techniques of his profession, and it is our belief that this approach is more likely to provide this intuitive feel. For those who wish to pursue these topics in the technical literature, we have provided references and suggestions for further reading.

Most discussions start with familiar material, partly by way of review and partly to provide a point of departure, and then go on to topics which are not normally part of a first course in DSP. For example, the chapter on fast algorithms starts with a review of the fast Fourier transform, which should be well known to the reader, and then goes on to less-familiar versions of the FFT and finally to the fast algorithms of

Rader, Good-Thomas, and Winograd. Similarly, Section 6.1 starts with a least-squares fit of a straight line and uses this elementary example as the basis for the more advanced material to follow, for example, least-squares FIR filter design and adaptive and optimal filters. We have followed this plan repeatedly in this book.

The two chapters on applications of DSP to speech and to image processing build upon the theory developed in Part I. This material can serve as a coherent bridge to specialized texts dedicated to these subjects and to the current literature in these fields.

Another objective of this text is to touch on some of the hardware development which has supported the explosive growth of DSP. The chapter on hardware is intended to convey a notion of how these algorithms are implemented on representative systems. It provides a survey of digital VLSI chips with particular attention to the TI TMS 320 family. We expect the advanced components of this family to remain current for several years.

We have relegated reviews of prerequisite material and of special ancillary topics to the Appendices. In this respect, the text is mostly self-contained. There are exercises at the ends of most of the chapters; these are intended to complement the textual material and to provide the instructor with problems whose solutions further elucidate the theory.

Chapter 1 and the supplementary material in the appendices provide a succinct review of discrete-time signal and system theory. Digital representation of signals is the topic of Chapter 2; after treatment of non-uniform sampling and quantization, we introduce orthogonal expansions as a prelude to transform coding.

We have devoted Chapter 3 to two-dimensional (2D) signals and spectra. Starting with the 2D sampling theorem, we develop the basic system theory in both spectral and frequency domains, followed by examples of FIR and IIR filter-design procedures. The chapter ends with a section on 2D random fields, which is subsequently referenced in Chapter 9 on image processing.

Fast transforms beyond the conventional radix-2 FFT constitute the subject matter of Chapter 4. The sections on Rader's prime-number transforms, and the brief digression into fast convolution algorithms provide the basis for the Winograd Fourier transform; the chapter ends with an introduction to number-theoretic transforms. Such background in number theory as this chapter requires is provided in an appendix.

Following a review of traditional FIR and IIR filter design in Chapter 5, we go on to describe multirate and polyphase filters, multiplierless filters, and median filters.

Chapter 6, is intended to provide a link between traditional DSP and optimal estimation theory. We employ the least-squares method to derive both fixed and adaptive FIR filters. Spectral and parameter estimation are the topics of Chapter 7. After a review of Markov processes, we present both traditional nonparametric spectral estimation and more recent parametric approaches, including the methods of Capon, Burg, and Pisarenko. The background in probability and stochastic processes is provided in an appendix.

The theory developed in Part I is used in the two chapters on applications. The preceding one-dimensional theory provides the underpinning for speech analysis and synthesis in Chapter 8. Chapter 9 similarly applies the 2D theory of Chapter 3, and other theory developed previously, to topics in image coding, compression, restoration, and enhancement. We also touch on topics of current research interest.

Chapter 10, on hardware, is a survey of special DSP chips, taking the TI TMS 320 family as representative. Examples of the hardware implementation of a linear FIR filter and of a median filter are described.

As this summary suggests, the text provides flexibility in the choice of topics for a one-semester course. Core topics might include signal representations, fast transforms, and spectral estimation. These could be followed by selections from among multirate signal processing, two-dimensional processing, adaptive filters, and one or both of the applications chapters.

We acknowledge our debt to Professor Stanley H. Smith of Stevens Institute of Technology, who wrote the chapter on DSP hardware. A word of thanks also goes to the reviewers who provided us with valuable criticism and suggestions. We are particularly grateful to Professor Raymond L. Pickholtz, the editor of the CSP Series on Signal Processing, who provided the initial impetus for the writing of this text. We also express our appreciation to the people at W. H. Freeman and Company, particularly to William Gruener, Nola Hague, and Stephen Wagley, Nancy Singer, and Julia De Rosa for their support and help, and to Carol Loomis, our copy editor, for her eagle eyes and unfailing patience.

Digital Signal Processing
Theory, Applications, and Hardware

Part I

THEORY

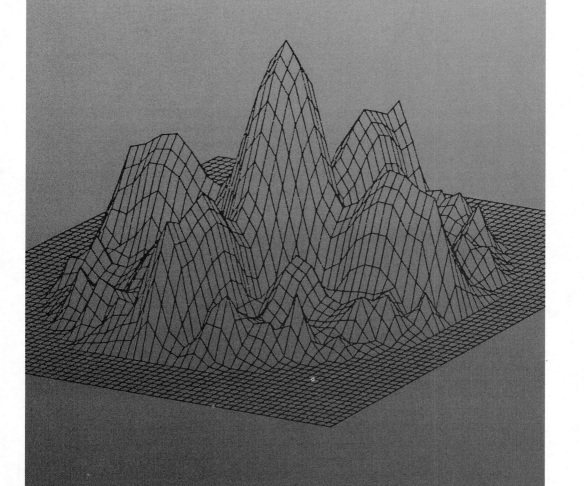

CHAPTER 1

■ ■

NUMERICAL OPERATIONS AND SYSTEM REPRESENTATION

This chapter is intended as a brief review of the material normally covered in a first course in digital signal processing. It is provided to familiarize the reader with our conventions of notation and usage and to establish a few principles to which we can refer later as they arise.

1.1 DISCRETE-TIME SIGNALS

A signal is a representation of information in a form that permits it to be stored, transmitted from one place to another, or otherwise manipulated. It can be thought of as a model of the information, and we obtain the information from the signal by observing its parameters in the light of our knowledge of the model. Historically, this model has generally taken the form of a function of time, although in applications such as image processing, the model may be a function of space (or, occasionally, of both space and time). In the first part of this book, we will for the most part assume that the signal is a function of time. The information conveyed by the signal may be evident, or it may be latent, masked by other, irrelevant data or by noise. Signal processing is the modification of the model, usually to extract or enhance the information it contains. Analog signal processing works with models in the form of continuous functions and typically applies them to systems made up of such things as inductors, capacitors, amplifiers, and delay lines. Digital signal processing uses models that are sequences of numbers and performs arithmetic operations on the numbers.

There are two principal problems in any kind of signal processing. The *analysis problem* is to evaluate the response of a signal processing system to a given input or class of inputs. A related problem is to ana-lyze the signal to extract information from it or about it. The *synthesis problem* is to design or specify a processing system which will produce a desired output from a given input.

As a starting point, we will consider signals themselves and how they can be represented for analysis purposes and then consider elementary processing operations and how they can be represented as numerical operations.

For analytical purposes, we will define a signal $f(t)$ as a real or com-plex function of the variable t. A signal is said to be discrete-time if it can mapped onto an ordered sequence of numbers $\{f(t_n)\}$. A discrete-time signal is thus completely defined by the corresponding sequence. Such a signal can be considered a time function defined on a set of discrete sample times $t_0, t_1, t_2, ...$, as shown in Fig. 1-1. We will use the terms "discrete-time signal" and "sequence" interchangeably, and we will often employ the notation $f_n \triangleq f(t_n)$.

In most practical applications, these sample times are equally spaced, and we will assume they are throughout this book, unless explicitly stated to the contrary. In this case, $t_n = nT$, where T is the sampling interval, and $f(t_n) = f_n = f(nT)$. Where no ambiguity will result, the interval T is often suppressed, or in effect, normalized to unity.

Discrete-time signals frequently result from sampling a continuous signal $f(t)$; in other cases, the signal may be inherently a sequence of numbers in time—for example, daily stock-market quotations. These two points of view coalesce when the signal is stored in a computer, where the sequence $\{f_n\}$ is a set of digital words stored in (usually) sequential memory locations.

It is occasionally important to distinguish between an entire se-quence and the value of the sequence at some particular time. In such cases, we will write $\{x(n)\}$ for the entire sequence and $x(n)$ for the value at time n. We will also use $x(n)$ for the entire sequence, however, when it is clear from context which is meant.

Discrete-time signals should be distinguished from discrete-ampli-tude, or *quantized*, signals, where the signal amplitude can assume only a discrete number of levels. Signals represented in a digital computer are always quantized, discrete-time signals. The term "digital signal" is used for such signals, although it is often used for any discrete-time signal whether quantized or not, and we will accept this usage when no

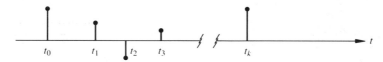

Figure 1-1   *Representation of discrete-time signal.*

confusion results. Similarly, an analog signal can be any continuous-time signal, whether quantized or not.

Quantization of an analog signal results in errors wherever its true amplitude has values falling between available quantization levels, and a complete analysis of a digital signal processing system requires an analysis of the effects of quantization errors. In practice, it is convenient to break this process into two steps, first designing and analyzing the system as if there were no quantization error and then estimating the effects of quantization errors. In addition, most processing techniques can be considered in isolation from quantization effects, and we will do so here, for the most part. Our justification for this is, that ignoring quantization errors simplifies the description greatly and that the analysis of quantization effects is well covered in most elementary texts. We will address quantization effects explicitly only in such cases as nonuniform quantization and bit allocation for maximum signal-to-quantizing-noise ratio.

1.2   OPERATIONS ON SEQUENCES

The processing of discrete-time signals can be described in terms of combinations of certain fundamental types of numerical operations on time sequences. In such a description, we assume a stream of time-indexed data fed into a device that performs certain operations on the data and outputs another data stream. We will assume that the operations are done instantaneously and with infinite precision. These operations are described in depth in App. A; we discuss only the most important ones here.

As a preliminary, we note that two sequences are equal if they have the same length and their corresponding samples are equal: $\{x(k)\} = \{y(k)\}$ if $x(k) = y(k)$ for all applicable values of k.

The simplest operations performed on sequences are: addition of two sequences, multiplication by a constant, shifts in time, multiplication of two sequences, and nonlinear operations.

1. *Addition.* Two sequences are summed by adding the corresponding values for each sample time. Formally, the sum of two sequences $\{x(n)\}$ and $\{y(n)\}$ is the sequence $\{z(n)\}$ where $z(n) = x(n) + y(n)$. Subtraction is similarly defined: $\{x(n)\} - \{y(n)\}$ is the sequence $\{z(n)\}$ where $z(n) = x(n) - y(n)$.

2. *Scalar multiplication.* This operation multiplies every sample of a sequence by a constant multiplier. If a is the multiplier, we may write

$$\{z(n)\} = a\{x(n)\} = \{ax(n)\} \tag{1-1}$$

3. *Time shifts.* A sequence may be delayed in time by storing the samples in a register or a delay line. The unit delay operator, E^{-1}, is defined as follows: If

$$\{y(n)\} = E^{-1}\{x(n)\} \qquad (1\text{-}2)$$

then each sample of $\{y(n)\}$ is equal to the previous sample of $\{x(n)\}$. That is,

$$y(n) = x(n-1) \qquad (1\text{-}3)$$

For example, if

$$x(n) = \begin{cases} w^n & n \geq 0 \\ 0 & n < 0 \end{cases}$$

then $y(n) = E^{-1}x(n) = x(n-1) = \begin{cases} w^{n-1}, & n-1 \geq 0 \\ 0, & n-1 < 0 \end{cases}$ $(1\text{-}4)$

Notice that a delay element provides us with a way of remembering a previous sample from a sequence. Systems with delay operators are said to possess *memory*.

4. *Time advance.* This is defined similarly to the delay operator. If

$$\{y(n)\} = E\{x(n)\}$$

then each y sample is equal to the forthcoming x sample:

$$y(n) = x(n+1)$$

or, more concisely,

$$E\{x(n)\} = \{x(n+1)\} \qquad (1\text{-}5)$$

5. *Multiplication.* Two sequences are multiplied by multiplying the corresponding elements. We write $\{z(n)\} = \{x(n)\} \cdot \{y(n)\} = \{x(n) \cdot y(n)\}$; that is, $z(n) = x(n) \cdot y(n)$.

6. *Arbitrary nonlinear operations.* Let $\phi(\cdot)$ be any nonlinear operator. Then we define

$$\phi\{x(n)\} = \{y(n)\} \qquad (1\text{-}6)$$

where $y(n) = \phi[x(n)]$

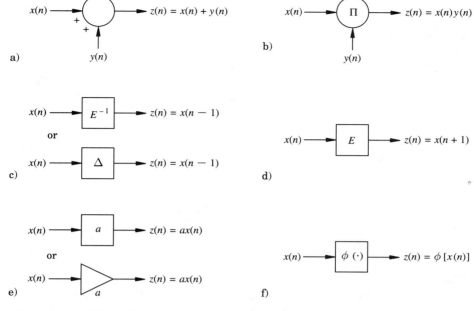

Figure 1-2   *Block-diagram representations of fundamental operations: (a) addition; (b) multiplication; (c) unit time delay; (d) unit time advance; (e) scalar multiplication (gain); (f) nonlinear operation. The use of Δ for delay (instead of difference) is potentially confusing but widespread.*

That is, the output sequence is obtained by applying the nonlinear operation to each sample of the input sequence.

These six operations are represented in block diagrams by the symbols shown in Fig. 1-2.

1.3 DISCRETE-TIME SYSTEMS

A *discrete-time system* is one in which some or all of the signals in the system are discrete-time signals. A digital computer is inherently a discrete-time system; but beyond this, a computer program for processing discrete-time signals can itself be thought of as a discrete-time system performing numerical operations on one or more input sequences $\{x_n\}$ to produce one or more output sequences $\{y_n\}$.

Such a system processes a signal (or, in some cases, a set of signals) by performing some combination of the basic operations on the signal. Discrete-time systems are classified in various ways:

Linear or nonlinear

Causal or noncausal

Time-dependent or time-invariant

In what follows, we will represent what the system does to its input by the operator $H[\cdot]$. Thus if the input is $\{x(n)\}$, then the output is given by $\{y(n)\} = H[\{x(n)\}]$, or, for any specific sample, $y(n) = H[x(n)]$.

A linear system has the property that the response to a linear combination of inputs is a corresponding linear combination of the responses to the individual inputs. In particular, for any constant multipliers a and b, if

$$H[x_1(n)] = y_1(n)$$

and

$$H[x_2(n)] = y_2(n)$$

then

$$H[ax_1(n) + bx_2(n)] = ay_1(n) + by_2(n) \qquad (1\text{-}7)$$

This means that the responses to two different inputs can be added, or superposed, to find the response to the composite input. This result can easily be generalized to any number of inputs; see Prob. 1-5.

This is the *superposition principle*; it is an important property, for we will see presently that any arbitrary signal can be represented as a weighted sum of elementary functions; then it must be possible to apply the superposition principle to represent the response to any arbitrary signal as a similarly weighted sum of the responses to these elementary functions.

A system is causal, or "physical," if its output cannot anticipate a change in its input. Formally, let $H[x]$ represent the output of the system for an input x. Then if x_1 and x_2 are two possible inputs and

$$x_1 = x_2 \qquad \text{for } t \leq t_0$$

then in a causal system,

$$H[x_1] = H[x_2] \qquad \text{for } t \leq t_0 \qquad (1\text{-}8)$$

Another way of stating this property is that the output $y(n)$ depends only on present and past values of the input. As the term "physical" implies, this is an inevitable constraint in physical hardware which must process the signal as it appears—that is, in real time. If immediacy of response is not necessary, however, we may choose to store the entire input signal in computer memory, in which case it is quite feasible to look ahead of the sample currently under consideration and take future samples into consideration. Furthermore, since image processing takes place in spatial dimensions rather than in time, the issue of causality does not arise in such applications. Various overlapping techniques are available for use when the signal is too long to fit in available memory.

A system is time-invariant if its response to a given input is independent of when the input arrives. Formally, if

$$H[x(n)] = y(n)$$

and the input is delayed (or advanced) by some number of samples k, then

$$H[x(n-k)] = y(n-k) \qquad\qquad (1\text{-}9)$$

The most important systems in DSP are linear, time-invariant (LTI) systems. Such a system can be described by a linear, constant-coefficient difference equation

$$y(n) = a_1 y(n-1) + a_2 y(n-2) + \cdots + a_N y(n-N)$$
$$+ b_0 x(n) + b_1 x(n-1) + b_2 x(n-2) + \cdots + b_M x(n-M) \quad (1\text{-}10)$$

where the coefficients $\{a_i\}$ and $\{b_j\}$ are all constant.

An equivalent representation is by means of a block diagram. Let

$$P(n) = b_0 x(n) + b_1 x(n-1) + b_2 x(n-2) + \cdots + b_M x(n-M)$$

Then $y(n) = a_1 y(n-1) + a_2 y(n-2) + \cdots + a_N y(n-N) + P(n)$

We can form $P(n)$ by passing the input sequence through a chain of M delays and computing the weighted sum of the delayed samples, as shown in Fig. 1-3.

We can similarly form $y(n)$ by passing the output sequence through a chain of N delays and adding $P(n)$ to the weighted sum of the delayed output samples, again as shown in Fig. 1-3.

Given the difference equation of any LTI system, it is possible to draw the block diagram; similarly, given the block diagram, we can construct the difference equation by noting the inputs to the delays and the adders.

A system in which the a coefficients in Eq. (1-10) are all zero has an impulse response of limited duration; this is because the impulse makes its way through the M input delays, after which it disappears and is lost forever so that the impulse response is identically zero from that time on. Such a system is called a finite-duration impulse-response (FIR) system. Any system in which any of the a coefficients is nonzero will have an impulse response which goes on forever, since output samples are recycled in the output delay chain and will continue to contribute to the impulse response. (There are exceptions to this; see Prob. 1-6 for one.) Such a system is called an infinite-duration impulse-response (IIR) system.

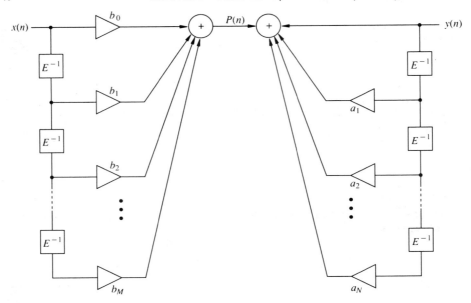

Figure 1-3 *Block-diagram representation of general difference equation.*

Using our delay operators, we may rewrite the difference equation as

$$(1 + a_1 E^{-1} + a_2 E^{-2} + \cdots + a_N E^{-N})y(n)$$
$$= (b_0 + b_1 E^{-1} + b_2 E^{-2} + \cdots + b_M E^{-M})x(n) \qquad (1\text{-}11)$$

Its solution has the operational form

$$y(n) = \frac{\sum_{k=0}^{M} b_k E^{-k}}{1 - \sum_{i=1}^{N} a_i E^{-i}} x(n) \qquad (1\text{-}12)$$

1.3.1 UNIT SIGNALS; CONVOLUTION

Signal analysis in the continuous domain has traditionally represented signals as linear combinations of certain elementary constituents. These elementary signals often serve as useful probe signals in characterizing the response of linear systems. The two most important elementary signals are the Dirac impulse function $\delta(t)$ and the complex exponential $e^{j\omega t}$. The response of a linear system to the delta function is termed the impulse response, while the response to the complex sinusoid yields the frequency response of the system.

    The digital impulse function corresponding to the Dirac delta is the sequence

$$\delta(n) = \begin{cases} 1 & n = 0 \\ 0 & \text{otherwise} \end{cases} \tag{1-13}$$

We may generalize this by noting that $\delta(n - k)$ is similarly zero every-where except at $n = k$, as shown in Fig. 1-4(a). We may write the corre-sponding sequence as $\{\delta(n-k)\} = \{...,0,0,1,0,0,...\}$ where the 1 appears in the kth time slot.

   Note: The terminology and notation used to describe the unit im-pulse both vary. The Dirac impulse takes its name from physics, and it can be thought of as the limit of a rectangular pulse of unit area as its width approaches zero. A rigorous and illuminating discussion of the Dirac impulse may be found in A. Papoulis (1962), App. I. The corresponding discrete-time function is sometimes called the Kronecker impulse, since we may write

$$\delta(n - k) = \delta_{nk}$$

where δ_{nk} is the Kronecker delta. Some authors prefer to reserve "unit impulse" and $\delta(t)$ for continuous-time systems and to call the discrete-time equivalent the "unit pulse," written $d(n)$. In this case, the response to $d(n)$ is called the "unit-pulse response." We find these terms verbose and misleading, since to most engineers a pulse is a rectangular function of nonzero duration. Hence we will use "impulse" and δ in both cases; it will usually be clear from context which is meant, and when it isn't, we will specify. In particular, $\delta(t)$ will be the Dirac impulse and $\delta(n)$ will be the Kronecker impulse. A sequence of the form, $\{...,0,0,1,0,0,...\}$ can also be termed a Kronecker sequence.

   The usefulness of this elementary function may be seen by noting that any sequence can be represented as a sum, or superposition, of

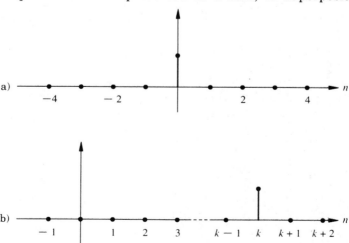

Figure 1-4   *Kronecker sequences: (a)* $\delta(n)$; *(b)* $\delta(n - k)$.

digital impulses. Consider any arbitrary sequence

$$\{f(n)\} = \{f(0), f(1), ...\}$$

defined for all $n \geq 0$. We may rewrite this as

$$
\begin{aligned}
\{f(n)\} &= \{f(0), 0, 0, ...\} + \{0, f(1), 0, ...\} + \cdots \\
&= f(0)\{1, 0, 0, ...\} + f(1)\{0, 1, 0, ...\} + \cdots \\
&= \{f(0)\delta(n)\} + \{f(1)\delta(n-1)\} + \cdots \\
&= \{\sum_{j=0}^{\infty} f(j)\delta(n-j)\}
\end{aligned}
\tag{1-14}
$$

This result can readily be extended to the case where $\{f(n)\}$ is defined over all time by changing the lower limit of the sum to $-\infty$:

$$f(n) = \sum_{j=-\infty}^{\infty} f(j)\delta(n-j) \tag{1-15}$$

Alternatively, by a change of dummy variables, we may write

$$f(n) = \sum_{j=-\infty}^{\infty} f(n-j)\delta(j) \tag{1-16}$$

Hence any arbitrary sequence is represented as a weighted sum of Kronecker sequences.

We can use the Kronecker sequence to find yet another characterization of linear systems. If we apply $\{\delta(n)\}$ to the input of any linear system, the output sequence is termed the impulse response of the system, analogously to of continuous-time systems. We reserve the symbol, $h(n)$ for the impulse response of an LTI system.

If any arbitrary function can be written as a weighted sum of unit impulses, it must be possible to represent the output of any LTI system as a similarly weighted sum of impulse responses. For if the system's impulse response is $h(n)$ and

$$x(n) = \sum_{j=-\infty}^{\infty} x(j)\delta(n-j)$$

then, by the principles of superposition and time invariance, we can substitute $h(n-j)$ for $\delta(n-j)$ to get

$$y(n) = \sum_{j=-\infty}^{\infty} x(j)h(n-j) \tag{1-17}$$

This operation is known as the *convolution* of $x(n)$ and $h(n)$; it is commonly written as $x(n) * h(n)$, and we may abbreviate Eq. (1-17) as

$$y(n) = x(n) * h(n)$$

The operation of convolution has a number of important properties, which we summarize here.

1. Convolution is commutative. For any a and b,

$$a(n) * b(n) = b(n) * a(n) \tag{1-18}$$

This can be seen by a simple change of variables in the convolution equation.

2. Convolution is associative. For any a, b, and c,

$$[a(n) * b(n)] * c(n) = a(n) * [b(n) * c(n)] \tag{1-19}$$

3. Not surprisingly, convolution is linear.

$$h(n) * [ax_1(n) + bx_2(n)] = ah(n) * x_1(n) + bh(n) * x_2(n) \tag{1-20}$$

This can be seen by substituting $ax_1(n) + bx_2(n)$ in Eq. (1-17). It follows from the linearity property that we may interchange the operations of convolution and summing.

The importance of convolution goes beyond this, however. All the elementary operations on a single sequence (except nonlinear operations) can be represented as convolutions with a suitable combination of elementary signals. It is left to the reader to verify that

Convolution with $\delta(n)$ leaves the sequence unchanged;
Scalar multiplication by a is equivalent to convolution with $a\delta(n)$;
The unit delay is equivalent to convolution with $\delta(n-1)$; that is,

$$\{x(n-1)\} = \{x(n) * \delta(n-1)\}$$

Similarly, an m-step delay is equivalent to convolution with $\delta(n-m)$;
An m-step time advance is equivalent to convolution with $\delta(n+m)$.
Summation is equivalent to convolution with the unit step

$$u(n) = \begin{cases} 1 & n \geq 0 \\ 0 & n < 0 \end{cases}$$

1.3.2 STABILITY

A system is said to be stable if its output is bounded for every bounded input. An LTI system is stable iff

$$\sum_{n=-\infty}^{\infty} |h(n)| < \infty \tag{1-21}$$

We say that $h(n)$ must be *absolutely summable*.

1.4 TRANSFORMS

Continuous-time systems are commonly analyzed with the aid of the Laplace transform or the Fourier transform. For discrete-time systems, the transform corresponding to the Laplace transform is the Z transform. There are two Fourier transforms used in the analysis of discrete-time systems; these will be described below.

1.4.1 Z TRANSFORMS

The two-sided Z transform is defined by

$$X(z) = \sum_{n=-\infty}^{\infty} x(n)z^{-n} \tag{1-22}$$

We will write the Z transform of a sequence $x(n)$ as $Z\{x(n)\}$ or occasionally simply $Z\{x\}$. Where feasible, we will also use lowercase letters for time-domain sequences and uppercase letters for transforms: thus $Z\{x(n)\} = X(z)$, as in the equation above. The sum, Eq. (1-22), may or may not converge for all values of z. The set of values for which the sum converges is known as the *region of convergence* (ROC) of the transform. For example, the sequence

$$x(n) = a^n u(n)$$

has the Z transform

$$X(z) = \sum_{n=0}^{\infty} a^n z^{-n}$$

$$= \frac{1}{1 + az^{-1}} \tag{1-23}$$

This sum converges for $|z| > a$; hence in the complex z plane, the ROC is the region outside the circle $|z| = a$.

The one-sided Z transform is defined similarly:

$$X(z) = \sum_{n=0}^{\infty} x(n)z^{-n} \tag{1-24}$$

The one-sided Z transform is of interest primarily in analyzing the transient response of a system. Since we are interested in general characterization of linear systems and discrete-time signals, we will use the two-sided transform. In particular, most of the properties listed below apply only to the two-sided transform.

Properties. These are discussed in detail in App. B. Here we review the most important ones briefly.

Linearity: Let $X(z) = Z\{x(n)\}$ and $Y(z) = Z\{y(n)\}$. Then it can be shown, from the defining equation, that for any constants a and b,

$$Z\{ax(n) + by(n)\} = aX(z) + bY(z) \tag{1-25}$$

As with the superposition principle for linear systems, this property can be generalized to linear combinations of any number of inputs.

Convolution: Let $y(n) = x(n) * h(n)$ and let the transforms of $y(n)$, $x(n)$, and $h(n)$ be $Y(z)$, $X(z)$, and $H(z)$, respectively. Then

$$Y(z) = X(z)H(z) \tag{1-26}$$

From the defining equation it follows that the Z transform of the unit impulse is 1:

$$Z\{\delta(n)\} = \sum_{n=-\infty}^{\infty} \delta(n)z^{-n} = 1 \tag{1-27}$$

since all terms in the sum vanish except z^0.

Similarly, the Z transform of a shifted impulse $\delta(n - k)$ is given by

$$Z\{\delta(n - k)\} = \sum_{n=-\infty}^{\infty} \delta(n - k)z^{-n} \tag{1-28}$$

$$= z^{-k}$$

$$\delta(n) = \begin{cases} 1 & n = 0 \\ 0 & \text{otherwise} \end{cases}$$

System function. From the definitions and properties of the Z transform, we can immediately rewrite the difference equation Eq. (1-10) in Z-transform notation. If $Z\{x(n)\} = X(z)$ and $Z\{y(n)\} = Y(z)$, then by the convolution property of the delta function, $x(n - m) = x(n) * \delta(n - m)$. But then by the convolution property of the Z transform,

$$Z\{x(n-m)\} = z^{-m}X(z) \qquad (1\text{-}29)$$

Hence the Z transform of our difference equation (1-10) becomes

$$Y(z) = (a_1 z^{-1} + a_2 z^{-2} + \cdots + a_N z^{-N})Y(z)$$
$$+ (b_0 + b_1 z^{-1} + b_2 z^{-2} + \cdots + b_M z^{-M})X(z) \qquad (1\text{-}30)$$

The solution can be written immediately as

$$Y(z) = \frac{b_0 + b_1 z^{-1} + b_2 z^{-2} + \cdots + b_M z^{-M}}{1 - (a_1 z^{-1} + a_2 z^{-2} + \cdots + a_N z^{-N})} X(z) \qquad (1\text{-}31)$$

Notice the similarity of this equation to Eq. (1-12). In the Z transform, z^{-1} takes the place of the delay operator, E^{-1}. Equation (1-31) may also be written as

$$Y(z) = H(z)X(z) \qquad (1\text{-}32)$$

where $H(z)$ is the fraction in Eq. (1-31). $H(z)$ is called the *system function*. The notation in Eq. (1-32) suggests that $H(z)$ is the Z transform of the impulse response, and so it is: for if $x(n)$ is the unit impulse, then $X(z) = 1$ and $Y(z)$ is simply $H(z)$.

The system function of an LTI system will always be a ratio of polynomials in z. The roots of the numerator polynomial are those values of z for which $H(z)$ will be zero; accordingly, they are the *zeroes* of $H(z)$. The roots of the denominator polynomial are those values of z for which $H(z)$ will be infinite; these are the *poles* of $H(z)$. In a causal system, since the polynomials are all in negative powers of z, the numerator will also provide M poles at $z = 0$ and the denominator will provide N zeroes at $z = 0$. If $M = N$, these poles and zeroes will cancel; otherwise there will be $M - N$ poles if $M > N$ or $N - M$ zeroes if $N > M$. In a noncausal system, there will be one or more positive powers of z in the numerator, which give rise to zeroes at the origin and poles at ∞. In all cases, the number of poles of $H(z)$ will be equal to the number of zeroes if singularities at 0 and ∞ are included.

If the numerator is a constant, then we speak of the filter as *all-pole*, and if the denominator is a constant, we call it *all-zero*, ignoring in both cases singularities at 0 and ∞. From the definitions above, it should be clear that an all-zero filter is an FIR filter, also called a *moving-average* (MA) filter. An all-pole filter is also called *autoregressive* (AR), and the general form Eq. (1-30) is also called an *autoregressive, moving-average* (ARMA) system.

Stability: The complex exponential

$$h(n) = a^n u(n) \qquad (1\text{-}33a)$$

has the Z transform

$$H(z) = \frac{1}{1 - az^{-1}} \qquad (1\text{-}33b)$$

Hence any single pole will give rise to such an exponential in the impulse response. The impulse response can thus be analyzed into a sum of sequences of the form of Eq. (1-33). For these sequences to be absolutely summable, $|a|$ must be less than unity. But $|a|$ gives the distance of the pole from the origin; hence for stability, all the poles must lie within the unit circle. (This formulation applies to causal systems; a more general requirement, which embraces noncausal systems as well, is that the ROC include the unit circle.)

1.4.2 FOURIER TRANSFORMS

The discrete-time Fourier transform (DTFT) of a sequence $\{x(n)\}$ is defined by

$$X(e^{j\omega}) = \sum_{n=-\infty}^{\infty} x(n)e^{-jn\omega} \qquad (1\text{-}34)$$

where $\omega = \Omega T_s$ is the normalized frequency in radians per sample. Note that this transform goes between a discrete-time domain and a continuous-frequency domain. We will write the Fourier transform of $x(n)$ as $X(e^{j\omega})$ or $\mathfrak{F}\{x(n)\}$ or, more concisely, $x(n) \leftrightarrow X(e^{j\omega})$. Note that for sequences having a Z transform whose ROC includes the unit circle, the Fourier transform is the Z transform evaluated on the unit circle.

We may motivate the use of the DTFT by recalling the use of the complex exponential $e^{j\omega t}$ as a probe signal in analyzing continuous-time systems. The corresponding discrete-time sequence is $e^{j\omega n}$. As with continuous systems, the response of a discrete-time system to a complex exponential is another complex exponential of the same frequency, changed only in magnitude and phase. The nature of these changes is given by the discrete-time Fourier transform. For if $x(n) = e^{j\omega_0 n}$, then by the convolution property,

$$y(n) = \sum_{k=-\infty}^{\infty} h(n)e^{j\omega_0(n-k)}$$

$$= e^{j\omega_0 n} \sum_{k=-\infty}^{\infty} h(n)e^{-j\omega_0 k}$$

$$= x(n)H(e^{j\omega_0})$$

$H(e^{j\omega_0})$ is a complex number which depends only on ω_0; hence the output exponential will indeed be the same as the input, but with its amplitude multiplied by the magnitude of $H(e^{j\omega_0})$ and with its phase shifted by the argument.

The inverse transform is given by

$$x(n) = \frac{1}{2\pi} \int_{-\pi}^{\pi} X(e^{j\omega})e^{jn\omega}\, d\omega \tag{1-35}$$

Properties.   These are discussed in detail in App. B. Here we review briefly the most important ones.

Periodicity: The transform is periodic in ω with period 2π. That is, for any integer n,

$$X(e^{j(\omega+2n\pi)}) = X(e^{j\omega}) \tag{1-36}$$

Linearity: Let $X(e^{j\omega}) \leftrightarrow x(n)$ and let $Y(e^{j\omega}) \leftrightarrow y(n)$. Then for any constants a and b,

$$ax(n) + by(n) \leftrightarrow aX(e^{j\omega}) + bY(e^{j\omega}) \tag{1-37}$$

Convolution: Let $x(n) \leftrightarrow X(e^{j\omega})$ and $h(n) \leftrightarrow H(e^{j\omega})$. Then

$$x(n) * h(n) \leftrightarrow X(e^{j\omega})H(e^{j\omega}) \tag{1-38}$$

Impulse functions: From the defining equation, we have

$$\mathcal{F}\{\delta(n)\} = 1 \tag{1-39a}$$

and
$$\mathcal{F}\{\delta(n-k)\} = e^{-jk\omega} \tag{1-39b}$$

Shifted sequences: If $x(n) \leftrightarrow X(e^{j\omega})$, then

$$x(n-m) \leftrightarrow e^{-jm\omega}X(e^{j\omega}) \tag{1-40}$$

This follows from the convolution properties of the unit impulse and the DTFT. Note that shifting simply imposes a linear phase ramp on the transform of the unshifted sequence.

Even-odd symmetry: If $x(n)$ is real, the magnitude and phase of $X(e^{j\omega})$ are respectively even- and odd-symmetric about $\omega = 0$ and $\omega = \pi$. In particular, if $x(n)$ is real and even, then $X(e^{j\omega})$ is likewise real and even. This has an important application in the design of digital filters. Any system for which $H(e^{j\omega})$ is real will have zero phase shift. Hence we have from this property the fact that a sufficient condition for zero phase shift is that the impulse response $h(n)$ be real and even.

If it is required that the filter be causal, then a real, even impulse response can be shifted in time to make it causal, provided its duration is finite. This shift means only that the output of the filter will be delayed by the amount of the shift. We have seen that a time shift imposes a linear phase ramp on the previously real Fourier transform; hence such filters are termed *linear-phase* filters. Thus if $h(n)$ is real and even, with a finite duration from $-N$ to N, the causal system

$$g(n) = h(n - N) \leftrightarrow e^{-j\omega N} H(e^{j\omega})$$

has the linear-phase term up front.

The discrete Fourier transform (DFT). This is obtained from the DTFT by sampling the latter at N equally spaced points on the unit circle:

$$X(k) = X(e^{j\omega})\Big|_{\omega = 2\pi k/N} \tag{1-41}$$

We may then write the defining equation of the DFT as follows: Let $\{x(n)\}$ be a sequence $\{x(0), x(1), ..., x(N-1)\}$. Then the DFT is given by

$$X(k) = \sum_{n=0}^{N-1} x(n)e^{-j2\pi nk/N} \qquad k = 0, 1, ..., N-1 \tag{1-42a}$$

Because of the ubiquity of the factor $e^{-j2\pi nk/N}$, it is common practice to use the abbreviation

$$W_N = e^{j2\pi/N}$$

Then the defining equation may be written

$$X(k) = \sum_{n=0}^{N-1} x(n)W_N^{-nk} \tag{1-42b}$$

The subscript N is frequently omitted when it is understood from context. Note that W_N is the principal Nth root of unity.

The inverse transform is given by

$$x(n) = \frac{1}{N} \sum_{k=0}^{N-1} X(k)W_N^{nk} \qquad n = 0, 1, ..., N-1 \tag{1-43}$$

We use the $\mathfrak{F}$ and double-arrow notations for the DFT as well: if $X(k)$ is the DFT of $x(n)$, then $X(k) = \mathfrak{F}[x(n)]$, or $X(k) \leftrightarrow x(n)$.

It is important to note that the DFT is cyclic with period N in both domains. This follows from the sampled nature of $x(n)$ and $X(k)$. We can account for this property in our equations most simply by restricting our domains to $(0, N - 1)$ and taking the results of all operations on indices modulo N.

Properties. As with the other transforms, the properties of the DFT are given in detail in App. B; we summarize only the most important here.

Linearity: Let $X(k) \leftrightarrow x(n)$ and $Y(k) \leftrightarrow y(n)$. Then for any constants a and b,

$$aX(k) + bY(k) \leftrightarrow ax(n) + by(n) \tag{1-44}$$

Shifting: When the sequence is shifted to the right, the vacated position at $n = 0$ is filled by the sample which was at $n = -1$ prior to the shift. But by the periodicity property, this is identical to the sample which was at $n = N - 1$ before the shift. Hence the effect of the shift can be viewed as a rotation in which the sample at $N - 1$ moves into position 0. Again, doing all arithmetic on indices and subscripts modulo N will build this property into our equations. If a sequence is shifted to the right by p samples, then the sample at point n will be moved to location $(n + p) \bmod N$.

With this understood, we may write the shifting property as follows: If $x(n) \leftrightarrow X(k)$, then

$$x(n + p) \leftrightarrow W^{-pk} X(k) \tag{1-45}$$

Convolution: The convolution of two sequences of length N is likewise cyclical, since it entails shifting; hence we must again take all indices and subscripts modulo N. Then cyclical convolution of the time functions corresponds to multiplication of the transforms: if $X(k) \leftrightarrow x(n)$ and $Y(k) \leftrightarrow y(n)$, then

$$x(n) * y(n) \leftrightarrow X(k)Y(k) \tag{1-46}$$

Symmetry: If $X(k) \leftrightarrow x(n)$, then

$$X(n) \leftrightarrow Nx(-k) \tag{1-47}$$

1.5 SUMMARY

This completes our review of elementary discrete-time signals and systems. The material presented here represents the bare minimum of background in these topics. For more details, the reader is referred to Apps. A and B and to the many books which have been written on the subject, some of which are listed in the References.

PROBLEMS

1.1. Let $x(n) = nu(n)$. Sketch:
(a) $x(n) + u(n)$
(b) $2x(n)$
(c) $E^{-1}\{x(n)\}$
(d) $E\{x(n)\}$
(e) $\sqrt{\{x(n)\}}$

1.2. Given the following systems, defined by their input-output relations:
(a) $y(n) = ax(n) + ax(n/2) + y(n-1)$
(b) $y(n) = a^2 x(n) + abx(n-1) - b^2 y(n-1)$
(c) $y(n) = x(2n) - x(n-1)$
(d) $y(n) = [x(n)]^2 + y(n-2)$
Which of these systems are linear? Which are time-invariant? Which are causal?

1.3. Draw the block diagrams corresponding to the following difference equations:
(a) $y(n) = x(n) - 3x(n-1) + 2x(n-2)$
(b) $y(n) = x(n) + y(n-1) - y(n-2) + 2y(n-3)$
(c) $y(n) = x(n) + 2x(n-1) - x(n-3) - 4y(n-4)$

1.4. Write the difference equations for the systems illustrated by the block diagrams in Fig. 1-P1.

1.5. Using the principle of mathematical induction, show that the superposition property can be extended to cover any number of inputs.

1.6. Given the system

$$y(n) = x(n) - x(n-8) + y(n-1)$$

show that, in spite of its appearance, this system has a finite-duration impulse response.

1.7. (a) Show that the system $h(n) = (-1)^n u(n)$ is unstable.
(b) For what values of a is the system $h(n) = a^n u(n)$ stable?

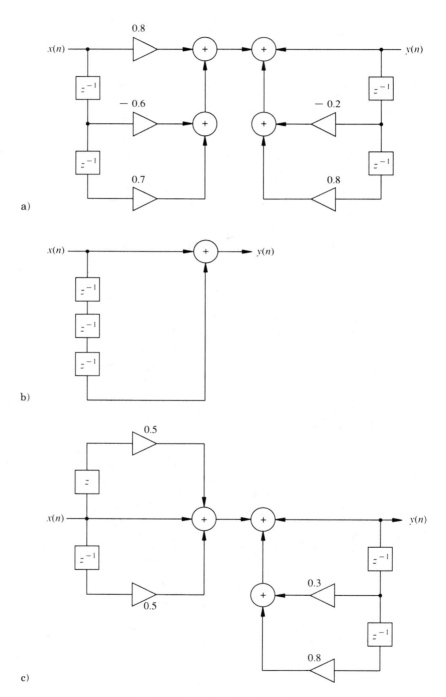

Figure 1-P1   *Problem 1.4.*

    (c) Show that any system with a bounded, finite-duration impulse response is stable.

1.8.   Prove the associativity property for convolution, Eq. (1-19).

1.9.   Write the system functions for each of the systems in Prob. 1.3.

1.10.   Use Z transforms to compute the convolution of the sequences, $x(n) = \{1, 3, 2, 4, 5\}$ and $y(n) = \{1, 0, 0, -1\}$. (Both of these sequences begin at $n = 0$.)

1.11.   Find the DTFT of the sequence,

$$x(n) = \begin{cases} 1 & n = 2 \\ -1 & n = -2 \\ 0 & \text{elsewhere} \end{cases}$$

1.12.   Let $x(n)$ be 1 for $0 \le n < N$ and 0 everywhere else. Show that

$$|X(e^{j\omega})| = \frac{\sin N\omega/2}{\sin \omega/2}$$

1.13.   Prove that if $x(n)$ is a real, odd sequence, then $X(e^{j\omega})$ is an imaginary, odd function of ω.

1.14.   Prove the symmetry property, Eq. (1-47).

1.15.   Let

$$y(n) = \sum_{k=n-N}^{n} a^{n-k} x(k)$$

    (a) Evaluate $h(n)$ and sketch. What type of response is this?
    (b) Obtain the difference equation relating $\{y(n)\}$ and $\{x(n)\}$.
    (c) Show a block-diagram realization of this difference equation.
    (d) Evaluate the system function $H(z)$ in closed form. Show the pole-zero pattern.
    (e) Show that

$$H(e^{j\omega}) = e^{-(N/2)(b+j\omega)} \frac{\sinh(M/2)(b+j\omega)}{\sinh(1/2)(b+j\omega)}$$

    where $M = N + 1$ and $a = e^{-b}$.
    (f) Let $a = 1$. Evaluate $H(e^{j\omega})$ and sketch.
    (g) Comment on the stability of this system.

1.16.   Let

$$y(n) = \sum_{k=\infty}^{n} a^{n-k} x(k)$$

Repeat (a) through (g) of Prob. 1.15 and compare.

1.17.   Let $H(z) = 1 - z^{-N}$. (This is a comb filter.)
     (a) Show a block-diagram realization of $H(z)$; show the pole-zero pattern.
     (b) Evaluate and sketch $h(n)$.
     (c) Show that $H(e^{j\omega}) = 2je^{-jN\omega/2} \sin N\omega/2$.
     (d) Sketch $|H(e^{j\omega})|$ *versus* ω and comment on the name "comb filter."
     (e) Sketch $\arg[H(e^{j\omega})]$ *versus* ω.

1.18.   Given an ideal, low-pass digital filter

$$H(e^{j\omega}) = \begin{cases} 1 & |\omega| \leq a \\ 0 & a < |\omega| \leq \pi \end{cases}$$

evaluate and sketch $h(n)$; comment on causality.

1.19.   Using Z transforms, solve the difference equation for this Fibonacci-number generator:

$$y(k) = y(k-1) + y(k-2)$$

starting values $y(0) = 1$ and $y(1) = 1$.

1.20.   Let

$$x(n) = \left(\frac{1}{2}\right)^{|k|} \quad \text{and} \quad h(n) = a^n u(n)$$

solve for $y(n)$ (a) by convolution and (b) by transforms.

1.21.   Let $y(n) = 1.6y(n-1) - 0.64y(n-2) + x(n-1)$.
     (a) Evaluate $H(z)$ and show a block-diagram realization.
     (b) Evaluate and sketch $h(n)$.
     (c) Evaluate the steady-state response if $x(n) = 4\cos(\pi/8)n$.

1.22.   (a) Given $x(n) = \cos L(2\pi/N)n$, $L < N/2$; evaluate and sketch the DFT.

(b) Given that a periodic sequence has the DFT

$$X(k) = \frac{\sin[(2p+1)k\pi]/N}{\sin k\pi/N}$$

evaluate $\sum_{k=0}^{N-1} |X(k)|^2$.

1.23. A system with the system function

$$H(z) = \frac{1}{1 - \frac{1}{2}z^{-1}} \qquad |z| > \frac{1}{2}$$

is excited by the input

$$x(n) = 2\cos(\pi n/20 + \pi/6) + 2\delta(n-2)$$

Evaluate the response $y(n)$ for all n.

1.24. The sequence

$$x(n) = \sum_{p=-\infty}^{\infty} \delta(n - pN) \qquad N = 8$$

is applied to the system with the system function shown in Fig. 1-P2. Evaluate and sketch the DFT of the periodic output sequence y(n). From this DFT, evaluate $y(0)$ and the sum

$$\sum_{n=0}^{7} |y(n)|^2$$

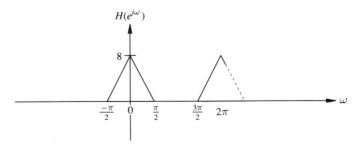

Figure 1-P2 *Problem 1.24.*

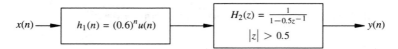

Figure 1-P3. *Problem 1.25.*

1.25.   Given the tandem connection of filters shown in Fig. 1-P3; if the input is as given in Prob. 1.24, evaluate the DFT of the output sequence $y(n)$.

REFERENCES

N. Ahmad and T. Natarajan. *Discrete-Time Signals and Systems.* Reston, Va.: Reston, 1983.

J. A. Cadzow. *Discrete-Time Systems.* Englewood Cliffs, N. J.: Prentice-Hall, 1973.

H. Freeman. *Discrete-Time Systems.* New York: John Wiley, 1965.

K. Ogata. *Discrete-Time Control Systems.* Englewood Cliffs, N. J.: Prentice-Hall, 1987.

A. V. Oppenheim and R. W. Schafer. *Discrete-Time Signal Processing.* Englewood Cliffs, N. J.: Prentice-Hall, 1989.

R. A. Schwarz and B. Friedland. *Linear Systems.* New York: McGraw-Hill, 1965.

CHAPTER 2

■ ■

DIGITAL REPRESENTATION
OF SIGNALS

Data for digital signal processing may occasionally be inherently discrete series of numbers. In many applications, however, they are quantized samples of a continuous signal. In this chapter we will consider how these samples are obtained and the various ways in which quantization may be done. Continuous data are normally sampled at uniform time intervals and at a high rate. We consider these conditions first, and then examine nonuniform sampling and, briefly, sampling at low rates. Finally, we will discuss various ways in which the samples, once obtained, may be represented in forms which can be handled more conveniently or transmitted (or stored) more economically.

2.1 SAMPLING

The starting point for analog-to-digital conversion systems is obtaining a sequence of sample values. For the moment we will assume that these samples are exact amplitudes, and we will address quantization later. The risk of such a sampling process is, of course, that some important event may occur between sample times and thus be missed. The sampling theorem (Shannon, 1949) states that if $f(t)$, the function being sampled, is band-limited—that is, if it contains no frequency components above some bound f_c (in hertz)—then it can be completely described by uniformly spaced samples taken at a rate of at least $2f_c$ samples per second.

In order to prove this theorem, we must define what we mean by a function being completely described by its samples. We will say that a function is completely described by its samples if an exact reconstruction of the continuous function can be obtained from the samples.

Let the function be real, with a continuous Fourier transform $F(\omega)$, as shown in Fig. 2-1(a). Then

$$f(t) = \frac{1}{2\pi} \int_{-\infty}^{\infty} F(\omega)e^{j\omega t}\, d\omega \qquad (2\text{-}1)$$

Let us consider what happens when we sample $f(t)$ at uniformly spaced times. If the sampling frequency is f_s, then we can model this as a multiplication of $f(t)$ by a train of Dirac impulses spaced $T_s = 1/f_s$ seconds apart:

$$\hat{f}(t) = f(t)T_s \sum_k \delta(t - kT_s) \qquad (2\text{-}2)$$

(We are still in the analog world here, and so must use Dirac impulses.) But multiplication by an impulse train in the time domain corresponds

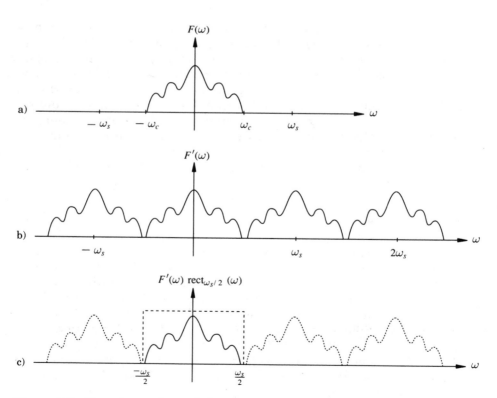

Figure 2-1 *Transforms of sampled signals: (a) signal before sampling; (b) sampled signal; (c) recovered signal.*

to convolution with an impulse train in the frequency domain: if the transform of the sampled sequence is $\hat{F}(\omega)$, then

$$\hat{F}(\omega) = F(\omega) * \sum_{k=-\infty}^{\infty} \delta(\omega - k\omega_s) = \sum_{k=-\infty}^{\infty} F(\omega - k\omega_s) \qquad (2\text{-}3)$$

Hence the Fourier transform of our samples is an infinitely repeated replication of $F(\omega)$ at intervals of $\omega_s = 2\pi f_s$, as shown in Fig. 2-1(b). We will have occasion later to speak of these replications at some length; we will call the portion of the transform between $-\omega_s/2$ and $\omega_s/2$ the *baseband* and all other replications *images*. If $f(t)$ is band-limited so that $F(\omega)$ is zero for $|\omega| > 2\pi f_c$, and if $f_s \geq 2f_c$ (Shannon's rate), then there is no overlap between successive replications. We have lost no information in the sampling process; if anything, we have picked up a lot of spurious "information" at frequencies outside the range of interest.

To recover the original signal, we must remove the images. But $F(\omega)$ can be obtained from $\hat{F}(\omega)$ by multiplying it by a rectangle:

$$\text{rect}_{\omega_s/2}(\omega) = \begin{cases} 1 & |\omega| < \omega_s/2 \\ 0 & |\omega| > \omega_s/2 \end{cases}$$

as shown in Fig. 2-1(c). (This is done by an analog filter known as an interpolation filter.) We are now back to $F(\omega)$, and the time function can be recovered in accordance with Eq. (2-1). But multiplication by a rectangle of width ω_s in the frequency domain is equivalent to convolution with $(\omega_s/2\pi)\sin(\omega_s t/2)/(\omega_s t/2)$ in the time domain:

$$\hat{F}(\omega)\,\text{rect}_{\omega_s/2}(\omega) \leftrightarrow \hat{f}(t) * \frac{\omega_s}{2\pi}\frac{\sin(\omega_s t/2)}{(\omega_s t/2)} \qquad (2\text{-}4)$$

Hence we conclude:

1. The signal can be sampled at a rate of $f_s \geq 2f_c$ samples per second with no loss of information.

2. The original signal can be reproduced exactly by convolving the sample sequence with $(\omega_s/2\pi)\sin(\omega_s t/2)/(\omega_s t/2)$.

The kernel, $\sin \omega t/\omega t$, is of considerable importance; it is frequently represented by the abbreviation $\text{sinc}(\omega t)$ [or alternatively $\text{sinc}_\omega(t)$], and we will use this notation here. (If the subscript is omitted, it is assumed to be 1.) Similarly, a rectangular pulse centered about the origin and of half-width T, is abbreviated $\text{rect}_T(t)$. The functions, $\text{rect}_T(t)$ and $2T\,\text{sinc}(\omega T)$, are a Fourier transform pair.

The notion that a sampling rate of $2f_c$ is necessary and sufficient was published by H. Nyquist (1928), and this minimum rate is commonly known as the Nyquist rate. Subsequently Shannon, in his 1949 paper, put this notion on a rigorous basis and further stated (without proof) that an *average* rate of $2f_c$ was sufficient. If the sampling rate is less than the Nyquist rate, the signal is said to be *under-sampled*; if the rate is significantly greater than the Nyquist rate, the signal is said to be *oversampled*.

If the signal is under-sampled, then the spacing of the replications of $F(\omega)$ is less than their width, and they overlap, as shown in Fig. 2-2(a). The consequence of this is that, in the output of the interpolation filter, frequency components lying outside the range $(-f_s/2, f_s/2)$ are reflected back into the range, as shown in Fig. 2-2(c). As an example of this, Fig. 2-3 shows a sinusoid of frequency $1.1f_s$, with a second sinusoid of frequency $0.9f_s$ superimposed; it will be seen that the sampled values of the two functions are identical. This apparent falsification of frequency is known as *aliasing*. In most practical applications, it is not possible to guarantee that the signal being processed will be strictly band-limited, and hence before every analog-to-digital (A/D) converter there is always a low-pass filter, known as an *anti-aliasing filter*, which, ideally, removes components above $f_s/2$. (Practical anti-aliasing filters attenuate these

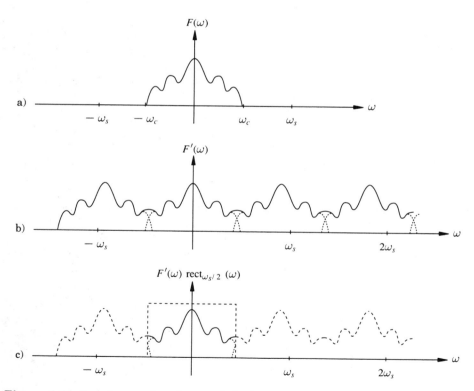

Figure 2-2 *Tranforms of undersampled signals: (a) signal before sampling; (b) sampled signal, showing overlaps between images; (c) recovered signal.*

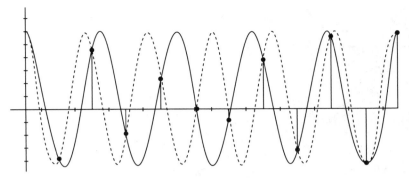

Figure 2-3   *Example of aliasing.*

components to the point that they are less than the errors introduced by the quantization process.)

We see, then, that any digital signal-processing system is normally bracketed by low-pass analog filters: an anti-aliasing filter at the input from the analog world and an interpolation filter at the output to the analog world.

An ideal interpolation filter of the sort described above is not physically realizable, since its impulse response is nonzero for all $t < 0$ (except at isolated points). In practice, a high-quality low-pass filter is used. But there is an additional problem, for when a digital output signal is converted back to analog form, in a practical system, the output is most commonly a "boxcar" function, as shown in Fig. 2-4. The boxcar effect results from the fact that the D/A converter holds each converted value until the next sampling instant, when a new value is converted. This effect can be most conveniently modelled as an impulse sequence convolved with a rectangular pulse whose width is the sampling interval T_s:

$$y(t) = \left[\sum_{n=-\infty}^{\infty} f(nT_s)\delta(t - nT_s) \right] * \text{rect}_{T_s/2}(t) \qquad (2\text{-}5)$$

(This formulation omits the half-sample delay, which we ignore here.) By the convolution property, this means that the actual frequency spectrum is the product of the spectrum of the sampled signal and the transform of $\text{rect}_{T_s/2}(t)$, which is $T \, \text{sinc}(\omega T/2)$. This causes a high-frequency rolloff which is approximately 4 dB down at one half the sampling frequency. This rolloff must be corrected by a compensating high-frequency adjustment. It may be feasible to introduce a high-frequency boost in the digital domain before the output digital-to-analog (D/A) conversion, or the boost may be built into the output interpolation filter.

In some applications, over-sampling may be used to reduce the high-frequency rolloff. In this technique, the output conversion rate is made to be some multiple L of the sampling rate, and each output sample

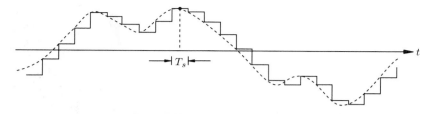

Figure 2-4　*Recovered signal from d/a converter. T_s is the sampling interval.*

is followed by $L - 1$ zero samples, as shown in Fig. 2-5. In this case, the ideal impulse sequence is convolved with a rectangular pulse whose width is $1/L$ times the sampling interval. Narrowing the rectangular pulse widens the sinc function and reduces the high-frequency rolloff correspondingly. For example, if $L = 4$, the rolloff is only 0.2 dB at one half the original sampling frequency.

2.1.1　SAMPLING AND HETERODYNING

We have seen that as a consequence of Eqs. (2-1) and (2-2), the Fourier transform of $\hat{f}(t)$ contains images centered about $k\omega_s$, $k = ..., -1, 0, 1, 2, ...$ If we had performed a heterodyning operation on $f(t)$, for example, if we had multiplied $f(t)$ by $\cos \omega_s t$, the transform would have been convolved with $1/2[\delta(\omega+\omega_s)+\delta(\omega-\omega_s)]$. This operation gives us a pair of replications of $F(\omega)$ centered about $-\omega_s$ and ω_s. Hence the sampling process could be thought of as an infinite sum of heterodyning operations. Alternatively, we can accomplish the equivalent of heterodyning to any desired integer multiple of f_s by sampling the signal and selecting the images centered about $-k\omega_s$ and $k\omega_s$. We will have more to say about this in Chapter 5, when we discuss multirate signal processing.

2.1.2　NONUNIFORM SAMPLING

If the spacing between samples is nonuniform, the situation is by no means as straightforward. A large part of the problem arises from the

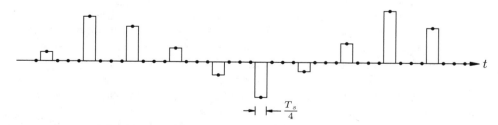

Figure 2-5　*Recovered signal narrowed by insertion of zero samples. T_s is the sampling interval.*

fact that we usually assume a signal of infinite duration, which implies that we must have the exact locations of an infinite set of nonuniformly-spaced samples. In practice, we never have this much information. In a practical reconstruction, we can back down from this requirement in several ways. These ways are covered in a paper by J. L. Yen (1956), on which the following discussion is based.

First, we can assume that all the samples are uniformly spaced except for a finite number of irregular samples; second, we can assume that there is a gap of indefinite size in an otherwise uniform series of samples; third, we can assume that the nonuniformity in sampling times is itself periodic; and finally, we can assume that our data are of finite length. Since we must partition an infinite sequence into finite-length blocks, this is a particularly realistic assumption. These cases are illustrated in Fig. 2-6. We will discuss only the last of these here.

In this case, we have a finite number N of nonuniformly-spaced samples, $f(\tau_i), i = 1, ..., N$. The time function is assumed band limited to f_c and the average sampling rate is at least f_c.

Suppose the sampling had been uniform. Then our given nonuniform sequence could be obtained by (1) interpolating with a sinc function to obtain the corresponding continuous function and then (2) sampling this continuous function at the given sampling moments $\{\tau_i\}$. The process of going from nonuniform to uniform sampling must, in effect, reverse these two steps.

Let $T = 1/2f_c = \pi/\omega_c$. Then if we had uniformly spaced samples, we could write, as before,

$$f(t) = \sum_{n=-\infty}^{\infty} f(nT)\,\mathrm{sinc}_{\omega_c}(t - nT) \qquad (2\text{-}6)$$

This is simply the usual reconstruction formula for uniformly-sampled data. We require that $f(t)$ pass through the available sample points:

$$f(\tau_i) = \sum_{n=-\infty}^{\infty} f(nT)\,\mathrm{sinc}_{\omega_c}(\tau_i - nT) \qquad i = 1, ..., N \qquad (2\text{-}7)$$

This illustrates how we could have gotten our nonuniform samples from a set of uniform ones. Reversing these steps involves inverting the infinite sum in Eq. (2-7)—not an attractive prospect. Suppose we could obtain, from $\{f(\tau_i)\}$, a finite set of N weights $\{b_i\}$ such that

$$f(nT) = \sum_{n=-\infty}^{\infty} b_i\,\mathrm{sinc}_{\omega_c}(nT - \tau_i) \qquad (2\text{-}8)$$

It is not immediately obvious whether, given these weights, we can get rid of the uniform sampling and generalize this by simply writing t for

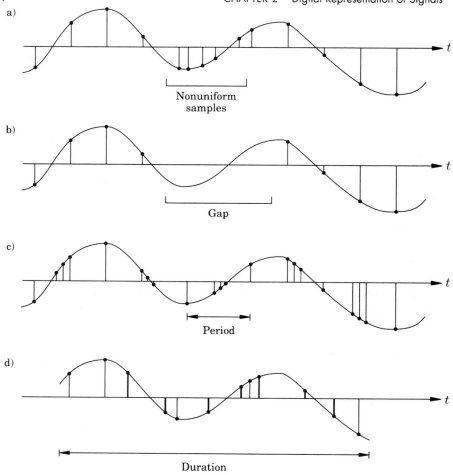

Figure 2-6   *Types of nonuniform sampling.*

nT. For the time being, let us assume that we can; we will justify this step later. Then, writing t for nT in Eq. (2-8), we have

$$f(t) = \sum_{i=1}^{N} b_i \, \text{sinc}_{\omega_c} (t - \tau_i) \qquad (2\text{-}9)$$

We must now find the weights b_i. Again we invoke the requirement that $f(t)$, as given by Eq. (2-9), pass through the sample points. Substituting τ_j for t in Eq. (2-9) gives us

$$f(\tau_j) = \sum_{i=1}^{N} b_i \, \text{sinc}_{\omega_c} (\tau_j - \tau_i), \qquad j = 1, ..., N \qquad (2\text{-}10)$$

This is a set of simultaneous equations. To solve it for $\{b_i\}$, let $\mathbf{S}$ be the $N \times N$ matrix $[s_{ij}]$, where $s_{ij} = \mathrm{sinc}_{\omega_c}(\tau_i - \tau_j)$ and let $\mathbf{f}^T$ be $[f(\tau_1), f(\tau_2), ..., f(\tau_N)]$. Then Eq. (2-10) becomes the matrix equation

$$\mathbf{Sb} = \mathbf{f}$$

Solving this system yields the desired weights, and we can then obtain the equivalent uniformly spaced sequence by using these weights in Eq. (2-8).

Using the same weights in Eq. (2-9) gives us the reconstructed continuous function. To obtain an explicit expression for this, we start by observing that

$$\mathbf{b} = \mathbf{S}^{-1}\mathbf{f}$$

For simplicity in notation, let the inverse of $\mathbf{S}$ be $\mathbf{A}$; then our desired expression is

$$f(t) = \sum_{j=1}^{N} f(\tau_j) \sum_{i=1}^{N} a_{ij}\,\mathrm{sinc}_{\omega_c}(t - \tau_i) \qquad (2\text{-}11)$$

It remains to establish that we can get Eq. (2-9) from Eq. (2-8) simply by writing t for nT. To show that we can, we substitute Eq. (2-8) into Eq. (2-6) to get

$$f(t) = \sum_{n=-\infty}^{\infty} \mathrm{sinc}_{\omega_c}(t - nT) \sum_{i=1}^{N} b_i\,\mathrm{sinc}_{\omega_c}(nT - \tau_i)$$

$$= \sum_{i=1}^{N} b_i \sum_{n=-\infty}^{\infty} \mathrm{sinc}_{\omega_c}(t - nT)\,\mathrm{sinc}_{\omega_c}(nT - \tau_i) \qquad (2\text{-}12)$$

We still have an infinite sum. We can get rid of it by the following subterfuge: Suppose we sample *any* band-limited function $\phi(t)$ at times $nT - \tau_i$ instead of nT; then, substituting $\phi(\cdot)$ for $f(\cdot)$ and $(nT - \tau_i)$ for nT in our reconstruction formula, Eq. (2-6), we have

$$\phi(t) = \sum_{n=-\infty}^{\infty} \phi(nT - \tau_i)\,\mathrm{sinc}_{\omega_c}[t - (nT - \tau_i)]$$

Since the sum is on n, τ_i is a constant, and in fact we have a simple shift in our sample times by a constant increment τ_i. We handle the shift by means of a translation of the time axis which changes t to $t - \tau_i$ and get

$$\phi(t - \tau_i) = \sum_{n=-\infty}^{\infty} \phi(nT - \tau_i)\,\mathrm{sinc}_{\omega_c}(t - nT)$$

But this is true for any band-limited $\phi(t)$; so it must be true in particular for $\phi(t) = \text{sinc}_{\omega_c}(t)$, which is certainly band-limited. So we can substitute $\text{sinc}_{\omega_c}(\cdot)$ for $\phi(\cdot)$, and

$$\text{sinc}_{\omega_c}(t - \tau_i) = \sum_{n=-\infty}^{\infty} \text{sinc}_{\omega_c}(nT - \tau_i)\,\text{sinc}_{\omega_c}(t - nT) \qquad (2\text{-}13)$$

But the right-hand side of Eq. (2-13) is the infinite sum found in Eq. (2-12). Hence Eq. (2-12) becomes

$$f(t) = \sum_{i=1}^{N} b_i\,\text{sinc}_{\omega_c}(t - \tau_i)$$

But this is Eq. (2-9), and Eq. (2-8) is now simply the special, sampled case where $t = nT$.

Nothing has been said about just how random these sampling times are. In principle, as long as there are, on the average, f_c points per second of time, they could be anywhere. Thus [as Shannon (1949) points out] we could imagine them bunched up toward the beginning of a long interval, in which case our reconstruction could be used to predict the behavior of the function over the rest of the interval. In practice, such a gross nonuniformity would require extreme accuracy, both in the locations and values of the sample points and in the computations. Methods such as the one derived here work best when the samples are "almost" uniform.

2.2 QUANTIZATION

Data samples must be converted to numerical values before we can subject them to the operations outlined in Chap. 1. This process is known as quantization. In practice, the samples are converted to binary integers. Suppose the integers are k bits long. A k-bit word can be used to represent 2^k different levels and thus can cover a range of either 0 to $2^k - 1$ or, in the case of bipolar signals, -2^{k-1} to $2^{k-1} - 1$. In uniform quantization, these levels correspond to multiples of a uniform step size; in nonuniform quantization, the corresponding levels are a monotonic function of the integer values but may have any spacing. We will start by considering uniform quantization.

2.2.1 UNIFORM QUANTIZATION

The input-output function of a typical uniform quantizer is shown in Fig. 2-7; as the input varies smoothly from large negative to large positive values, the quantizer's output moves in steps. The error resulting from this stepwise motion is shown below the input-output function. From this error function, we see that the quantizers are liable to two types of

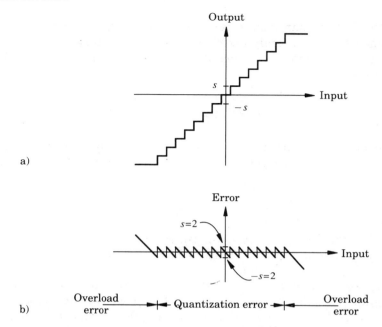

Figure 2-7 *Characteristics of uniform sampler: (a) input-output relation; (b) sampling errors.*

error. While the signal is within the operating range of the quantizer, the error cycles uniformly through $(-S/2, S/2)$. When the signal falls outside this range, the error increases without limit.

In the first case, where the sample value falls in between the available quantization levels, the error is termed a quantization error. If the number of quantization levels is small, the error will generally be correlated with the signal. If the number of levels is large, and the step size is uniform, it is usually safe to assume that the quantization errors are uncorrelated with the signal. We will assume uniform quantization and uncorrelated errors, unless specifically stated otherwise.

If the signal falls outside the range of the quantizer, then an overload error occurs. This overload error is similar to hard limiting in the analog domain: if the quantizer has a maximum value of 2047 and the true value of the sample is 2900, the quantizer will output 2047. In nonreal-time applications, as when the signal has been recorded on magnetic tape, the playback gain can be adjusted so that the maximum excursions of the signal fit neatly within the range of the quantizer; in real time, this is usually out of the question and the input has to be subjected to a fixed scaling, usually on the basis of statistical observations.

Although the quantization error is a deterministic function of the input signal, as is clear from Fig. 2-7, it is frequently modelled as a random noise signal. If the true value is x_t and the nearest quantization level is x_q, then the error is obviously $x_q - x_t$. Let the quantization step size be S. The quantization error then ranges from $-S/2$ to $S/2$ and is assumed to be uniformly distributed over the interval, as can be seen from Fig. 2-7. Hence we conclude that the probability density function of the error is

$$f(z) = \begin{cases} 1/S & |z| < S/2 \\ 0 & \text{elsewhere} \end{cases}$$

If we model the quantization error as uniformly distributed additive noise, then the noise power will be the variance associated with this density function; this is

$$\sigma_n^2 = S^2/12 \tag{2-14}$$

The signal-to-quantization-noise ratio (snr) is then the variance of the signal divided by σ_n^2. This calculation is complicated by the fact that the signal level depends not on its variance, but on its maximum values, since the maximum must be scaled so that it will pass through the quantizer without overload errors. Hence we cannot estimate the snr without some knowledge of the statistics of the signal. It is customary to assume a ratio, called the *loading factor*, that relates the rms amplitude of the signal to its maximum value:

$$p = x_{\max}/\sigma_x \tag{2-15}$$

Assume a bipolar signal whose range is $(-x_{\max}, x_{\max})$ and a k-bit bipolar quantizer whose range is approximately $(-2^k/2, 2^k/2)$. Then if the signal level is set so that $x_{\max}$ corresponds to $2^k/2$, $\sigma_x = x_{\max}/p$ and the step size is $x_{\max}/2^{k-1}$. The signal power is $\sigma_x^2 = (x_{\max}/p)^2$, and the quantization noise variance is $(x_{\max}/2^{k-1})^2/12$. In that case, the signal-to-noise ratio is

$$\text{snr} = \sigma_x^2/\sigma_n^2$$
$$= 3 \cdot 2^{2k}/p^2 \tag{2-16}$$

In decibels, this is

$$\text{snr}(dB) = 4.77 + 6k - 20\log_{10} p \tag{2-17}$$

Notice that the signal-to-noise ratio increases by 6 dB for each additional bit.

2.2.2 NONUNIFORM QUANTIZATION

Up to this point, we have given no consideration to the probability density of the signal being quantized. But it would seem, intuitively, that if the probability density function (PDF) of the signal were not rectangular, and its statistics were stationary, that we could achieve better results by tailoring the step size to the PDF, using small step sizes, which would produce a small quantization error for the more probable input values at the expense of larger step sizes for the less probable ones. J. Max (1960) put this on a rigorous basis.

Suppose we consider a range of input values from x_i to x_{i+1}; we want our quantizer to represent any input falling in this range by some output value y_i. We call the $\{x_i\}$ quantization levels and the $\{y_i\}$ representation levels. Then if x has the probability density $p(x)$, the mean-squared quantization error will be

$$D = \sum_{i=1}^{N} \int_{x_i}^{x_{i+1}} (x - y_i)^2 p(x)\, dx \tag{2-18}$$

where N is the number of quantization levels. To minimize this, we differentiate with respect to x_i and y_i. These derivatives involve only those terms of the sum involving x_i. Using the rule,

$$\frac{d}{dx} \int_{\alpha(x)}^{\beta(x)} \phi(x, v)\, dv = \int_{\alpha(x)}^{\beta(x)} \frac{d}{dx} \phi(x, v)\, dv + \phi(x, \beta(x)) \frac{d\beta}{dx} - \phi(x, \alpha(x)) \frac{d\alpha}{dx}$$

we get

$$\frac{\partial D}{\partial x_1} = (x_i - y_{i-1})^2 p(x_i) - (x_i - y_i)^2 p(x_i) \tag{2-19}$$

$$\frac{\partial D}{\partial y} = -2 \int_{x_i}^{x_{i+1}} (x - y_i) p(x)\, dx \tag{2-20}$$

If $p(x) > 0$, setting the partial derivative in Eq. (2-19) to zero gives

$$(x_i - y_{i-1})^2 = (x_i - y_i)^2$$

If we assume that x_i lies between y_i and y_{i-1}, then as our first requirement we have,

$$y_i - x_i = x_i - y_{i-1} \tag{2-21}$$

This means that x_i lies midway between y_{i-1} and y_i. Setting the partial derivative in Eq. (2-20) to zero gives

$$\int_{x_i}^{x_{i+1}} xp(x)\, dx = y_i \int_{x_i}^{x_{i+1}} p(x)\, dx$$

or

$$y_i = \left[\int_{x_i}^{x_{i+1}} xp(x)\, dx \right] \Big/ \left[\int_{x_i}^{x_{i+1}} p(x)\, dx \right] \tag{2-22}$$

This means that y_i is at the centroid of that portion of $p(x)$ between x_i and x_{i+1}.

These requirements do not lend themselves to an easy solution. Max used an iterative numerical approach, assuming an initial y_1, computing the succeeding $\{x_i\}$ and $\{y_i\}$ according to the requirements set forth above, seeing whether the final y_N landed in the right place, and adjusting the starting guess for the next iteration accordingly. In this way, he found and tabulated the optimum quantization and representation levels for a zero-mean Gaussian input signal for various numbers of quantization levels up to a maximum of 36. His tables show that the quantization levels do indeed bunch up for small values of $|x|$ and spread out as $|x|$ becomes large.

P. F. Panter and W. Dite (1951) showed that when the number of quantization levels is large, the analysis can be simplified by assuming that the PDF does not change significantly over the interval from x_i to x_{i+1}. Then we can write $p(x) \approx p(x_i), x_i \leq x < x_{i+1}$. Subject to this assumption, the mean-squared error for any step i will be

$$D_i = p(x_i) \int_{x_i}^{x_{i+1}} (x - y_i)^2 \, dx$$

It is easy to verify that this error will be a minimum when y_i is in the middle of the step:

$$y_i = (x_{i+1} + x_i)/2 \tag{2-23}$$

This minimum error can now be found as follows. Let $u = x - y_i$ and let the step size be $S_i = x_{i+1} - x_i$. Then

$$D_i = p(x_i) \int_{-S_i/2}^{S_i/2} u^2 \, du = \frac{1}{12} S_i^3 p(x_i)$$

Then the total quantization error is the sum of D_i over all steps:

$$D = \frac{1}{12} \sum_{i=1}^{N} S_i^3 p(x_i). \tag{2-24}$$

For brevity, write $\mu_i = S_i[p(x_i)]^{1/3}$. Then we wish to minimize

$$D = \frac{1}{12} \sum_{i=1}^{N} \mu_i^3$$

subject to some reasonable constraint. Suppose we wish to cover a range from $-x_{max}$ to x_{max}. Since we have many steps and since each step will therefore tend to be small, we will write $S_i \approx dx$. Then we have

$$\sum_{i=1}^{N} \mu_i \approx \int_{-x_{max}}^{x_{max}} [p(x)]^{1/3} \, dx$$

This integral is a constant, since its limits are constant and the probability density is given. Then we can minimize the distortion subject to the constraint $\sum \mu_i = K$, using Lagrange's method:

$$v = \frac{1}{12} \sum_{i=1}^{N} \mu_i^3 + \lambda \sum_{i=1}^{N} \mu_i$$

and for a minimum,

$$\frac{\partial v}{\partial \mu_i} = \frac{1}{4}\mu_i^2 + \lambda = 0 \tag{2-25}$$

or $\qquad\qquad \mu_i = C \qquad$ where C is a constant. $\qquad\qquad$ (2-26)

This leads to two important results. First, if all the quantities μ_i are equal, then so are their cubes; hence every step makes an equal contribution to the mean-squared error. Second, the step size is inversely proportional to the cube root of $p(x)$ for that step:

$$S_i = C/[p(x_i)]^{1/3} \tag{2-27}$$

The constant is determined by the requirement that the sum of all step sizes must span the range from $-x_{max}$ to x_{max}.

It is sometimes convenient to model a nonuniform quantizer as a nonlinear distortion function, commonly known as a compressor, followed by a uniform quantizer, as shown in Fig. 2-8. The compression function $\phi(x)$ maps the desired nonuniform levels of the optimum quantizer onto the available uniform levels of the practical quantizer. At the receiving end, a complementary nonlinear function $\phi^{-1}(x)$, called the expansion function, restores the original levels. Let us assume that $p(x)$ is symmetrical about 0. If the uniform quantizer's step size is Δ, then

$$\phi'(x) \approx \Delta/S_i = \Delta[p(x_i)]^{1/3}/C \tag{2-28}$$

Again, if the number of steps is large enough, we can approximate this by integrating the step sizes: this gives us

Figure 2-8 *Nonuniform quantizer modelled as uniform quantization enclosed within compandor.*

$$\phi(x) \approx \Delta/C \int_0^x [p(u)]^{1/3} \, du \qquad (2\text{-}29)$$

In practice, we replace Δ/C by a constant which makes $\phi(x_{\max}) = x_{\max}$.

2.3 ORTHONORMAL SEQUENCES AND EXPANSIONS

We turn now to the problem of finding alternative representations of sampled, quantized signals. The expansion of an arbitrary sequence $f(k)$ into a weighted sum of Kronecker delta sequences is shown in Eqs. (1-15) and (2-30) below,

$$f(n) = \sum_{j=-\infty}^{\infty} f(j)\delta(n - j) \qquad (2\text{-}30)$$

is but one of many useful ways of representing a discrete-time signal as a superposition of component sequences. There are families of other probe signals—the sine and exponential functions, for example—which also provide useful signal representations. The representation of sequences as a superposition of the members of an *orthogonal* family of functions is discussed in this section.

Certain geometric concepts borrowed from the theory of linear vector spaces are useful in establishing the properties of the functions to be introduced here. To this end, suppose a sequence $\{f(k)\}$ is defined on the interval, $0 \le k \le 2$. We can portray this sequence of samples, $f(0)$, $f(1), f(2)$ as a point or vector in a three-dimensional space as shown in Fig. 2-9, with the projections along the coordinate axes as indicated. The coordinate system for this space is determined by the unit vectors in the three mutually orthogonal directions. Clearly, these *basis vectors*

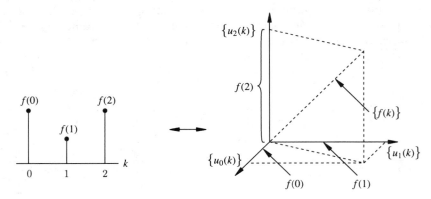

Figure 2-9   *Sequence of three samples depicted as a vector in three-dimensional space.*

are the Kronecker delta sequences $\{1, 0, 0\}$, $\{0, 1, 0\}$, and $\{0, 0, 1\}$, and the sequence $\{f(k)\}$ is expressible as the linear combination

$$f(k) = f(0)\{1, 0, 0\} + f(1)\{0, 1, 0\} + f(2)\{0, 0, 1\}$$
$$= f(0)\{u_0(k)\} + f(1)\{u_1(k)\} + f(2)\{u_2(k)\} \qquad (2\text{-}31)$$

which is a special case of the superposition of Eq. (2-30). The set of values $f(0)$, $f(1)$, $f(2)$ are said to be the components of $\{f(k)\}$ relative to the given basis. Now with the sequence $\{f(k)\}$ treated as a vector, we can say that the norm, or length of $\{f(k)\}$ is

$$\text{norm}\{f(k)\} = \left| \sum_{k=0}^{2} |f(k)|^2 \right|^{1/2} \qquad (2\text{-}32)$$

It is clear that any other vector (sequence) in the space has a similar decomposition in terms of the basis vectors $\{u_0(k)\}$, $\{u_1(k)\}$, and $\{u_2(k)\}$.

The three unit vectors possess two general properties, which were tacitly assumed in writing Eq. (2-31):

1. *Linear independence:* if c_0, c_1, c_2 are constants, the linear combination,

$$c_0\{u_0(k)\} + c_1\{u_1(k)\} + c_2\{u_2(k)\}$$

can vanish over $0 \le k \le 2$ only if $c_0 = c_1 = c_2 = 0$. Another way of saying this is that no one basis vector can be represented as a linear combination of the other two.

2. *Orthogonality:* again, borrowing from the terminology of linear vector spaces, we say that two sequences $\{g(k)\}$ and $\{f(k)\}$ are orthogonal if the inner product,

$$\sum_{k=0}^{2} g(k)f(k)$$

vanishes.

The Kronecker delta sequences are orthogonal since, obviously,

$$\sum_{k=0}^{2} u_r(k)u_s(k) = 0, \qquad r \ne s$$

Moreover, the norm of each basis vector is unity:

$$\sum_{k=0}^{2} u_r^2(k) = 1$$

Any such set of orthogonal unit-norm discrete functions is said to be *orthonormal*.

Generalization of the preceding concepts to sequences defined on any interval, finite or infinite, leads directly to the superposition formula of Eq. (2-30) and to the interpretation of the Kronecker delta sequence $\{\delta(n - k)\}, -\infty < k < \infty$, as an orthonormal basis for the space of all sequences $\{f(k)\}$ defined on $-\infty < k < \infty$. Our interest here, however, is the extension of this principle to other families of functions.

Suppose $\{f(k)\}$ is defined on the interval $N_1 \leq k \leq N_2$; then $\{f(k)\}$ can be regarded as a vector in an M-dimensional space, where $M = N_2 + 1 - N_1$. Let $\{x_r(k), r = 1, 2, ..., M\}$ be a family of M linearly independent sequences; this family is orthogonal on the interval $[N_1, N_2]$ if

$$\sum_{k=N_1}^{N_2} x_r(k) x_s^*(k) = c_r^2 \delta_{r-s} = \begin{cases} 0, & r \neq s \\ c_r^2, & r = s \end{cases} \tag{2-33}$$

where c_r is the norm of $\{x_r(k)\}$. (We will write the complex conjugate with an asterisk throughout this chapter.) The orthonormal family is obtained by dividing each member $x_r(k)$ by c_r. Thus, with

$$\phi_r(k) \equiv \frac{1}{c_r} x_r(k) \qquad N_1 \leq k \leq N_2 \tag{2-34}$$

we have

$$\sum_{k=N_1}^{N_2} \phi_r(k) \phi_s^*(k) = \delta_{r-s} \tag{2-35}$$

Any set of M functions satisfying Eq. (2-35) constitutes an orthonormal basis for the space. Hence $\{f(k)\}$ can be *uniquely* represented as

$$f(k) = \sum_{r=1}^{M} \alpha_r \phi_r(k) \qquad N_1 \leq k \leq N_2 \tag{2-36}$$

where

$$\alpha_s = \sum_{k=N_1}^{N_2} f(k) \phi_s^*(k) \qquad 1 \leq s \leq M \tag{2-37}$$

The validity of this last equation is established by multiplying both sides of Eq. (2-36) by $\phi_s^*(k)$ and then summing over the index k. Upon interchanging the order of summation and invoking the orthonormal property, we find

$$\sum_k f(k) \phi_s^*(k) = \sum_r \alpha_r \sum_k \phi_r(k) \phi_s^*(k) = \sum_r \alpha_r \delta_{r-s} = \alpha_s$$

The set of coefficients $\{a_s\}, s = 1, 2, ..., M$, are the components of $\{f(k)\}$ relative to the orthonormal basis. They are sometimes called the spectral coefficients of the signal relative to the given family.[†]

Multiplying both sides of Eq. (2-36) by their conjugates and summing over k gives the square of the norm of $\{f(k)\}$ as

$$\sum_k f(k)f^*(k) = \sum_r \sum_s \alpha_r \alpha_s^* \sum_k \phi_r(k)\phi_s^*(k)$$

$$= \sum_r \sum_s \alpha_r \alpha_s^* \delta_{r-s}$$

or

$$\sum_{k=N_1}^{N_2} |f(k)|^2 = \sum_{r=1}^{M} |\alpha_r|^2 \qquad (2\text{-}38)$$

The latter is termed the Parseval theorem, which in continuous-time systems relates the energy of the signal, $\int |f(t)|^2 \, dt$, to the sum of the squared magnitude of the spectral coefficients. Clearly, this representation makes sense only for signals with finite norms (or energy). On a finite interval, we require only that all samples $f(k)$ be bounded on that interval. On the other hand, for signals defined on a semi-infinite $(0, \infty)$ or doubly-infinite $(-\infty, \infty)$ interval, the convergence of the norm[‡] requires much more stringent conditions—the obvious necessary one being $|f(k)| \to 0$ as $k \to \pm\infty$.

The set of coefficients determined by Eq. (2-37) is said to give the best least-squares approximation to $f(k)$. The meaning of this is as follows: Suppose $f(k)$ is approximated by a linear combination of the first L of the M basis functions, using some arbitrary weighting coefficients $\beta_1, \beta_2, ..., \beta_L$. We will show that the best choice for the coefficients is $\beta_j = \alpha_j, j = 1, ..., L$, where α_j is the spectral coefficient of Eq. (2-37). The approximant is

$$\hat{f}(k) = \sum_{r=1}^{L} \beta_r \phi_r(k) \qquad (2\text{-}39)$$

so that the instantaneous error is

$$\epsilon(k) = f(k) - \sum_{r=1}^{L} \beta_r \phi_r(k) \qquad (2\text{-}40)$$

[†]We can also think of these as generalized discrete Fourier coefficients, even when $\{\phi_r(k)\}$ is not sinusoidal.

[‡]Such sequences with bounded norms are termed square-summable sequences, or L_2.

We wish to choose the $\{\beta_r\}$ so that the following sum-squared error is minimized:

$$J_L = \sum_{k=N_1}^{N_2} |\epsilon(k)|^2 \tag{2-41}$$

Expanding the latter and invoking orthonormality[†] gives

$$J_L = \sum_k |f(k)|^2 - 2 \sum_k f(k) \sum_r \beta_r \phi_r(k) + \sum_r (\beta_r)^2$$

Setting the partial derivative of J_L with respect to β_s to zero gives

$$\frac{\partial J_L}{\partial \beta_s} = -2 \sum_k f(k)\phi_s(k) + 2\beta_s = 0$$

From this we find

$$\beta_s = \sum_{k=N_1}^{N_2} f(k)\phi_s(k) \equiv \alpha_s \qquad \text{the spectral coefficient}$$

and when $L = M$, we see that $J_M = 0$. Thus the sum-squared error in using the partial sum of Eq. (2-39) as an approximant to $\{f(k)\}$ is minimized by selecting the spectral coefficients themselves as the weights.

A geometric interpretation of the resulting error sequence can be reconstructed from

$$\epsilon(k) = f(k) - \sum_{r=1}^{L} \alpha_r \phi_r(k) = \sum_{r=L+1}^{M} \alpha_r \phi_r(k) \tag{2-42}$$

Thus, the minimized error sequence lies in the space spanned by the remaining basis functions $\{\phi_{L+1}(k), ..., \phi_M(k)\}$. Since each of these is orthogonal to $\{\phi_s(k)\}, 1 \leq s \leq L$, and since $\{f(k)\}$ lies in the space spanned by the first L basis functions, it follows that $\{\epsilon(k)\}$ is orthogonal to $\{\hat{f}(k)\}$. That is,

$$\sum_{k=N_1}^{N_2} \epsilon(k)\hat{f}(k) = 0 \qquad \text{for real sequences}$$

$$\sum_{k=N_1}^{N_2} \epsilon(k)\hat{f}^*(k) = 0 \qquad \text{for complex sequences} \tag{2-43}$$

[†]For convenience only, we assume $f(k)$, β_j, and $\phi_s(k)$ are all real.

It can be shown that the orthogonality of error and approximant is both necessary and sufficient to minimize the sum-squared error. Hence the latter can serve as the principle from which the spectral coefficients are derived. This rule, called the *orthogonality principle*, has far-reaching implications in optimal estimation theory, as we shall see in Chap. 7.

A diagram depicting this orthogonal relationship is shown in Fig. 2-10 for the case $M = 3$ and $L = 2$. Note that $\{\hat{f}(k)\}$, the least-squares optimal approximation to $\{f(k)\}$, is the orthogonal projection of $\{f(k)\}$ onto the two-dimensional subspace generated by basis vectors $\{\phi_1(k)\}$ and $\{\phi_2(k)\}$.

One final point: *all* the results and theorems of this section are valid for infinite as well as finite-dimensional spaces or intervals so long as the norms of the sequences are bounded.

Before ending this section, we present two additional theorems which find application in later chapters. The first is the Cauchy-Schwarz inequality which asserts that

$$\left|\sum_k x_k y_k^*\right|^2 \le \sum_k |x_k|^2 \sum_k |y_k|^2 \tag{2-44}$$

That is, the square of the magnitude of the inner product of two sequences is less than or equal to the product of the sums of the squares of the norms of the sequences. The equality holds whenever y_k is proportional to x_k, that is, whenever $y_k = ax_k$.

The second theorem reveals the equality between the inner product of two sequences with the inner product of their corresponding spectral coefficients. With α_r the spectral coefficient of $x(n)$ and β_r that of $y(n)$, we have

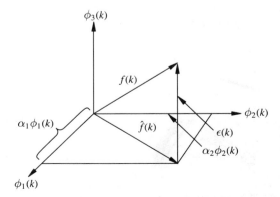

Figure 2-10 *Illustrating the orthogonality theorem.*

$$\sum_{n=N_1}^{N_2} x_n y_n^* = \sum_{r=1}^{M} \alpha_r \beta_r^* \tag{2-45}$$

as a generalization of the Parseval theorem. Proofs of these last two theorems are left as exercises for the reader.

2.3.1 INTERPOLATION AND SMOOTHING

We have proved that the generalized Fourier coefficients minimize the sum of squared errors between a sequence $\{f(k)\}$, defined on $[0, N]$, and its approximant

$$\hat{f}(k) = \sum_{j=0}^{L} \alpha_j \phi_j(k) \tag{2-46}$$

Clearly, for the case $L = N$, the basis vectors $\phi_0, ..., \phi_N$ span the $(N+1)$-dimensional vector space, and hence

$$\epsilon(k) = f(k) - \hat{f}(k) = 0$$

at all points $0 \le k \le N$. In this case, the envelope obtained by replacing k by t, a continuous variable, gives a function

$$\hat{f}(t) = \sum_{j=0}^{N} \alpha_j \phi_j(t) \tag{2-47}$$

which can be used to estimate or *interpolate* the values of $f(t)$ between sampling instants. The interpolation formula Eq. (2-47) thus gives a continuous curve fitting the data at all points, as shown in Fig. 2-11(a). Next, suppose that $L+1$, the number of basis functions employed, is less than $N+1$, the number of data points. In this case, $f(k) \ne \hat{f}(k)$, and a nonzero sum-squared error results. The envelope $\hat{f}(t)$, again obtained

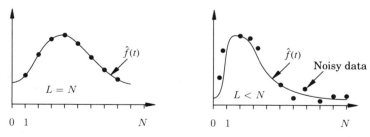

Figure 2-11 *Use of orthogonal expansions for interpolation and smoothing: (a) interpolation; (b) smoothing.*

by replacing k with t, is said to provide a smoothed least-squares fit to the data, as shown in Fig. 2-11(b). Smoothing is required when the data points are subject to random errors. In this instance, the redundancy of the data is used to reduce the effects of randomness in the raw data. This matter is taken up in much greater detail in Chap. 6.

2.4 ORTHOGONAL EXPANSIONS AND TRANSFORM CODING

The orthonormal expansions developed in Sec. 2-3 provide the mathematical underpinnings for signal classification and identification, and for *transform coding* of signals, particularly speech and images. The relationship between orthonormal expansions and transform coding is developed here.

We begin by reformulating the orthogonal expansions of Eqs. (2-36) and (2-37) in vector-matrix form. We define the signal and spectral vectors by

$$\mathbf{f}^T = [f_1, ..., f_M]$$
$$\boldsymbol{\alpha}^T = [\alpha_1, ..., \alpha_M] \tag{2-48}$$

The rows of the transformation matrix $\boldsymbol{\Phi}$ are the orthonormal sequences, $\phi_r(k) = \phi_{rk}$; that is,

$$\boldsymbol{\Phi} = [\phi_{rk}] \tag{2-49}$$

Now it is easy to recognize that

$$\boldsymbol{\alpha} = \boldsymbol{\Phi} f \tag{2-50}$$

and
$$\mathbf{f} = \boldsymbol{\Phi}^{-1}\boldsymbol{\alpha} = \boldsymbol{\Phi}^T \boldsymbol{\alpha} \tag{2-51}$$

Thus
$$\boldsymbol{\Phi}^{-1} = \boldsymbol{\Phi}^T \tag{2-52}$$

a property which identifies $\boldsymbol{\Phi}$ as an *orthogonal* matrix. For complex signals and matrices, these equations are

$$\boldsymbol{\alpha} = \boldsymbol{\Phi}^* \mathbf{f} \tag{2-53}$$

$$\mathbf{f} = \boldsymbol{\Phi}^T \boldsymbol{\alpha} \tag{2-54}$$

and
$$\boldsymbol{\Phi}^{-1} = (\boldsymbol{\Phi}^*)^T \tag{2-55}$$

This last equation says that the inverse of $\boldsymbol{\Phi}$ is its conjugate transpose; thus $\boldsymbol{\Phi}$ is a *unitary* matrix. It is also evident that the Parseval relation is

$$\boldsymbol{\alpha}^T \boldsymbol{\alpha} = \mathbf{f}^T \mathbf{f} \qquad \text{real signals}$$
$$\boldsymbol{\alpha}^T \boldsymbol{\alpha}^* = \mathbf{f}^T \mathbf{f}^* \qquad \text{complex signals} \tag{2-56}$$

In pulse-code modulation (PCM), the signal (for example, speech) is sampled, quantized, and encoded at the transmitter, as indicated in Fig. 2-12. The formatted signal is then transmitted via some transmission channel to a receiver which reconstructs the signal.

In transform coding (TC), illustrated in Fig. 2-13, the orthonormal spectrum of a batch of N signal samples is first evaluated. These spectral coefficients are then quantized, encoded, and transmitted. The receiver performs inverse operations to reconstruct the signal. The purpose of the transformation is to convert the data vector $\mathbf{f}$ into a spectral-coefficient vector α which can be optimally quantized.

Suppose that $f(n)$ is a sequence of zero-mean, gaussian, correlated random variables, each with the same variance σ_x^2. The orthogonal transformation is intended to decorrelate the components of the coefficient vector α. In effect, the transformation tries to "whiten" the sequence $\{\alpha_r\}$. Moreover, the variances of the individual components of α will generally be different, which simply indicates that the signal $\mathbf{f}$ has a nonuniform spectrum under the transformation. We can exploit this fact by choosing the quantizer to allocate bits to each coefficient in accordance with its variance. Thus some coefficients can be quantized more finely than others. In fact, the nonuniform quantizer of Section 2-2 gives an explicit formulation for the design of the optimal quantizer, which depends on the probability density of the coefficient and in particular on its variance.

The justification for this added complexity is based on the observation that in the absence of channel noise, the mean-squared signal reconstruction error equals the mean-squared quantization error. With reference to Fig. 2-13,

$$\tilde{\mathbf{f}} = \mathbf{f} - \hat{\mathbf{f}} \qquad \text{reconstruction error}$$
$$\tilde{\alpha} = \alpha - \hat{\alpha} \qquad \text{coefficient quantization error}$$

(2-57)

It can be shown (Prob. 2.7) that

$$E\{\tilde{\mathbf{f}}^T \tilde{\mathbf{f}}\} = E\{\tilde{\alpha}^T \tilde{\alpha}\} \qquad (2\text{-}58)$$

$x(t) \rightarrow \boxed{\text{Sampler}} \xrightarrow{x(n)} \boxed{\text{Quantizer}} \rightarrow \boxed{\text{Encoder}} \dashrightarrow \boxed{\text{Decoder}} \xrightarrow{\hat{x}(n)} \boxed{\text{D/A}} \rightarrow \hat{x}(t)$

Channel

Figure 2-12 *Elements of a PCM system.*

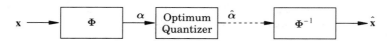

Figure 2-13 *Transform coding system elements. (Sampler, encoder, decoder, and d/a converter omitted for clarity.)*

Hence we can optimally code the data stream by an orthogonal transformation followed by an optimal quantizer. This can be done by

1. A fixed transformation and quantization based on an a priori model for the input signal, or

2. An adaptive transformation and quantization which adjust on the basis of the observed data (Jayant and Noll, 1984).

Zonal sampling is a term used to indicate an approximation wherein only a subset of the N spectral coefficients are used to represent the signal **f**. This is nothing but the least-squares approximation addressed in Sec. 2-3, Eqs. (2-39) through (2-43). The truncation error was

$$\epsilon(k) = f(k) - \sum_{r=1}^{L} \alpha_r \phi_r(k) = \sum_{r=L+1}^{M} \alpha_r \phi_r(k) \qquad (2\text{-}42)$$

from which it is easy to show that

$$\sum_{k=N_1}^{N_2} |\epsilon(k)|^2 = \sum_{r=L+1}^{M} |\alpha_r|^2 \qquad (2\text{-}59)$$

Thus the best zonal sampler is one that packs the maximum energy into the first L coefficients. The Karhunen-Loève transform (KLT), a signal-dependent transform discussed below, has this property.

The Fourier transform is a non-signal-dependent transform. Based on an a priori signal model (for example, a low-frequency signal), the optimum fixed quantizer would allocate bits to spectral coefficients based on the expected strengths or variances of these coefficients. Zonal sampling would simply discard those coefficients that the signal model suggested were small.

2.5 SOME ORTHONORMAL SEQUENCES

In this section, we introduce various examples of orthonormal transformations which have applications in signal processing. Some of these are taken up in detail in subsequent chapters.

2.5.1 THE DISCRETE FOURIER TRANSFORM (DFT)

The topic of discrete Fourier analysis is taken up in considerable detail in App. B. Here we mention only that the set $\{x_r(k) = e^{j2\pi rk/N}\}$, $r = 0, 1, ..., N - 1$, is orthogonal (but not normalized) on the interval $0 \le k \le N - 1$. Consequently, the relations

$$f(k) = \sum_{r=0}^{N-1} \alpha_r e^{j2\pi rk/N}$$

$$\alpha_r = \frac{1}{N} \sum_{k=0}^{N-1} f(k)e^{-j2\pi rk/N}$$

(2-60)

constitute the DFT pair. The set of coefficients $\{\alpha_r\}$ constitutes the frequency spectrum of the sampled signal. Since $x_r(k) = x_r(k+N)$ is periodic in k with period N, the series gives a periodic repetition of $f(k)$ outside the interval $[0, N-1]$, and thus is perfectly suited for functions which are periodic with this period. The coefficient corresponding to $r = 0$ is the "d-c" value of the sequence, while α_1 is the amplitude of the fundamental $\{x_1(k)\}$. Note that the fundamental $\{x_1(k)\}$ sequence corresponds to the N roots of unity, as shown in Fig. 2-14. The second harmonic $\{x_2(k)\}$ can then be imagined as being generated by a unit vector in the complex plane, rotating twice as fast as the fundamental. The higher harmonics are generated similarly.

The DFT is the most important orthogonal transformation in signal analysis, with far-reaching implications in almost every phase of signal processing. The Fast Fourier transform (FFT), reviewed in Chap. 4, is a computationally efficient algorithm for the calculation of α.

2.5.2 THE DISCRETE COSINE TRANSFORM (DCT)

This transform has become almost an industry standard in speech and image processing. It has near-optimal properties in transform coding of speech signals (Ahmed and Rao, 1975) and can be computed via an FFT-like algorithm.

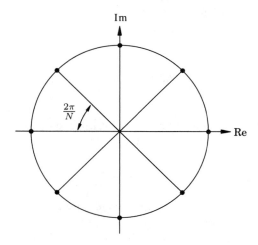

Figure 2-14   *Sketch of $\{x_1(k)\}$ in the complex plane for $N = 8$.*

The orthogonal sequences are defined on [0, N - 1] by

$$x_r(n) = \cos \frac{(2n+1)r\pi}{2N} \qquad r = 0, 1, ..., N-1$$

One can verify that

$$\sum_{n=0}^{N-1} x_r(n) x_s(n) = c_r^2 \delta_{r-s} \tag{2-61}$$

where

$$c_r^2 = \begin{cases} N & r=0 \\ N/2 & r \neq 0 \end{cases} \tag{2-62}$$

This set is normalized by

$$\phi_r(n) = \frac{1}{c_r} x_r(n)$$

so that

$$f(n) = \sum_{r=0}^{N-1} \alpha_r \left(\frac{1}{c_r}\right) \cos \frac{(2n+1)r\pi}{2N} \tag{2-63}$$

$$\alpha_r = \frac{1}{c_r} \sum_{n=0}^{N-1} f(n) \cos \frac{(2n+1)r\pi}{2N} \tag{2-64}$$

Figure 2-15 shows the discrete cosine orthonormal family for $N = 8$.

2.5.3 THE DISCRETE LAGUERRE FAMILY

This set of sequences is orthogonal on $[0, \infty)$ and is given by

$$\phi_r(k) = A_r p_r(k) \lambda^k = L_r(k) \lambda^k \tag{2-65}$$

where λ is a constant, $0 < \lambda < 1$, A_r is a normalizing factor,

$$A_r = \lambda^r \sqrt{1 - \lambda^2} \tag{2-66}$$

and $p_r(k)$ is a degree-r polynomial in k given explicitly as

$$p_r(k) = \sum_{m=0}^{r} (-1)^m \binom{r}{m} \alpha^m \frac{k^{(m)}}{m!} \tag{2-67}$$

where $\binom{r}{m}$ is the binomial coefficient, $\alpha = (1 - \lambda^2)/\lambda^2$, and

$$k^{(m)} \equiv k(k-1)\cdots(k-m+1) \qquad k^{(0)} \equiv 1 \tag{2-68}$$

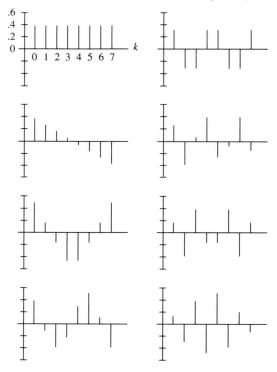

Figure 2-15 *Discrete cosine family, $N = 8$.*

The latter is called the forward factorial function and is a degree-m polynomial in k with zeroes at $k = 0, 1, ..., m - 1$.

The proof of the orthogonality of these sequences is greatly facilitated by Z transform techniques and is left as an exercise for the reader (Prob. 2-8). We can show that

$$\sum_{k=0}^{\infty} \phi_r(k)\phi_s(k) = \delta_{r-s} \qquad (2\text{-}69)$$

These sequences are useful for the representation of certain classes of signals defined on $[0, \infty)$. The coefficients of the expansion are then termed the Laguerre spectrum. Digital filters using these sequences have found some application in signal analysis (Morrison, 1969).

The polynomial multipliers $L_r(k)$ appearing in Eq. (2-65) are the discrete Laguerre polynomials

$$L_r(k) = A_r p_r(k) \qquad (2\text{-}70)$$

When this form is substituted into the orthonormality condition, there results

$$\sum_{k=0}^{\infty} L_r(k)L_s(k)\lambda^{2k} = \sum_{k=0}^{\infty} L_r(k)L_s(k)\theta^k = \delta_{r-s} \qquad (2\text{-}71)$$

where $\theta = \lambda^2$. These polynomials are then said to be orthonormal over $[0, \infty)$ with respect to the weighting function θ^k. Again with the aid of Z transforms, it can be shown that these polynomials (without the normalizing factors A_r) satisfy the recurrence relation,

$$p_r(k) = p_{r-1}(k) - \alpha \sum_{j=0}^{k-1} p_{r-1}(j), \qquad p_0(k) = 1 \qquad (2\text{-}72)$$

The latter can be expressed as the constant-coefficient difference equation,

$$p_r(k) = p_r(k-1) + p_{r-1}(k) - (1-\alpha)p_{r-1}(k-1) \qquad (2\text{-}73)$$

with the solution taking the form of Eq. (2-67).

For $k \geq 0$, the first five polynomials are

$$p_0(k) = 1$$

$$p_1(k) = 1 - \alpha k$$

$$p_2(k) = 1 - 2\alpha k + \frac{\alpha^2 k(k-1)}{2!}$$

$$p_3(k) = 1 - 3\alpha k + \frac{3\alpha^2 k(k-1)}{2!} - \frac{\alpha^3 k(k-1)(k-2)}{3!} \qquad (2\text{-}74)$$

$$p_4(k) = 1 - 4\alpha k + \frac{6\alpha^2 k(k-1)}{2!} - \frac{4\alpha^3 k(k-1)(k-2)}{3!}$$

$$+ \frac{\alpha^4 k(k-1)(k-2)(k-3)}{4!}$$

The form of Eq. (2-73) suggests that the Laguerre polynomials can be generated by the tandem arrangement shown in Fig. 2-16.

In a similar manner, it can be established that the $\phi_r(k)$ satisfy the difference equation,

$$\phi_r(n) = \lambda\phi_r(n-1) + \lambda\phi_{r-1}(n) - \phi_{r-1}(n-1)$$

$$\phi_0(n) = \sqrt{1 - \lambda^2}\lambda^n \qquad (2\text{-}75)$$

and $\phi_0(n)$ satisfies

$$\phi_0(n) = \lambda\phi_0(n-1) + \sqrt{1 - \lambda^2}\delta(n)$$

$$\phi_0(-1) = 0 \qquad (2\text{-}76)$$

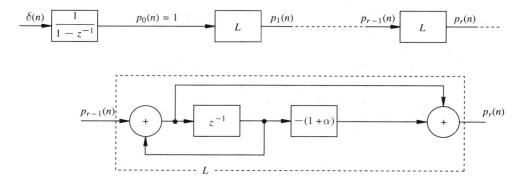

Figure 2-16   Laguerre polynomial generator.

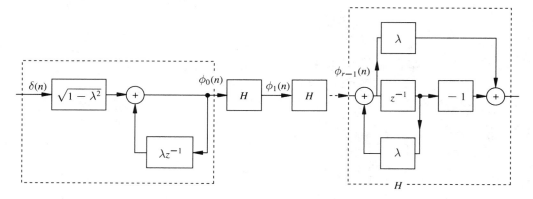

Figure 2-17   Generation of the orthonormal Laguerre sequences.

The orthonormal sequence can thus be generated by the digital network of Fig. 2-17.

2.5.4 THE BINOMIAL AND DISCRETE HERMITE SEQUENCES

The discrete counterparts (R. Haddad, 1971) to the continuous-time orthogonal Hermite family are generated by successive differences of the binomial sequence, defined on the interval $[0, N]$ as

$$x_0(k) = \binom{N}{k} = \frac{N!}{(N-k)!k!} \qquad 0 \le k \le N \qquad (2\text{-}77)$$

The other members of this family are

$$x_r(k) \equiv \Delta^r \binom{N-r}{k-r} \qquad r = 0, 1, ..., N$$

$$= \binom{N}{k} \sum_{m=0}^{r} (-2)^m \binom{r}{m} \frac{k^{(m)}}{N^{(m)}}$$

$$= \binom{N}{k} H_r(k) \qquad r = 0, 1, ..., N; k = 0, 1, ..., N \qquad (2\text{-}78)$$

where $k^{(m)}$ is the forward factorial function, Eq. (2-68).

The polynomials appearing in Eq. (2-78) are the discrete Hermite polynomials. They are symmetric with respect to index and argument:

$$H_r(k) = H_k(r) \qquad 0 \le r, k \le N \qquad (2\text{-}79)$$

The other members of the binomial family are generated by the recurrence relation

$$x_{r+1}(k) = -x_{r+1}(k-1) + x_r(k) - x_r(k-1) \qquad 0 \le r, k \le 1 \quad (2\text{-}80)$$

with initial values $x_r(-1) = 0, 0 \le r \le N$, and initial sequence $x_0(k) = \binom{N}{k}$.

In transform notation, the binomial sequence $x_0(k)$ is

$$X_0(z) = \sum_{k=0}^{N} \binom{N}{k} z^{-k} = (1 + z^{-1})^N \qquad (2\text{-}81)$$

From the recurrence Eq. (2-80), the other members are

$$X_r(z) = \left(\frac{1 - z^{-1}}{1 + z^{-1}} \right) X_{r-1}(z) = \left(\frac{1 - z^{-1}}{1 + z^{-1}} \right)^r X_0(z) \qquad (2\text{-}82)$$

These last two equations suggest the digital network shown in Fig. 2-18 for generating $x_r(k)$. Note that there are no multiplications in either the recurrence Eq. (2-80) or the network in the figure. Hence this family can be realized using only adders and delay elements. The topic of filters without multipliers is taken up in Chap. 5; there we show that the binomial network can serve as an efficient low-pass, band-pass, or high-pass filter with linear phase and almost Gaussian magnitude response.

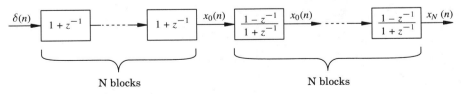

Figure 2-18 *Digital filters for generation of the binomial family.*

The Hermite and binomial sequences are orthogonal on $[0, N]$ with respect to the weighting sequences $\binom{N}{k}$ and $\binom{N}{k}^{-1}$, respectively.

$$\sum_{k=0}^{N} H_r(k) H_s(k) \binom{N}{k} = \sum_{k=0}^{N} \frac{x_r(k) x_s(k)}{\binom{N}{k}} = 2^N \binom{N}{k}^{-1} \delta_{r-s} \quad (2\text{-}83)$$

The associated Hermite and binomial transformation matrices are

$$\mathbf{H} = [H_{rk}]$$
$$\mathbf{X} = [X_{rk}]$$

where we use the notation $H_{rk} = H_r(k)$ and $X_{rk} = x_r(k)$. $\mathbf{H}$ is real and symmetric, and the rows and columns of $\mathbf{X}$ are orthogonal:

$$\mathbf{H} = \mathbf{H}^T$$
$$\mathbf{X}^2 = 2^N \mathbf{I} \quad (2\text{-}84)$$

The orthogonality expressed in Eq. (2-83) suggests that by proper normalization, the Hermite transform can provide an orthonormal matrix suitable for signal coding. This *modified* Hermite transform (MHT) is defined by

$$\Phi(r, k) = \frac{\binom{N}{r}^{1/2} \binom{N}{k}^{1/2}}{2^{N/2}} H_r(k) \quad (2\text{-}85)$$

The matrix $\mathbf{\Phi} = [\Phi(r, k)]$ has the property

$$\mathbf{\Phi}\mathbf{\Phi}^T = \mathbf{I} \quad (2\text{-}86)$$

The key advantage of the MHT for signal coding is the efficient three-step process by which the coefficients are calculated (Haddad and Akansu, 1988). The MHT algorithm requires 2N real multiplications compared with approximately $2N \log_2 N$ for the DCT algorithm. The penalty paid for this efficiency is a slightly degraded performance measure of 1 to 2 dB for the MHT, as compared with that of the DCT, for positively correlated signals. For negatively correlated signals, the MHT is slightly better than the DCT.

2.5.5 WALSH FUNCTIONS AND TRANSFORMS

Another orthogonal set used for signal processing is due to J. L. Walsh (1923). We will define the discrete Walsh functions as follows:

$$\text{wal}^{(N)}(j, k) = \prod_{i=0}^{p-1} (-1)^{j_i k_{p-i}} \quad (2\text{-}87)$$

where $N = 2^p$; j_i is the ith bit of the binary representation of j and k_i is similarly the ith bit of k. The function takes on values of 1 and -1 only; its value for any j and k depends on the parity (that is, the number of 1-bits) of the bitwise product (logical AND) between j and the p-bit bit reversal of k. From this it can be seen that the Walsh functions are symmetric in j and k:

$$\mathrm{wal}^{(N)}(j, k) = \mathrm{wal}^{(N)}(k, j) \qquad (2\text{-}88)$$

For any N, there are exactly N Walsh functions, distinguished by j, which ranges from 0 to $N - 1$. The Walsh functions for $N = 16$ are shown in Fig. 2-19(a). This figure shows values only for $k = 0, ..., 15$; but Walsh functions are generally taken to be infinite in extent, repeating with period N; alternatively, for values outside the interval $(0, N - 1)$, the k indices are taken modulo N. We will omit the superscript (N) whenever no ambiguity will result from doing so. If we replace each sample of the discrete Walsh function with a rectangular pulse of unity width, we obtain the continuous Walsh functions, shown in Fig. 2-19(b).

The first two Walsh functions should be singled out for special consideration. For any N,

$$\mathrm{wal}^{(N)}(0, k) = 1 \qquad \text{for all } k$$

and
$$\mathrm{wal}^{(N)}(1, k) = \begin{cases} 1 & k < N/2 \\ -1 & \text{otherwise} \end{cases}$$

There are a number of recurrence relations among the Walsh functions; three of the more useful are

$$\mathrm{wal}(i \oplus j, k) = \mathrm{wal}(i, k)\, \mathrm{wal}(j, k) \qquad (2\text{-}89a)$$

$$\mathrm{wal}(j, k) = \mathrm{wal}(j/2, 2k) \qquad \text{for } j \text{ even} \qquad (2\text{-}89b)$$

$$\mathrm{wal}^{(N/2)}(j, k) = \mathrm{wal}^{(N)}(j, 2k) \qquad (2\text{-}89c)$$

The symbol $\oplus$ indicates addition without carries—that is, the result of an exclusive OR of the corresponding bits of i and j.

The first relation, Eq. (2-89a), follows from the defining equation:

$$\mathrm{wal}(i, k)\, \mathrm{wal}(j, k) = \prod_{b=0}^{p-1} (-1)^{i_b k_{p-b}} \prod_{b=0}^{p-1} (-1)^{j_b k_{p-b}}$$

$$= \prod_{b=0}^{p-1} (-1)^{(i_b \oplus j_b) k_{p-b}}$$

The second relation, Eq. (2-89b), can be seen by examining the binary forms of j and k. Dividing j by 2 shifts j right by one bit; multiplying k

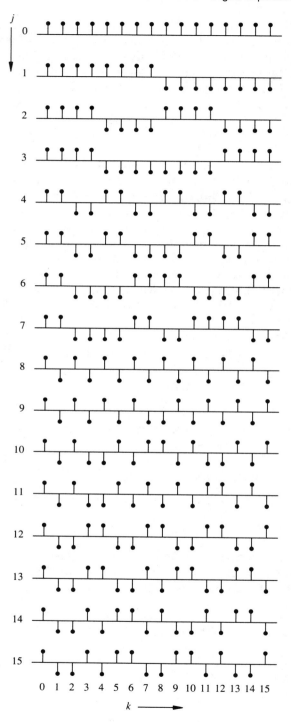

Figure 2-19   (a) *Discrete Walsh functions*

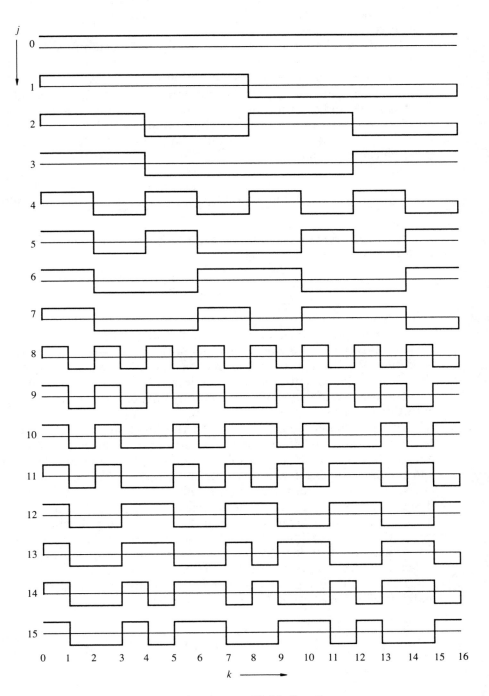

Figure 2-19 (b) Continuous Walsh functions.

by 2 shifts k left by one bit and hence shifts the bit reversal of k right by one bit. The corresponding bits thus line up as before and the product is unchanged.

The third relation, Eq. (2-89c), is also seen by examining the bits of j and k. If k is multiplied by 2, the bit reversal of k is shifted one bit to the right, resulting in the alignment found in $\text{wal}_{N/2}(j, k)$.

In deriving any $\text{wal}(j, k)$ from a predecessor, the total number of positive samples and the total number of negative samples never changes, with the sole exception of the step which obtains $\text{wal}(1, k)$ from $\text{wal}(0, k)$. Since $\text{wal}(1, k)$ has exactly $N/2$ positive samples and $N/2$ negative samples, we conclude that

$$\sum_{j=0}^{N-1} \text{wal}(j, k) = \begin{cases} N & k = 0 \\ 0 & \text{otherwise} \end{cases} \tag{2-90}$$

It follows from Eq. (2-90) and Eq. (2-89a) that the Walsh functions for any N are an orthogonal set:

$$\sum_{k=0}^{N-1} \text{wal}(i, k)\,\text{wal}(j, k) = N\delta_{i-j} \tag{2-91}$$

Alternative definitions. The definition of the Walsh functions that we have used thus far is only one of three more-or-less equivalent definitions that have appeared in the literature. The definitions all yield the same functions, but in different orders. The ordering resulting from our definition is known as the Paley, or dyadic, order. We chose to start with this because it has a simple defining equation.

Walsh's original definition ordered the functions by the number of zero crossings—that is, sign changes between consecutive points. This number is the nearest analog to the frequency of sinusoidal functions, and is commonly expressed by the term, *sequency*. The sequency (Harmuth, 1969) is one half the average number of zero crossings per second. If the interval, $(0, N - 1)$, is normalized to a duration of one second, then the sequency is given as one half the zero crossings in this interval. (Since the domain is periodic in N, the zero crossing between sample $N - 1$ and sample 0, if it occurs, must be counted.) Walsh's ordering is called, not surprisingly, the Walsh, or sequency, order.

The third ordering derives from the Hadamard matrix. Hadamard matrices of order 2^p are defined recursively:

$$\mathbf{H}_1 = \begin{bmatrix} 1 & 1 \\ 1 & 1 \end{bmatrix} \tag{2-92}$$

and for all other powers of 2,

$$\mathbf{H}_{2n} = \begin{bmatrix} \mathbf{H}_n & \mathbf{H}_n \\ \mathbf{H}_n & -\mathbf{H}_n \end{bmatrix} = \mathbf{H}_1 \times \mathbf{H}_n \tag{2-93}$$

where $\times$ indicates the Kronecker product. For example,

$$\mathbf{H}_2 = \begin{bmatrix} 1 & 1 & 1 & 1 \\ 1 & -1 & 1 & -1 \\ 1 & 1 & -1 & -1 \\ 1 & -1 & -1 & 1 \end{bmatrix}$$

In the Hadamard, or natural, order, the jth function is found in row j of the appropriately-sized Hadamard matrix.

Sequency-ordered Walsh functions have been compared to trigonometric functions, and it has become customary to name the even functions $\mathrm{cal}(j,k)$ and the odd functions $\mathrm{sal}(j,k)$, on the analogy of cos and sin:

$$\begin{aligned} \mathrm{cal}(j,k) &= \mathrm{wal}(2j,k) & 0 < j \le N/2 \\ \mathrm{sal}(j,k) &= \mathrm{wal}(2j-1,k) & 0 \le j < N/2 \end{aligned} \tag{2-94}$$

Here j is equal to the sequency. The close correspondence between sequency and frequency can be seen from Fig. 2-20, in which sine and cosine functions are superimposed on continuous Walsh functions in sequency order. For $\mathrm{cal}(j,k)$, the superimposed sinusoid is $\cos 2\pi jk$; for $\mathrm{sal}(j,k)$, the sinusoid is $\sin 2\pi jk$. Notice that each flat part of the function contains exactly one extremum of the sinusoid.

Transforms.   The Walsh transform (WT) of a sequence $\{x(n)\}$ of N samples is defined as follows:

$$X(k) = \sum_{n=0}^{N-1} x(n)\,\mathrm{wal}^{(N)}(n,k) \tag{2-95}$$

By using the orthogonality property, it is easy to show that the inverse transform is given by

$$x(n) = \frac{1}{N}\sum_{k=0}^{N-1} X(k)\,\mathrm{wal}^{(N)}(n,k) \tag{2-96}$$

We will write the WT of $x(n)$ as $W[x(n)]$.

It is easy to show that the WT is linear. If $X(k) = W[x(n)]$ and $Y(k) = W[y(n)]$, then for any constants a and b,

$$W[ax(n) + bx(n)] = aX(k) + bY(k)$$

There is no shifting property of the kind found in the Z transform and the DFT, and no convolution property. To see why this is so, consider the WT of the sequence $x(n-t)$:

$$W[x(n-t)] = \sum_{t=0}^{N_1} x(n-t)\,\mathrm{wal}(n,k)$$

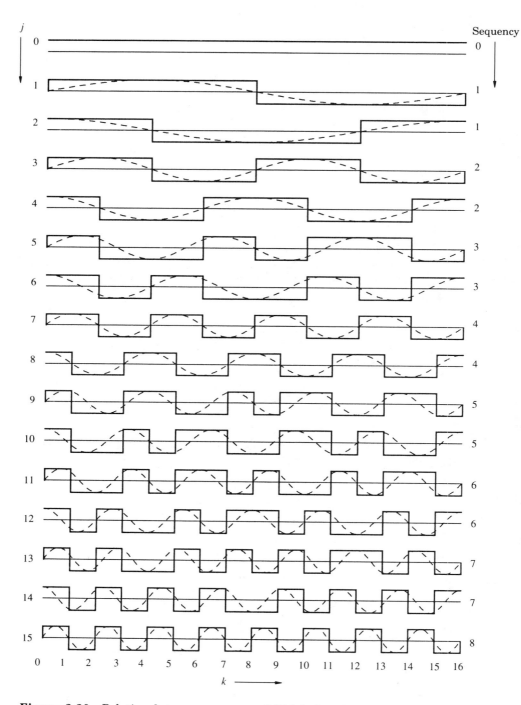

Figure 2-20 *Relation between sequency of Walsh functions and frequency of sinusoids.*

The next step would ordinarily be to make a change of variables, letting $q = n - t$ and obtaining

$$W[x(n - t)] = \sum_{t=0}^{N_1} x(q)\,\text{wal}(q + t, k)$$

but here we reach a dead end, because there is no arithmetic shifting relation for the Walsh functions. We meet a similar problem in convolution. Kennett (1970) describes a logical convolution, which we may write as

$$x(n) \otimes y(n) = \sum_{t=0}^{N_1} x(n)y(t \oplus n)$$

It is fairly clear that a logical convolution property follows immediately from (2-89a) and the orthogonality property:

$$W[x(n) \otimes y(n)] = X(k)Y(k)$$

Part of the appeal of the WT is the simplicity of its hardware implementation; since all multiplications are by ± 1, simple inversion, gating, and summing hardware is sufficient. Because of the absence of a conventional convolution property, however, the WT does not lend itself readily to digital filtering.

The WT has also been proposed as an intermediate step to the DFT. If the time sequence is represented as a sum of Walsh functions, then the DFT can be found from the WT by summing the DFTs of the Walsh functions: if $X(k)$ is the WT of $x(n)$, then

$$x(n) = \sum_{k=0}^{N-1} X(k)\,\text{wal}(n, k)$$

then
$$\mathfrak{F}\{x(n)\} = \sum_{p=0}^{N-1} \sum_{k=0}^{N-1} X(k)\,\text{wal}(p, k)W^{-pk}$$

$$= \sum_{k=0}^{N-1} X(k)\mathfrak{F}\{\text{wal}(n, k)\}$$

For small-sized transforms, it may not be unreasonable to maintain tables of the DFT coefficients of the Walsh functions, in which case the conversion from the WT to the DFT is a matter of a small number of table lookups and multiplications.

2.5.6 THE DISCRETE LEGENDRE POLYNOMIALS

The Laguerre functions introduced earlier provide a means of representing discrete signals in terms of weighted polynomials on the semi-infinite
interval $[0, \infty)$. The discrete Hermite polynomials weighted by the binomial sequence, on the other hand, are suitable for the representation
of signals which are Gaussian-like on a finite interval. Typically, such
signals fall off rapidly near the two endpoints of the $[0, N]$ interval. The
discrete Legendre polynomials are used to represent signals in terms
of uniformly weighted polynomials on a finite interval; these have been
used by Morrison (1969) in connection with finite-memory polynomial
filters.

Let $L_r(k)$ be a degree-r polynomial in k on the interval $0 \leq k \leq N-1$,
of the form

$$L_r(k) = 1 + \sigma_{r1}k + \sigma_{r2}k^{(2)} + \cdots + \sigma_{rs}k^{(r)} \qquad r = 0, ..., N - 1 \quad (2\text{-}97)$$

When the σ_{rs} are chosen to satisfy the uniformly weighted orthogonality
relation,

$$\sum_{k=0}^{N-1} L_r(k)L_s(k) = C_r^2 \delta_{r-s} \qquad (2\text{-}98)$$

the result is

$$\sigma_{rs} = \begin{cases} (-1)^s \binom{r}{s} \binom{r+s}{s} \frac{1}{(N-1)^{(s)}} & 0 \leq s \leq r \\ 0, & s > r \end{cases} \qquad (2\text{-}99)$$

and the norms are

$$C_r^2 = \frac{(N + r + 1)^{(r+1)}}{(2r + 1)(N)^{(r)}} \qquad (2\text{-}100)$$

Thus the discrete Legendre polynomials are

$$L_r(k) = \sum_{s=0}^{r} \sigma_{rs}k^{(s)} \qquad (2\text{-}101)$$

Scaling $L_r(k)$ by C_r gives the orthonormal discrete Legendre transform
(DLT),

$$\Phi(r, k) = \frac{L_r(k)}{C_r} \qquad (2\text{-}102)$$

If $\{\Phi\}$ is the matrix $[\Phi(r, k)]$, then

$$\Phi\Phi^T = I$$

The rows of the DLT matrix for $N = 8$ are shown in Fig. 2-21. These sequences are symmetrical about $N/2$ for even-indexed rows and skew-symmetric for odd-indexed rows. The similarity to the DCT is evident from these sketches; however, a fast DLT has yet to be developed.

2.5.7 THE KARHUNEN-LOÈVE TRANSFORM

The transformations studied up to this point are independent of the input signal statistics. The Karhunen-Loève transform (KLT, Netravali and Limb, 1980) is a unitary transform matched to the statistics of the input signal in such a way that it decorrelates the signal vector **f** and packs a maximum amount of energy into the first $L \leq M$ components of the transformed vector α (where M is the length of **f**). The KLT is thus the optimum least-squares decorrelating transformation.

Let **f** be the signal vector, whose components $\{f(1), f(2), \cdots, f(M)\}$ constitute a sequence of zero-mean, real random variables characterized by the wide-sense stationary (see App. C) correlation function,

$$r(k) = \langle f(n) \cdot f(n + k) \rangle \tag{2-103}$$

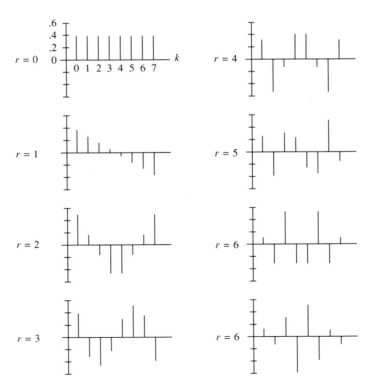

Figure 2-21   *Basis sequences for the discrete Legendre polynomials, $N = 8$.*

or covariance matrix

$$\mathbf{R}_f = \langle \mathbf{ff}^T \rangle = \begin{bmatrix} r_0 & r_1 & \cdots & r_{M-1} \\ r_1 & r_0 & \cdots & r_{M-2} \\ \vdots & \vdots & \vdots & \\ r_{M-1} & r_{M-2} & \cdots & r_0 \end{bmatrix} \tag{2-104}$$

The latter is a real, symmetric, Toeplitz matrix.

We seek a unitary matrix $\mathbf{A}$ which transforms $\mathbf{f}$ into a coefficient vector α with uncorrelated components and which minimizes the truncation error Eq. (2-59). That is, find $\mathbf{A}$ such that the transformation

$$\alpha = \mathbf{Af} \tag{2-105}$$

$$\mathbf{f} = \mathbf{A}^T \alpha \tag{2-106}$$

results in a diagonal covariance matrix

$$\mathbf{R}_\alpha = \langle \alpha \alpha^T \rangle = \Lambda = \mathrm{diag}(\lambda_1, \lambda_2, \cdots, \lambda_M) \tag{2-107}$$

and minimizes a mean-squared error measure.

The basic result of the KLT theory is that the *rows* of the optimal matrix $\mathbf{A}$ are the orthonormalized eigenvectors of $\mathbf{R}_f$; that is,

$$\mathbf{A} = \begin{bmatrix} \phi_1^T \\ \phi_2^T \\ \vdots \\ \phi_M^T \end{bmatrix} \tag{2-108}$$

where

$$\mathbf{R}_f \phi_i = \lambda_i \phi_i \tag{2-109}$$

and λ_i are the eigenvalues of $\mathbf{R}_f$, that is, the roots of the characteristic polynomial

$$p(\lambda) = \det[\lambda \mathbf{I} - \mathbf{R}_f] = \prod_{i=1}^{M} (\lambda - \lambda_i) \tag{2-110}$$

Since $\mathbf{R}_f$ is a covariance matrix, its eigenvalues are ≥ 0.

From Eq. (2-107), these eigenvalues may be recognized as just the variances of the spectral coefficients $\{\alpha_i\}$. If we order these in nonincreasing order, $\lambda_1 \geq \lambda_2 \geq \cdots \geq \lambda_M \geq 0$, we see that the maximum energy is compacted into the first $L \leq M$ coefficients corresponding to the eigenvectors with the largest eigenvalues.

Rewriting Eq. (2-105) and Eq. (2-106) as

$$\mathbf{f} = \sum_{i=1}^{M} \alpha_i \phi_i \tag{2-111}$$

$$\alpha_i = \mathbf{f}^T \boldsymbol{\phi}_i \tag{2-112}$$

and noting that
$$\hat{\mathbf{f}} = \sum_{i=1}^{L} \alpha_i \boldsymbol{\phi}_i \tag{2-113}$$

we conclude that the KLT diagonalizes the covariance matrix of α and minimizes the mean-squared truncation error,

$$J = \langle (\mathbf{f} - \hat{\mathbf{f}})^T (\mathbf{f} - \hat{\mathbf{f}}) \rangle \tag{2-114}$$

It is left as an exercise (Prob. 2-11) to show that Eq. (2-114) reduces to

$$J = \sum_{i=L+1}^{M} \lambda_i \tag{2-115}$$

The foregoing is readily extended to the case where $\mathbf{f}$ is non-zero-mean and possibly complex valued. In this instance, we take the mean and covariance of the input as

$$\langle \mathbf{f} \rangle = \mathbf{m}$$

$$\mathbf{C}_f = \langle (\mathbf{f} - \mathbf{m})(\mathbf{f} - \mathbf{m})^T \rangle$$

The transformation is now

$$\alpha = \mathbf{A}(\mathbf{f} - \mathbf{m})$$

and the *rows* of $\mathbf{A}$ are the *conjugates* of the eigenvectors of $\mathbf{C}_f$.

While the KLT is the optimum transform, it suffers from two serious practical drawbacks:

1. The KLT is input-dependent, as we have seen, and must be calculated for each signal model on which it operates. (A closed-form result for a Markov-1 stationary process was derived by U. Grenander and G. Szegö, 1969. See also A. K. Jain, 1989.)

2. Even after the KLT basis vectors are determined for a given source model, the transformation of Eq. (2-105) requires $O(M^2)$ multiply/add operations. As we shall see in Chap. 4, certain fixed transforms possess fast algorithms which allow computation of the coefficient vector in far fewer operations. The FFT and the DCT, in particular, require $O(M \log_2 M)$ multiplications.

Additionally, it has been observed that for ρ large (on the order of 0.9), the fixed but suboptimal DCT performs quite comparably to the KLT. See discussions in A. K. Jain (1981), A. N. Netravali and J. O. Limb (1980), and N. S. Jayant and P. Noll (1984).

2.6 SUMMARY

We have considered two questions relating to the representation of a continuous signal in discrete form. The first is the simple matter of how to accomplish the conversion so that the corresponding analog signal can be reconstructed to the desired degree of accuracy; the fundamental result—the sampling theorem—was extended to the case of nonuniformly-spaced samples. The second question is how the discrete form can be represented economically. This second problem has many different types of solution: probably the simplest is the use of nonuniform quantization levels; other solutions rely mostly on various types of orthogonal expansion, which have been explored in depth in this chapter. One possibility which was omitted is that of characterizing the signal in terms of a model of its source; this is touched on in App. E, which reviews the use of linear prediction as a way of deriving an all-pole model of the signal source.

PROBLEMS

2.1. Suppose a band-limited signal of bandwidth $W/2$ is amplitude-modulated by a carrier at some frequency $f_0 > W/2$. Determine the sampling rate necessary to represent this signal.

2.2. An analog signal is sampled at twice its Nyquist rate.
      (a) Sketch the spectra of the analog signal and of the sampled signal.
      (b) Suppose every other sample is discarded. Sketch the spectrum of the resulting sequence $x_0, x_2, x_4, \ldots$.
      (c) Suppose, instead of discarding samples, we insert a sample with value 0 in between every pair of existing samples. Sketch the spectrum of the resulting sequence $x_0, 0, x_1, 0, x_2, 0, x_3, \ldots$.

2.3. A real, band-limited analog signal has a bandwidth extending from $4f_0$ to $5f_0$ and from $-4f_0$ to $-5f_0$. The signal is sampled at a frequency of $2f_0$. Sketch the spectra of the analog signal and the sampled signal. Suggest how such a process could be extended to yield a baseband signal from $-f_0$ to f_0 without heterodyning.

2.4. Compact-disc audio systems use a sampling rate of 44.1 kHz and 16-bit samples. (a) Assuming a loading factor of 0 dB, what is the dynamic range of the system? (b) If it is desired to cover a frequency range out to 20 kHz, sketch the specifications of the anti-aliasing filter that must precede the A/D converter.

2.5. Determine the compression and expansion functions for a Panter-Dite quantizer given the PDF

$$f(x) = \begin{cases} 1 - |x| & |x| < 1 \\ 0 & \text{otherwise} \end{cases}$$

2.6. Prove Eqs. (2-44) and (2-45).

2.7. Prove Eq. (2-58).

2.8. Prove the orthogonality of the Laguerre sequences, Eq. (2-65).

2.9. Show that the Walsh transform of the unit impulse is $n = 1$.

2.10. Show that for any Hadamard matrix, $\mathbf{H}_N \mathbf{H}_N = N\mathbf{I}$, where $\mathbf{I}$ is the order-N identity matrix.

2.11. Show that the mean-squared truncation error of the KLT, Eq. (2-114), is given by Eq. (2-115).

REFERENCES

N. Ahmed, et al. "On notation and definitions of terms related to a class of complete orthogonal functions." *IEEE Trans.*, vol. EMC-15, pp 75–80, May 1973.

N. Ahmed and K. R. Rao. *Orthogonal Transforms for Digital Signal Processing*. Berlin: Springer-Verlag, 1975.

N. Ahmed, et al. "A generalized discrete transform." *Proc. IEEE*, vol. 59, pp. 1360–1362, Sept. 1961.

D. A. Bell. "Walsh functions and Hadamard matrices." *Electron. Lett.*, vol. 2, pp. 340–341, Sept. 3, 1966.

N. M. Blackman. "Spectral analysis with sinusoids and Walsh functions, *IEEE Trans.*, vol. AES-7, pp. 900–905, Sept. 1971.

J. E. Gibbs and F. O. Pichler, "Comments on transformation of Fourier power spectra into Walsh power spectra." *Proc. 1971 Symp. Applications of Walsh Functions*, pp. 51–54.

U. Grenander and G. Szegö. *Toeplitz Forms and Their Applications*. Berkeley, Cal., Univ. of California Press, 1958.

R. A. Haddad. "A class of orthogonal non-recursive binomial filters." *IEEE Trans.*, vol. AU-19, no. 4, pp. 296–304, Dec. 1971.

R. A. Haddad and A. N. Akansu, "A new orthogonal transform for signal coding." *IEEE Trans.*, vol. ASSP-36, no. 9, pp. 1404–1411, Sept. 1988.

H. F. Harmuth. "A generalized concept of frequency and some applications". *IEEE Trans.*, vol. IT-14, pp. 375–382, May 1968.

K. W. Henderson. "Some notes on the Walsh functions." *IEEE Trans.*, vol. EC-13, pp. 50–52, Feb. 1964.

N. S. Jayant, ed. *Waveform Quantization and Coding.* New York: IEEE Press, 1976.

N. S. Jayant and P. Noll. *Digital Coding of Waveforms.* Englewood Cliffs, N. J.: Prentice-Hall, 1984.

B. L. N. Kennett. "A note on the finite Walsh transform." *IEEE Trans.*, vol. IT-16, pp. 489–491, July 1970.

R. Kitai and K.-H. Siemens. "Discrete Fourier transform via Walsh transform." *IEEE Trans.*, vol. ASSP-27, no.3, p. 288, June 1979.

J. Max. "Quantizing for minimum distortion." *IRE Trans. Inform. Theory*, vol. IT-6, pp. 7–12, Mar. 1960.

W. E. Milne. *Numerical Calculus.* Princeton: Princeton Univ. Press, 1949.

N. Morrison. *Introduction to Sequential Smoothing and Prediction.* New York: McGraw-Hill, 1969.

A. N. Netravali and J. O. Limb. "Picture coding: a review." *Proc. IEEE*, vol. 68, no. 3, pp. 366–406, March 1980.

H. Nyquist. "Certain topics in telegraph communication theory." *AIEE Trans.*, p. 617, April 1928.

R. E. A. C. Paley. "A remarkable series of orthogonal functions." *Proc. London Math. Soc.*, vol. 34, pp. 241–279, 1932.

P. F. Panter and W. Dite. "Quantization distortion in pulse-count modulation with nonuniform spacing of levels." *Proc. IRE*, vol. 39, no. 1, pp. 44–48, Jan. 1951.

Proc. Symposium on Applications of Walsh Functions, 1971–1978

G. S. Robinson. "Logical convolution and discrete Walsh and Fourier power spectra." *IEEE Trans.*, vol. AU-20, no. 4, pp. 271–280, Oct. 1972.

J. L. Shanks. "Computation of the fast Walsh-Fourier transform." *IEEE Trans.*, vol. C-18, pp. 457–459, May 1969.

C. E. Shannon. "Communication in the presence of noise." *Proc. IRE*, vol. 37, no. 1, pp. 10–21, Jan. 1949.

K.-H. Siemens and R. Kitai. "Digital Walsh-Fourier analysis of periodic waveforms." *IEEE Trans.*, vol. IM-18, pp. 316–320, Dec. 1969.

———. "A nonrecursive equation for the Fourier transform of a Walsh function." *IEEE Trans.*, vol. EMC-15, pp. 81–83, May 1973.

Y. Tadakoro and T. Higuchi. "Discrete Fourier transform computation via the Walsh transform." *IEEE Trans.*, vol. ASSP-26, no. 3, p 236–240, June 1978.

———. "Comments on 'Discrete Fourier transforms via Walsh Transform'." *IEEE Trans.*, vol. ASSP-27, no. 3, pp. 295–296, June 1979.

J. L. Walsh. "A closed set of orthogonal functions." *Am. J. Math.*, vol. 45, pp. 5–24, 1923.

J. L. Yen. "On nonuniform sampling of band-limited signals." *IRE Trans. Circ. Theory*, pp. 251–257, Dec. 1956.

CHAPTER 3

■ ■

TWO-DIMENSIONAL
SIGNAL PROCESSING

In this chapter we extend the one-dimensional (1D) signal and system analysis to two dimensions. The extrapolation to dimensions greater than two is then implicit.

While a 1D signal $v(t)$ has one independent variable (usually time, as in speech signals), a two-dimensional (2D) signal $v(x, y)$ has two independent variables, usually spatial. Typical 2D signals are photographic images, from visible or infrared light or from x-rays. A time-varying image would be characterized by three independent variables, two spatial and one temporal.

At an abstract conceptual level, multidimensional signals and systems are no different from the 1D variety. For example, the concepts of linearity, spectra, and low-pass filtering or smoothing carry over. The concept of causality, however, is basically related to time as the independent variable. An image is a function of space rather than time, so the concept of causality is not meaningful. We are free to select the origin at any convenient point and to process the image pixels in any direction. On an implementation level and on a test basis, multidimensional systems can be far more complex. For example, we can factor 1D polynomials into a product of first-order polynomials and thereby study stability and system response. This statement cannot be made for a 2D polynomial. The stability test for a 1D system can be readily determined from the location of the system poles. For a 2D system, poles are surfaces rather than points, so the stability question becomes problematical.

2D signal-processing algorithms offer considerably more flexibility in implementation, however. For example, one can process image data in a formally non-causal way, and the image can be sampled in various geometric patterns, not necessarily in a rectangular grid.

In this chapter, we extend 1D results to 2D wherever possible, indicate where significant differences lie, and provide caveats for the reader whenever extrapolations are hazardous. For a more detailed exposition of the subject, the reader is referred to texts dedicated to multidimensional systems (Dudgeon and Mersereau, 1984; Jain, 1989; Lim, 1990; Gonzales and Wintz, 1987; and Papoulis, 1968). The application of 2D signal theory to image processing is taken up in Chapter 9. In this chapter, first we define 2D analog signals and systems following essentially a Fourier optics approach (Papoulis, 1968). Next, we sample the analog signal to create the 2D discrete-variable signal and the attendant processor, the 2D discrete-variable filter. Finally, we describe this filter by its transfer function, impulse response, or difference equations.

3.1 TWO-DIMENSIONAL ANALOG SIGNALS AND SYSTEMS

This section introduces and summarizes the properties of 2D analog signals and systems. It is a codification and 2D extrapolation of the 1D analog Fourier theory developed in App. B. Sampling these analog signals leads us to the 2D discrete sequences that are the main concern of this chapter.

3.1.1 ANALOG SIGNALS

A 2D signal is simply a function $f(x, y)$ of two variables. We say it is analog when x and y range over a continuum of values, as shown in Fig. 3-1; it is simply a surface.

A fundamental signal is the pulse, defined by

$$p_A(x, y) = \begin{cases} 1, & x, y \in A \\ 0, & x, y \notin A \end{cases} \qquad (3\text{-}1)$$

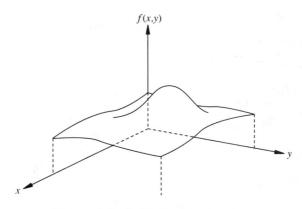

Figure 3-1   *A 2D analog signal.*

We can define pulses over certain domains by selecting the region A. For example, A could be a rectangle or a quarter circle, as in Fig. 3-2(a) and (b).

The 2D (Dirac) impulse $\delta(x, y)$ is defined by the property,

$$\int_{-\infty}^{\infty} \int_{-\infty}^{\infty} f(x, y)\delta(x - a, y - b)\, dx\, dy = f(a, b) \qquad (3\text{-}2)$$

The impulse is commonly drawn as an arrow, as shown in Fig. 3-2(c). For manipulations, $\delta(x, y)$ can be represented as the product of two 1D impulses:

$$\delta(x, y) = \delta(x) \cdot \delta(y)$$

With $f(x, y) = 1$ in Eq. (3-2), we see that the 2D impulse has unit "volume."

The 2D impulse is the probe signal used for linear, spatially invariant systems. The response of such a system to an impulse is the 2D impulse response or *point-spread function*.

3.1.2 LINEAR SYSTEMS AND CONVOLUTION

A 2D system is a mapping of a set of inputs $f(x, y)$ into a set of outputs $g(x, y)$, denoted by

$$g(x, y) = T[f(x, y)] \qquad (3\text{-}3)$$

The mapping is linear if superposition holds, i.e., if

$$T[af_1(x, y) + bf_2(x, y)] = aT[f_1(x, y)] + bT[f_2(x, y)] \qquad (3\text{-}4)$$

where a and b are arbitrary constants and f_1 and f_2 are arbitrary allowable inputs.

The system is shift-invariant if

$$T[f(x - c_1, y - y_1)] = g(x - c_1, y - y_1) \qquad (3\text{-}5)$$

when

$$T[f(x, y)] = g(x, y) \qquad (3\text{-}6)$$

And a system is linear, shift-invariant if both Eqs. (3-4) and (3-5) hold.

The point-spread function for a linear, shift-invariant system is the 2D system response to a Dirac delta function:

$$T[\delta(x, y)] \triangleq h(x, y) \qquad (3\text{-}7)$$

By superposition and the shift-invariant property, since

$$f(x, y) = \int_{-\infty}^{\infty} \int_{-\infty}^{\infty} f(\xi, \eta)\delta(x - \xi, y - \eta)\, d\xi\, d\eta$$

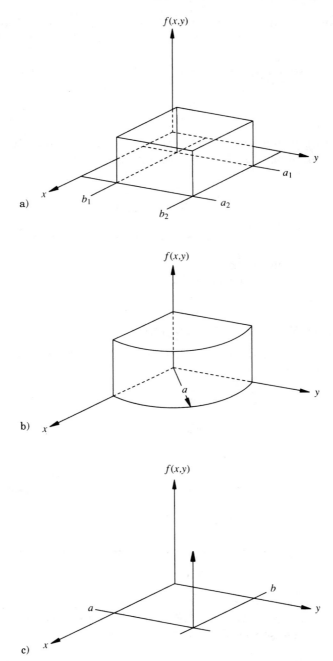

Figure 3-2 *Elementary 2D analog signals: (a) pulse over rectangular region*
$a_1 < x \le a_2,\ b_1 < y \le b_2;$ *(b) pulse over quarter-circle region* $x^2 + y^2 \le a;\ x$ *and* y
positive; (c) impulse $\delta(x - a, y - b).$

we have

$$g(x, y) = \int_{-\infty}^{\infty} \int_{-\infty}^{\infty} f(\xi, \eta) T[\delta(x - \xi, y - \eta)] \, d\xi \, d\eta$$

$$= \int_{-\infty}^{\infty} \int_{-\infty}^{\infty} f(\xi, \eta) h(x - \xi, y - \eta) \, d\xi \, d\eta \qquad (3\text{-}8)$$

$$= f ** h \qquad (3\text{-}9)$$

Equation (3-8) defines the 2D convolution of the functions $f(x, y)$ and $h(x, y)$, and the double asterisk in Eq. (3-9) indicates a 2D convolution.

Thus input and output are related by a 2D convolution of the input with the 2D impulse response. These properties are all direct extensions of the corresponding 1D properties.

Several other properties can be derived, the proofs of which are left as homework exercises; these are summarized in Table 3.1.

The system is separable if

$$h(x, y) = h_1(x) \cdot h_2(y)$$

For this important special case, the convolution reduces to

$$g(x, y) = \int_{-\infty}^{\infty} \int_{-\infty}^{\infty} f(\xi, \eta) h_1(x - \xi) h_2(y - \eta) \, d\xi \, d\eta$$

$$= \int_{-\infty}^{\infty} h_2(y - \eta) \left[\int_{-\infty}^{\infty} f(\xi, \eta) h_1(x - \xi) \, d\xi \right] d\eta$$

$$= h_2(y) * [h_1(x) * f(x, y)] \qquad (3\text{-}10)$$

Thus the 2D convolution reduces to two 1D convolutions. For a fixed but arbitrary value of η, one performs the 1D convolution over the first dummy variable ξ

Table 3.1.   Properties of 2D Linear, Shift-Invariant Systems

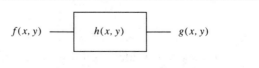

Impulse response: $T[\delta(x, y)] = h(x, y)$
Convolution: $T[f(x, y)] = f(x, y) ** h(x, y)$
Separable systems: $h(x, y) = h_1(x) \cdot h_2(y)$
                                $g(x, y) = h_2(y) * [h_1(x) * f(x, y)]$
                                $= h_1(x) * [h_2(y) * f(x, y)]$

$$\phi(x,\eta) = \int_{-\infty}^{\infty} f(\xi,\eta)h_1(x-\xi)\,d\xi \qquad (3\text{-}11)$$

Then $\phi(x,\eta)$ is convolved with $h_2(y)$ via

$$\int_{-\infty}^{\infty} \phi(x,\eta)h_2(y-\eta)\,d\eta = g(x,y) \qquad (3\text{-}12)$$

Conceptually, this extension is evident. Computationally, it is an order of magnitude more tedious to execute than the 1D case.

3.1.3 SYSTEM FUNCTION AND FOURIER TRANSFORM

As in the 1D case, we define the sinusoidal probe signal by the separable complex exponential,

$$f(x,y) = e^{j(\omega_1 x + \omega_2 y)} = e^{j\omega_1 x} \cdot e^{j\omega_2 y} \qquad (3\text{-}13)$$

The response to this probe signal is

$$g(x,y) = \int_{-\infty}^{\infty}\int_{-\infty}^{\infty} e^{j[\omega_1(x-\xi)+\omega_2(y-\eta)]}h(\xi,\eta)\,d\xi\,d\eta \qquad (3\text{-}14)$$

$$= H(\omega_1,\omega_2)e^{j(\omega_1 x + \omega_2 y)} \qquad (3\text{-}15)$$

where $H(\omega_1,\omega_2) = \mathfrak{F}\{h(x,y)\}$

$$= \int_{-\infty}^{\infty}\int_{-\infty}^{\infty} h(\xi,\eta)e^{-j(\omega_1\xi+\omega_2\eta)}\,d\xi\,d\eta \qquad (3\text{-}16)$$

is the Fourier transform of $h(x,y)$. Conversely, the inversion formula is

$$h(x,y) = \left(\frac{1}{2\pi}\right)^2 \int_{-\infty}^{\infty}\int_{-\infty}^{\infty} H(\omega_1,\omega_2)e^{j(\omega_1 x + \omega_2 y)}\,d\omega_1\,d\omega_2 \qquad (3\text{-}17)$$

The notation for the 2D Fourier transform and its inverse is similar to the 1D notation,

$$h(x,y) \leftrightarrow H(\omega_1,\omega_2) \qquad (3\text{-}18)$$

Since the complex sinusoidal signal is separable, the output can be obtained by successive 1D convolutions. In the transform domain, this translates into successive 1D transforms. Specifically, the two stages are:

1. Take the 1D transform of $f(x,y)$ with respect to x to obtain

$$\Gamma(\omega_1,y) = f(x,y)e^{-j\omega_1 x}\,dx \qquad (3\text{-}19)$$

2. Do the 1D transform of $\Gamma(\omega_1, y)$ with respect to y:

$$F(\omega_1, \omega_2) = \int_{-\infty}^{\infty} \Gamma(\omega_1, y) e^{-j\omega_2 y} \, dy \qquad (3\text{-}20)$$

The inversion formula, Eq. (3-17), is similarly derived in two stages.

The most important result in linear system theory carries over intact to the 2D case. The Fourier transform of the 2D convolution of two functions is equal to the product of the respective 2D transforms: If

$$f_1(x, y) \leftrightarrow F_1(\omega_1, \omega_2)$$
$$f_2(x, y) \leftrightarrow F_2(\omega_1, \omega_2)$$

then

$$\boxed{f_1(x, y) ** f_2(x, y) \leftrightarrow F_1(\omega_1, \omega_2) \cdot F_2(\omega_1, \omega_2)} \qquad (3\text{-}21)$$

Table 3.2.   Theorems and Properties of the 2D Fourier Transform[†]

Let $f(x, y) \leftrightarrow F(\omega_1, \omega_2)$.

1. $\bar{f}(x, y) \leftrightarrow \bar{F}(-\omega_1, -\omega_2)$.	conjugate time function
2. $F(x, y) \leftrightarrow (2\pi)^2 f(-\omega_1, -\omega_2)$.	symmetry property
3. $f(x - x_0, y - y_0) \leftrightarrow e^{-j(\omega_1 x_0 + \omega_2 y_0)} F(\omega_1, \omega_2)$.	spatial shift
4. $f(x, y) e^{j(\alpha_1 x + \alpha_2 y)} \leftrightarrow F(\omega_1 - \alpha_1, \omega_2 - \alpha_2)$.	frequency shift
5. $f(ax, by) \leftrightarrow \dfrac{1}{|ab|} F\left(\dfrac{\omega_1}{a}, \dfrac{\omega_2}{b}\right)$.	scaling
6. $\dfrac{\partial}{\partial x} f(x, y) \leftrightarrow j\omega_1 F(\omega_1, \omega_2)$.	spatial differentiation
$\quad x f(x, y) \leftrightarrow j \dfrac{\partial}{\partial \omega_1} F(\omega_1, \omega_2)$.	frequency differentiation
7. $xy f(x, y) \leftrightarrow -\dfrac{\partial^2}{\partial \omega_1 \partial \omega_2} F(\omega_1, \omega_2)$.	
8. If $f(x, y)$ is real, then $F(\omega_1, \omega_2) = \bar{F}(-\omega_1, -\omega_2)$. real signals	
$\quad$ If $f(x, y)$ is real and even, $F(\omega_1, \omega_2)$ is real and even.	
9. If $f(x, y) = f(x) \cdot f(y)$, then $F(\omega_1, \omega_2) = F(\omega_1) \cdot F(\omega_2)$	
10. If $\rho(x, y) = f(x, y) ** f(-x, -y)$, then $\qquad \mathfrak{F}\{\rho(x, y)\} = |F(\omega_1, \omega_2)|^2$	

[†] Overbar indicates the complex conjugate.

Table 3.3.   Two-Dimensional Fourier Transform Pairs

$f(x, y)$	$F(\omega_1, \omega_2)$
$\delta(x, y)$	1
$\delta(x - a, y - b)$	$e^{-j(a\omega_1 + b\omega_2)}$
$e^{j(\beta_1 x + \beta_2 y)}$	$(2\pi)^2 \delta(\omega_1 - \beta_1, \omega_2 - \beta_2)$
$e^{-(ax^2 + by^2)}$	$\dfrac{\pi}{\sqrt{ab}} e^{-(\omega_1^2/a + \omega_2^2/b)}$
$\text{rect}(x/a, y/b)$	$ab \; \text{sinc}(a\omega_1/2) \; \text{sinc}(b\omega_2/2)$
$\text{tri}(x/a, y/b)$	$ab/4 \; \text{sinc}^2(a\omega_1/4) \; \text{sinc}^2(b\omega_2/4)$
$\displaystyle\sum_m \sum_n \delta(x - m\,\Delta x, y - n\,\Delta y)$	$\dfrac{(2\pi)^2}{\Delta x\, \Delta y} \displaystyle\sum_m \sum_n \delta\left(\omega_1 - \dfrac{m 2\pi}{\Delta x}, \omega_2 - \dfrac{n 2\pi}{\Delta y}\right)$

Definitions:

$$\text{rect}(x/a) = \begin{cases} 1, & |x/a| < 1/2 \\ 0, & \text{otherwise} \end{cases}$$

$$\text{tri}(x/a) = \begin{cases} 1 - 2|x|/a, & |x/a| < 1/2 \\ 0, & \text{otherwise} \end{cases}$$

$$\text{rect}(u, v) = \text{rect}(u) \cdot \text{rect}(v)$$

$$\text{tri}(u, v) = \text{tri}(u) \cdot \text{tri}(v)$$

From this equation, it is easy to establish Parseval's theorem,

$$\int_{-\infty}^{\infty} \int_{-\infty}^{\infty} |f(\xi, \eta)|^2 \, d\xi \, d\eta = \left(\frac{1}{2\pi}\right)^2 \int_{-\infty}^{\infty} \int_{-\infty}^{\infty} |F(\omega_1, \omega_2)|^2 d\omega_1 \, d\omega_2 \quad (3\text{-}22)$$

Other theorems and properties of the 2D Fourier transform can be found in Papoulis (1968). These are summarized in Table 3.2; Table 3.3 gives a selection of 2D transform pairs.

3.2 SAMPLING 2D SIGNALS

Consider a band-limited signal $f(x, y)$, that is, one in which the Fourier transform vanishes outside a finite region R, as in Fig. 3-3:

$$F(\omega_1, \omega_2) = 0 \qquad (\omega_1, \omega_2) \notin R \tag{3-23}$$

The two-dimensional sampling function $i(x, y)$ shown in Fig. 3-4 is defined by[†]

[†]Summation signs without explicit upper and lower limits run from $-\infty$ to $+\infty$.

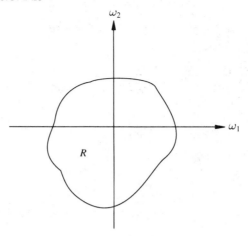

Figure 3-3   *Band-limited spectrum: outside region R, $F(\omega_1, \omega_2) = 0$.*

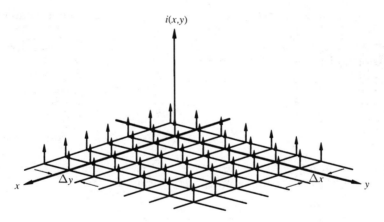

Figure 3-4   *2D impulse sampling function (the "bed-of-nails" function).*

$$i(x, y) = \sum_m \sum_n \delta(x - m \, \Delta x, y - n \, \Delta y) \tag{3-24}$$

The Fourier transform of $i(x, y)$ is shown (Papoulis, 1977) to be

$$I(\omega_1, \omega_2) = \frac{2\pi}{\Delta x} \frac{2\pi}{\Delta x} \sum_m \sum_n \delta \left(\omega_1 - m \frac{2\pi}{\Delta x}, \omega_2 - n \frac{2\pi}{\Delta x} \right) \tag{3-25}$$

Observe that, just as a 1D impulse train transforms into another 1D impulse train, a 2D periodic repetition of impulses transforms into another 2D periodic repetition of impulses. The impulse-sampled function is

$$f_s(x, y) = f(x, y) \cdot i(x, y) \tag{3-26}$$

$$= \sum_m \sum_n f(m\,\Delta x, n\,\Delta y)\delta(x - m\,\Delta x, y - n\,\Delta y) \qquad (3\text{-}27)$$

Since a product in the spatial domain transforms to a convolution in the frequency domain and $I(\omega_1, \omega_2)$ is a train of impulses, the convolution is easy and

$$F_s(\omega_1, \omega_2) = \mathfrak{F}\{f(x, y) \cdot i(x, y)\}$$

$$= \left(\frac{1}{2\pi}\right)^2 F(\omega_1, \omega_2) ** I(\omega_1, \omega_2) \qquad (3\text{-}28)$$

$$= \frac{1}{\Delta x\,\Delta y} \sum_m \sum_n F(\omega_1 - m\frac{2\pi}{\Delta x}, \omega_2 - n\frac{2\pi}{\Delta y}) \qquad (3\text{-}29)$$

Equation (3-29) asserts that the transform of the sampled signal is the periodic repetition of $F(\omega_1, \omega_2)$ at a spacing of $2\pi/\Delta x, 2\pi/\Delta y$, as shown in Fig. 3-5.

Now assume that the spacings Δx and Δy are small enough that the lobes in Fig. 3-5 do not overlap—that is to say, there is no aliasing. Then we can recover the lobe centered at the origin by multiplying the transform of the sampled signal by an ideal low-pass filter,

$$H(\omega_1, \omega_2) = \begin{cases} 1 & (\omega_1, \omega_2) \in R \\ 0 & (\omega_1, \omega_2) \notin R \end{cases} \qquad (3\text{-}30)$$

Thus $\qquad H(\omega_1, \omega_2) \cdot F_s(\omega_1, \omega_2) = F(\omega_1, \omega_2) \qquad (3\text{-}31)$

In the spatial domain, we can invert Eq. (3-31) to obtain an expression for $f(x, y)$ in terms of its samples $f(m\,\Delta x, n\,\Delta y)$, via

$$f(x, y) = h(x, y) ** f_s(x, y) \qquad (3\text{-}32)$$

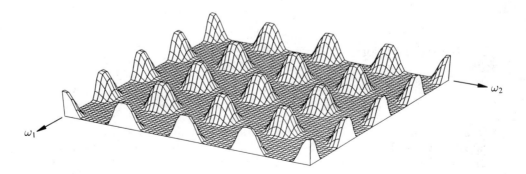

Figure 3-5   *Spectrum of sampled band-limited signal.*

To obtain an explicit expression, we choose $H(\omega_1, \omega_2)$ to be the rectangular pulse function

$$H(\omega_1, \omega_2) = \begin{cases} 1 & |\omega_1| < \pi/\Delta x, |\omega_2| < \pi/\Delta y \\ 0 & \text{otherwise} \end{cases}$$

The resulting impulse response is the 2D sinc function

$$h(x, y) = \left(\frac{\sin \theta_1 x}{\theta_1 x} \right) \left(\frac{\sin \theta_2 y}{\theta_2 y} \right) \tag{3-33}$$

where $\theta_1 \triangleq \pi/\Delta x$ and $\theta_2 \triangleq \pi/\Delta y$.

Now it is straightforward to convolve Eq. (3-33) with Eq. (3-27) to obtain the 2D version of the Shannon reconstruction formula,

$$f(x, y) = \sum_m \sum_n f(m \, \Delta x, n \, \Delta y) h(x - m \, \Delta x, y - n \, \Delta y) \tag{3-34}$$

One form of the Fourier spectrum of the sampled 2D signal is the periodic repetition form of Eq. (3-29). Another form is obtained by taking the term-by-term Fourier transform of Eq. (3-27). Since

$$\mathfrak{F}\{\delta(x - m \, \Delta x, y - n \, \Delta y)\} = e^{-j(m\Delta x + n\Delta y)} \tag{3-35}$$

we have

$$F_s(\omega_1, \omega_2) = \sum_m \sum_n f(m\Delta x, n\Delta y) e^{-j(m\Delta x \omega_1 + n\Delta y \omega_2)} \tag{3-36}$$

Equations (3-29) and (3-36) are different but equivalent representations of the spectrum of the sampled 2D signal. In fact, Eq. (3-36) can be recognized as the 2D Fourier series expansion of the periodic function $F_s(\omega_1, \omega_2)$. The spectrum of the sampled signal is thus given in terms of the original spectrum $F(\omega_1, \omega_2)$ by Eq. (3-29), or in terms of the spatial samples $f(m\Delta x, n\Delta y)$ by Eq. (3-36). These results are also direct extensions of the 1D frequency-domain Poisson sum formula, App. B.

The 2D sampling theorem provides the theoretical bound for image reconstruction from its samples. Of course, an image that is spatially limited cannot be band-limited in frequency. Hence the Shannon formula is regarded as a guide and an upper bound rather than as an explicit formula for image reconstruction.

3.3 2D DISCRETE SIGNALS AND OPERATIONS

Two-dimensional sequences and operations are obvious and simple extensions of the 1D versions encountered in Chap. 1. In this section, we present a succinct listing of these corresponding definitions and properties.

3.3.1 SIGNAL REPRESENTATION

The previous section dealt with sampling a 2D spatially-continuous signal. In that context, it was convenient (as in the 1D case) to use the
artifice of the 2D Dirac impulse train as the sampling function. The
information, however, lies in the samples, not in the impulses, and the
samples can be represented simply as 2D sequences without recourse to
Dirac delta functions. This is the view adopted from here on. Furthermore, it is convenient to normalize the Δx and Δy spacings to unity.
Thus our discrete-variable sequence is $x(n_1, n_2)$, shown in the the three-
dimensional sketch of Fig. 3-6. The value of x at each grid point is
represented by the height of the sample at that point. Another representation is given in Fig. 3-7(a), which shows the extent of the image by
heavy dots, and the value of x is indicated alongside each dot. A similar representation shows the dots, but not the sample values, and the
tabular representation shown in Fig. 3-7(b) gives the numerical values
themselves as a 2D array.

For monochrome image representation, each dot, or *pixel*, is represented by a theoretically infinite number of gray levels. In practice, the
gray scale would be quantized to 2^B levels, where B is the number of
bits per pixel.

3.3.2 A COLLECTION OF SIGNALS

It is convenient at this point to define the region of support of a signal as
the region in the (n_1, n_2) plane outside of which the sequence is zero. The
support (or extent) of a 2D signal is the spatial equivalent of the temporal
duration of a 1D signal. Later in this chapter we regard 2D FIR and IIR
filters as those with finite- (or infinite-) support impulse responses. We
now provide a listing of 2D versions of standard 1D discrete signals.

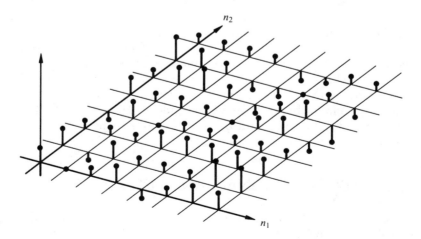

Figure 3-6   *2D discrete-variable sequence.*

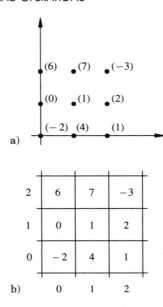

Figure 3-7 *(a) 2D sequence represented as values on a grid; (b) 2D sequence represented as an array.*

1. The *2D* (Kronecker) *impulse* is

$$\delta(n_1, n_2) = \delta(n_1) \cdot \delta(n_2) = \begin{cases} 1 & n_1 - n_2 = 0 \\ 0 & \text{otherwise} \end{cases} \tag{3-37}$$

and a 2D *line impulse* is just

$$x(n_1, n_2) = \delta(n_1) \cdot 1 = \begin{cases} 1 & n_1 = 0 \\ 0 & \text{otherwise} \end{cases} \tag{3-38}$$

These are shown in Fig. 3-8.

2. The unit *step function*, shown in Fig. 3-8(c), is

$$u(n_1, n_2) = \begin{cases} 1 & n_1 \geq 0, n_2 \geq 0 \\ 0 & \text{otherwise} \end{cases} \tag{3-39}$$

This step function is said to have first-quadrant support. It should be clear that

$$u(n_1, n_2) = \sum_{k_2=-\infty}^{n_1} \sum_{k_1=-\infty}^{n_2} \delta(k_1, k_2) \tag{3-40}$$

3. Another important sequence is the exponential

$$x(n_1, n_2) = a^{n_1} b^{n_2} \tag{3-41}$$

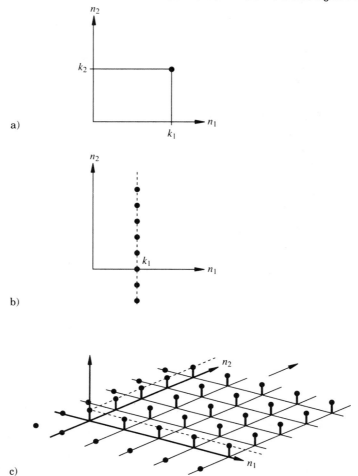

Figure 3-8   (a) Impulse $\delta(n_1 - k_1, n_2 - k_2)$; (b) line impulse $\delta(n_1 - k_1)$; (c) unit step function $u(n_1, n_2)$.

where a and b can be complex. Exponential sequences are the eigenfunctions of linear, shift-invariant systems. For the case where $a = e^{j\omega_1}$, $b = e^{j\omega_2}$, we obtain the exponential sinusoid,

$$x(n_1, n_2) = e^{j(\omega_1 n_1 + \omega_2 n_2)} \tag{3-42}$$

The step, exponential, and complex sinusoid may be generated by sampling their continuous precursors. Thus for example,

$$e^{j(\Omega_1 x + \Omega_2 y)}\bigg|_{\substack{x=n_1\Delta x \\ y=n_2\Delta y}} = e^{j(\Omega_1 n_1 \Delta x + \Omega_2 n_2 \Delta y)}$$

$$= e^{j(\omega_1 n_1 + \omega_2 n_2)}$$

In our context, Ω_1 and Ω_2 are spatial frequencies of dimension rad/unit length. Then $\omega_1 = \Omega_1 \Delta x$ and $\omega_2 = \Omega_2 \Delta y$ are normalized dimensionless frequencies. Similarly, the real exponential signal arises from

$$e^{\alpha x} e^{\beta y} \Big|_{\substack{x = n_1 \Delta x \\ y = n_2 \Delta y}} = (e^{\alpha \Delta x})^{n_1} (e^{\beta \Delta y})^{n_2}$$

$$= a^{n_1} b^{n_2}$$

Separable Sequences.   A signal $x(n_1, n_2)$ is separable if it can be expressed as a product of a function of n_1 alone and a function of n_2 alone:

$$x(n_1, n_2) = x_1(n_1) \cdot x_2(n_2) \tag{3-43}$$

In this respect, it is similar to the property of a bivariate probability density function when the random variables are independent. These separable sequences considerably reduce the computational burden in signal processing. All the signals mentioned above—the impulse, exponential, and complex sinusoid—are separable.

Periodic Sequences.   Sequences have a rectangular periodicity if

$$x(n_1, n_2) = x(n_1 + N_1, n_2 + N_2) \qquad \forall n_1, n_2 \tag{3-44}$$

The integers N_1 and N_2 are the horizontal and vertical periods, respectively. Other periodicities are possible; sequences with hexagonal sampling and periodicity are described by Mersereau (1979). The advantages of this form of signal representation and filtering are storage and computational efficiency and a higher degree of circular symmetry.

3.3.3 OPERATIONS ON SEQUENCES

The elementary operations on sequences are direct extensions of the 1D operators described in Chap. 1; these are summarized in Table 3.4. A linear, shift-invariant system can be represented by combinations of operations 1, 2, and 5 in Table 3.4.

We now look to establishing the properties of the 2D (scalar) system shown symbolically in Fig. 3-9. The system operator is represented by $T[\cdot]$. This operator is linear if

$$T[a x_1(n_1, n_2) + b x_2(n_1, n_2)] = a T[x_1(n_1, n_2)] + b T[x_2(n_1, n_2)] \tag{3-45}$$

for all x_1 and x_2 and arbitrary constants a and b. For a linear system, we define the impulse response or discrete point-spread function as

$$T[\delta(n_1 - k_1, n_2 - k_2)] = h(n_1, k_1; n_2, k_2) \tag{3-46}$$

Table 3.4. Operations on 2D Sequences

1. Addition of two sequences	$y(n_1, n_2) = x_1(n_1, n_2) + x_2(n_1, n_2)$
2. Multiplication by scalar constant	$y(n_1, n_2) = ax(n_1, n_2)$
3. Multiplication by spatially varying gain	$y(n_1, n_2) = a(n_1, n_2)x(n_1, n_2)$
4. Memoryless nonlinear function ϕ	$y(n_1, n_2) = \phi[x(n_1, n_2)]$
5. Shift by k_1, k_2	$y(n_1, n_2) = x(n_1 - k_1, n_2 - k_2)$
6. Superposition property	$x(n_1, n_2) = \sum_{k_1} \sum_{k_2} x(k_1, k_2) \cdot$ $\delta(n_1 - k_1, n_2 - k_2)$

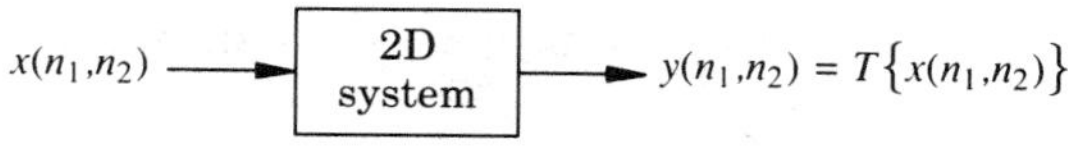

$$x(n_1, n_2) \longrightarrow \boxed{\begin{array}{c} \text{2D} \\ \text{system} \end{array}} \longrightarrow y(n_1, n_2) = T\{x(n_1, n_2)\}$$

Figure 3-9 *A 2D discrete-variable system.*

This is the response observed at (n_1, n_2) to a Kronecker impulse at (k_1, k_2). From the superposition property for linear systems, we have

$$T[x(n_1, n_2)] = T \left[\sum_{k_1} \sum_{k_2} x(k_1, k_2)\delta(n_1 - k_1, n_2 - k_2) \right]$$

$$= \sum_{k_1} \sum_{k_2} x(k_1, k_2)T[\delta(n_1 - k_1, n_2 - k_2)]$$

$$= \sum_{k_1} \sum_{k_2} x(k_1, k_2)h(n_1, k_1; n_2, k_2) \qquad (3\text{-}47)$$

The operator is shift-invariant if

$$T[x(n_1, n_2)] = y(n_1, n_2)$$

implies $\qquad T[x(n_1 - m_1, n_2 - m_2)] = y(n_1 - m_1, n_2 - m_2)] \qquad (3\text{-}48)$

$$\forall m_1, m_2$$

Consequently, for a linear, shift-invariant (LSI) system, the impulse response reduces to

$$h(n_1, k_1; n_2, k_2) = h(n_1 - k_1, n_2 - k_2) \qquad (3\text{-}49)$$

and the superposition summation Eq. (3-47) reduces to the 2D discrete convolution

$$y(n_1, n_2) = \sum_{k_1} \sum_{k_2} x(k_1, k_2) h(n_1 - k_1, n_2 - k_2) \qquad (3\text{-}50)$$

$$= x ** h = h ** x \qquad (3\text{-}51)$$

Henceforth, in this chapter, we deal only with LSI systems.

Fig. 3-10 shows two systems connected in cascade and in parallel. It is clear that the composite impulse response h_c is the convolution of the two individual impulse responses for the cascade connection and that h_c is the sum of the individual responses for the parallel connection.

The 2D convolution of Eq. (3-50) simplifies considerably when the signal (or the system, for that matter) is separable. In that case as we said previously,

$$x(n_1, n_2) = x_1(n_1) \cdot x_2(n_2) \qquad (3\text{-}52)$$

and the convolution can be expressed as

$$y(n_1, n_2) = \sum_{k_1} x_1(k_1) \sum_{k_2} x_2(k_2) h(n_1 - k_1, n_2 - k_2) \qquad (3\text{-}53)$$

Equation Eq. (3-53) asserts that the (separable) 2D convolution can be done in two stages:

1.
$$g(n_1, n_2) = \sum_{k_2} x_2(k_2) h(n_1, n_2 - k_2)$$

$$= x_2 * h \qquad (3\text{-}54)$$

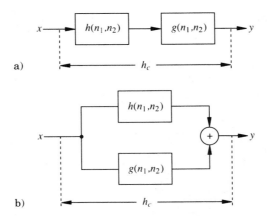

a)

b)

Figure 3-10 *Linear, shift-invariant systems: (a) cascaded $h_c = h ** g$; (b) in parallel $h_c = h + g$.*

In Eq. (3-54), n_1 is held fixed and a 1D convolution is performed in the second (vertical) dimensional variable. This is done for each n_1 to obtain an intermediate array $g(n_1, n_2)$.

2. This intermediate array is 1D-convolved with $x_1(k_1)$ in the first variable, using

$$y(n_1, n_2) = \sum_{k_1} x_1(k_1)g(n_1 - k_1, n_2) \qquad (3\text{-}55)$$

The variable n_2 is fixed for each convolution; then it is incremented by 1 and the convolution is repeated. This process in continued until each row of $g(n_1, n_2)$ has been convolved with $h_2(k_2)$ and the final array $y(n_1, n_2)$ is obtained.

These steps can be summarized by

$$g(n_1, n_2) = h_1(n_1, n_2) * x_2(n_2) \qquad (3\text{-}56)$$
$$\text{(repeated 1D convolutions along } N \text{ columns of } h)$$
$$y(n_1, n_2) = g(n_1, n_2) * x_1(n_1) \qquad (3\text{-}57)$$
$$\text{(repeated convolutions along } M \text{ rows of } g)$$

Thus a total of $M \cdot N$ 1D convolutions are needed to do the separable 2D convolution. An even simpler case arises when h and x are each separable; then y reduces to the product of two 1D convolutions.

A detailed example of a (nonseparable) 2D convolution is given as Example 3.4 later in this chapter.

3.4 FREQUENCY RESPONSE AND THE FOURIER TRANSFORM

When the complex sinusoid $e^{j(\omega_1 n_1 + \omega_2 n_2)}$ is applied to the LSI system, the result is

$$y(n_1, n_2) = \sum_{k_1} \sum_{k_2} e^{j[\omega_1(n_1 - k_1) + \omega_2(n_2 - k_2)]} h(k_1, k_2)$$

$$= H(\omega_1, \omega_2)e^{j(\omega_1 n_1 + \omega_2 n_2)} \qquad (3\text{-}58)$$

where the sums are from $-\infty$ to ∞ and[†]

$$H(\omega_1, \omega_2) = \sum_{k_1} \sum_{k_2} h(k_1, k_2)e^{-j(\omega_1 k_1 + \omega_2 k_2)} \qquad (3\text{-}59)$$

[†]In the 1D case, we used $H(e^{j\omega})$ as the system function. A consistent notation for the 2D system would be $H(e^{j\omega_1}, e^{j\omega_2})$, which is clumsy, so we opt for $H(\omega_1, \omega_2)$.

is the discrete system function or frequency response of the filter (also sometimes called the *modulation transfer function*). The system function $H(\omega_1, \omega_2)$ is periodic in ω_1 and ω_2 with period 2π:

$$H(\omega_1, \omega_2) = H(\omega_1 + 2\pi, \omega_2) = H(\omega_1, \omega_2 + 2\pi) \qquad (3\text{-}60)$$

Consequently Eq. (3-59) is identified as the Fourier series expansion of a periodic function of two variables, and $h(k_1, k_2)$ is the Fourier-series coefficient. Hence, from 2D Fourier series theory, we evaluate

$$h(n_1, n_2) = \left(\frac{1}{2\pi}\right) \int_{-\pi}^{\pi} \int_{-\pi}^{\pi} H(\omega_1, \omega_2) e^{j(\omega_1 n_1 + \omega_2 n_2)} \, d\omega_1 \, d\omega_2 \qquad (3\text{-}61)$$

The pair Eqs. (3-59) and (3-61), constitute a 2D Fourier-transform pair, with notation

$$h(n_1, n_2) \leftrightarrow H(\omega_1, \omega_2) \qquad (3\text{-}62)$$

A sufficient condition for the existence of $H(\omega_1, \omega_2)$ is that $h(n_1, n_2)$ be stable in the sense that

$$\sum_{n_1} \sum_{n_2} |h(n_1, n_2)| < \infty \qquad (3\text{-}63)$$

More generally (and not surprisingly), the convolution sum can be transformed into a product of transforms: that is, if

$$y(n_1, n_2) = x(n_1, n_2) ** h(n_1, n_2)$$

then
$$Y(\omega_1, \omega_2) = X(\omega_1, \omega_2) \cdot H(\omega_1, \omega_2) \qquad (3\text{-}64)$$

In pattern-recognition studies, we want to determine the closest match between a test image and one of a set of standard or reference images. A time-honored approach to this problem, in both 1D and 2D, is to obtain the correlation between the test signal and each of the references and declare the closest match as the one with the largest correlation.

In 1D, the correlation of $h(n)$ and $g(n)$ is the convolution of $h(n)$ and $\overline{g}(-n)$. Using the symbol $\circ$ for correlation, we have

$$h(n) * \overline{g}(-n) = h(n) \circ g(n) = \sum_{n} \overline{g}(n) h(n + k) \qquad (3\text{-}65)$$

The corresponding transforms are

$$H(\omega) \cdot \overline{G}(\omega) \leftrightarrow h(n) \circ g(n) \qquad (3\text{-}66)$$

where the overbar indicates the complex conjugate.

In 2D, we define the correlation between two images by

$$f \circ\circ g = \sum_{n_1} \sum_{n_2} \overline{g}(n_1, n_2) h(n_1 + k_1, n_2 + k_2) = r(k_1, k_2) \qquad (3\text{-}67)$$

or,

$$r = f \circ\circ g \leftrightarrow F(\omega_1, \omega_2) \cdot \overline{G}(\omega_1, \omega_2) = R(\omega_1, \omega_2) \qquad (3\text{-}68)$$

The 2D FFT can be used for efficient computation of the correlation pattern between two images, as described in the next section concerning the DFT.

The DFT is a sampling of the Fourier transform $F(\omega_1, \omega_2)$, and the FFT is a computationally efficient algorithm for calculating the DFT. In the next section, we will discuss how the FFT can be used to compute correlations and convolutions. Before embarking on that, we provide examples of 2D Fourier transforms and signals, and a tabular summary of manipulations and properties of the transform pair defined by Eqs. (3-59) and (3-61).

EXAMPLE 3.1:   We start with a one-dimensional binomial-filter (Haddad, 1971) impulse response

$$h(n) = \begin{cases} 6 & n = 0 \\ 4 & n = \pm 1 \\ 1 & n = \pm 2 \\ 0 & \text{elsewhere} \end{cases} \qquad (3\text{-}69)$$

The corresponding 1D transform is

$$H(\omega) = 6 + 8 \cos \omega + 2 \cos 2\omega = (2 \cos \omega/2)^4 \qquad (3\text{-}70)$$

These are sketched in Fig. 3-11.

We form the separable, symmetric 2D impulse response

$$h(n_1, n_2) = h(n_1) \cdot h(n_2) \qquad (3\text{-}71)$$

and note that the frequency response is

$$H(\omega_1, \omega_2) = 2^8 (\cos \omega_1/2)^4 (\cos \omega_2/2)^4 \qquad (3\text{-}72)$$

The impulse response has the four-quadrant symmetry shown in Fig. 3-12(a). The corresponding separable frequency response is the low-pass characteristic sketched in Fig. 3-12(b). This filter is noteworthy in that it can be implemented using only add operations, as described in Section 5.3.1.

EXAMPLE 3.2:   The ideal low-pass filter shown in Fig. 3-13(a) is defined as

$$H(\omega_1, \omega_2) = \begin{cases} 1 & |\omega_1| < a, |\omega_2| < b \\ 0 & \text{elsewhere} \end{cases} \qquad (3\text{-}73)$$

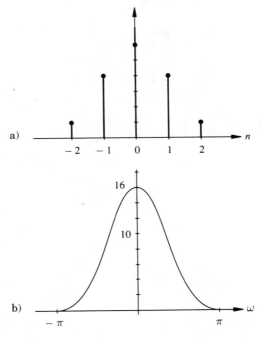

Figure 3-11 *(a) Impulse response, 1D binomial filter; (b) frequency response, 1D binomial filter.*

on the interval, $-\pi \le \omega_1 < \pi, -\pi \le \omega_2 < \pi$. This is clearly separable into

$$H(\omega_1, \omega_2) = H_1(\omega_1) \cdot H_2(\omega_2)$$

where

$$H_1(\omega_1) = 1, \qquad |\omega_1| < a \qquad (3\text{-}74)$$

$$H_2(\omega_2) = 1, \qquad |\omega_2| < b$$

The corresponding impulse response separates into

$$h(n_1, n_2) = h_1(n_1) \cdot h_2(n_2) = \frac{\sin an_1}{\pi n_1} \frac{\sin bn_2}{\pi n_2} \qquad (3\text{-}75)$$

and has infinite support in each dimension.

EXAMPLE 3.3: An ideal low-pass filter with circular symmetry is shown in Fig. 3-14(a):

$$H(\omega_1, \omega_2) = \begin{cases} 1 & \omega_1^2 + \omega_2^2 \le \omega_R^2, \qquad \omega_R < \pi \\ 0 & \text{elsewhere} \end{cases} \qquad (3\text{-}76)$$

This filter is *not* separable; the associated impulse response is

$$h(n_1, n_2) = \left(\frac{1}{2\pi}\right)^2 \int\!\!\int_C e^{j(\omega_1 n_1 + \omega_2 n_2)} \, d\omega_1 \, d\omega_2 \qquad (3\text{-}77)$$

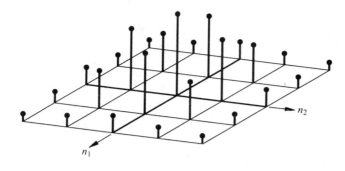

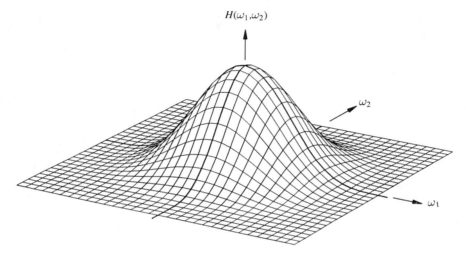

Figure 3-12 *(a) Impulse response, 2D binomial filter; (b) frequency response, 2D binomial filter.*

where C is the circular region, $\omega_1^2 + \omega_2^2 \leq \omega_R^2$. Transforming to polar coordinates, with

$$\rho = \sqrt{\omega_1^2 + \omega_2^2}$$

$$\phi = \tan^{-1}(\omega_2/\omega_1)$$

and letting $\theta = \tan^{-1}(n_2/n_1)$, we obtain

$$h(n_1, n_2) = \left(\frac{1}{2\pi}\right)^2 \int_0^{\omega_R} \int_0^{2\pi} \rho e^{j\rho\sqrt{n_1^2 + n_2^2}\cos(\theta - \phi)}\, d\phi\, d\rho$$

$$= \frac{1}{2\pi} \int_0^{\omega_R} \rho J_0(\rho\sqrt{n_1^2 + n_2^2})\, d\rho$$

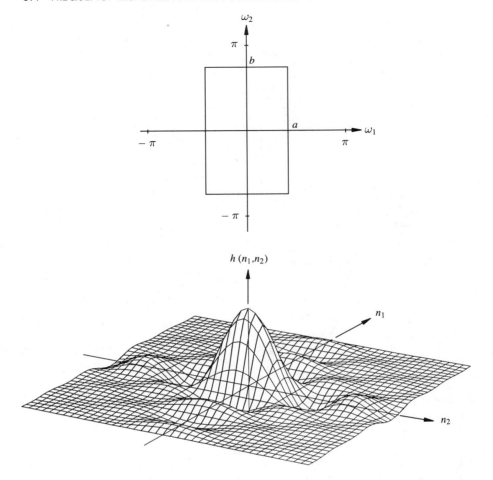

Figure 3-13 *Ideal low-pass filter, rectangular symmetry: (a) frequency response, (b) impulse response.*

$$= \frac{\omega_R}{2\pi} \frac{J_1(\omega_R \sqrt{n_1^2 + n_2^2})}{\sqrt{n_1^2 + n_2^2}}$$

$$= \frac{\omega_R}{2\pi} \frac{J_1(\omega_R m)}{m} \tag{3-78}$$

where
$$m = \sqrt{n_1^2 + n_2^2}$$

The functions J_0 and J_1 are Bessel functions of order zero and one, respectively. Tables of these functions may be found in, for example, CRC (1970). Where $\sin x/x$ is commonly called "sinc x," some authors have used the term, "jinc x" for $J_1(x)/x$. The impulse response is clearly not separable; rather, it has circular symmetry, as seen from Eq. (3-78) and the sketch in Fig. 3-14(b). Note that while

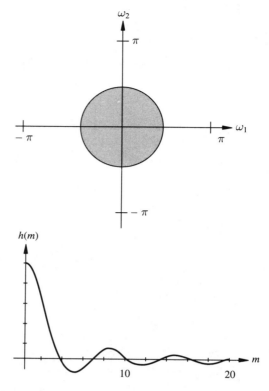

Figure 3-14 *Ideal low-pass filter, circular symmetry: (a) frequency response; (b) impulse response.*

it is true that a circularly symmetric $H(\omega_1, \omega_2)$ implies a circularly symmetric $h(n_1, n_2)$, as in this example, the converse is not necessarily true.

The properties and manipulations of the 2D Fourier transform are mostly extensions of the 1D versions. These are summarized in Table 3.5. The proofs of some of these properties are left as homework exercises.

3.4.1 THE DISCRETE FOURIER TRANSFORM (DFT)

A standard computational problem in Fourier analysis is: given a set of samples $\{x(n)\}$, compute the Fourier transform numerically. That is, for a given $\{x(n)\}$ and[†]

$$X(\omega) = \sum_{n=-\infty}^{\infty} x(n)e^{-j\omega n} = X(\omega + 2\pi) \qquad (3\text{-}79)$$

[†]In this chapter, we use $X(\omega)$ instead of $X(e^{j\omega})$ to denote the Fourier transform of the sampled signal.

Table 3.5.   2D Fourier Transform Properties

Given: $x(n_1, n_2) \leftrightarrow X(\omega_1, \omega_2)$

---

1. Spatial shift/linear phase

   $$x(n_1 - m_1, n_2 - m_2) \leftrightarrow e^{-j(\omega_1 m_1 + \omega_2 m_2)} X(\omega_1, \omega_2)$$

2. Frequency shift/spatial modulation

   $$x(n_1, n_2) e^{j(\theta_1 n_1 + \theta_2 n_2)} \leftrightarrow X(\omega_1 - \theta_1, \omega_2 - \theta_2)$$

3. Differentiation

   $$-jn_i x(n_1, n_2) \leftrightarrow \frac{\partial X(\omega_1, \omega_2)}{\partial \omega_i} \qquad i = 1, 2$$

   $$n_1 n_2 x(n_1, n_2) \leftrightarrow \frac{\partial^2 X(\omega_1, \omega_2)}{\partial \omega_1 \omega_2}$$

4. Transposition

   $$x(n_2, n_1) \leftrightarrow X(\omega_2, \omega_1)$$

5. Reflection

   $$x(-n_1, n_2) \leftrightarrow X(-\omega_1, \omega_2)$$

6. Conjugates

   $$\overline{x}(n_1, n_2) \leftrightarrow \overline{X}(-\omega_1, -\omega_2)$$

7. Real functions $x(n_1, n_2)$

   $$X(\omega_1, \omega_2) = \overline{X}(-\omega_1, -\omega_2)$$

   or $\mathrm{Re}\,\{X(\omega_1, \omega_2)\} = \mathrm{Re}\,\{X(-\omega_1, -\omega_2)\}$

   $\mathrm{Im}\,\{X(\omega_1, \omega_2)\} = -\mathrm{Im}\,\{X(-\omega_1, -\omega_2)\}$

8. Convolution and correlation

   $$x(n_1, n_2) ** y(n_1, n_2) \leftrightarrow X(\omega_1, \omega_2) \cdot Y(\omega_1, \omega_2)$$

   $$x(n_1, n_2) \circ\circ y(n_1, n_2) \leftrightarrow X(\omega_1, \omega_2) \cdot \overline{Y}(\omega_1, \omega_2)$$

9. Parseval's theorem

   $$\sum_{n_1} \sum_{n_2} |x(n_1, n_2)|2 = \left(\frac{1}{2\pi}\right)^2 \int_{-\pi}^{\pi} \int_{-\pi}^{\pi} |X(\omega_1, \omega_2)|^2 d\omega_1 \, d\omega_2$$

---

we wish to compute $X(\omega)$ at N points on the interval $[0, 2\pi)$, or on the interval $(-\pi, \pi]$. In effect, we are computing N *samples* of $X(\omega)$, denoted by

$$\tilde{X}(k) = X(\omega)\big|_{\omega = k 2\pi / N}$$

But we know that $\tilde{X}(k)$ is the DFT of the *periodic* sequence $\tilde{x}(n)$, that is,

$$\tilde{X}(k) = \sum_{n=0}^{N-1} \tilde{x}(n)e^{-j2\pi nk/N} = DFT\{\tilde{x}(n)\}$$

(3-80)

$$\tilde{x}(n) = 1/N \sum_{k=0}^{N-1} \tilde{X}(k)e^{j2\pi nk/N} = (DFT)^{-1}\{\tilde{X}(k)\}$$

where $\tilde{x}(n)$ is the periodic repetition of $x(n)$ with period N:

$$\tilde{x}(n) = \sum_{r=-\infty}^{\infty} x(n - rN) = \tilde{x}(n - N)$$ (3-81)

Equation (3-80) constitutes the discrete Fourier transform (DFT) pair. The interpretation of Eq. (3-80) is vitally important in understanding the DFT. If one were to obtain N uniformly spaced samples of $X(\omega)$ on $[0, 2\pi)$ and use the inverse DFT formula, one would get $\tilde{x}(n)$ rather than $x(n)$. Now if $x(n)$ were of extent $[0, N - 1]$ to start, then there would be no aliasing in Eq. (3-81) and we would obtain $\tilde{x}(n) = x(n)$ on $[0, N - 1]$.

The fast Fourier transform (FFT) is used to calculate $\tilde{X}(\omega)$ and $\tilde{x}(n)$, so care must be exercised in formatting the data for the FFT routine and in interpreting the results.

The 2D version is readily extrapolated from Eqs. (3-80) and (3-81). Thus with the Fourier transform pair

$$x(n_1, n_2) \leftrightarrow X(\omega_1, \omega_2)$$

and $$\tilde{X}(k_1, k_2) \triangleq X(\omega_1, \omega_2)\Big|_{\substack{\omega_1 = k_1 2\pi/N_1 \\ \omega_2 = k_2 2\pi/N_2}}$$ (3-82)

and $$\tilde{x}(n_1, n_2) = \sum_{r_2=-\infty}^{\infty} \sum_{r_1=-\infty}^{\infty} x(n_1 - r_1 N_1, n_2 - r_2 N_2)$$ (3-83)

it follows that $\tilde{x}(n_1, n_2)$ and $\tilde{X}(k_1, k_2)$ are a DFT pair denoted by

$$\tilde{x}(n_1, n_2) \leftrightarrow \tilde{X}(k_1, k_2).$$ (3-84)

Explicitly,

$$\tilde{X}(k_1, k_2) = \sum_{n_1=0}^{N_1-1} \sum_{n_2=0}^{N_2-1} \tilde{x}(n_1, n_2)e^{-j(2\pi n_1 k_1/N_1 + 2\pi n_2 k_2/N_2)}$$

(3-85)

$$\tilde{x}(n_1, n_2) = \frac{1}{N_1 N_2} \sum_{k_1=0}^{N_1-1} \sum_{k_2=0}^{N_2-1} \tilde{X}(k_1, k_2)e^{j(2\pi n_1 k_1/N_1 + 2\pi n_2 k_2/N_2)}$$

The DFT for the $N_1 \times N_2$ array $x(n_1, n_2)$ is computed in two stages:

1. Compute the 1D DFT for each row of the array:

$$\Gamma(k_1, n_2) = \sum_{n_1=0}^{N_1-1} x(n_1, n_2) e_1^{-j2\pi n_1 k_1/N_1} \qquad n_2 = 0, 1, ..., N_2 - 1 \quad (3\text{-}86)$$

For $n_2 = 0$, we scan the 0th row of $[x(n_1, n_2)]$ and perform a 1D N_1-point DFT, $\Gamma(k_1, 0)$ for $0 \le k_1 \le N_1 - 1$. This constitutes the 0th row of the intermediate array $\Gamma(k_1, n_2)$. The FFT routine (Chap. 4) achieves this with $(N_1/2) \log_2 N_1$ complex multiplications. The procedure is repeated N_2 times on all the rows of the input array to yield $\Gamma(k_1, 0)$, $\Gamma(k_1, 1), ..., \Gamma(k_1, N_2 - 1)$. The intermediate array is now complete after $(N_2 N_1/2) \log_2 N_1$ multiplications.

2. Next calculate

$$\tilde{X}(k_1, k_2) = \sum_{n_1=0}^{N_1-1} \Gamma(k_1, n_2) e^{-j2\pi n_2 k_2/N_2} \qquad\qquad (3\text{-}87)$$

by computing the 1D DFT of each *column* of the intermediate array. The result is the final transform array $\tilde{X}(k_1, k_2)$. The process is illustrated schematically in Fig. 3-15. Note that the FFT permits all computations to be done in place, just as in the 1D case. Stage 2 can be completed using N_1 FFTs each of length N_2, requiring another $(N_2 N_1/2) \log_2 N_2$ multiplications. The total for both stages is therefore $(N_2 N_1/2) \log_2(N_1 N_2)$ complex multiplications.

Since any image is a finite array, the aliasing described by Eq. (3-81) can be circumvented and efficient fast transforms can be used to transform, or code, any image. The DFT transformation of Eq. (3-85) can also be represented by the matrix product indicated in Fig. 3-16, where $\tilde{\mathbf{X}} \leftrightarrow \tilde{\mathbf{x}}$ and $\mathbf{W}_{N_i}$ is the matrix $[e^{-j2\pi nk/N_i}]$. Postmultiplication by $\mathbf{W}_{N_1}$ transforms the rows of $\tilde{\mathbf{x}}$ and premultiplication by $\mathbf{W}_{N_2}$ transforms

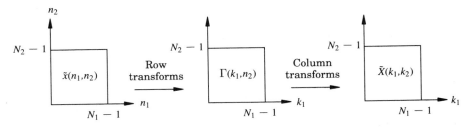

Figure 3-15   *Two-dimensional DFT as row and column one-dimensional DFTs.*

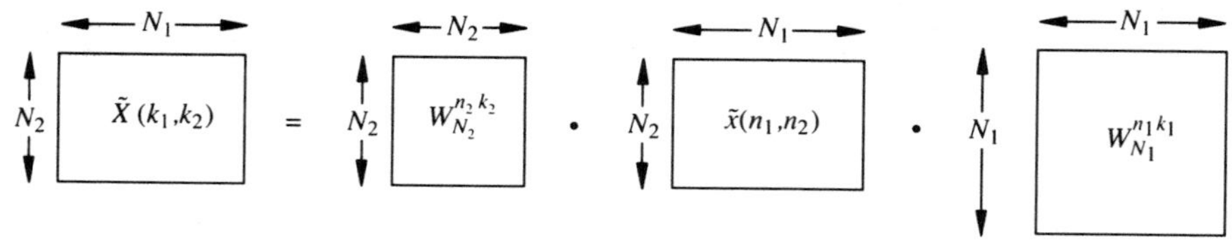

Figure 3-16 *The DFT array as product of three matrices.* $W_{N_1} = e^{-j2\pi/N_1}, W_{N_2} = e^{-j2\pi/N_2}$.

the columns. For example, for an array with three columns and two rows ($N_1 = 3$ and $N_2 = 2$), the DFT is given explicitly by

$$\begin{bmatrix} X(0,0) & X(1,0) & X(2,0) \\ X(0,1) & X(1,1) & X(2,1) \end{bmatrix}$$

$$= \begin{bmatrix} 1 & 1 \\ 1 & -1 \end{bmatrix} \begin{bmatrix} x(0,0) & x(1,0) & x(2,0) \\ x(0,1) & x(1,1) & x(2,1) \end{bmatrix} \begin{bmatrix} 1 & 1 & 1 \\ 1 & e^{-j2\pi/3} & e^{-j4\pi/3} \\ 1 & e^{-j4\pi/3} & e^{-j8\pi/3} \end{bmatrix}$$

(Tildes are omitted here for notational simplicity.) The general form is

$$[X] = [W_{N_2}][x][W_{N_1}] \tag{3-88}$$

Another important consequence of the DFT equations is the fact that if $\{\tilde{x}(n_1, n_2) \leftrightarrow \tilde{X}(k_1, k_2)\}$ and $\{\tilde{h}(n_1, n_2) \leftrightarrow \tilde{H}(k_1, k_2)\}$, then the product $\tilde{X}(k_1, k_2) \cdot \tilde{H}(k_1, k_2)$ is the DFT of the *cyclical* convolution of $\tilde{x}(n_1, n_2)$ and $\tilde{h}(n_1, n_2)$. In 1D, this property is simply

$$\tilde{H}(k) \cdot \tilde{X}(k) \leftrightarrow \sum_{k=0}^{N-1} \tilde{x}(n-k)\tilde{h}(k)$$

where $\tilde{x}(n)$ and $\tilde{h}(n)$ are periodic with period N.

As in the 1D case, we can fool the DFT into performing a linear convolution by suitably padding the time (or space) function with zeros. The linear convolution of a P_1-point sequence with a P_2-point sequence is a sequence whose extent (or support) is $P_1 + P_2 - 1$. Choose N as the least power of 2 greater than or equal to $P_1 + P_2 - 1$, and let

$$\tilde{x} = \begin{cases} x(n) & 0 \le n \le P_1 - 1 \\ 0 & P_1 \le n \le N - 1 \end{cases}$$

$$\tilde{h} = \begin{cases} h(n) & 0 \le n \le P_2 - 1 \\ 0 & P_2 \le n \le N - 1 \end{cases}$$

On the interval $[0, N-1]$, the linear convolution of $x(n)$ and $h(n)$ equals the cyclical convolution of $\tilde{x}(n)$ and $\tilde{h}(n)$. Hence we can compute $\tilde{X}(k)$

and $\tilde{H}(k)$, the N-point DFTs of $\tilde{x}(n)$ and $\tilde{h}(n)$, using the FFT. The product, $\tilde{Y}(k) = \tilde{H}(k) \cdot \tilde{X}(k)$ is then computed and then inverted by the inverse FFT. The resulting $\tilde{y}(n)$ equals $y(n)$, the linear convolution of $x(n)$ and $h(n)$.

The same technique of zero padding can be used to do 2D convolutions via the FFT, as shown in the following example.

EXAMPLE 3.4: Figure 3-17 illustrates the steps in the 2D convolution of the sequences shown in Fig. 3-17(a) using the form shown. In 1D convolution, $h(k)$ is reversed in time to form $h(-k)$, then shifted by n to obtain $h(n - k)$. This last sequence is multiplied point by point by the fixed $x(k)$ and summed to produce $y(n)$.

In the 2D case, we reverse the h sequence by flipping it over both coordinate axes to obtain $h(-k_1, -k_2)$ as in Fig. 3-17(b). Next we multiply corresponding slots in $h(-k_1, -k_2)$ and $x(k_1, k_2)$ and sum (noting that nonoverlapping entries yield zero). The result is $y(0, 0) = 5$, which is entered in the output array in slot $(0, 0)$. In (c), the flipped h array is shifted up one slot; corresponding entries in the shifted h and the fixed x are multiplied and added to give $y(0, 1)$, which is entered in the appropriate slot of the y array. In (d), the h array is shifted up again; in this case, the product of the shifted h and the fixed x is 0. So $y(0, 2) = 0$ is entered in the output array as shown. This completes the first column $y(0, n_2)$ of the output array.

To do the second column of the output array $y(1, n_2)$, the flipped $h(-k_1, -k_2)$ of Fig. 3-17(b) is shifted one slot to the *right*, as shown in Fig. 3-17(e). The products of the corresponding entries are taken and summed to give $y(1, 0)$. Next, in Fig. 3-17(f), the h array of Fig. 3-17(e) is shifted up one slot, multiplied, and summed as previously to give $y(1, 1)$. The next entry $y(1, 2)$ is obtained similarly, as shown in Fig. 3-17(g), and an additional upward shift (not shown) yields 0 for $y(1, 3)$. The second column is now complete. The third column is obtained similarly, and the final output array is shown in Fig. 3-17(h).

EXAMPLE 3.5: We now obtain the convolution of $x(n_1, n_2)$ and $h(n_1, n_2)$ of Example 3.4, using the DFT.

First, the support of $y(n_1, n_2)$ is determined to be 3×3. Hence we pad x and h with zeroes to form the 4×4 arrays

$$\mathbf{x} = \begin{bmatrix} 1 & 2 & 0 & 0 \\ 3 & 4 & 0 & 0 \\ 0 & 0 & 0 & 0 \\ 0 & 0 & 0 & 0 \end{bmatrix} \qquad \mathbf{h} = \begin{bmatrix} 5 & 6 & 0 & 0 \\ 0 & 7 & 0 & 0 \\ 0 & 0 & 0 & 0 \\ 0 & 0 & 0 & 0 \end{bmatrix}$$

Now we can compute the DFT arrays

$$[X(k_1, k_2)] = \mathbf{W}[x(n_1, n_2)]\mathbf{W} = DFT\{x(n_1, n_2)\}$$
$$[H(k_1, k_2)] = \mathbf{W}[h(n_1, n_2)]\mathbf{W} = DFT\{h(n_1, n_2)\}$$

where

$$\mathbf{W} = \begin{bmatrix} 1 & 1 & 1 & 1 \\ 1 & -j & 1 & j \\ 1 & -1 & 1 & -1 \\ 1 & j & 1 & -j \end{bmatrix}$$

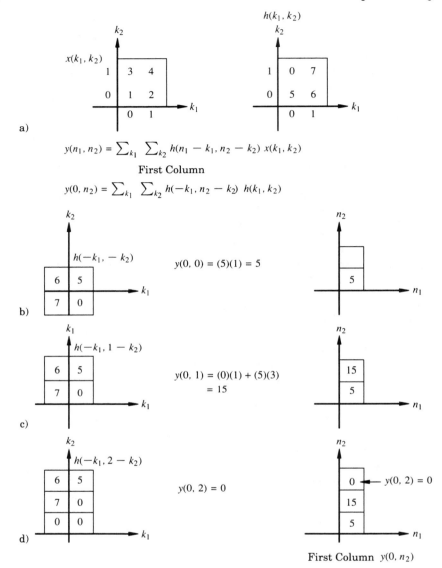

Figure 3-17 *Steps in 2D convolution.*

Next we multiply the corresponding entries in **X** and **H**, *element by element*, to obtain

$$Y(k_1, k_2) = H(k_1, k_2)X(k_1, k_2)$$

We take the inverse DFT of the **Y** matrix and get the **Y** matrix obtained in Example 3.4.

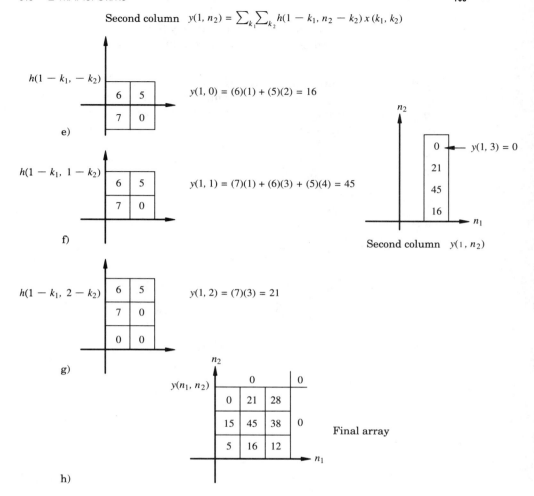

Second column $y(1, n_2) = \sum_{k_1}\sum_{k_2} h(1 - k_1, n_2 - k_2) x(k_1, k_2)$

$h(1 - k_1, -k_2)$

$y(1, 0) = (6)(1) + (5)(2) = 16$

e)

$h(1 - k_1, 1 - k_2)$

$y(1, 1) = (7)(1) + (6)(3) + (5)(4) = 45$

f)

Second column $y(1, n_2)$

$h(1 - k_1, 2 - k_2)$

$y(1, 2) = (7)(3) = 21$

g)

$y(n_1, n_2)$

Final array

h)

Figure 3-17   *(continued)*

3.5 Z TRANSFORMS

Thus far we have discussed signal processing from the standpoints of signal analysis and representation and have mentioned the convolution operation. We now turn to the description of signal-processing operations in the spatial domain by the difference equation and its associated Z transform. The difference equation itself is often used as the actual processing algorithm.

Again while the 2D concepts are similar to those of the 1D case, there are nevertheless fundamental limitations to the simple extrapolation of 1D properties. The two major difficulties are

1. There is no polynomial factorization property for 2D polynomials. The tests for stability and the concept of a pole are far more complex.

2. The difference equation for a 2D IIR filter must have a special property, *recursive computability*, to permit recursions on an image. Furthermore the boundary values that must be specified are also challenging, as we shall see.

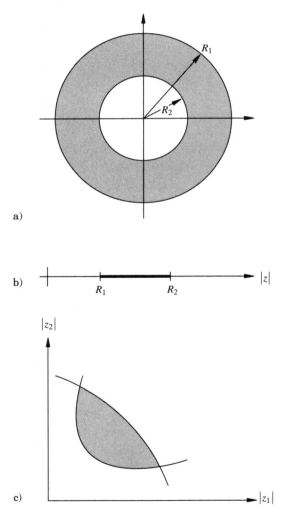

a)

b)

c)

Figure 3-18   *(a) Region of convergence in z plane for 1D transform; (b) region of convergence on $|z|$ line for 1D transform; (c) region of convergence on $|z_1||z_2|$ plane for 2D transform.*

The Z transform for a 2D sequence $x(n_1, n_2)$, a simple extension of the 1D case, is

$$X_z(z_1, z_2) = \sum_{n_1=-\infty}^{\infty} \sum_{n_2=-\infty}^{\infty} x(n_1, n_2) z_1^{-n_1} z_2^{-n_2} \qquad (3\text{-}89)$$

If $X_z(z_1, z_2)$ converges on $|z_1| = 1$ and $|z_2| = 1$, we can obtain the Fourier transform by

$$X(\omega_1, \omega_2) = X_z(z_1, z_2)\Big|_{\substack{z_1=e^{j\omega_1} \\ z_2=e^{j\omega_2}}} \qquad (3\text{-}90)$$

In 1D, we know that $X_z(z)$ converges in a ring $R_1 < |z| < R_2$ as shown in Fig. 3-18(a). An abbreviated version of this is the interval of convergence shown in Fig. 3-18(b). In the 2D case, the region of convergence is represented in the $|z_1||z_2|$ plane as indicated in Fig. 3-18(c). To illustrate these points, we will work some examples.

EXAMPLE 3.6: Let x be a separable exponential signal

$$x(n_1, n_2) = a^{n_1} b^{n_2} u(n_1, n_2) = a^{n_1} u(n_1) \cdot b^{n_2} u(n_2) \qquad (3\text{-}91)$$

$$X_z(z_1, z_2) = \sum_{n_1=0}^{\infty} \sum_{n_2=0}^{\infty} a^{n_1} b^{n_2} z_1^{-n_1} z_2^{-n_2}$$

$$= \sum_{n_1=0}^{\infty} (a z_1^{-1})^{n_1} \sum_{n_2=0}^{\infty} (b z_2^{-1})^{n_2}$$

$$= \frac{1}{1 - a z_1^{-1}} \cdot \frac{1}{1 - b z_2^{-1}} \qquad (3\text{-}92)$$

$X_z(z_1, z_2)$ is convergent for $|z_1| > |a|$ and $|z_2| > |b|$, as shown in Fig. 3-19.

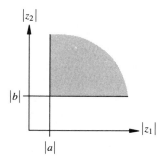

Figure 3-19   *Region of convergence for transform of exponential signal.*

EXAMPLE 3.7:   This is an exponential function along a line in the first quadrant, Fig. 3-20(a):

$$x(n_1, n_2) = a^{n_1} \underbrace{\delta(n_1 - n_2)}_{\text{along line}} \underbrace{u(n_1, n_2)}_{\text{first quadrant}} \tag{3-93}$$

Then

$$X_z(z_1, z_2) = \sum_0^\infty \sum_0^\infty a^{n_1} \delta(n_1 - n_2) u(n_1, n_2) z_1^{-n_1} z_2^{-n_2}$$

$$= \sum_0^\infty z_2^{-n_2} \sum_0^\infty (a z_1^{-1})^{n_1} \delta(n_1 - n_2)$$

$$= \sum_0^\infty z_2^{-n_2} (a z_1^{-1})^{-n_2} = \sum_0^\infty (a z_1^{-1} z_2^{-1})^{n_2}$$

$$= \frac{1}{1 - a z_1^{-1} z_2^{-1}} \tag{3-94}$$

convergent for $|z_1 z_2| > |a|$ as indicated in Fig. 3-20(b).

In 1D poles of $X_z(z)$ are points in the 2D complex z plane. In 2D the poles of $X_z(z_1, z_2)$ are surfaces in the 4D complex $z_1 z_2$ space. In Example 3.6, the pole surfaces are separable:

$$z_1 = a \qquad \text{for any } z_2$$

$$z_2 = b \qquad \text{for any } z_1$$

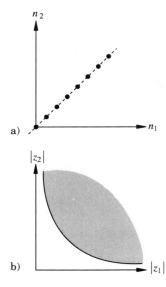

Figure 3-20   (a) Exponential sequence along a line; (b) region of convergence.

Table 3.6. Some Properties of the 2D Z Transform

$$\text{Let } x(n_1, n_2) \leftrightarrow X_z(z_1, z_2).$$

1. Finite support sequence: Z transform converges everywhere except $|z_1| = 0, \infty$ and $|z_2| = 0, \infty$.
2. Stable sequence: $\sum \sum |x(n_1, n_2)| < \infty$ has a region of convergence including $|z_1| = 1, |z_2| = 1$; the Fourier transform exists.
3. Convolution: $x(n_1, n_2) ** h(n_1, n_2) \leftrightarrow X_z(z_1, z_2) \cdot H_z(z_1, z_2)$.
4. Shift and modulation:
$$x(n_1 - m_1, n_2 - m_2) \leftrightarrow z_1^{-m_1} z_2^{-m_2} X_z(z_1, z_2)$$
$$a^{n_1} b^{n_2} x(n_1, n_2) \leftrightarrow X_z(a^{-1} z_1, b^{-1} z_2)$$
5. Differentiation:
$$-n_i x(n_1, n_2) \leftrightarrow z_i \frac{\partial}{\partial z_i} X(z_1, z_2) \qquad i = 1, 2$$
6. Conjugates and reversal:
$$\bar{x}(n_1, n_2) \leftrightarrow X(\bar{z}_1, \bar{z}_2)$$
$$x(-n_1, n_2) \leftrightarrow X(z_1^{-1}, z_2)$$
7. Inversion formula:
$$x(n_1, n_2) = \left(\frac{1}{2\pi j}\right)^2 \oint_{r_1} \oint_{r_2} X_z(z_1, z_2) z_1^{n_1-1} z_2^{n_2-1} dz_1 dz_2$$
8. Stability (Huang, 1972, 1978): For $h(n_1, n_2)$ a sequence with first-quadrant support and transform $H_z(z_1, z_2) = N(z_1, z_2)/D(z_1, z_2)$, the necessary and sufficient conditions for stability are
    a. $D(z_1, z_2) \neq 0, |z_1| \geq 1, z_2 = 1$.
    b. $D(z_1, z_2) \neq 0, |z_2| \geq 1, |z_1| = 1$.

The main difficulty in using the Z transform is that, in general, a 2D (bivariate) polynomial cannot be factored. Hence stability tests are very difficult and inversion via partial fractions cannot be done. These are far more complex than the 1D signals and offer relatively little insight, and we will not pursue their study further.

Some properties of the Z transform are summarized in Table 3.6.

3.6 DIFFERENCE EQUATIONS AND IIR FILTERS

The difference equation determines the structure of the filter prescribed by the frequency response or impulse response. The difference equation is the computational procedure that realizes the desired $H(\omega_1, \omega_2)$ or $H_z(z_1, z_2)$. Therefore, we impose the requirement of *recursive computability*, which permits the algorithm to calculate signal samples recursively, that is, in terms of adjacent, previously calculated signal values. To illustrate this point, we give a simple difference equation and its associated transfer function:

$$y(n_1, n_2) = ay(n_1 - 1, n_2) + by(n_1 + 1, n_2) + x(n_1, n_2) \qquad (3\text{-}95)$$

$$H_z(z_1, z_2) = \frac{1}{1 - az_1^{-1} - bz_1}$$

In the course of computing the output sequence, we must calculate

$$y(0, 0) = ay(-1, 0) + by(1, 0) + x(0, 0)$$

$$y(1, 0) = ay(0, 0) + by(2, 0) + x(1, 0)$$

This is impossible, for the computation of $y(0, 0)$ needs $y(1, 0)$, but $y(1, 0)$ needs $y(0, 0)$. There is no way around this difficulty, and therefore the equation is not recursively computable.

It is clear that all FIR filters are recursively computable, since their output samples depend only on inputs and not on other outputs.

Recall that in 1D, the general difference equation is

$$\sum_{k=0}^{N} a_k y(n - k) = \sum_{j=-M_1}^{M_2} b_j x(n - j)$$

or
$$H_z(z) = \frac{N(z)}{D(z)} \qquad (3\text{-}96)$$

$$= \frac{\sum_{j=-M_1}^{M_2} b_j z^{-j}}{\sum_{k=0}^{N} a_k z^{-k}}$$

and if $M_1 = 0$, we say the system is causal. In imaging, causality as such is not an issue, since we can process the image pixels in any spatial direction—up, down, right, or left—and time is not an issue.

Also, with a_0 normalized to unity, we can write the (causal) difference equation as

$$y(n) = -[a_1 y(n - 1) + \cdots + a_N y(n - N)]$$
$$+ b_0 x(n) + b_1 x(n - 1) + \cdots + b_M x(n - M) \qquad n > 0 \quad (3\text{-}97)$$

If the difference equation is to be used as the explicit vehicle for filtering the signal, a set of initial conditions on the output $\{y(-1), ..., y(-N)\}$ must be specified.

The 2D version is more complex as we have seen from the simple example of a non-recursively-computable equation. The general 2D equation is

$$\sum_{k_1} \sum_{k_1} a(k_1, k_2) y(n_1 - k_1, n_2 - k_2)$$

$$= \sum_{r_1} \sum_{r_2} b(r_1, r_2) y(n_1 - r_1, n_2 - r_2) \qquad (3\text{-}98)$$

The finite coefficient arrays $\{b(r_1, r_2)\}$ and $\{a(k_1, k_2)\}$ are given over appropriate regions R_B and R_A, typically rectangles. Then the transfer function is

$$H_z(z_1, z_2) = \frac{\displaystyle\sum_{k_1, k_2 \in R_B} b(k_1, k_2) z_1^{-k_1} z_2^{-k_2}}{\displaystyle\sum_{r_1, r_2 \in R_A} a(r_1, r_2) z_1^{-r_1} z_2^{-r_2}} \qquad (3\text{-}99)$$

$$= \frac{N(z_1, z_2)}{D(z_1, z_2)}$$

To make this equation recursive, we normalize so that $a(0,0) = 1$ and rewrite Eq. (3-98) as

$$y(n_1, n_2) = - \sum_{\substack{k_1, k_2 \, /= n_1, n_2 \\ k_1, k_2 \in R_a}} a(n_1 - k_1, n_2 - k_2) y(k_1, k_2)$$

$$+ \sum_{k_1, k_2 \in R_B} b(n_1 - k_1, n_2 - k_2) x(k_1, k_2) \qquad (3\text{-}100)$$

$$= y_a(n_1, n_2) + y_b(n_1, n_2)$$

The coefficient arrays $\{a(k_1, k_2)\}$ and $\{b(k_1, k_2)\}$ in Eq. (3-100) constitute windows or *masks* in the (k_1, k_2) plane shown in Fig. 3-21, which depicts the implementation of Eq. (3-100). For the example illustrated, the equation is

$$y(n_1, n_2) = [x(n_1 + 1, n_2 + 1) + x(n_1 + 1, n_2) + x(n_1, n_2 + 1)$$

$$+ x(n_1, n_2) + x(n_1 - 1, n_2) + x(n_1, n_2 - 1)]$$

$$+ [y(n_1 - 1, n_2) + y(n_1 - 2, n_2) + y(n_1, n_2 - 1)$$

$$+ y(n_1 - 1, n_2 - 1), + y(n_1 - 2, n_2 - 1)] \qquad (3\text{-}101)$$

The recursions are performed as follows:

1. At any point (n_1, n_2), the input mask is positioned over the input array. The input values are multiplied by the b values and summed.

2. Simultaneously the output mask covers the already computed output values, weights these values by the a coefficients, and sums them.

3. The results of steps 1 and 2 are added and the result is entered as $y(n_1, n_2)$.

4. The masks are then shifted by one position, and steps 1 through 3 are repeated.

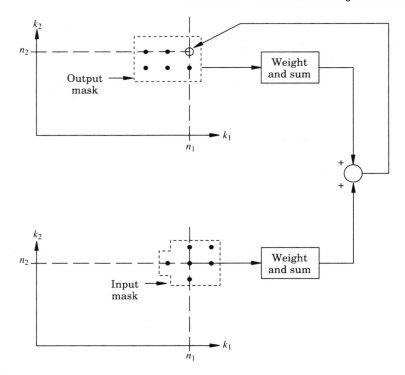

Figure 3-21   *Input and output masks for Eq. (1-102).*

Conditions for recursive computability are described in Dudgeon and Mersereau (1984) and in Lim and Oppenheim (1988). There it is shown that for a finite-support input, the system is recursively computable if the output mask has wedge support. [Wedge support means that the support lies in a region bounded by lines emanating from the origin and making an angle of less than 180°, as in Fig. 3-22(a).] A simpler sufficient condition is that $a(k_1, k_2)$ have first-quadrant support, as shown in Fig. 3-22(b).

A recursive equation is not complete unless initial conditions are specified. The mask for the coefficient array of Fig. 3-22(b) is shown in Fig. 3-22(c) for $n_1 = 0$ and $n_2 = 0$. It is clear that initial conditions within the L-shaped region of N_1 columns and N_2 rows must be specified to allow recursive computation of $y(n_1, n_2)$ for $n_1 \geq 0$ and $n_2 \geq 0$. For example, the output mask for Eq. (3-105) has first-quadrant support since $\{a(k_1, k_2)\}$ is nonzero in $(0 \leq k_1 \leq 2, 0 \leq k_2 \leq 1)$. Therefore the required initial conditions are

$$y(n_1, -1) \qquad \text{for } n_1 = -2, -1, 0, 1, \ldots$$

$$\left. \begin{array}{l} y(-1, n_2) \\ y(-2, n_2) \end{array} \right\} \qquad \text{for } n_2 = 0, 1, \ldots$$

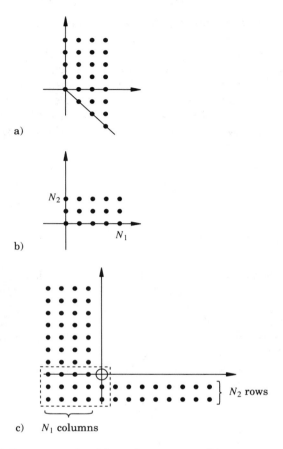

Figure 3-22 *(a) Output mask with wedge support; (b) output mask with first-quadrant support; (c) initial conditions for first-quadrant output mask.*

Our discussion of boundary conditions here is necessarily brief, and we consider only the simple case of first-quadrant output mask support. We refer the reader to the texts mentioned above for a complete treatment of the more general initial conditions for arbitrary wedge masks.

The question of the stability of an IIR filter is quite complex. We noted earlier that a bivariate polynomial cannot, in general, be factored into lower-order polynomials. Furthermore, the poles of $H_z(z_1, z_2)$ of Eq. (3-100) are surfaces in the four-dimensional complex $z_1 z_2$ space, so stability statements and tests are not simple extrapolations of 1D results.

The spatial domain requirement for BIBO (bounded-input bounded-output) stability is, however, a straightforward extension of the 1D case. It is easy to show that the necessary and sufficient condition is

$$\sum_{n_1} \sum_{n_2} |h(n_1, n_2)|^2 < \infty \qquad (3\text{-}102)$$

But the translation of this requirement to the transform domain is far from obvious.

We are concerned with transfer functions of the form of Eq. (3-99), where the output mask has wedge support. A sequence with wedge support can be mapped into a mask with first-quadrant support by a linear transformation of variables (Lim, 1990). Hence it suffices to consider stability theorems for IIR filters with first-quadrant support. In fact, we need not be concerned with the numerator of Eq. (3-99) but need only consider

$$H_z(z_1, z_2) = \frac{1}{A(z_1, z_2)}$$

where
$$A(z_1, z_2) = \sum_{k_2=0}^{N_2} \sum_{k_1=0}^{N_1} a(k_1, k_2) z_1^{-k_1} z_2^{-k_2} \qquad (3\text{-}103)$$

Stability tests for these IIR filters have been developed by Shanks, extended by Huang, and recodified as the DeCarlo-Strintzis theorem, which we quote here. First, we note that the necessary and sufficient condition for BIBO stability is that $A(z_1, z_2) \neq 0$ outside $|z_1| = 1$ and $|z_2| = 1$. This is an extrapolation of the 1D stability requirement that the zeroes of $A(z)$ lie within $|z| = 1$.

For the 2D case, the filter is stable if and only if:

1. $A(z_1, z_2) \neq 0$ for $|z_1| = 1, |z_2| = 1$

2. $A(z_1, 1) \neq 0$ for $|z_1| \geq 1$ (3-104)

3. $A(1, z_2) \neq 0$ for $|z_2| \geq 1$

Condition 1 asserts that the frequency-response function $A(e^{j\omega 1}, e^{j\omega 2}) \neq 0$ for any ω_1 and ω_2, while conditions 2 and 3 require $A(z_1, z_2)$ to be nonzero in the space indicated. Problem 3.18 demonstrates the application of these theorems in stability testing.

3.7 FIR FILTERS IN TWO DIMENSIONS

This class of filters is far simpler than the IIR variety and can be exercised using either a spatial domain or an FFT frequency-domain implementation.

The FIR equation is a special case of the general difference equation of Eq. (3-100) with $a(0, 0) = 1$ and $a(k_1, k_2) = 0$ otherwise. Hence there is just one mask, the input mask, and the $y_a(n_1, n_2)$ branch shown in Fig. 3-21 is absent. The questions of stability and recursive computability are now moot, and the equation reduces to

$$y(n_1, n_2) = \sum_{k_2} \sum_{k_1} b(k_1, k_2) x(n_1 - k_1, n_2 - k_2) \qquad (3\text{-}105)$$

and $b(k_1, k_2)$ is recognized as the FIR finite-support impulse response

$$h(n_1, n_2) = b(n_1, n_2) \qquad (3\text{-}106)$$

$$H_z(z_1, z_2) = \sum_{n_1} \sum_{n_2} b(n_1, n_2) z_1^{-n_1} z_2^{-n_2} \qquad (3\text{-}107)$$

The FIR filter can be readily designed to have zero phase by requiring $h(n_1, n_2)$ to be an even function of its arguments. This is so because elementary Fourier transform theory asserts that if $h(n_1, n_2)$ is real and if

$$h(-n_1, -n_2) = h(n_1, n_2) \qquad (3\text{-}108)$$

then $H(\omega_1, \omega_2)$ is a real, even function of its arguments:

$$H(\omega_1, \omega_2) = H(-\omega_1, -\omega_2) \qquad (3\text{-}109)$$

Since spatial causality is not a constraint, Eq. (3-108) is easily satisfied.

An example of a zero-phase FIR filter was given in Example 3.1 and Fig. 3-12. Two additional examples are given here.

EXAMPLE 3.8: . This very simple filter, with impulse and frequency responses shown in Fig. 3-23, is nonseparable, since

$$h(n_1, n_2) = \delta(n_1 - 1, n_2) + \delta(n_1 + 1, n_2)$$
$$+ \delta(n_1, n_2 - 1) + \delta(n_1, n_2 + 1) \qquad (3\text{-}110)$$

$$H_z(z_1, z_2) = (z_1 + z_1^{-1}) + (z_2 + z_2^{-1}) \qquad (3\text{-}111)$$

$$H(\omega_1, \omega_2) = 2(\cos \omega_1 + \cos \omega_2) \qquad (3\text{-}112)$$

EXAMPLE 3.9: . A separable FIR transfer function is

$$H_z(z_1, z_2) = (z_1 + 2 + z_1^{-1})(z_2 + 2 + z_2^{-1}) \qquad (3\text{-}113)$$

Expansion of this product gives

$$H_z(z_1, z_2) = z_1 z_2 + 2 z_1 z_2^0 + z_1 z_2^{-1} + 2 z_1^0 z_2 + 4 z_1^0 z_2^0$$
$$+ 2 z_1^0 z_2^{-1} + z_1^{-1} z_2 + 2 z_1^{-1} z_2^0 + z_1^{-1} z_2^{-1} \qquad (3\text{-}114)$$

from which the impulse response $h(n_1, n_2)$ is identified and plotted in Fig. 3-24(a). The associated frequency response [Fig. 3-24(b)] is just

$$H(\omega_1, \omega_2) = (2 + 2 \cos \omega_1)(2 + 2 \cos \omega_2)$$

$$= \left(4 \cos \frac{\omega_1}{2} \cos \frac{\omega_2}{2} \right)^2 \qquad (3\text{-}115)$$

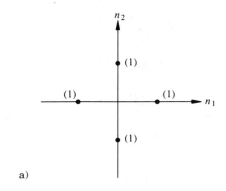

a)

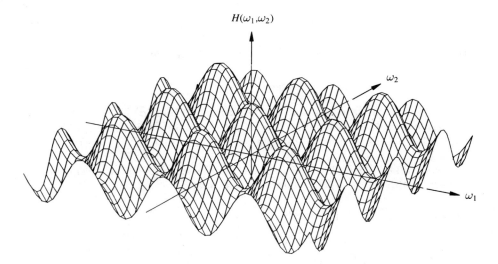

Figure 3-23   (a) Weighting sequence of Eq. (3-108); (b) frequency response from Eq. (3-110).

These last two examples show that the determination of separability from the impulse-response pattern is not a simple matter. The converse, however, is quite trivial. One simply starts with an $H_z(z_1, z_2)$ written as $H_1(z_1) \cdot H_2(z_2)$ and expands the finite polynomials as in Eq. (3-114). Separable FIR filters are therefore easy to handle, since they reduce to two 1D FIR filters in tandem. In fact, the filter can be realized by the series form shown in Fig. 3-25. The first filter performs row operations on the input array, followed by column operations in the second block. The design of separable 2D filters thus reduces to the design of two 1D FIR filters. The latter can be synthesized by the techniques to be developed in Chap. 5—the window method, the frequency-sampling method, or the Parks-McClellan algorithm.

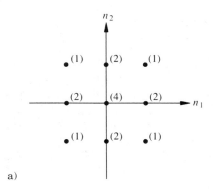

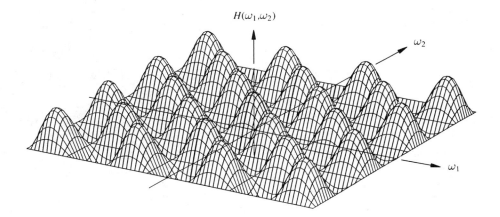

Figure 3-24   *(a) Separable FIR impulse response for Eq. (3-110); (b) frequency response from Eq. (3-112).*

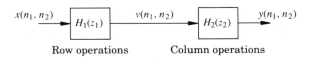

Row operations   Column operations

Figure 3-25   *Separable FIR filter as tandem connection of 1D FIR filters.*

Even nonseparable FIR filters can be designed by a conceptually simple extension of the 1D technique. For example, once the desired frequency response $H(\omega_1, \omega_2)$ is specified, we can:

1. Compute $h(n_1, n_2)$ by the inverse FFT.

2. Truncate $h(n_1, n_2)$ by a tapered window $w(n_1, n_2)$ of certain support:

$$\tilde{h}(n_1, n_2) = h(n_1, n_2)w(n_1, n_2)$$

Typically, the window itself is separable.

3. Pad $\tilde{h}(n_1, n_2)$ with zeroes and compute $\tilde{H}(k_1, k_2)$ with the FFT.

This procedure is used iteratively until a satisfactory solution is obtained. In the foregoing, $H(\omega_1, \omega_2)$ usually has zero phase, so $h(n_1, n_2)$ will have the even symmetry of Eq. (3-108); the window is necessarily even.

Another carryover from the 1D domain is the frequency-sampling method. The 2D version is a straightforward extrapolation of the 1D method outlined in Chap. 5.

Another useful and easily applied design technique is the transformation method due to McClellan (1973). Simply stated, first we design a 1D prototype zero-phase FIR filter of the form

$$H(\omega) = \sum_{n=-M}^{M} h(n)e^{-jn\omega} = \sum_{n=0}^{M} a_n \cos n\omega$$

$$= \sum_{n=0}^{M} b_n (\cos \omega)^n \tag{3-116}$$

Next, we substitute a 2D function, $G(\omega_1, \omega_2)$ for $\cos \omega$ in Eq. (3-116):

$$\cos \omega \rightarrow G(\omega_1, \omega_2) \tag{3-117}$$

This leaves us with the 2D filter,

$$H(\omega_1, \omega_2) = \sum_{n=0}^{M} b_n [G(\omega_1, \omega_2)]^n \tag{3-118}$$

The constraints on $G(\omega_1, \omega_2)$ are that it be a zero-phase, real, even function and that $|G(\omega_1, \omega_2)| \leq 1$. Hence from Eq. (3-114), the resulting $H(\omega_1, \omega_2)$ is a 2D zero-phase filter. The representation of Eq. (3-118) suggests the realization shown in Fig. 3-26, which resembles a tapped delay line with delays replaced by the transformation $G(\omega_1, \omega_2)$.

The principle behind the transformation method is the mapping of a desirable 1D frequency response into two dimensions. Suppose, for example, that we have a 1D bandpass filter with acceptable properties as illustrated in Fig. 3-27(a). We want to map values of $H(\omega)$ at points ω_1', ω_2' into contours in the $H(\omega_1, \omega_2)$ plane, as indicated in Fig. 3-27(b). A transformation that achieves this is

$$G(\omega_1, \omega_2) = \tfrac{1}{2}\cos\omega_1 + \tfrac{1}{2}\cos\omega_2 + \tfrac{1}{2}\cos\omega_1\cos\omega_2 - \tfrac{1}{2} \tag{3-119}$$

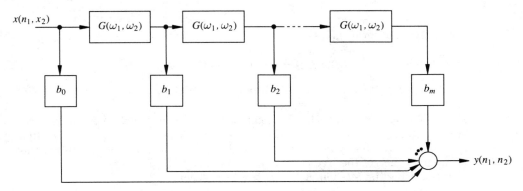

Figure 3-26 *Tapped transformation line realization of Eq. (3-118).*

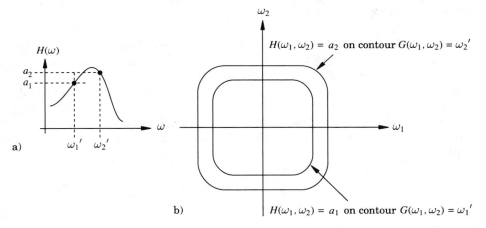

Figure 3-27 *(a) 1D bandpass prototype; (b) desired mapping in ω_1, ω_2 plane.*

This transformation is circularly symmetric, and thus low-pass, band-pass, and high-pass 1D filters map into circularly symmetric 2D counterparts.

EXAMPLE 3.10: We apply the transformation method to the binomial filter of Example 3.1, where the 1D low-pass filter was

$$H(\omega) = 6 + 8 \cos \omega + 2 \cos 2\omega$$

$$= 4 + 8 \cos \omega + 4 \cos^2 \omega \qquad (3\text{-}120)$$

$$H(\omega_1, \omega_2) = 4 + 8G(\omega_1, \omega_2) + 4[G(\omega_1, \omega_2)]^2 \qquad (3\text{-}121)$$

where $G(\omega_1, \omega_2)$ is as given in Eq. (3-119). Expansion of Eq. (3-121) leads to

$$H(\omega_1, \omega_2) = 1 + 2 \cos \omega_1 + 2 \cos \omega_2 + 4 \cos \omega_1 \cos \omega_2 + \cos^2 \omega_1$$

$$+ \cos^2 \omega_2 + 2 \cos \omega_1 \cos^2 \omega_2 + 2 \cos^2 \omega_1 \cos \omega_2$$

$$+ \cos^2 \omega_1 \cos^2 \omega_2 \qquad (3\text{-}122)$$

A regrouping of terms gives

$$H(\omega_1, \omega_2) = (1 + \cos \omega_1)^2 (1 + \cos \omega_2)^2$$

$$= 16 (\cos \omega_1 / 2)^4 (\cos \omega_2 / 2)^4 \qquad (3\text{-}123)$$

This last result is seen to be identical (within a scale factor of 2^4) to the 2D binomial filter described in Example 1. In the earlier example, the 2D filter was obtained by simply imposing a separability condition and extrapolating the 1D filter to each of the frequency axes in the 2D case. Here the same result is obtained using the McClellan 1D-2D transformation.

3.8 IIR FILTER DESIGN

The problem in IIR filter design is the determination of filter coefficients $\{a(k_1, k_2)\}$ and $\{b(k_1, k_2)\}$ in $\hat{H}_z(z_1, z_2)$ of Eq. (3-99) that approximate some desired filter characteristic. Typically one starts with some ideal frequency response function $H(\omega_1, \omega_2)$ and the associated impulse response $h(n_1, n_2)$, which can be obtained from $H(\omega_1, \omega_2)$ by FFT routines. A least-squares design philosophy seeks to find coefficients in $\hat{H}_z(z_1, z_2) \leftrightarrow \hat{h}(n_1, n_2)$ such that the following sum-squared error is minimized:

$$J = \frac{1}{2} \sum_{n_2} \sum_{n_1} |h(n_1, n_2) - \hat{h}(n_1, n_2)|^2 \qquad (3\text{-}124)$$

The first step is the choice of a recursively computable structure with input and output masks consistent with the spatial support of $h(n_1, n_2)$. In effect we specify the form of $\hat{H}_z(z_1, z_2)$ (or of the difference equation), including the order of the numerator and denominator, but not the parameter values. Generally the minimization of J with respect to the pole (denominator) parameters leads to a set of nonlinear algebraic equations that can only be solved iteratively. Here, we describe an approximate, linearized solution due to Shanks (1972), using the method of least squares. (See also Chap. 6.)

A direct approach to Eq. (3-124) involves the formulation of an error,

$$E(z_1, z_2) = H_z(z_1, z_2) - \hat{H}_z(z_1, z_2)$$

or $\qquad e(n_1, n_2) = h(n_1, n_2) - \hat{h}(n_1, n_2) \qquad (3\text{-}125)$

where $\hat{H}_z(z_1, z_2)$ has the form of Eq. (3-99) and $\hat{h}(n_1, n_2)$ is the solution of Eq. (3-100) when $x(n_1, n_2) = \delta(n_1, n_2)$, or

$$\hat{h}(n_1, n_2) = - \sum_{k_1, k_2 \in R} a(k_1, k_2) \hat{h}(n_1 - k_1, n_2 - k_2) + b(n_1, n_2) \qquad (3\text{-}126)$$

Subtracting Eq. (3-126) from $h(n_1, n_2)$ still leaves the unknown $\hat{h}(\cdot, \cdot)$ in the error equation by virtue of its presence inside the summation in Eq. (3-126). To circumvent this, we approximate $\hat{h}(n_1, n_2)$ by using $h(n_1, n_2)$ in the right-hand side of Eq. (3-126). That is, we define

$$\hat{h}(n_1, n_2) \approx \tilde{h}(n_1, n_2)$$

$$\triangleq -\sum_{k_1 \neq n_1} \sum_{k_2 \neq n_2} a(k_1, k_2)h(n_1 - k_1, n_2 - k_2) + b(n_1, n_2) \quad (3\text{-}127)$$

Now we can subtract $\tilde{h}(n_1, n_2)$ from $h(n_1, n_2)$ to get

$$\tilde{e}(n_1, n_2) = h(n_1, n_2) - \tilde{h}(n_1, n_2)$$

$$= \sum_{k_1} \sum_{k_2} a(k_1, k_2)h(n_1 - k_1, n_2 - k_2) - b(n_1, n_2) \quad (3\text{-}128)$$

where $a(0, 0) = 1$.

In this format, the modified error $\tilde{e}(n_1, n_2)$ is a *linear* function of the unknown a's and b's, and

$$\tilde{J} = \frac{1}{2} \sum_{n_1} \sum_{n_2} |\tilde{e}(n_1, n_2)|^2 \quad (3\text{-}129)$$

is quadratic in a and b. Setting the partial derivatives of $\tilde{J}$ with respect to $a(k_1, k_2)$ and $b(k_1, k_2)$ to zero leads to a set of linear algebraic equations with a closed form solution.

Before embarking on the details of the solution, we observe that Eq. (3-128) can be expressed as

$$\tilde{E}(z_1, z_2) = A(z_1, z_2)H_z(z_1, z_2) - B(z_1, z_2) \quad (3\text{-}130)$$

This in turn suggests the block diagram of Fig. 3-28. This diagram is used by Shynk (1989) in deriving a quadratic performance measure for an adaptive 1D IIR filter. It is also suggestive of the first step in the iterative solution of the nonlinear equations resulting from a minimization of J in Eq. (3-124).

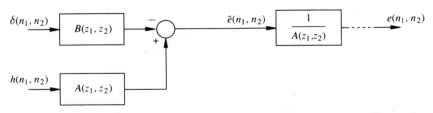

Figure 3-28 *Modified error signal for quadratic measure.*

To complete the problem formulation, we take $\{a(k_1, k_2)\}$ and $\{b(k_1, k_2)\}$ each to be first-quadrant rectangular support sequences shown in Fig. 3-29(a) and (b); we also assume $h(n_1, n_2)$ to be first-quadrant support, extending from $(0, 0)$ to $(L_1 - 1, L_2 - 1)$ as in Fig. 3-29(c).

The error measure $\tilde{J}$ can be expressed as

$$\tilde{J} = \frac{1}{2} \sum_{n_2=0}^{L_2-1} \sum_{n_1=0}^{L_1-1} |\tilde{e}(n_1, n_2)|^2$$

$$= \frac{1}{2} \sum_{n_1, n_2 \notin R} |\tilde{e}(n_1, n_2)|^2 + \frac{1}{2} \sum_{n_1, n_2 \in R} |\tilde{e}(n_1, n_2)|^2$$

$$= \tilde{J}_1(a, b) + \tilde{J}_2(a) \tag{3-131}$$

Note that in the region R, $b(n_1, n_2) = 0$ so that $\tilde{J}_2$ depends only on the a's, and $\tilde{e}(n_1, n_2)$ reduces to

$$\tilde{e}(n_1, n_2) = \sum_{k_2=0}^{M_2-1} \sum_{k_1=0}^{M_1-1} a(k_1, k_2) h(n_1 - k_1, n_2 - k_2) \tag{3-132}$$

The procedure is now straightforward, but messy. First we find $\{a(k_1, k_2)\}$ to minimize $\tilde{J}_2$. Then we note that when these a's are substituted into Eq. (3-128), we can force $\tilde{e}(n_1, n_2)$ to 0 for $(n_1, n_2) \in R_b$ by making

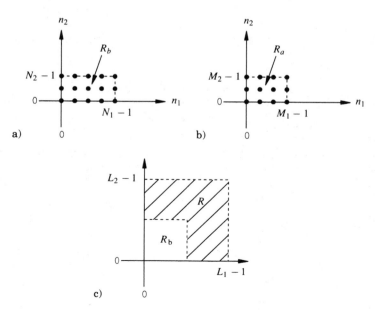

Figure 3-29   *(a), (b) Support regions for numerator and denominator of* $\hat{H}_z(z_1, z_2)$. *(c) Regions used in* $\tilde{J}_1$ *and* $\tilde{J}_2$ *error measures.*

$$b(n_1, n_2) = \sum_{k_2=0}^{M_2-1} \sum_{k_1=0}^{M_1-1} a(k_1, k_2) h(n_1 - k_1, n_2 - k_2) \qquad (3\text{-}133)$$

for $(n_1, n_2) \in R_b$.

The details are as follows:

$$\tilde{J}_2 = \frac{1}{2} \sum_{(n_1, n_2) \in R} \sum [\tilde{e}(n_1, n_2)]^2$$

$$0 = \frac{\partial \tilde{J}_2}{\partial a(p_1, p_2)} = \sum_{(n_1, n_2) \in R} \sum [\tilde{e}(n_1, n_2)] \frac{\partial}{\partial a(p_1, p_2)} \tilde{e}(n_1, n_2)$$

$$= \sum_{(n_1, n_2) \in R} \sum \tilde{e}(n_1, n_2) h(n_1 - p_1, n_2 - p_2)$$

$$= \sum_{k_1} \sum_{k_2} a(k_1, k_2) \sum_{(n_1, n_2) \in R} \sum h(n_1 - p_1, n_2 - p_2) h(n_1 - k_1, n_2 - k_2)$$

Expressing the second sum as $r(p_1, p_2; k_1, k_2)$ and noting that $a(0, 0) = 1$, we can write this last equation as

$$\sum_{\substack{k_1 \\ (k_1, k_2) \neq (0,0)}}^{M_1-1} \sum_{k_2}^{M_2-1} a(k_1, k_2) r(p_1, p_2; k_1, k_2) = -r(p_1, p_2; 0, 0) \qquad (3\text{-}134)$$

for
$$0 \leq p_1 \leq M_1 - 1$$
$$0 \leq p_2 \leq M_2 - 1$$
$$(p_1, p_2) \neq (0, 0)$$

This last equation has the form of the *normal equations* (see Chap. 6) and can be solved for the $(M_1 M_2 - 1)$ unknowns $\{a(k_1, k_2)\}$. These in turn yield the b's when substituted into Eq. (3-133).

   This method can also be used as the basis for an iterative minimization of the original sum-squared error Eq. (3-124). The problem with Shanks's method and the iterative procedure arising from it is that there is no a priori assurance that the Shanks filter is stable or that the iterations converge. Nevertheless, IIR filters designed in the spatial domain as described here tend to have better stability properties than those designed using frequency-domain procedures.

   Many other design procedures are described in the literature, some of which incorporate implementations with the design (for example, Lee and Woods, 1986). Earlier works due to Aly and Fahmy (1978, 1980) provide alternative views on both frequency- and spatial-domain recursive designs.

3.9 TWO-DIMENSIONAL RANDOM FIELDS

The two-dimensional equivalent to the 1D random sequence is the *random field*, a 2D grid of random variables. The properties of 2D random sequences are for the most part direct extrapolations of the 1D progenitor. Perhaps the main conceptual difference between 1D and 2D is the permissible noncausality of the 2D structure. As we shall see, this leads to different dynamic representations for the same field. Our presentation here is necessarily abbreviated. Jain (1981) developed a more detailed exposition, on which we will draw heavily in what follows.

Each pixel $x(n_1, n_2)$ is a random variable with some probability density function (pdf). The collection of pixels $\{x(n_1, n_2)\}$ and the probability relations among them constitute the random field. We are primarily concerned with wide-sense properties of the field—the means and correlations. We define mean and autocorrelation by

$$E\{x(n_1, n_2)\} = \mu(n_1, n_2) \tag{3-135}$$

$$E\{x(n_1, n_2)x(k_1, k_2)\} = R(n_1, n_2; k_1, k_2) \tag{3-136}$$

The field is *wide-sense stationary* (WSS) (or *spatially invariant* or *homogeneous*) when the mean is constant and the autocorrelation shift-invariant, that is,

$$E\{x(n_1, n_1)\} = \mu = \text{constant} \tag{3-137}$$

$$R\{n_1, n_2; k_1, k_2\} = R(n_1 - k_1, n_2, -k_2) \tag{3-138}$$

We assume a WSS field in this chapter, and we can often take the mean to be zero for convenience.

A white-noise field is defined by zero mean and uncorrelated pixels, or

$$R(n_1, n_2) = \sigma^2 \delta(n_1, n_2) \tag{3-139}$$

where σ^2 is the variance in each pixel.

For WSS fields, we define the 2D power spectral density (psd) as the 2D Fourier transform of the correlation function. Thus[†]

$$\mathcal{S}(\omega_1, \omega_2) = \mathfrak{F}\{R(k_1, k_2)\} = \sum_{k_1}\sum_{k_1} R(k_1, k_2)e^{-j(\omega_1 k_1 + \omega_2 k_2)} \tag{3-140}$$

or $$S(z_1, z_2) = Z\{R(k_1, k_2)\} = \sum_{k_1}\sum_{k_2} R(k_1, k_2)z_1^{-k_1}z_2^{-k_2} \tag{3-141}$$

[†]In this section, a script capital $\mathcal{S}$ is used for the Fourier transform while a simple capital S denotes the Z transform.

The inverse formulae are the standard ones given in Eq. (3-62) and Table 3.6.

The basic filtering property of a 2D linear system is illustrated in Fig. 3-30. It is left as an exercise for the reader to show that the following extrapolations from 1D are correct:

$$R_y(k_1, k_2) = [h(k_1, k_2) \circ\circ h(k_1, k_2)] ** R_x(k_1, k_2) \qquad (3\text{-}142)$$

$$\mathcal{S}_y(\omega_1, \omega_2) = |H(\omega_1, \omega_2)|^2 \mathcal{S}_x(\omega_1, \omega_2) \qquad (3\text{-}143)$$

$$S_y(z_1, z_2) = H_z(z_1, z_2) H_z(z_1^{-1}, z_2^{-1}) S_x(z_1, z_2) \qquad (3\text{-}144)$$

$$\mu_y = H(0, 0)\mu_x \qquad (3\text{-}145)$$

Equations Eq. (3-142) through Eq. (3-145) are the fundamental relationships for spatially invariant random fields and algorithms.

Now we pose another problem: For a given $R_y(m, n)$, find a realization of the form shown in Fig. 3-30. From Eq. (3-144), we see that there are an infinite number of filters $H_z(z_1, z_2)$ and source psds $S_x(z_1, z_2)$ which can be combined to yield a given $S_y(z_1, z_2)$. By putting certain constraints on $H_z(z_1, z_2)$, however, we can reduce the representations to three broad classes: causal, semicausal, and noncausal.

This classification can be illustrated by example for a simple, separable, exponential correlation function. Here

$$R_y(k_1, k_2) = \sigma^2 \rho_1^{|k_1|} \rho_2^{|k_2|} \qquad (3\text{-}146)$$

and the corresponding psd is

$$S_y(z_1, z_2) = \sigma^2 \frac{(1 - \rho_1^2)}{(1 - \rho_1 z_1)(1 - \rho_1 z_1^{-1})} \frac{(1 - \rho_2^2)}{(1 - \rho_2 z_2)(1 - \rho_2 z_2^{-1})} \qquad (3\text{-}147)$$

A causal system can generate the specified autocorrelation function from white noise via the causal difference equation

$$y(n_1, n_2) = \rho_1 y(n_1 - 1, n_2) + \rho_2 y(n_1, n_2 - 1)$$
$$\qquad\qquad - \rho_1 \rho_2 y(n_1 - 1, n_2 - 1) + x(n_1, n_2) \qquad (3\text{-}148)$$

$$x(n_1, n_2) \,\forall\, n_1, n_2 \longrightarrow \boxed{\begin{array}{c} h(n_1, n_2) \leftrightarrow H_z(z_1, z_2) \\ \text{Algorithm} \end{array}} \longrightarrow y(n_1, n_2)$$

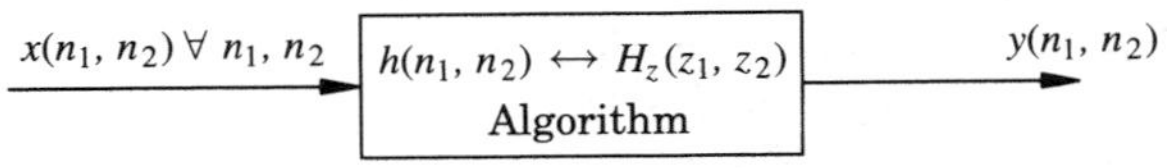

Figure 3-30   2D linear system as a filter.

The input sequence is white noise,

$$R_x(k_1, k_2) = \sigma^2(1 - \rho_1^2)(1 - \rho_2^2)\delta(k_1, k_2) \tag{3-149}$$

$$S_x(z_1, z_2) = \sigma^2(1 - \rho_1^2)(1 - \rho_2^2) \qquad \text{a constant}$$

and the filter transfer function is causal and separable.

$$H_z(z_1, z_2) = \frac{1}{1 - \rho_1 z_1^{-1} - \rho_2 z_2^{-1} + \rho_1 \rho_2 z_1^{-1} z_2^{-1}}$$

$$= \frac{1}{(1 - \rho_1 z_1^{-1})(1 - \rho_2 z_2^{-1})} \tag{3-150}$$

Equation Eq. (3-148) is a recursive, causal equation. It is said to be a first-order 2D Markov field, and it is used in predictive coding algorithms.

A semicausal model is causal in one direction or variable and non-causal in the other. For our example the representation is

$$y(n_1, n_2) = \alpha_1[y(n_1 - 1, n_2) + y(n_1 + 1, n_2)] + \rho_2 y(n_1, n_2 - 1)$$
$$- \rho_2 \alpha_1[y(n_1 + 1, n_2 - 1) + y(n_1 - 1, n_2 - 1)] + x(n_1, n_2) \tag{3-151}$$

where $\alpha_1 = \dfrac{\rho_1}{1 + \rho_1^2}$

From Eq. (3-151), the transfer function can be evaluated and factored into

$$H_z(z_1, z_2) = \frac{1}{[1 - \alpha_1(z_1 + z_1^{-1})](1 - \rho_2 z_2^{-1})} \tag{3-152}$$

This transfer function is noncausal in the z_1 variable, and now the input correlation is also nonwhite in the corresponding spatial variable. It is straightforward to show that

$$R_x(k_1, k_2) = A[\delta(k_1, k_2) - \alpha_1\delta(k_1 - 1, k_2) - \alpha_1\delta(k_1 + 1, k_2)] \tag{3-153}$$

$$S_x(z_1, z_2) = A[1 - \alpha_1(z_1^{-1} + z_1)] \tag{3-154}$$

$$A = \frac{\sigma^2(1 - \rho_1^2)(1 - \rho_2^2)}{1 + \rho_1^2} \tag{3-155}$$

The product of $S_x(z_1, z_2)$ in Eq. (3-154) with $H_z(z_1, z_2)H_z(z_1^{-1}, z_2^{-1})$ defined by Eq. (3-152) yields the desired output spectral density. This semicausal representation lends itself to hybrid coding in applications described by Jain, where the processing is recursive in one direction but noncausal in the other. (See also Sec. 9.2.3.)

The third representation is noncausal in both directions. For our desired output spectral density, we can have

$$y(n_1, n_2) = \alpha_1[x(n_1 - 1, n_2) + x(n_1 + 1, n_2)]$$
$$+ \alpha_2[x(n_1, n_2 - 1) + x(n_1, n_2 + 1)]$$
$$- \alpha_1\alpha_2[x(n_1 - 1, n_2 - 1) + x(n_1 + 1, n_2 - 1)$$
$$+ x(n_1 - 1, n_2 + 1) + x(n_1 + 1, n_2 + 1)]$$
$$+ x(n_1, n_2) \tag{3-156}$$

Therefore
$$H_Z(z_1, z_2) = \frac{1}{[1 - \alpha_1(z_1 + z_1^{-1})][1 - \alpha_2(z_2 + z_2^{-1})]} \tag{3-157}$$

and
$$S_x(z_1, z_2) = B[1 - \alpha_1(z_1 + z_1^{-1})][1 - \alpha_2(z_2 + z_2^{-1})] \tag{3-158}$$

where
$$B = \sigma^2 \frac{(1 - \rho_1^2)(1 - \rho_2^2)}{(1 + \rho_1^2)(1 + \rho_2^2)}$$

The correlation function for the input is determined by expanding Eq. (3-157) and identifying each term with a Kronecker delta grid point in the (k_1, k_2) plane.

$$\frac{S_x(z_1, z_2)}{B} = 1 - \alpha_1 z_1^{-1} - \alpha_1 z_1 - \alpha_2 z_2^{-1} - \alpha_2 z_2$$
$$+ \alpha_1\alpha_2[z_1^{-1}z_2^{-1} + z_1 z_2^{-1} + z_1^{-1}z_2 + z_1 z_2] \tag{3-159}$$

Then
$$\frac{R_x(k_1, k_2)}{B} = \delta(k_1, k_2) - \alpha_1\delta(k_1 - 1, k_2) - \alpha_1\delta(k_1 + 1, k_2)$$
$$- \alpha_2\delta(k_1, k_2 - 1) - \alpha_2\delta(k_1, k_2 + 1)$$
$$+ \alpha_1\alpha_2[\delta(k_1 - 1, k_2 - 1) + \delta(k_1 + 1, k_2 - 1)$$
$$+ \delta(k_1 - 1, k_2 + 1) + \delta(k_1 + 1, k_2 + 1)] \tag{3-160}$$

This noncausal representation lends itself to transform coding applications; see Sec. 9.1.1.

The source autocorrelations for the causal, semicausal, and noncausal representations of Eqs. (3-149), (3-151), and (3-160) are shown in Fig. 3-31(a), (b), and (c), respectively.

EXAMPLE 3.11: This example illustrates the use of one of the foregoing models in signal smoothing. The problem is posed in Fig. 3-32. The image $x(n_1, n_2)$ is corrupted by white noise $\eta(n_1, n_2)$, and $G(z_1, z_2)$ is to be selected for optimal filtering of the corrupted image. We assume that the image and noise are uncorrelated, with zero means and with psds given by

$$S_x(z_1, z_2) = \frac{\beta^2}{1 - \alpha(z_1 + z_1^{-1} + z_2 + z_2^{-1})} \tag{3-161}$$

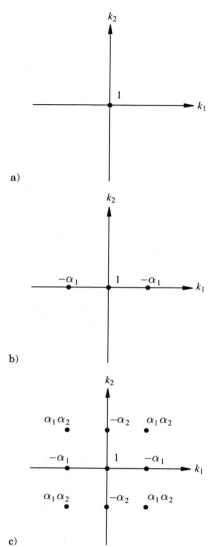

Figure 3-31   *Source autocorrelation functions for causal, semicausal, and non-causal models.*

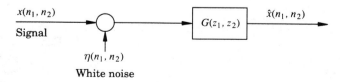

Figure 3-32   *Image-smoothing configuration.*

$$S_\eta(z_1, z_2) = \sigma_\eta^2 \qquad \text{(white noise)} \tag{3-162}$$

By the orthogonality principle (Chap. 5), the best noncausal linear mean-squared estimator is known to satisfy

$$G(z_1, z_2) = \frac{S_x(z_1, z_2)}{S_x(z_1, z_2) + S_\eta(z_1, z_2)} \tag{3-163}$$

Hence

$$G(z_1, z_2) = \frac{1}{1 + \frac{\sigma_\eta^2}{\beta^2}[F(z_1, z_2)]} \tag{3-164}$$

where

$$F(z_1, z_2) = 1 - \alpha(z_1 + z_1^{-1} + z_2 + z_2^{-1})$$

For large signal-to-noise ratios, $\sigma_\eta^2/\beta^2 << 1$, we can approximate G by the noncausal FIR filter

$$G(z_1, z_2) \approx 1 - \frac{\sigma_\eta^2}{\beta^2} F(z_1, z_2) \tag{3-165}$$

$$= \left(1 - \frac{\sigma_\eta^2}{\beta^2}\right) + \alpha\left(\frac{\sigma_\eta^2}{\beta^2}\right)(z_1 + z_1^{-1} + z_2 + z_2^{-1}) \tag{3-166}$$

If $\sigma_\eta^2/\beta^2 = 1/5$ and $\alpha = 1/4$, this becomes

$$G(z_1, z_2) = \tfrac{4}{5}\left[1 + \tfrac{1}{16}(z_1 + z_1^{-1} + z_2 + z_2^{-1})\right] \tag{3-167}$$

a simple, non-causal, nonrecursive spatial averaging filter which weights pixels in an approximately optimal way. The scale factor in Eq. (3-167) can be adjusted for any desired calibration.

EXAMPLE 3.12: . An IIR filter that has been used (Natravali and Limb, 1980; and Jung, 1979) as an inverse predictive filter in differential pulse-code modulation (DPCM), and also as a model for a Markov random field (Jain, 1981), is given by

$$H_z(z_1, z_2) = \frac{1}{1 - bz_1^{-1} - cz_2^{-1} - dz_1^{-1}z_2^{-1}} = \frac{1}{D(z)} \tag{3-168}$$

$$y(n_1, n_2) = by(n_1 - 1, n_2) + cy(n_1, n_2 - 1) + dy(n_1 - 1, n_2 - 1) + x(n_1, n_2) \tag{3-169}$$

Some characteristics of this filter are

1. Stability (Huang, 1972)

$$|b| < 1, \qquad \left|\frac{c + d}{1 - b}\right| < 1, \qquad \left|\frac{c - d}{1 + b}\right| < 1 \tag{3-170}$$

2. Separability: $d = -bc$ and

$$D(z) = (1 - bz_1^{-1})(1 - cz_2^{-1}) \tag{3-171}$$

3. Power:

$$\sum \sum |h(n_1, n_2)|^2 = \left(\frac{1}{2\pi}\right)^2 \int_{-\pi}^{\pi} \int_{-\pi}^{\pi} |H(\omega_1, \omega_2)|^2 d\omega_1 \, d\omega_2$$

$$= \frac{1}{[(1+b)^2 - (c-d)^2]^{\frac{1}{2}} [(1-b)^2 - (c+d)^2]^{\frac{1}{2}}} \tag{3-172}$$

For a random-field model, the input $x(n_1, n_2)$ is represented by a 2D stationary white-noise sequence

$$E\{x(n_1, n_2)\} = 0$$
$$R_x(k_1, k_2) = \sigma_x^2 \delta(k_1, k_2) \tag{3-173}$$

This is an autoregressive Markov model for an image. In this case, the coefficients are chosen to be

$$b = \rho_1$$
$$c = \rho_2$$
$$d = -bc = -\rho_1 \rho_2$$

The coefficients ρ_1 and ρ_2 represent the first-order correlation coefficients between adjacent horizontal and vertical pixels, respectively. In this case, the filter $H_z(z_1, z_2)$ is separable and the input is white noise and hence also separable. It follows that the output correlation sequence is likewise separable and is given by

$$R_y(k_1, k_2) = \sigma_y^2 \rho_1^{|n_1|} \rho_2^{|n_2|} \tag{3-174}$$

where

$$\sigma_y^2 = \left(\sum \sum |h(n_1, n_2)|^2\right) \sigma_x^2$$

In our case, the noise power term reduces to

$$\left(\sum |h_1(n)|^2\right)\left(\sum |h_2(n)|^2\right) = \frac{1}{(1-\rho_1^2)(1-\rho_2^2)} \tag{3-175}$$

hence

$$\sigma_y^2 = \frac{\sigma_x^2}{(1-\rho_1^2)(1-\rho_2^2)} \tag{3-176}$$

In a DPCM application (see Sec. 9.1.2), as shown in Fig. 3-33, the source signal would be modelled by Eq. (3-169) and the predictor represents the source-model coefficients ρ_1 and ρ_2 as best it can:

$$\hat{P}(z_1, z_2) = \hat{\rho}_1 z_1^{-1} + \hat{\rho}_2 z_2^{-1} - \hat{\rho}_1 \hat{\rho}_2 z_1^{-1} z_2^{-1} \tag{3-177}$$

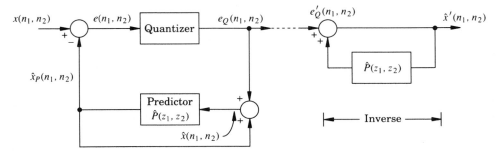

Figure 3-33   *DPCM transmitter and receiver.*

At the receiver, the inverse filter is simply

$$H_Z(z_1, z_2) = \frac{1}{1 - \hat{P}(z_1, z_2)} \frac{1}{1 - \hat{\rho}_1 z_1^{-1} - \hat{\rho}_2 z_2^{-1} + \hat{\rho}_1 \hat{\rho}_2 z_1^{-1} z_2^{-1}}$$

$$= \frac{1}{(1 - \hat{\rho}_1 z_1^{-1})(1 - \hat{\rho}_2 z_2^{-1})} \tag{3-178}$$

The subject of DPCM is taken up in greater detail in Sect. 9.1.

3.10 SUMMARY

In this chapter, we laid the groundwork for the representation, analysis, and design of 2D signals and filters. One objective was to extend 1D results to two dimensions wherever possible and also to point out the pitfalls in blindly extrapolating 1D theory. Starting from 2D analog signals, we outlined an analog 2D linear system theory along with the 2D analog Fourier transform. The sampling theorem provided the link between this analog world and the discrete-space linear world that is the subject of this book. We saw that the 2D Fourier transform, DFT, FIR filters, and difference equations are relatively straightforward extrapolations of their 1D counterparts, with the expected attendant complexity involved in going from one independent variable to two. Some of the more vexing problems arose in IIR filters—in their design, implementation, recursion, and stability. The 1D progenitors in these cases did not provide much help and in fact were misleading. Special care has to be taken, as noted in the text.

We attempted to demonstrate a few of these techniques by way of example. Our treatment here is not exhaustive; the reader can pursue the subject in greater depth by studying some of the references, particularly Dudgeon and Mersereau (1984), Jain (1989), Huang (1981), and Lim (1990). We will resume our consideration of two-dimensional systems in Chap. 9, where image-processing applications are developed.

PROBLEMS

3.1.   (a) Derive the inversion formula for the 2D Fourier transform, Eq. (3-17).
        (b) Derive Parseval's theorem, Eq. (3-22), from Eq. (3-21).

3.2.   Derive entries 3, 5, and 7 in Table 3.2.

3.3.   Derive entry 10 in Table 3.2.

3.4.   In Table 3.5, derive entry 9 from entry 8.

3.5.   Given the 2D sequence

$$x(n_1, n_2) = \sum_{p=0}^{N-1} \delta(n_1 - pm_1, n_2 - pm_2)$$

(a) Sketch and label this sequence in terms of parameters m_1 and m_2.
(b) Evaluate and sketch the Fourier transform of $x(n_1, n_2)$.

3.6.   Evaluate and sketch the *linear* convolution of the pair of 2×2 arrays shown in Fig. 3-P1.

3.7.   When a linear mapping $y = L\{x\}$ results in $y = kx$, the input x is an eigenfunction of the transformation and k is the eigenvalue. Show that $e^{j(\omega_1 n_1 + \omega_2 n_2)}$ is an eigenfunction of the linear shift-invariant operator and that $H(\omega_1, \omega_2) = \mathcal{F}\{h(n_1, n_2)\}$ is the associated eigenvalue.

3.8. $H(\omega_1, \omega_2) = (1 - \cos\omega_1)(1 - \cos\omega_2)$ defines the frequency response of a FIR filter.
        (a) Evaluate $H_z(z_1, z_2)$ and show the difference equation relating the input sequence $x(n_1, n_2)$ and the output sequence $y(n_1, n_2)$.
        (b) Evaluate and plot the 2D impulse response $h(n_1, n_2)$.

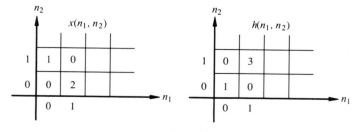

Figure 3-P1   *Problem 3.6.*

(c) Show a cascaded, separable realization of this filter.

3.9.   A FIR filter has the mask shown in Fig. 3-P2. Evaluate $h(n_1, n_2)$,
       $H_z(z_1, z_2)$, and $H(\omega_1, \omega_2)$. What operation is this filter performing?

3.10. The input to a recursive filter is a 128×128 picture $x(n_1, n_2)$, where
       $0 \le n_1 \le 127, 0 \le n_2 \le 127$. The filter is defined by

$$y(n_1, n_2) = 0.1y(n_1 - 1, n_2) + 0.2y(n_1 - 1, n_2 - 1)$$
$$+ x(n_1, n_2) + 0.5x(n_1 - 1, n_2 - 1)$$

       (a) Evaluate $H_z(z_1, z_2)$, $H(\omega_1, \omega_2)$.
       (b) Sketch the output mask for this filter. Describe how you recur-
           sively compute the 128×128 output. Describe the minimum
           set of (n_1, n_2) values outside the input square where initial
           values of the output must be defined in order to carry out the
           recursions.
       (c) Evaluate the output $y(n_1, n_2)$ for

$$(n_1, n_2) = (0, 0), (0, 1), (1, 0), (1, 1)$$

       Assume zero for all necessary initial conditions, and let the
       input be

$$x(n_1, n_2) = \begin{cases} 2 & n_1 = n_2 = 0 \\ 0 & \text{otherwise} \end{cases}$$

3.11. (a) Compute the Fourier transform of the first-quadrant signal

$$x(n_1, n_2) = \begin{cases} a^{n_1} b^{n_2} & 0 \le n_1, n_2 \le \infty \\ 0 & \text{otherwise} \end{cases}$$

Figure 3-P2   Problem 3.9.

(b) This transform is sampled at points

$$\omega_1 = k_1 \frac{2\pi}{8} \quad \text{and} \quad \omega_2 = k_2 \frac{2\pi}{8}, \quad 0 \le k_1, k_2 \le 7$$

to yield $X(k_1, k_2)$, and then the inverse DFT is taken. Evaluate the spatial-domain function. (*Hint:* Use the Poisson sum formula.)

3.12. Evaluate $H(\omega_1, \omega_2)$ and $H_z(z_1, z_2)$ for
(a) $y(n_1, n_2) = a_1 y(n_1 - 1, n_2) + a_2 y(n_1, n_2 - 1) + x(n_1, n_2)$.
(b) $y(n_1, n_2) = a_1 y(n_1 - 1, n_2) + a_2 y(n_1, n_2 - 1)$
$\quad -a_1 a_2 y(n_1 - 1, n_2 - 1) + x(n_1, n_2)$.
(c) Explain the differences between parts (a) and (b).

3.13. Evaluate and plot $h(n_1, n_2)$ for a system with

$$H(\omega_1, \omega_2) = 1 - 2a \cos \omega_1 - 2a \cos \omega_2$$

3.14. $H_z(z_1, z_2) = 1 + a z_1^{-1} + b z_1^{-1} z_2^{-1} + c z_2^{-1} + d z_1 z_2^{-1}$. Evaluate the impulse response.

3.15. A spatially invariant filter $h(n_1, n_2)$ is driven by a wide-sense stationary input $x(n_1, n_2)$ as in Fig. 3-30. Derive Eq. (3-142) through Eq. (3-145) starting with the convolution $y(n_1, n_2) = h(n_1, n_2) ** x(n_1, n_2)$.

3.16. See Fig. 3-33. Suppose that $\hat{\rho}_1 = \rho_1, \hat{\rho}_2 = \rho_2$, that is, the vertical and horizontal correlation coefficients are known and are implemented in the predictor block. Show that
(a) $\hat{X}(z_1, z_2) = \dfrac{1}{1 - \hat{P}(z_1, z_2)} E_Q(z_1, z_2)$
(b) In the absence of channel errors,

$$e_Q'(n_1, n_2) = e_Q(n_1, n_2)$$

Show that

$$\hat{x}_Q'(n_1, n_2) = \hat{x}(n_1, n_2)$$

and consequently the reconstruction error $\tilde{x}(n_1, n_2)$ is equal to the quantization error, that is,

$$\tilde{x}(n_1, n_2) \triangleq x(n_1, n_2) - \hat{x}'(n_1, n_2)$$

$$= e(n_1, n_2) - e_Q(n_1, n_2) \triangleq q(n_1, n_2)$$

3.17. See Example 3.11 and Fig. 3-32. Evaluate the power spectral density for the estimation error $e(n_1, n_2) \triangleq x(n_1, n_2) - \hat{x}(n_1, n_2)$. Show that, for $.\sigma_\eta^2/\beta^2) \ll 1$.

$$S_{ee}(z_1, z_2) \approx \sigma_n^2 \left[1 - \frac{\sigma_n^2}{\beta^2} F(z_1, z_2) \right]$$

Evaluate the resulting mean-squared error, $E\{|e(n_1, n_2)|^2\} = R_e(0, 0)$.

3.18. Use the DeCarlo-Strintzis test to determine whether the following are stable IIR filters:

(a) $H(z_1, z_2) = \dfrac{1}{1 - az_1^{-1} - 0.5z_2^{-1}}$,

Find the range of values of a for which this filter is stable.

(b) $H(z_1, z_2) = \dfrac{1}{1 - \frac{1}{2}z_1^{-1} - \frac{1}{4}z_2^{-1}}$

REFERENCES

S. A. H. Aly and M. M Fahmy. "Design of two dimensional recursive digital filters with specified magnitude and group delay characteristics." *IEEE Trans Circ. and Sys.*, vol. CAS–25, pp. 908–916, Nov. 1978.

———. "Spatial domain design of two-dimensional recursive digital filters." *IEEE Trans. Circ. and Sys.*, vol. CAS-27, pp. 892–901, Oct. 1980.

N. K. Bose, *Applied Multidimensional Systems Theory*. New York: Van Nostrand Reinhold, 1982.

Computer, vol. 7, no. 5, 1974. Special issue on digital image processing.

R. A. DeCarlo *et al.* "Multivariable Nyquist theory." *Int. J. Control*, vol. 25, pp. 657–675, 1976.

D. D. Dudgeon and R. M. Mersereau, *Multidimensional Signal Processing*. Englewood Cliffs, N.J.: Prentice-Hall, 1984.

R. C. Gonzales and P. Wintz. *Digital Image Processing*. Reading, Mass.: Addison-Wesley, 1987.

J. Goodman, *Introduction to Fourier Optics*. New York: McGraw-Hill, 1968.

R. A. Haddad. "A class of orthogonal nonrecursive binomial filters." *IEEE Trans. Audio and Electroacous.*, vol. AU-19, no. 4, pp 296–304, Dec. 1971.

T. S. Huang, ed. *Two Dimensional Signal Processing. (Topics in Applied Physics, vol. 42, 43)* Berlin: Springer-Verlag, 1981.

T. S. Huang. "Stability of two-dimensional recursive filters." *IEEE Trans. Audio and Electroacous.*, vol. AU-20, no. 2, pp. 158–163, June, 1972.

A. K. Jain. "Advances in mathematical models for image processing." *Proc. IEEE*, vol. 69, no. 5, pp 502–528, May, 1981.

————, *Fundamentals of Digital Image Processing*. Englewood Cliffs, N. J.: Prentice-Hall, 1989.

N. S. Jayant and P. Noll, *Digital Coding of Waveforms*. Englewood Cliffs, N.J.: Prentice-Hall, 1984.

P. Jung. "Statistical comparison of additive error effects in DPCM image coding with two-dimensional prediction" [in German], *NTZ Archiv*, vol. 12, pp. 254–262, 1979.

J. H. Lee and J. W. Woods. "Design and implementation of two-dimenstional fully recursive digital filters." *IEEE Trans. ASSP*, vol. ASSP-34, no. 1, pp. 178–191, Feb. 1986.

J. S. Lim and A. V. Oppenheim (Eds.), *Advanced Topics in Signal Processing*. Englewood Cliffs, N.J.: Prentice-Hall, 1988.

J. S. Lim. *Two-Dimensional Signal and Image Processing*. Englewood Cliffs, N. J.: Prentice-Hall, 1990.

J. H. McClellan. "The design of two-dimensional digital filters by transformations." *Proc. 7th Annu. Princeton Conf. Inform. Sciences and Syst.*, pp 247–251, 1973.

R. M. Mersereau. "Processing of hexagonally sampled two dimensional signals." *Proc. IEEE*, vol. 67, no. 6, pp 930–949, June 1979.

W. F. G. Mecklenbrauker and R. M. Mersereau. "McClellan transformation for 2D digital filtering. II–Implementation." *IEEE Trans. Circ. and Syst.*, vol. CAS-23, no. 7, pp 414–422, July 1976.

R. M. Mersereau et al. "McClellan transformation for 2D digital filtering. I–Design." *IEEE Trans. Circ. and Syst.*, vol. CAS-23, no. 7, pp 405–414, July 1976.

A. N. Netravali and J. O. Limb. "Picture coding: a review." *Proc. IEEE*, vol. 68, no. 3, pp. 366–406, Mar. 1980.

B. T. O'Connor and T. S. Huang. "Stability of general two-dimensional recursive digital filters." *IEEE Trans. ASSP*, vol. ASSP-26, no. 6, pp 550–560, Dec. 1978.

A. Papoulis, *Systems and Transforms with Applications in Optics*. New York: McGraw-Hill, 1968.

————, *Signal Analysis*. New York: McGraw-Hill, 1977.

Proc. IEEE, vol. 60, no. 7, 1972. Special issue on digital image processing.

A. Rosenfeld and A. Kak, *Digital Picture Processing*. New York: Academic Press, 1976.

J. L. Shanks et al. "Stability and synthesis of two dimensional recursive filters." *IEEE Trans. Audio and Electroacous.*, vol. AU-20, no. 2, pp 115–128, June, 1972.

J. J. Shynk. "Adaptive IIR filtering." *IEEE ASSP Magazine*, vol. 6, no. 2, pp. 4–21, Apr. 1989.

Standard Mathematical Tables. Cleveland: Chemical Rubber Co., 1970, pp. 518–524.

M. G. Strentzis. "Test of stability of multidimensional filters." *IEEE Trans. on Circ. and Systs.*, vol. CAS-24, pp. 432–437, Aug. 1977.

E. Vanmarke, *Random Fields: Analysis and Synthesis*. Cambridge, Mass.: MIT Press, 1983.

CHAPTER 4

■ ■

THE FFT AND OTHER FAST TRANSFORMS

In this chapter we will start by reviewing the theory of the fast Fourier transform (FFT); then we will consider other high-speed algorithms for Fourier and other transforms. In so doing, we will digress briefly to examine methods for doing fast convolutions. One of the traditional uses for fast Fourier transforms is fast convolution, and it is appropriate to consider algorithms that implement fast convolution directly. In addition, these fast convolution algorithms are themselves the basis for the most important alternative to the FFT. We will conclude by considering number-theoretic transforms.

4.1 THE FOURIER TRANSFORM

Recall the defining equation of the DFT: let

$$\{x(n)\} = \{x(0), x(1), ..., x(N-1)\}$$

Then its transform is given by

$$X(k) = \sum_{n=0}^{N-1} x(n) W_N^{-nk} \qquad k = 0, 1, ..., N-1 \qquad (4\text{-}1)$$

where W_N is the principal Nth root of unity:

$$W_N = \exp(j2\pi/N)$$
$$= \cos(2\pi/N) + j\sin(2\pi/N)$$

We will normally use lowercase letters for the input sequence and uppercase for the transform; we will also write $X(k) = \mathfrak{F}\{x(n)\}$ to indicate that x and X are a transform pair.

It is customary to regard $X(k)$ as defined only for $k = 0, 1, ..., N - 1$, as shown. From App. B, however, we have for any integer p,

$$X(k + Np) = X(k). \tag{4-2}$$

Alternatively, we can use modular arithmetic and agree that all indices n and k are taken modulo N.

To compute this transform as defined, we must do N^2 complex multiplications. This requirement makes DFT algorithms based on the defining equation prohibitively expensive except for very small N. Virtually the only shortcut the defining equation offers is replacing multiplications by $1 + j0$ and by $0 + j1$ with the appropriate data transfers.

We can reduce the number of multiplications by 25 percent using a well-known trick for multiplying complex numbers. Although we are looking for much better speedups than this, we will review this technique here, since it the first, and simplest, example of a fast algorithm and presents some features of which we should take note. The product of $(a + jb)(c + jd)$ is normally written, $(ac - bd) + j(ad + bc)$. Let $P = d(a - b)$, $Q = a(c - d)$, and $R = b(c + d)$. Finding P, Q, and R requires only three multiplications, and the product is $(P + Q) + j(P + R)$.

This device illustrates in miniature two characteristics that will recur again and again in the fast algorithms to be described in this chapter. The first is that the saving in multiplications is bought at the price of an increase in the number of additions; this trade-off is typical of these fast algorithms. The theory of computational complexity has had its greatest successes in the minimization of multiplications, and these successes are always paid for by extra additions. The second characteristic illustrated by this shortcut is its structure: a set of preliminary additions (and subtractions), followed by a minimal number of multiplications, followed by some more additions. We will find this structure in most of the fast algorithms we will examine.

There is one more lesson to be learned from this simple example. Let us look at some execution times. In the 8086, a well-known 16-bit microprocessor, a 16-bit multiplication takes roughly 125 clock pulses and a 32-bit add takes 6 clocks. Then the conventional algorithm requires 512 clocks for a complex multiplication, and the alternative requires only 405 clocks. On the other hand, the 8087, a floating-point coprocessor used with the 8086, requires approximately 138 clocks for a long (64-bit) multiply and 85 clocks for an add. This is not unreasonable: the need to normalize floating-point sums usually slows down floating-point addition and subtraction. In this case, the conventional method takes roughly 722 clocks and our "fast" alternative requires 839 clocks; here using the fast algorithm is actually counterproductive.

In general, let t_a be the time required for an addition and t_m the time required for a multiplication, and assume that these two execution times dominate the execution time for the algorithm as a whole. Let M_o and M_n be the number of multiplications for the old and new methods, respectively, and similarly let A_o and A_n be the number of additions. Then if

$$t_{\text{old}} = M_o t_m + A_o t_a$$

and
$$t_{\text{new}} = M_n t_m + A_n t_a$$

Then the difference in execution time will be

$$t_{\text{old}} - t_{\text{new}} = t_m \Delta m + t_a \Delta a$$

where
$$\Delta m = M_o - M_n \qquad \Delta a = A_o - A_n$$

Under these assumptions, there will be a net reduction in execution time if $t_{\text{old}} > t_{\text{new}}$ or if

$$t_m \Delta m > t_a \Delta a$$

The lesson to be drawn from this is that the trade-off between adds and multiplies must always be carefully examined. In digital signal processing, fixed-point arithmetic is used wherever possible, but not everywhere. In this chapter we will always assume that the cost of additions is negligible in comparison with the cost of multiplications; nevertheless, in an actual application this assumption must be tested whenever speed is important.

4.1.1 THE FAST FOURIER TRANSFORM

This is the name given to a family of algorithms for computing the DFT with far fewer multiplications than are required by the defining equation. We will review this algorithm briefly by way of setting the stage for what is to come. In its simplest form the FFT (Cooley and Tukey, 1965) proceeds as follows: suppose N is a power of 2. We will see shortly that we can find the Fourier transform of a sequence by combining the transforms of two sub-sequences of length $N/2$, obtained from the given sequence. Then instead of applying the defining equation to x, let us transform its subsequences and combine the transforms. In this way, we need perform only $\frac{1}{2}N^2$ multiplications, plus the cost of the combining operation.

There is obviously no reason to stop here: we can transform our two subsequences by dividing each of them into two sub-sub-sequences of length $N/4$ and combining their transforms. At this point, it is clear that we can continue this process until we are down to sub-sub-$\cdots$- sub-sequences of length 1. By Eq. (4-1), however, a sequence of length 1 is its own transform. We have therefore repeatedly postponed the transformation process until the need to transform disappears, and all we are left with is the series of combining operations. Programmers will recognize this approach as an example of the well-known "divide-and-conquer" strategy.

It takes $\log_2 N$ iterations to get down to a sequence of length 1. (In this chapter, all logarithms are to the base 2 unless stated otherwise.) This means that we require $\log N$ combining operations. We will see that a combining operation requires N complex multiplications; hence we have reduced this computational burden from N^2 to $N \log N$. For $N = 1,024$, we have a 100-to-1 reduction in the number of multiplications needed. Since multiplications are normally the most time-consuming operations, this results in a significant speedup and justifies the term "fast."

The key to this process is clearly the combining operation, which we will now derive. Let

$$\{x_e(n)\} = \{x(0), x(2), ..., x(N-2)\}$$

so that

$$x_e(n) = x(2n)$$

and

$$\{x_o(n)\} = \{x(1), x(3), ..., x(N-1)\}$$

so that

$$x_o(n) = x(2n+1)$$

Thus $\{x_e(n)\}$ and $\{x_o(n)\}$ are the even-numbered and odd-numbered points of $\{x(n)\}$, considered as $N/2$-vectors, as shown in Fig. 4-1. Let their transforms be $X_e(k)$ and $X_o(k)$, respectively.

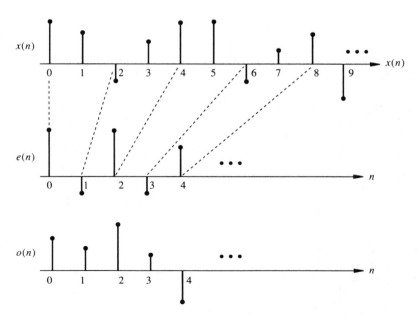

Figure 4-1 *Even and odd subsequences used in the FFT.*

We may rewrite our defining equation in terms of the even- and odd-numbered points as follows:

$$X(k) = \sum_{n=0}^{N/2-1} x(2n)W_N^{-2nk} + \sum_{n=0}^{N/2-1} x(2n+1)W_N^{-(2n+1)k}$$

$$= \sum_{n=0}^{N/2-1} x_e(n)W_N^{-2nk} + \sum_{n=0}^{N/2-1} x_o(n)W_N^{-(2n+1)k}$$

Now observe that for even N, $W_N^2 = W_{N/2}$. Therefore

$$X(k) = \sum_{n=0}^{N/2-1} x_e(n)W_{N/2}^{-nk} + W_N^{-k} \sum_{n=0}^{N/2-1} x_o(n)W_{N/2}^{-nk}$$

$$= X_e(k) + W_N^{-k}X_o(k) \tag{4-3}$$

For values of $k > N/2$, we will rely on the periodic property given above for DFTs in general:

$$X_e(k + N/2) = X_e(k)$$
$$X_o(k + N/2) = X_o(k)$$

Notice also that for any even N, $W_N^{k+N/2} = -W_N^k$. Hence it is convenient to rewrite Eq. (4-3) as follows

$$X(k) = \begin{cases} X_e(k) + W_N^{-k}X_o(k) & 0 \le k < N/2 \\ X_e(k - N/2) - W_N^{-(k-N/2)} X_o(k - N/2) & \text{otherwise} \end{cases} \tag{4-4}$$

This gives us the combining method we are after: we find the transforms of the even-numbered points and of the odd-numbered points and then combine them as specified in Eq. (4-4). One way of looking at the W^{-k} factor is to note that a simple DFT of the odd points has built into it the assumption that these points start at $n = 0$. Actually, they start at $n = 1$, however, and the W^{-k} factor can be thought of as an application of the shifting property of the DFT.

The combining operation requires N complex multiplications, as promised. We may represent this operation by the flow diagram in Fig. 4-2(a), where it is shown for $N = 8$. The operation defined by Eq. (4-4) is commonly called the butterfly, because of the appearance of its flow diagram. The transforms $X_e(k)$ and $X_o(k)$ are obtained similarly; we take the even- and odd-numbered points of $\{x_e(n)\}$ and combine their transforms to obtain $X_e(k)$; similarly for $X_o(k)$. The entire process, consisting of three iterations for $N = 8$, is shown in Fig. 4-2(b).

If the FFT is computed in the manner suggested by Fig. 4-2(b), the computation can be done in place. In each iteration, the newly computed

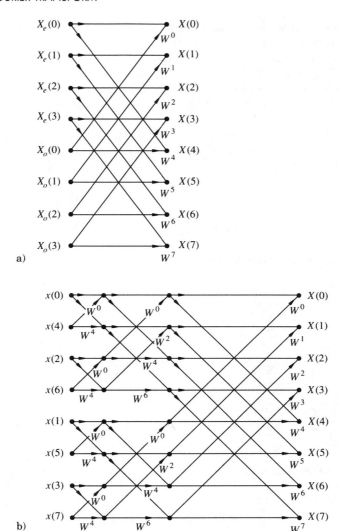

Figure 4-2   *FFT flow diagrams: (a) one stage in 8-point FFT transform; (b) entire 8-point FFT.*

values of the sequence replace the old ones. This is clearly an advantage when transforming long sequences in environments with limited storage, but as the relative orderliness of the diagram suggests, it is also the easiest way to program the FFT. The price one pays for this straightforward procedure is that the input sequence must be presented to the algorithm in a permuted order. It can be shown (Cooley and Tukey, 1965) that the address of the ith input sample can be found by writing i as an N-bit binary number with the bits in reversed order. Thus in the figure, it will be seen that $x(3)$ (011_2) is in location 6 (110_2) and $x(6)$ is in location

3. Samples whose indices are palindromes in binary are placed in their natural locations: $x(2)$ is in location 2 (010_2). This input permutation normally consumes a negligible amount of processing time.

If the same graph is redrawn with the input sequence in natural order, as in Fig. 4-3, then we obtain an alternative form for which in-place computation causes the transform sequence to appear in bit-reversed order.

An alternative combining method divides the input sequence into two sequences consisting of the first $N/2$ points and the second $N/2$ points. Let these sequences be $\{x_f(n)\}$ and $\{x_s(n)\}$, where

$$x_f(n) = x(n) \qquad\qquad n = 0, 1, ..., N/2 - 1$$

$$x_s(n) = x(n + N/2) \qquad n = 0, 1, ..., N/2 - 1$$

We can now examine the contributions of these two subsequences to $X(k)$ by writing,

$$X(k) = \sum_{n=0}^{N/2-1} x(n)W_N^{-nk} + \sum_{n=N/2}^{N-1} x(n)W_N^{-nk}$$

$$= \sum_{n=0}^{N/2-1} x(n)W_N^{-nk} + \sum_{n=0}^{N/2-1} x(n+N/2)W_N^{-(n+N/2)k}$$

$$= \sum_{n=0}^{N/2-1} x_f(n)W_N^{-nk} + W_N^{-Nk/2}\sum_{n=0}^{N/2-1} x_s(n)W_N^{-nk}$$

If k is even, $k = 2p$ and

$$X(2p) = \sum_{n=0}^{N/2-1} x_f(n)W_N^{-2np} + W_N^{-Np}\sum_{n=0}^{N/2-1} x_s(n)W_N^{-2np} \qquad (4\text{-}5)$$

Since $W_N^2 = W_{N/2}$ and $W_N^N = 1$, this is the transform of $x_f(n) + x_s(n)$.

If k is odd, $k = 2p + 1$, and

$$X(2p + 1) = \sum_{n=0}^{N/2-1} x_f(n)W_N^{-n(2p+1)}$$

$$+ W_N^{-Np}W_N^{-N/2}\sum_{n=0}^{N/2-1} x_s(n)W_N^{-n(2p+1)}$$

Hence $\quad X(2p + 1) = \sum_{n=0}^{N/2-1} W_N^{-n}[x_f(n) - x_s(n)]W_{N/2}^{-np}$

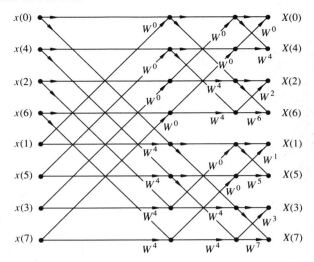

Figure 4-3   *8-point FFT with inputs in natural order and outputs in permuted order.*

This is the transform of $W_N^{-n}[x_f(n) - x_s(n)]$. We may summarize as follows:

$$X(k) = \begin{cases} \mathfrak{F}\{x_f(n) + x_s(n)\} & k \text{ even} \\ \mathfrak{F}\{W_N^{-n}[x_f(n) - x_s(n)]\} & k \text{ odd} \end{cases} \qquad (4\text{-}6)$$

This combining operation is shown in Fig. 4-4(a).

Notice the symmetries between this method, known as *decimation in frequency*, and the previous method, known as *decimation in time*. In the first method, the first and last halves of the transform were obtained from the even and odd points in the input sequence. In the second method, the even and odd points are obtained from the transforms of the first and second halves of the input sequence. These symmetries show up in the flow diagrams as well, as may be seen by comparing Fig. 4-2(b) with Fig. 4-4(b). There is virtually nothing to choose between these two methods; normally one is not faster or more efficient than the other, and in both cases a permutation pass is required. Decimation in frequency, as described here, requires permutation of the output samples after the iterations. As with decimation in time, however, the graph can be rearranged to require permutation of the input instead.

4.1.2 TRANSFORMS OF REAL FUNCTIONS

If $x(n)$ is real, then $X(N - k) = \bar{X}(k)$, where the overbar indicates the complex conjugate. In this case, half the transform is redundant, since it can be obtained from the other half. Cooley et al. (1970) provide the

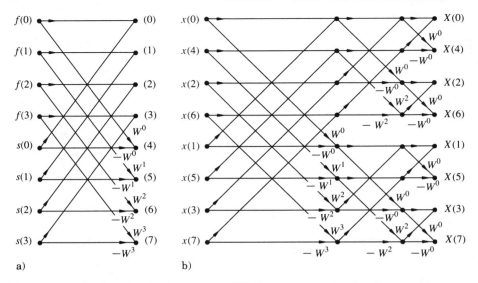

Figure 4-4 *Decimation-in-frequency FFT: (a) one stage in 8-point transform;*
(b) entire 8-point FFT.

following technique for finding the DFT of a sequence of $2N$ points. For
clarity we will call the input sequence $\{y(n)\}$. Then we let

$$p(n) = y(2n) \qquad n = 0, 1, ..., N - 1$$

and

$$q(n) = y(2n + 1) \qquad n = 0, 1, ..., N - 1$$

and form the x sequence

$$x(n) = p(n) + jq(n)$$

In many programming languages, the real and imaginary parts of complex arrays are stored in alternating locations; hence this can be done
simply by making x and y share the same address space. The transform
of this sequence will be

$$X(k) = P(k) + jQ(k)$$

To separate these parts, we note that since $P(k)$ and $Q(k)$ are each
transforms of the real sequences $p(n)$ and $q(n)$

$$P(N - k) = \overline{P}(k), Q(N - k) = \overline{Q}(k)$$

Hence we have

$$X(k) = P(k) + jQ(k)$$

$$\overline{X}(N - k) = P(k) - jQ(k)$$

in which case we can solve for $P(k)$ and $Q(k)$:

$$P(k) = [X(k) + \overline{X}(N - k)]/2 \tag{4-7}$$

$$Q(k) = j[\overline{X}(N - k) - X(k)]/2$$

We can now obtain $Y(k)$ by combining $P(k)$ and $Q(k)$ as in Eq. (4-3).

4.1.3 NON-RADIX-2 TRANSFORMS

In all the FFTs considered so far, N has been assumed a power of 2; such FFTs are termed radix-2. Suppose N were a power of 3. Then, parallelling the derivation above, we can define three sequences of length $N/3$, as follows:

$$s_0(n) = x(3n)$$

$$s_1(n) = x(3n + 1) \tag{4-8}$$

$$s_2(n) = x(3n + 2)$$

Let these sequences have transforms $S_0(k)$, $S_1(k)$, and $S_2(k)$, respectively. Then decimation in time gives

$$X(k) = \sum_{n=0}^{N/3-1} x(3n)W_N^{-3nk} + W_N^{-k} \sum_{n=0}^{N/3-1} x(3n + 1)W_N^{-3nk}$$

$$+ W_N^{-2k} \sum_{n=0}^{N/3-1} x(3n + 2)W_N^{-3nk} \tag{4-9}$$

Since for N divisible by 3, $W_N^3 = W_{N/3}$, this gives us

$$X(k) = S_0(k) + W_N^{-k}S_1(k) + W_N^{-2k}S_2(k) \tag{4-10}$$

This process is the basis for a radix-3 FFT; it is illustrated in Fig. 4-5(a) for $N = 9$. Where we had repeated butterflies in the radix-2 case, we now have a much more complicated structure [Fig. 4-5(b)]—possibly a cat's cradle. The completed graph is shown in Fig. 4-5(c). The input sequence must now be permuted in digit-reversed order: if the samples are numbered in base 3, then their locations are given by the values obtained by writing the digits in reverse order: $x(5)$ (12_3) is found in location 7 (21_3).

This method can clearly be generalized to any desired radix. Most practical systems use radix 2, because addressing tends to be simpler, and also because the available transform sizes are more closely spaced.

4.1.3.1 Mixed-Radix Transforms   If N has more than one prime factor, then a mixed-radix transform (Singleton, 1969) can be used; for example

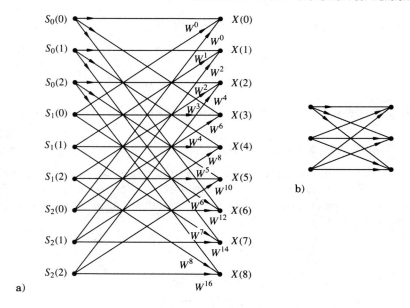

a)

b)

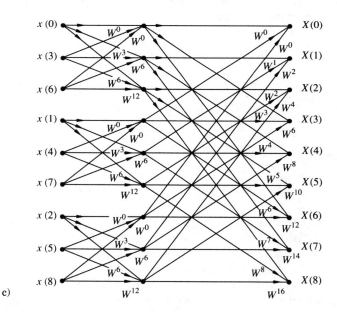

c)

Figure 4-5 *Radix-3 FFT flow diagrams: (a) one stage in 9-point transform;* (b) 3-point cat's cradle; (c) entire 9-point transform.*

if $N = 6$ we can decimate in time by first combining the even and odd points and, on the next iteration, combining as in Eq. (4-10). An example for $N = 12$ is shown in Fig. 4-6; the first iteration uses a radix of 3 and yields four groups of three points each; the last two passes are radix 2 and combine these four groups to produce the 12-point transform. (The multipliers next to the arrows are powers of W_{12} throughout.) The permutation must be found by writing the indices in mixed-radix form (see for example Knuth, 1969) and reversing the order of digits and radices: for example, element 8 (102 in the 2-2-3 radix system) is placed in position 13 (201 in the 3-2-2 system).

4.1.3.2 Split-radix transforms

An additional speedup can be obtained by handling one of the subtransforms more cleverly. This tack has been taken in various forms by a number of researchers (Vetterli and Nussbaumer, 1984; Martens, 1984; and Duhamel, 1986); we follow Duhamel here.

Consider a radix-4 transform. Let us form the subsequences

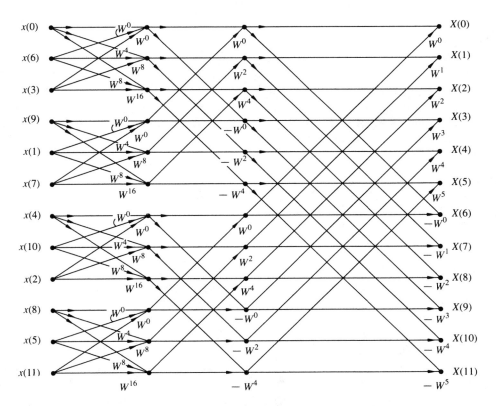

Figure 4-6   *Flow diagram of 12-point mixed-radix FFT.*

$$s_0(n) = x(n)$$
$$s_1(n) = x(n + N/4)$$
$$s_2(n) = x(n + N/2) \qquad n = 0, 1, ..., N/4 - 1$$
$$s_3(n) = x(n + 3N/4)$$

with transforms $S_0(k)$, $S_1(k)$, $S_2(k)$, and $S_3(k)$, respectively. Then

$$X(k) = \sum_{n=0}^{N/4-1} x(n)W_N^{-nk} + \sum_{n=N/4}^{N/2-1} x(n)W_N^{-nk} + \sum_{n=N/2}^{3N/4-1} x(n)W_N^{-nk}$$

$$+ \sum_{n=3N/4}^{N-1} x(n)W_N^{-nk}$$

$$= \sum_{n=0}^{N/4-1} s_0(n)W_N^{-nk} + \sum_{n=0}^{N/4-1} s_1(n)W_N^{-(n+N/4)k}$$

$$+ \sum_{n=0}^{N/4-1} s_2(n)W_N^{-(n+N/2)k} + \sum_{n=0}^{N/4-1} s_3(n)W_N^{-(n+3N/4)k}$$

Up to this point, we have a conventional decimation-in-frequency radix-4 transform. However, there is no need to follow this through consistently. We can combine the transforms for the even numbers $X(k)$, $k = 4p$, and $X(k)$, $k = 4p+2$, and the result will be indistinguishable from the radix-2 form of Eq. (4-5). For the odd-numbered transform points, we write (Duhamel, 1986),

$$X(4p + 1) = \sum_{n=0}^{N/4-1} s_0(n)W_N^{-n(4p+1)} + \sum_{n=0}^{N/4-1} s_1(n)W_N^{-n(4p+1)}W_N^{(N/4+Np)}$$

$$+ \sum_{n=0}^{N/4-1} s_2(n)W_N^{-n(4p+1)}W_N^{(N/2+2Np)}$$

$$+ \sum_{n=0}^{N/4-1} s_3(n)W_N^{-n(4p+1)}W_N^{(3N/4+3Np)}$$

Noting that $W_N^{Np} = 1$, $W_N^{N/4} = j$, $W_N^{N/2} = -1$ and $W_N^{3N/4} = -j$, we have

$$X(4p + 1) = \sum_{n=0}^{N/4-1} \{[s_0(n) - s_2(n)] + j[s_1(n) - s_3(n)]\}W_N^{-n(4p+1)}$$

Reasoning similarly, we have

$$X(4p + 3) = \sum_{n=0}^{N/4-1} \{[s_0(n) - s_2(n)] - j[s_1(n) - s_3(n)]\}W_N^{-n(4p+3)}$$

This hybrid, which seems unable to make up its mind whether it is radix-2 or radix-4, is called the split-radix transform. A typical butterfly for this transform is shown in Fig. 4-7. Notice that the even-numbered points require no complex multiplications at all; hence we obtain a four-point butterfly at the cost of two complex multiplications. Duhamel tabulates the number of multiplications and additions required for split-radix transforms of various sizes and shows that these are fewer than required for a conventional FFT and comparable to the methods of Vetterli and Nussbaumer (1984) and Martens (1984). (Martens' approach is essentially equivalent to Duhamel's; this equivalence is explored in Prob. 4-4.)

4.1.3.3 Prime-Radix transforms   Rader (1968) gives a way of adapting the FFT for use when N is a prime number greater than 2. The method consists of transforming the defining equation of the DFT into a cyclical correlation of two sequences of length $N - 1$. Since $N - 1$ is not prime, this correlation can be carried out using FFT techniques.

Consider the defining equation for $N = 7$. In vector-matrix notation this is

$$\begin{bmatrix} X(0) \\ X(1) \\ X(2) \\ X(3) \\ X(4) \\ X(5) \\ X(6) \end{bmatrix} = \begin{bmatrix} 1 & 1 & 1 & 1 & 1 & 1 & 1 \\ 1 & W^1 & W^2 & W^3 & W^4 & W^5 & W^6 \\ 1 & W^2 & W^4 & W^6 & W^1 & W^3 & W^5 \\ 1 & W^3 & W^6 & W^2 & W^5 & W^1 & W^4 \\ 1 & W^4 & W^1 & W^5 & W^2 & W^6 & W^3 \\ 1 & W^5 & W^3 & W^1 & W^6 & W^4 & W^2 \\ 1 & W^6 & W^5 & W^4 & W^3 & W^2 & W^1 \end{bmatrix} \begin{bmatrix} x(0) \\ x(1) \\ x(2) \\ x(3) \\ x(4) \\ x(5) \\ x(6) \end{bmatrix}$$

For notational simplicity, it is convenient to use positive exponents here. We have partitioned off the W^0 factors and written all the expo-

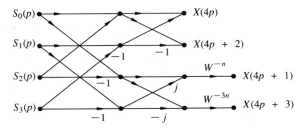

Figure 4-7   *Butterfly for split-radix FFT.*

nents modulo 7. The border elements are easily handled, and we direct our attention to the inner square.

Let g be a primitive root of N. Primitive roots have the property that $j = g^i \bmod N$ produces a cyclical mapping of the integers onto $(1, 2, ..., N - 1)$. For example, 3 is a primitive root of 7, and for $g = 3$, we have

$$i: \qquad 1\ 2\ 3\ 4\ 5\ 6\ 7\ 8\ 9\ 10\ 11\ \cdots$$

$$g^i \bmod N:\ 3\ 2\ 6\ 4\ 5\ 1\ 3\ 2\ 6\ \ \ 4\ \ \ 5\ \cdots$$

If we now replace n and k by $g^n \bmod N$ and $g^k \bmod N$, respectively, this permutes the x and X sequences. That is, for all $k > 0$, in the kth position of the X array, we will place $X(g^k \bmod N)$, and likewise for all $n > 0$, in the nth position of the x array we will place $x(g^n \bmod N)$. Hence the rows and columns of the $\mathbf{W}$ matrix are also permuted, and the defining equation becomes

$$X(0) = \sum_{n=0}^{N-1} x(n) \tag{4-11a}$$

$$X(g^k \bmod N) - X(0) = \sum_{n=0}^{N-1} x(g^n \bmod N)W^{(g^{k+n} \bmod N)} \qquad k > 0 \tag{4-11b}$$

The right-hand side of Eq. (4-11b) is a cyclical correlation of the sequences $\{x(g^n \bmod N)\}$ and $\{W^{(g^n \bmod N)}\}$, as can be seen from inspecting the matrix that results when the x and X vectors are permuted

$$
\begin{bmatrix} X(0) \\ X(3) \\ X(2) \\ X(6) \\ X(4) \\ X(5) \\ X(1) \end{bmatrix} =
\begin{bmatrix}
1 & 1 & 1 & 1 & 1 & 1 & 1 \\
1 & W^2 & W^6 & W^4 & W^5 & W^1 & W^3 \\
1 & W^6 & W^4 & W^5 & W^1 & W^3 & W^2 \\
1 & W^4 & W^5 & W^1 & W^3 & W^2 & W^6 \\
1 & W^5 & W^1 & W^3 & W^2 & W^6 & W^4 \\
1 & W^1 & W^3 & W^2 & W^6 & W^4 & W^5 \\
1 & W^3 & W^2 & W^6 & W^4 & W^5 & W^1
\end{bmatrix}
\begin{bmatrix} x(0) \\ x(3) \\ x(2) \\ x(6) \\ x(4) \\ x(5) \\ x(1) \end{bmatrix}
$$

Note that the actual transform values are not changed by this. For example, $X(3) = x(0) + W^2 x(3) + W^6 x(2) + W^4 x(6) + W^5 x(4) + W^1 x(5) + W^3 x(1)$; this is no different from the value of $X(3)$ given by the defining equation. But since this mapping has transformed the DFT computation into a cyclical correlation, by the convolution property of the DFT,

$$\sum_n a(n)b(n - k) = \{a(n)\} * \{b(-n)\}$$

$$= \mathfrak{F}^{-1}\{A(k)B(-k)\} \tag{4-12}$$

Hence the cyclical correlation can be accomplished with the aid of three other DFTs that would normally be implemented by FFTs. In his paper, Rader points out that computation speed could be further increased if $\mathfrak{F}\{\exp[j2\pi(g^n \bmod N)/N]\}$ were precomputed and stored in memory. We will find a still better way to proceed below.

4.2   FAST CONVOLUTION

Convolution is clearly of interest as a device for doing digital filtering, but it is also important to us as a step on the way to the Winograd Fourier transform algorithm, which we will describe below. Convolution is normally noncyclical, and the first fast convolution algorithm we shall consider is noncyclical. The fastest algorithms, however, are cyclical. It is always possible to use cyclical convolution algorithms for noncyclical convolution, however, by making the cycle length longer than the length of the result of the convolution.

The noncyclical algorithm we will consider is due to Toom (1963) and Cook (1966); the cyclical algorithm is due primarily to Winograd (1975); both algorithms are summarized in a classic paper by Agarwal and Cooley (1977), on which our discussion here is based.

We know from Z transform theory that convolution is equivalent to multiplying polynomials. In the present case, it is convenient to use polynomials in z rather than in z^{-1}; these polynomials are commonly known as generating polynomials. It is trivial to show that most of the properties of Z transforms, and in particular the convolution property, carry over to generating polynomials.

4.2.1   THE COOK-TOOM ALGORITHM

If we wish to find $\{y(n)\} = \{h(n)\} * \{x(n)\}$, we will find $Y(z) = H(z)X(z)$, as described above. If $\{a\}$ and $\{b\}$ are of length N, then their polynomials will be of order $N-1$ and $Y(z)$ will be of order $2N-2$. Let us evaluate $Y(z)$ for any convenient set of $2N - 1$ values of z. [For brevity, we will follow Agarwal and Cooley in using m_i to represent the value of $Y(z)$ at z_i.] These evaluations cost us one multiplication per point; they will give us $2N - 1$ points through which $Y(z)$ passes, and we can then write $Y(z)$ using the Lagrange interpolation formula:

$$Y(z) = \sum_{j=0}^{2N-2} m_j \prod_{i \neq j} \frac{z - z_i}{z_j - z_i} \tag{4-13}$$

Substituting the appropriate values of m_i and z_i in this formula gives us Y as a polynomial in z; the coefficients of this polynomial are the desired terms in the convolution.

An example is necessary to make this clear. Suppose we wish to convolve two 3-point sequences. Then

$$H(z) = h_0 + h_1 z + h_2 z^2$$

and
$$X(z) = x_0 + x_1 z + x_2 z^2$$

Then
$$Y(z) = (h_0 + h_1 z + h_2 z^2)(x_0 + x_1 z + x_2 z^2)$$

We can pick any $2N - 1 = 5$ points at which to evaluate $Y(z)$; we try to select points for which Y is easy to evaluate. Here we will choose $z_i = -2, -1, 0, 1,$ and 2:

$$m_0 = Y(-2) = (h_0 - 2h_1 + 4h_2)(x_0 - 2x_1 + 4x_2)$$
$$m_1 = Y(-1) = (h_0 - h_1 + h_2)(x_0 - x_1 + x_2)$$
$$m_2 = Y(0) = h_0 x_0$$
$$m_3 = Y(1) = (h_0 + h_1 + h_2)(x_0 + x_1 + x_2)$$
$$m_4 = Y(2) = (h_0 + 2h_1 + 4h_2)(x_0 + 2x_1 + 4x_2)$$

Then the Lagrange interpolation formula gives us

$$
\begin{aligned}
Y(z) = {} & m_0(z + 1)z(z - 1)(z - 2)/24 \\
& - m_1(z + 2)z(z - 1)(z - 2)/6 \\
& + m_2(z + 2)(z + 1)(z - 1)(z - 2)/4 \\
& - m_3(z + 2)(z + 1)z(z - 2)/6 \\
& + m_4(z + 2)(z + 1)z(z - 1)/24 \\
= {} & z^4(m_0 - 4m_1 + 6m_2 - 4m_3 + m_4)/24 \\
& + z^3(-m_0 + 2m_1 - 2m_3 + m_4)/12 \\
& + z^2(-m_0 + 16m_1 - 30m_2 + 16m_3 - m_4)/24 \\
& + z(m_0 - 8m_1 + 8m_3 - m_4)/12 \\
& + m_2 \qquad\qquad\qquad\qquad\qquad\qquad\qquad\qquad (4\text{-}14)
\end{aligned}
$$

It is not yet clear how this is going to save us any multiplications. Note that in practice, we are usually interested in convolving a long x sequence with a relatively short h sequence, presumably some finite-duration impulse response. Thus while the x samples will be continually changing, the h samples will not. Hence in the expansion of the coefficients of $Y(z)$ in terms of m, and hence ultimately in terms of h and x, we can move many of the constant multipliers into the h values, which are then pre-computed.

As an example of how this might be done, let us form the quantities

$$a_0 = x_0 - 2x_1 + 4x_2$$
$$a_1 = x_0 - x_1 + x_2$$

$$a_2 = x_0$$
$$a_3 = x_0 + x_1 + x_2$$
$$a_4 = x_0 + 2x_1 + 4x_2$$

and

$$b_0 = (h_0 - 2h_1 + 4h_2)/24$$
$$b_1 = (h_0 - h_1 + h_2)/6$$
$$b_2 = h_0/4$$
$$b_3 = (h_0 + h_1 + h_2)/6$$
$$b_4 = (h_0 + 2h_1 + 4h_2)/24$$

Then the convolution can be written,

$$w_0 = 4b_2 a_2$$
$$w_1 = 2(b_0 a_0 - 2b_1 a_1 + 2b_3 a_3 - b_4 a_4)$$
$$w_2 = -b_0 a_0 + 4b_1 a_1 - 5b_2 a_2 + 4b_3 a_3 - b_4 a_4$$
$$w_3 = 2b_0 a_0 + b_1 a_1 + b_3 a_3 + 2b_4 a_4$$
$$w_4 = b_0 a_0 + b_1 a_1 + b_2 a_2 + b_3 a_3 + b_4 a_4 \qquad (4\text{-}15)$$

Computation of the b values is done once, before the convolution commences; hence the divisions by 4, 6, and 24 are a one-time setup cost and need not be counted. The a values must be computed afresh for each new set of input points. Note, however, that multiplication of the x samples by 2 and by 4 can be done by shifting their binary representations. Since digital filtering commonly uses fixed-point arithmetic, the time for shifts and adds is normally negligible in comparison with the time required for multiplications.

For each new set of input points, the products $b_0 a_0, ..., b_4 a_4$ must be computed: these are our five multiplications. The five output points are obtained by repeated use of these products and require only shifts and adds, with the possible exception of the term, $5b_2 a_2$, which in most systems is still more quickly done by two shifts and an add than by a multiply.

Because we have had to choose a short convolution to keep the size of the example within bounds, the saving in multiplications is not large. For longer sequences, however, the savings become significant. Since we need only $2N - 1$ multiplications to get the reference points for the Lagrange interpolation, the number of multiplications required is in general $O(N)$, whereas conventional convolution requires $O(N^2)$.

On the other hand, the amount of computation involved in *deriving* the algorithm, even for so short a convolution as this one, is considerable. Agarwal and Cooley made use of an algebraic-symbol-manipulating program to obtain their convolution formulae, and this course is highly recommended for deriving any of the fast algorithms to be considered here.

4.2.2 THE WINOGRAD ALGORITHM

Winograd's (1975) algorithm is specifically for cyclical convolutions. If we are to do cyclic convolution with a cycle length of N, then we must multiply the polynomials modulo $(z^N - 1)$. There are a number of ways of seeing why this is so; perhaps the simplest is to let the product of the polynomials be $Y(z)$ and consider $Y(z)/(z^N - 1)$. Since $1/(z^N - 1) = z^{-N} + z^{-2N} + \cdots$, we can think of $1/(z^N - 1)$ as the generating function of a train of Kronecker impulses spaced N points apart, starting from 0; hence $Y(z)/(z^N - 1)$ is the generating function of the convolution of $Y(z)$ and that impulse train. This convolution (after a certain startup transient, depending on the order of Y) is clearly periodic with period N; we can isolate the periodic part by writing out the division in full:

$$Y(z) = Q(z) + R(z)/(z^N - 1)$$

The quotient $Q(z)$ corresponds to the transient part; the remainder $R(z)$, whose order will be less than N, gives one cycle of the repetitive part. Since the remainder is $Y(z) \bmod (z^N - 1)$, we have our result. Hence to convolve two sequences $\{h\}$ and $\{x\}$, we find

$$Y(z) = H(z)X(z) \bmod (z^N - 1)$$

With $Y(z)$ in this form, we can apply the polynomial version of the Chinese remainder theorem (CRT), proved in App. F. This theorem states: Let $P(z)$ have m distinct irreducible polynomial factors:

$$P(z) = P_1(z)P_2(z) \cdots P_m(z)$$

Then the congruences,

$$Y_j(z) \equiv Y(z) \bmod P_j(z) \qquad j = 1, 2, ..., m \tag{4-16}$$

can be solved for Y as follows: let $Q_i(z) = P(z)/P_i(z)$; then there exists an $A_i(z)$ for which

$$Q_i(z)A_i(z) \equiv 1 \bmod P_i(z) \tag{4-17}$$

Then the solution is given by

$$Y(z) = \sum_{i=1}^{m} Q_i(z)A_i(z)Y_i(z) \bmod P(z) \tag{4-18}$$

To apply the CRT to the convolution problem, we work it backward. We already have $Y(z)$, which in our case is $H(z)X(z)$, and the theorem tells us that we can express $Y(z)$ as the sum of the terms on the right-hand side. $P(z)$ is $(z^N - 1)$. This polynomial can always be factored; the factors are called *cyclotomic polynomials*. We can summarize Winograd's method as follows:

Factor $P(z)$ into $P_1(z)P_2(z)\cdots P_m(z)$.
For $i = 1, ..., m$:
    Form $Y_i(z) = H_i(z)X_i(z)\bmod\ P_i(z)$.
    Find $A_i(z)$ satisfying Eq. (4-17).
    Find $A_i(z)Q_i(z)$.
Find $Y(z) = \sum_i Q_i(z)A_i(z)Y_i(z)$.

The bulk of the work to be done is finding A_i and Y_i; as with the Cook-Toom algorithm, there is a lot of groundwork to be done to set up an algorithm that, at run time, is very simple.

The virtue of this approach is that it breaks a large convolution into a number of smaller ones—another divide-and-conquer technique. We started with $X(z)H(z)$, where X and H are of order $N - 1$; applying the CRT reduces this to a series of smaller products $H_i(z)X_i(z)$. Since the number of multiplications required for a convolution of length N is $O(N^2)$, this results in a significant reduction in the number of multiplications.

Let us consider the cyclical convolution of two 4-point sequences. In this case, $N - 1 = 3$, and we wish to find

$$Y(z) = H(z)X(z)\bmod (z^4 - 1)$$

We may factor $(z^4 - 1)$ as

$$(z^4 - 1) = P_1(z)P_2(z)P_3(z) = (z^2 + 1)(z + 1)(z - 1)$$

Then $H_1(z) = H(z)\bmod (z^2 + 1) = h_{1,1}z + h_{1,0}$

where $h_{1,1} = h_1 - h_3$

             $h_{1,0} = h_0 - h_2$

             $H_2(z) = H(z)\bmod (z + 1) = h_{2,0}$

where $h_{2,0} = h_0 - h_1 + h_2 - h_3$

and $H_3(z) = H(z)\bmod (z - 1) = h_{3,0}$

where $h_{3,0} = h_0 + h_1 + h_2 + h_3$

and similarly for $X_1(z)$, $X_2(z)$, and $X_3(z)$. Then the products (mod P_i) of the corresponding $\{X_i\}$ and $\{H_i\}$ are

$$H_1(z)X_1(z) = (h_{1,0}x_{1,1} + h_{1,1}x_{1,0})z + (h_{1,0}x_{1,0} - h_{1,1}x_{1,1})$$

$$H_2(z)X_2(z) = h_{2,0}x_{2,0}$$

$$H_3(z)X_3(z) = h_{3,0}x_{3,0}$$

These are the only multiplications we shall need. To emphasize this fact, let us write

$$m_1 = h_{1,0}x_{1,1}$$
$$m_2 = h_{1,1}x_{1,0}$$
$$m_3 = h_{1,0}x_{1,0} \qquad\qquad\text{(4-19)}$$
$$m_4 = h_{1,1}x_{1,1}$$
$$m_5 = h_{2,0}x_{2,0}$$
$$m_6 = h_{3,0}x_{3,0}$$

so that
$$H_1(z)X_1(z) = (m_1 + m_2)z + (m_3 - m_4)$$
$$H_2(z)X_2(z) = m_5$$
$$H_3(z)X_3(z) = m_6$$

Then $Q(z)$ and $A(z)$ are as follows:

$$Q_1(z) = P(z)/P_1(z) = (z+1)(z-1)$$
$$Q_2(z) = P(z)/P_2(z) = (z^2+1)(z-1)$$
$$Q_3(z) = P(z)/P_3(z) = (z^2+1)(z+1)$$
$$A_1(z) = -1/2$$
$$A_2(z) = -1/4$$
$$A_3(z) = 1/4$$

Hence
$$A_1(z)Q_1(z) = -(z^2-1)/2$$
$$A_2(z)Q_2(z) = -(z^3 - z^2 + z - 1)/4$$
$$A_3(z)Q_3(z) = (z^3 + z^2 + z + 1)/4$$

Putting these together and using the $\{m\}$ notation for the six products, we have

$$\begin{aligned}
Y(z) &= H_1(z)X_1(z)A_1(z)Q_1(z) + H_2(z)X_2(z)A_2(z)Q_2(z) \\
&\quad + H_3(z)X_3(z)A_3(z)Q_3(z) + H_4(z)X_4(z)A_4(z)Q_4(z) \\
&= [(m_6 - m_5 - 2m_1 - 2m_2)z^3 + (m_6 + m_5 - 2m_3 + 2m_4)z^2 \\
&\quad + (m_6 - m_5 + 2m_1 + 2m_2)z + (m_6 + m_5 + 2m_3 - 2m_4)]/4 \quad\text{(4-20)}
\end{aligned}$$

This is a straightforward application of the Winograd algorithm. The desired terms in the convolution are, as usual, the coefficients of the corresponding powers of z. Again because we have chosen a rather short convolution for brevity, the reduction in the number of multiplications is not striking. We can do better, however. Having broken a (relatively) large convolution into three relatively small convolutions, we contented ourselves with a straightforward computation of the smaller ones. In the computational complexity business, however, the struggle to reduce multiplications is never-ending, and by streamlining one of the smaller convolutions, we can do a little better than the above example suggests.

Recall our trick for computing a complex product with only three multiplications, and notice that the product $X_1(z)H_1(z) \bmod P_1(z)$ has a similar form. If we multiply $(az + b)(cz + d) \bmod (z^2 + 1)$, we will get $z(ad + bc) + (bd - ac)$. If we now let $P = c(b - a)$, $Q = a(c + d)$, and $R = b(d - c)$, the product is $z(P + Q) + (P + R)$. Hence let us redefine our multiplications as follows: Let

$$m_1 = x_{1,1}(h_{1,0} - h_{1,1})$$
$$m_2 = h_{1,1}(x_{1,0} + x_{1,1})$$
$$m_3 = h_{1,0}(x_{1,0} - x_{1,1}) \qquad (4\text{-}21)$$
$$m_4 = h_{2,0}x_{2,0}$$
$$m_5 = h_{3,0}x_{3,0}$$

This saves us one multiplication, at the expense of some extra additions as usual. It is left to the reader to verify that in this case,

$$H_1(z)X_1(z) = (m_1 + m_2)z + (m_1 + m_3)$$

and
$$Y(z) = [(m_5 - m_4 - 2m_1 - 2m_2)z^3$$
$$+ (m_5 + m_4 - 2m_1 - 2m_3)z^2$$
$$+ (m_5 - m_4 + 2m_1 + 2m_2)z$$
$$+ (m_5 + m_4 + 2m_1 + 2m_3)]/4 \qquad (4\text{-}22)$$

4.3 OTHER FAST FOURIER TRANSFORM ALGORITHMS

We now return to the problem of computing Fourier transforms. The following methods all depend, in one way or another, on the CRT. We will start with a classic algorithm for converting a mixed-radix transform into a multidimensional form that allows us to use smaller transforms separately on the several dimensions; then we will show how the Rader prime-N transform can be combined with Winograd's fast convolution algorithms to produce a transform requiring the minimum number of multiplications.

4.3.1 THE GOOD-THOMAS ALGORITHM

Another mixed-radix transform, devised independently by Good (1958, 1960) and Thomas (1963), maps the input vector $\mathbf{x}$ into a multidimensional array $\mathbf{z}$, which is then transformed by conventional methods; the desired transform $\mathbf{X}$ is obtained from the transformed array by a second mapping.

This transform requires that we be able to factor N into pairwise coprime factors. Let this factorization be $N = N_1 N_2 \cdots N_m$, and let the dimensions of $\mathbf{z}$ be $(0 : N_1 - 1), (0 : N_2 - 1), ..., (0 : N_m - 1)$. We will map $\mathbf{x}$ into $\mathbf{z}$ as follows: Let the ith subscript of this array be

$$n_i = n \bmod N_i \qquad i = 1, 2, ..., m; n = 0, 1, ..., N - 1 \qquad (4\text{-}23)$$

then
$$z(n_1, n_2, ..., n_m) = x(n) \qquad (4\text{-}24)$$

For example, for $m = 2$, $N_1 = 5$, and $N_2 = 7$, the mapping would be

$$\mathbf{z} = \begin{bmatrix} x(0) & x(15) & x(30) & x(10) & x(25) & x(5) & x(20) \\ x(21) & x(1) & x(16) & x(31) & x(11) & x(26) & x(6) \\ x(7) & x(22) & x(2) & x(17) & x(32) & x(12) & x(27) \\ x(28) & x(8) & x(23) & x(3) & x(18) & x(33) & x(13) \\ x(14) & x(29) & x(9) & x(24) & x(4) & x(19) & x(34) \end{bmatrix}$$

Next let $\mathbf{Z}$ be the transform of $\mathbf{z}$. We can find this transform by first transforming the rows of $\mathbf{z}$ with a seven-point transform and second transforming the columns of the result with a five-point transform. If we represent these transforms by $\mathbf{W}_7$ and $\mathbf{W}_5$, respectively, then

$$\mathbf{Z} = \mathbf{W}_5 \mathbf{z} \mathbf{W}_7$$

We can then obtain the $\mathbf{X}$ transform from

$$X(k) = Z(k_1, k_2, ..., k_m) \qquad (4\text{-}25)$$

where
$$k = (Q_1 k_1 + Q_2 k_2 + \cdots + Q_m k_m) \bmod N$$
$$Q_i = N/N_i$$

In our example, $Q_1 = 7$, $Q_2 = 5$, and the desired X values are to be found in $\mathbf{Z}$ as follows:

$$\mathbf{Z} = \begin{bmatrix} X(0) & X(5) & X(10) & X(15) & X(20) & X(25) & X(30) \\ X(7) & X(12) & X(17) & X(22) & X(27) & X(32) & X(2) \\ X(14) & X(19) & X(24) & X(29) & X(34) & X(4) & X(9) \\ X(21) & X(26) & X(31) & X(1) & X(6) & X(11) & X(16) \\ X(28) & X(33) & X(3) & X(8) & X(13) & X(18) & X(23) \end{bmatrix}$$

To justify this procedure, we must write the defining equation in terms of **z**. The key is to see what happens to W_N^{-nk} under this mapping. The CRT states that Eq. (4-23) can be solved for n:

$$n = (Q_1 a_1 n_1 + Q_2 a_2 n_2 + \cdots + Q_m a_m n_m) \bmod N \qquad (4\text{-}26)$$

where the a_i satisfy

$$Q_i a_i \bmod N_i = 1$$

Using Eqs. (4-25) and (4-26),

$$nk = \sum_{i=1}^{m} Q_i a_i n_i \sum_{j=1}^{m} Q_j k_j \bmod N$$

Let T_{ij} be a typical term in this product. Then

$$T_{ij} = Q_i n_i Q_j a_j k_j$$

If $i \neq j$, $Tij = (N/N_i N_j) N n_i a_j k_j$; since this contains a factor of N, $T_{ij} \bmod N = 0$. Hence all terms for which $i \neq j$ drop out of the sum. If $i = j$, then the factor N_i is missing from T_{ii}, and T_{ii} is not divisible by N. Hence nk reduces to

$$nk = \sum_{i=0}^{m} \frac{N}{N_i} Q_i a_i k_i n_i$$

and W_N^{-nk} becomes a product of terms,

$$W_N^{-(N/N_i)Q_i a_i k_k n_k} = W_{N_i}^{-Q_i a_i k_i n} \qquad i = 1, 2, \ldots, m$$

and since $Q_i a_i = 1 \bmod N_i$, this is simply

$$W_{N_i}^{-k_i n}$$

Therefore, we can write the DFT as

$$Z(k_1, k_2, \ldots, k_m) = \sum_n \left\{ \sum_n \left\{ \cdots \right. \right.$$

$$\sum_n \left[\sum_n z(n_1, n_2, \ldots, n_m) W_N^{-n_i k_i} \right]$$

$$\left. \left. \cdots W_N^{-n_i k_i} \right\} W_N^{-n_i k_i} \right\} \qquad (4\text{-}27)$$

which is a multidimensional DFT.

The multidimensional DFT requires $N \sum_i N_i$ complex multiplications; in our example this is $35 \cdot 12 = 420$, as compared with $352 = 1{,}225$.

4.3.2 THE WINOGRAD FOURIER TRANSFORM ALGORITHM (WFTA)

Winograd (1976, and also Silverman, 1977, 1978a, 1978b) gave an algorithm that reduces the number of multiplications below the number required for the FFT. His algorithm builds on Rader's prime-N transform and the Good-Thomas mixed-radix transform.

We saw from Rader's method that the DFT could be rewritten so that the $(N-1) \times (N-1)$ core of the transform became a cyclical convolution. It is clumsy and unattractive to have to do the convolution using three FFTs, and there ought to be a better way (although it should be noted that this was the best method available at the time Rader's paper appeared). Winograd's fast convolution algorithm provides us with that better way; it is a short step to applying these methods to Rader's algorithm.

This process has been explained with admirable clarity by Kolba and Parks (1977); we will draw on their explanation here. All the materials we need lie ready to hand. We will use Rader's algorithm to turn the transform into a cyclical correlation; we will turn one sequence backward to make this into a cyclical convolution; and we will use Winograd's algorithm to perform the convolution. In spite of this apparently simple summary, we have a fair amount of work ahead of us.

We will demonstrate the technique with the transform for $N = 5$. The defining equation gives us

$$\begin{bmatrix} X(0) \\ X(1) \\ X(2) \\ X(3) \\ X(4) \end{bmatrix} = \begin{bmatrix} 1 & 1 & 1 & 1 & 1 \\ 1 & W^1 & W^2 & W^3 & W^4 \\ 1 & W^2 & W^4 & W^1 & W^3 \\ 1 & W^3 & W^1 & W^4 & W^2 \\ 1 & W^4 & W^3 & W^2 & W^1 \end{bmatrix} \begin{bmatrix} x(0) \\ x(1) \\ x(2) \\ x(3) \\ x(4) \end{bmatrix}$$

(For notational simplicity we are again using positive exponents.) The primitive roots of 5 are 2 and 3. Using 3, we get the correspondence

$$n \qquad \quad 0\ 1\ 2\ 3\ 4\ 5\ 6\ 7\ \cdots$$

$$3^n \bmod 5 \quad 1\ 3\ 4\ 2\ 1\ 3\ 4\ 2\ \cdots$$

Using this root, Rader's method gives us

$$\begin{bmatrix} X(0) \\ X(1) \\ X(2) \\ X(3) \\ X(4) \end{bmatrix} = \begin{bmatrix} 1 & 1 & 1 & 1 & 1 \\ 1 & W^1 & W^3 & W^4 & W^2 \\ 1 & W^3 & W^4 & W^2 & W^1 \\ 1 & W^4 & W^2 & W^1 & W^3 \\ 1 & W^2 & W^3 & W^3 & W^4 \end{bmatrix} \begin{bmatrix} x(0) \\ x(1) \\ x(2) \\ x(3) \\ x(4) \end{bmatrix}$$

Ignoring the border elements for the moment, we see that we can get the rest of the transform by convolving

$$\{W, W^3, W^4, W^2\}$$

with $$\{x(1), x(2), x(4), x(3)\}$$

We now use the Winograd convolution which we developed above. Note that $X(1)$ corresponds to a shift of 0 and hence gets the constant term and that $X(3)$ corresponds to a shift of 1 and hence gets the coefficient of the z term. Similarly, $X(4)$ and $X(2)$ get the coefficients of the z^2 and z^3 terms, respectively. Including the border elements, we immediately obtain from Eq. (4-22)

$$X(0) = x(0) + x(1) + x(2) + x(3) + x(4)$$

$$X(1) = x(0) + (m_5 + m_4)/4 + (m_1 + m_3)/2$$

$$X(3) = x(0) + (m_5 - m_4)/4 + (m_1 + m_2)/2 \qquad \text{(4-28)}$$

$$X(4) = x(0) + (m_5 + m_4)/4 - (m_1 + m_3)/2$$

$$X(2) = x(0) + (m_5 - m_4)/4 - (m_1 + m_2)/2$$

where the m's are as given above in Eq. (4-21). If we let

$$\begin{aligned} h_0 &= W & x_0 &= x(1) \\ h_1 &= W^3 & x_1 &= x(2) \\ h_2 &= W^4 & x_2 &= x(4) \\ h_3 &= W^2 & x_3 &= x(3) \end{aligned}$$

then with $W = \cos 72° - j \sin 72°$ for the forward transform,

$$\begin{aligned} m_1 &= [(W - W^4) + (W^2 - W^3)][x(2) - x(3)] \\ &= -j2(\sin 72° + \sin 144°)[x(2) - x(3)] \\ m_2 &= (W^3 - W^2)[x(1) + x(2) - x(3) - x(4)] \\ &= j2 \sin 144°[x(1) + x(2) - x(3) - x(4)] \\ m_3 &= (W - W^4)[x(1) - x(2) - x(3) + x(4)] \qquad \text{(4-29)} \\ &= -j2 \sin 72°[x(1) - x(2) + x(3) - x(4)] \\ m_4 &= [(W + W^4) - (W^2 + W^3)][x(1) - x(2) - x(3) + x(4)] \\ &= 2(\cos 72° - \cos 144°)[x(1) - x(2) - x(3) + x(4)] \\ m_5 &= [(W + W^4) + (W^2 + W^3)][x(1) + x(2) + x(3) + x(4)] \\ &= 2(\cos 72° + \cos 144°)[x(1) + x(2) + x(3) + x(4)] \end{aligned}$$

We can now see the familiar pattern emerging: a set of preliminary additions, followed by the minimum number of multiplications, followed by another set of additions. Notice also that the WFTA very conveniently separates the real and imaginary multipliers, so each of the multiplications is, for all practical purposes, a multiplication by a real constant. Indeed, it is customary to compute each m as a real multiplication and move the multiplication by j (which is merely data shuffling) into the postadditions.

It is also customary to use temporary variables generously to minimize the rather large number of preadditions and postadditions. Combining all of these modifications, we get the following set of computations for a five-point WFTA:

$$a_1 = x(1) + x(4) \qquad a_5 = a_1 + a_2$$
$$a_2 = x(2) + x(3) \qquad a_6 = a_1 - a_2$$
$$a_3 = x(1) - x(4) \qquad a_7 = a_3 + a_4$$
$$a_4 = x(2) - x(3) \qquad a_8 = a_3 - a_4$$

$$b_1 = -1.538842a_4 \qquad c_1 = x(0) + b_5$$
$$b_2 = 0.587785a_7 \qquad c_2 = c_1 + b_4$$
$$b_3 = -0.951057a_8 \qquad c_3 = c_1 - b_4$$
$$b_4 = 0.559017a_6 \qquad c_4 = b_1 + b_2$$
$$b_5 = -0.25a_5 \qquad c_5 = b_1 + b_3 \qquad (4\text{-}30)$$
$$X(0) = x(0) + a_5$$
$$X(1) = c_2 + jc_5$$
$$X(2) = c_3 - jc_4$$
$$X(3) = c_3 + jc_4$$
$$X(4) = c_2 - jc_5$$

From our derivation, it would appear that we are limited to transforms whose lengths are prime numbers. McClellan and Rader (1976), however, showed that Rader's method can be extended to include lengths which are powers of primes. We will illustrate the method by an example.

For $N = 9$, the defining equation gives us

$$\begin{bmatrix} X(0) \\ X(1) \\ X(2) \\ X(3) \\ X(4) \\ X(5) \\ X(6) \\ X(7) \\ X(8) \end{bmatrix} = \begin{bmatrix} 1 & 1 & 1 & 1 & 1 & 1 & 1 & 1 & 1 \\ 1 & W^1 & W^2 & W^3 & W^4 & W^5 & W^6 & W^7 & W^8 \\ 1 & W^2 & W^4 & W^6 & W^8 & W^1 & W^3 & W^5 & W^7 \\ 1 & W^3 & W^6 & 1 & W^3 & W^6 & 1 & W^3 & W^6 \\ 1 & W^4 & W^8 & W^3 & W^7 & W^2 & W^6 & W^1 & W^5 \\ 1 & W^5 & W^1 & W^6 & W^2 & W^7 & W^3 & W^8 & W^4 \\ 1 & W^6 & W^3 & 1 & W^6 & W^3 & 1 & W^6 & W^3 \\ 1 & W^7 & W^5 & W^3 & W^1 & W^8 & W^6 & W^4 & W^2 \\ 1 & W^8 & W^7 & W^6 & W^5 & W^4 & W^3 & W^2 & W^1 \end{bmatrix} \begin{bmatrix} x(0) \\ x(1) \\ x(2) \\ x(3) \\ x(4) \\ x(5) \\ x(6) \\ x(7) \\ x(8) \end{bmatrix}$$

We start by eliminating those elements that have factors in common with N; they will be handled separately. In this example, these are $X(0)$, $X(3)$, $X(6)$, $x(0)$, $x(3)$, and $x(6)$. This leaves us with the reduced transform

$$
\begin{bmatrix} X'(1) \\ X'(2) \\ X'(4) \\ X'(5) \\ X'(7) \\ X'(8) \end{bmatrix} = \begin{bmatrix} W^1 & W^2 & W^4 & W^5 & W^7 & W^8 \\ W^2 & W^4 & W^8 & W^1 & W^5 & W^7 \\ W^4 & W^8 & W^7 & W^2 & W^1 & W^5 \\ W^5 & W^1 & W^2 & W^7 & W^8 & W^4 \\ W^7 & W^5 & W^1 & W^8 & W^4 & W^2 \\ W^8 & W^7 & W^5 & W^4 & W^2 & W^1 \end{bmatrix} \begin{bmatrix} x(1) \\ x(2) \\ x(4) \\ x(5) \\ x(7) \\ x(8) \end{bmatrix}
$$

We can now do a Rader mapping, using the correspondence,

$$n: \qquad 0\ 1\ 2\ 3\ 4\ 5\ 6\ 7 \cdots$$

$$2^n \bmod 9 \quad 1\ 2\ 4\ 8\ 7\ 5\ 1\ 2 \cdots$$

Note that the cycle repeats at 6 rather than 9, but since we have only six subscripts left after dropping the 0, 3, and 6 subscripts, these are sufficient. We permute the reduced transform, exactly as in Rader's method, to obtain

$$
\begin{bmatrix} X'(1) \\ X'(2) \\ X'(4) \\ X'(8) \\ X'(7) \\ X'(5) \end{bmatrix} = \begin{bmatrix} W^1 & W^2 & W^4 & W^8 & W^7 & W^5 \\ W^2 & W^4 & W^8 & W^7 & W^5 & W^1 \\ W^4 & W^8 & W^7 & W^5 & W^1 & W^2 \\ W^8 & W^7 & W^5 & W^1 & W^2 & W^4 \\ W^7 & W^5 & W^1 & W^2 & W^4 & W^8 \\ W^5 & W^1 & W^2 & W^4 & W^8 & W^7 \end{bmatrix} \begin{bmatrix} x(1) \\ x(2) \\ x(4) \\ x(8) \\ x(7) \\ x(5) \end{bmatrix} \qquad (4\text{-}31)
$$

Again we have a cyclical correlation, and we can proceed as before, using a Winograd six-point convolution to derive this part of the transform. It is not difficult to see that the omitted samples give us

$$
\begin{bmatrix} X(0) \\ X(3) \\ X(6) \end{bmatrix} = \begin{bmatrix} 1 & 1 & 1 \\ 1 & W^3 & W^6 \\ 1 & W^6 & W^3 \end{bmatrix} \begin{bmatrix} x(0) + x(3) + x(6) \\ x(1) + x(4) + x(7) \\ x(2) + x(6) + x(8) \end{bmatrix} \qquad (4\text{-}32)
$$

This is a three-point DFT, which can be found using a two-point cyclical convolution. To obtain the X values from the X' values, we note that

$$
\begin{bmatrix} X(1) \\ X(2) \\ X(4) \\ X(5) \\ X(7) \\ X(8) \end{bmatrix} = \begin{bmatrix} x(1) \\ x(2) \\ x(4) \\ x(5) \\ x(7) \\ x(8) \end{bmatrix} + \begin{bmatrix} Q(1) \\ Q(2) \\ Q(1) \\ Q(2) \\ Q(1) \\ Q(2) \end{bmatrix} \qquad (4\text{-}33)
$$

where

$$
\begin{bmatrix} Q(1) \\ Q(2) \end{bmatrix} = \begin{bmatrix} W^3 & W^6 \\ W^6 & W^3 \end{bmatrix} \begin{bmatrix} x(3) \\ x(6) \end{bmatrix}
$$

This is the core of yet another three-point DFT, which can also be computed with the aid of a cyclical convolution. A decomposition along these lines can be used for any N that is a power of a prime.

The derivation of prime-N transforms based on this method is clearly laborious and must essentially be done afresh for each new value of N; hence there is a limited repertoire of these transforms, and they do not include really large values of N. These transforms are therefore known as small-N transforms. Small-N transforms have been published (Silverman, 1977, 1978a, b, and Kolba and Parks, 1977) for $N = 2, 3, 4, 5, 7, 8, 9$, and 16. The number of multiplications required for a small-N transform is approximately equal to N. For large values of N, however, the Good-Thomas method can be pressed into service provided the factors of N are drawn from the available repertoire of small-N transforms.

4.3.2.1 The Nested Good-Thomas Algorithm It should be clear at this point that if N can be factored into numbers for which Winograd small-N transforms are known, the full transform can be computed by applying the small-N transforms to the Good-Thomas algorithm. The result will be a succession of adds, multiplies, and adds for each subtransform as it is applied to the two-dimensional array. Winograd has shown a method in which these operations can be grouped to form the structure with which we are by now familiar: preadditions, multiplications, and postadditions. More importantly, the total number of multiplications is less, using the nested form, than is required by a straightforward application of Winograd small-N transforms to the Good-Thomas algorithm.

To see how this rearrangement can be accomplished, it is convenient to represent the small-N transforms in matrix form. These transforms can be viewed as a factorization of the $\mathbf{W}$ matrix:

$$\mathbf{W} = \mathbf{SCT} \qquad (4\text{-}34)$$

where the $\mathbf{T}$ matrix contains the preadditions, the $\mathbf{C}$ matrix contains the multiplications, and the $\mathbf{S}$ matrix contains the postadditions. For example, if we return to Eqs. (4-28) and (4-29), we may rewrite these equations in matrix form as follows:

$$\mathbf{X} = \mathbf{SCTx} \qquad (4\text{-}35a)$$

where

$$\mathbf{T} = \begin{bmatrix} 1 & 1 & 1 & 1 & 1 \\ 1 & 0 & 0 & 0 & 0 \\ 0 & 1 & 1 & 1 & 1 \\ 0 & 1 & -1 & -1 & 1 \\ 0 & 1 & -1 & 1 & -1 \\ 0 & 1 & 1 & -1 & -1 \\ 0 & 0 & 1 & -1 & 0 \end{bmatrix} \qquad (4\text{-}35b)$$

$$\mathbf{C} = \text{diag}[1, 1, -1.538842, 0.587785, -0.951057,$$
$$0.559017, -0.25] \qquad (4\text{-}35c)$$

$$\mathbf{S} = \begin{bmatrix} 1 & 0 & 0 & 0 & 0 & 0 & 0 \\ 0 & 1 & 1 & 1 & j & 0 & j \\ 0 & 1 & 1 & -1 & 0 & -j & -j \\ 0 & 1 & 1 & -1 & 0 & j & j \\ 0 & 1 & 1 & 1 & -j & 0 & -j \end{bmatrix} \tag{4-35d}$$

Let us now apply this representation to the Good-Thomas algorithm. We will apply it directly first and then, using this form as a starting point, show how Winograd's nested form can be obtained from it. In our previous example, we transformed a 35-point sequence by forming the $\mathbf{z}$ matrix and then computing

$$\mathbf{Z} = \mathbf{W}_5 \mathbf{z} \mathbf{W}_7$$

If we have five-point and seven-point small-N transforms, then we may write

$$\mathbf{W}_5 = \mathbf{S}_5 \mathbf{C}_5 \mathbf{T}_5$$

and

$$\mathbf{W}_7 = \mathbf{S}_7 \mathbf{C}_7 \mathbf{T}_7$$

(Note that the subscripts here are for identification purposes and do not indicate order. $\mathbf{S}$ and $\mathbf{T}$ are not generally square, and the order of $\mathbf{C}$ is equal to the number of multiplications required, usually slightly greater than the length of the transform.) Since $\mathbf{W}_7$ is symmetric, we can post-multiply by $\mathbf{W}_7^T$ instead of by $\mathbf{W}_7$; then we have

$$\mathbf{Z} = \mathbf{S}_5 \mathbf{C}_5 \mathbf{T}_5 \mathbf{z} \mathbf{T}_7^T \mathbf{C}_7^T \mathbf{S}_7^T$$

and in general, for an rs-point transform, r and s coprime,

$$\mathbf{Z} = \mathbf{S}_r \mathbf{C}_r \mathbf{T}_r \mathbf{z} \mathbf{T}_s^T \mathbf{C}_s^T \mathbf{S}_s^T \tag{4-36}$$

The five-point Winograd transform requires five multiplications, as we have seen; a seven-point small-N transform requires nine multiplications. Hence the five row transforms require a total of 45 multiplications and the seven column transforms a total of 35 multiplications. It would be nice if we could reduce this total still further.

We can do this by manipulating the matrices in Eq. (4-36). In this equation $\mathbf{z}$ is surrounded by three layers, formed by the $\mathbf{T}$, $\mathbf{C}$, and $\mathbf{S}$ multiplications. We will peel these layers off one at a time and rearrange them in order to get (1) all the multipliers to the left of $\mathbf{z}$ and (2) all the $\mathbf{S}$, $\mathbf{C}$, and $\mathbf{T}$ multiplications together. Let

$$\mathbf{P} = \mathbf{T}_r \mathbf{z} \mathbf{T}_s^T$$

and

$$\mathbf{Q} = \mathbf{C}_r \mathbf{P} \mathbf{C}_s$$

(since the $\mathbf{C}$ matrices are diagonal, we can omit the transposition of $\mathbf{C}_s$) so that

$$\mathbf{Z} = \mathbf{S}_r \mathbf{Q} \mathbf{S}_s^T$$

Now we may write

$$\mathbf{Z} = \mathbf{S}_r (\mathbf{S}_s \mathbf{Q}^T)^T \tag{4-37}$$

and

$$\mathbf{P}^T = \mathbf{T}_s (\mathbf{T}_r \mathbf{z})^T \tag{4-38}$$

We now have the $\mathbf{T}$ matrices together on the left of $\mathbf{z}$ and the $\mathbf{S}$ matrices together on the left of $\mathbf{Q}$; it remains to find a way to bridge the gap from $\mathbf{Q}$ to $\mathbf{P}$. Let us write

$$\mathbf{Q}^T = \mathbf{C}_s \mathbf{P}^T \mathbf{C}_r$$

Since the $\mathbf{C}$ matrices are diagonal, we may write

$$\mathbf{Q}^T(i, j) = \mathbf{C}_s(i, i)\mathbf{P}^T(i, j)\mathbf{C}_r(j, j)$$

We can then move the $\mathbf{C}$'s to the left of $\mathbf{P}$ as follows: Define the matrix $\mathbf{D}$ as

$$D(i, j) = C_s(i, i)C_r(j, j)$$

Then if we represent element-by-element multiplication by $\circ$, we may write

$$\mathbf{Q}^T = \mathbf{D} \circ \mathbf{P}^T \tag{4-39}$$

This is the trick. Combining Eqs. (4-37), (4-38), and (4-39), we have

$$\mathbf{Z} = \mathbf{S}_r [\mathbf{S}_s \mathbf{D} \circ \mathbf{T}_s (\mathbf{T}_r \mathbf{z})^T]^T \tag{4-40}$$

which is our desired form.

Notice, however, what happened to our multiplications. When we formed $\mathbf{D}$, we consolidated all the multiplications in a single matrix. In our 5×7 example, $\mathbf{D}$ has 35 elements. Hence instead of 80 multiplications, we have only 35. In general, if the factors of N are r and s and the small-N multiplications required for the factors are m_r and m_s, respectively, then the total number of multiplications required is $m_r m_s$.

This technique of peeling off the layers and consolidating them can be generalized to any number of factors, but conventional matrix notation, being restricted to two dimensions, gives out on us. In the case of three dimensions, we can indicate the structure by writing

$$\mathbf{Z} = \mathbf{S}_q(\mathbf{S}_r\{\mathbf{S}_s\mathbf{D} \circ \mathbf{T}_s[\mathbf{T}_r(\mathbf{T}_q\mathbf{z})^T]^T\}^T)^T$$

provided we interpret the transpose signs as meaning something like "applied to the conformable dimension." In the general case, we may similarly write

$$\mathbf{Z} = \mathbf{S}_n(\mathbf{S}_n \cdots \mathbf{S}_n\{\mathbf{S}_n\mathbf{D} \circ \mathbf{T}_n[\mathbf{T}_n \cdots \mathbf{T}_n(\mathbf{T}_\mathbf{z})^T \cdots]^T\}^T \cdots)^T \qquad (4\text{-}41)$$

In programming this nested form, one never actually carries out these transpositions; instead, the programmer accesses the appropriate vectors of $\mathbf{z}$ by means of indexing techniques or, in languages which support it, the cross-section convention.

If each individual small-N transform requires mn multiplications, then the nested form requires $\prod_n m_n$ multiplications. To get an idea of the numbers involved, we may compare a 1,008-point WFTA with a 1,024-point FFT. The factors of 1,008 are 7, 9, and 16; small-N transforms for these factors require 9, 13, and 18 multiplications, respectively, so the WFTA requires 2,106 complex multiplications as compared with 10,240 for the FFT. Since a WFTA multiplication is a real multiplier times a complex multiplicand, while an FFT multiplication is the product of two complex factors, the WFTA requires 4,212 real multiplies and the FFT (ignoring shortcuts for trivial cases like $1+j0$ and $0+j1$) requires 40,960.

4.3.2.2 Winograd Transforms of Real Sequences We have seen that there is a great deal of redundancy in the DFTs of real sequences and that, in the case of the FFT, the amount of computation can be reduced by approximately one half by exploiting this fact. A similar saving can be accomplished in the WFTA (Parsons, 1979).

Let us have another look at the $\mathbf{S}$ matrix in Eq. (4-35d).

$$\mathbf{S} = \begin{bmatrix} 1 & 0 & 0 & 0 & 0 & 0 & 0 \\ \hline 0 & 1 & 1 & 1 & j & 0 & j \\ 0 & 1 & 1 & -1 & 0 & -j & -j \\ \hline 0 & 1 & 1 & -1 & 0 & j & j \\ 0 & 1 & 1 & 1 & -j & 0 & -j \end{bmatrix} \qquad (4\text{-}35d)$$

Here we have partitioned the matrix to show its structure. For a real input, the vertical partition separates those parts contributing to the real and imaginary parts of the transform. The horizontal partitions mark off the first and second halves of the transform and the term for which $n = 0$. (In the case of even N, we can also partition off a row for $n = N/2$.) The fact that

$$X(N - n) = \overline{X}(n) \tag{4-42}$$

is clearly reflected in the structure of the matrix. If we rewrite this as

$$\mathbf{S} = \begin{cases} \begin{bmatrix} \phi & 0 \\ \alpha & \beta \\ \gamma & \delta \end{bmatrix} & N \text{ odd} \\[2em] \begin{bmatrix} \phi & 0 \\ \alpha & \beta \\ \phi & 0 \\ \gamma & \delta \end{bmatrix} & N \text{ even} \end{cases}$$

We can see that partitions α and γ are redundant, and likewise β and δ. We therefore define a new matrix $\mathbf{S}'$ in which we discard β and γ and change the sign of δ. We will also remove the factors of j and take care of them in a subsequent step. Then

$$\mathbf{S}' = \begin{bmatrix} 1 & 0 & 0 & 0 & 0 & 0 & 0 \\ 0 & 1 & 1 & 1 & 0 & 0 & 0 \\ 0 & 1 & 1 & -1 & 0 & 0 & 0 \\ 0 & 0 & 0 & 0 & 0 & -1 & -1 \\ 0 & 0 & 0 & 0 & 1 & 0 & 1 \end{bmatrix} \tag{4-43}$$

This places the real parts of the first half of the transform into the first half of the output vector and the imaginary parts of the first half of the transform into the second half of the output vector. The transform $\mathbf{X}$ is now formed by combining elements from the first and second parts of this output vector as required. Then the transform is given by

$$\mathbf{X} = \mathbf{B}\mathbf{S}'\mathbf{C}\mathbf{T}\mathbf{x} \tag{4-44}$$

where

$$B = \begin{bmatrix} 1 & 0 & 0 & 0 & 0 \\ 0 & 1 & 0 & 0 & j \\ 0 & 0 & 1 & j & 0 \\ 0 & 0 & 1 & -j & 0 \\ 0 & 1 & 0 & 0 & -j \end{bmatrix}$$

For any small-N transform similar modifications are possible. $\mathbf{S}'$ is formed from Winograd's $\mathbf{S}$ matrix by identifying the α, β, γ, and δ partitions and modifying them as described above. The $\mathbf{B}$ matrix is defined as follows: For any element b_{rs}, if $r = s$ or $r = N - s$, $|b_{rs}| = 1$; otherwise $b_{rs} = 0$. If $2s > N$, b_{rs} is imaginary, and if $2r > N$, the imaginary elements are negative. The structure of $\mathbf{B}$ can be thought of as a paraphrase of Eq. (4-42). Because there are no complex values in Eq. (4-44) until the multiplication by $\mathbf{B}$, real storage can be used throughout until this point.

As with Winograd's basic factorization, the form Eq. (4-44) can be extended to large-N transforms by means of the Good-Thomas prime-factor algorithm, applying Eq. (4-44) to each dimension of the multidimensional real array $\mathbf{z}$. For two dimensions, the nested version of the process will be of the form

$$\mathbf{Z} = \mathbf{B}_r \{ \mathbf{B}_s \mathbf{S}'_s [\mathbf{S}'_r \mathbf{D}' \circ \mathbf{T}_r (\mathbf{T}_s \mathbf{z}^T)^T]^T \}^T \qquad (4\text{-}45)$$

When the process reaches the $\mathbf{B}$ multiplications, however, the working array will begin to contain complex values. For the algorithm to be practical, we must find a way to avoid this and, if possible, to obtain the transform $\mathbf{X}$ directly from the intermediate result at this point.

To do this, we exploit the relationship between pre- and post-multiplication and the Kronecker product. The Kronecker product of two matrices $\mathbf{S}$ and $\mathbf{B}$ is a partitioned matrix whose (i, j) partition is the product $a_{ij}\mathbf{B}$. For example, for $\mathbf{A}$ and $\mathbf{B}$ of order 3,

$$\mathbf{A} \times \mathbf{B} = \begin{bmatrix} a_{00}\mathbf{B} & a_{01}\mathbf{B} & a_{02}\mathbf{B} \\ a_{10}\mathbf{B} & a_{11}\mathbf{B} & a_{12}\mathbf{B} \\ a_{20}\mathbf{B} & a_{21}\mathbf{B} & a_{22}\mathbf{B} \end{bmatrix}$$

Following Nissen (1968), we define the stacking operator $v(\cdot)$ as follows: Let $\mathbf{S}_i$ be the ith column of the matrix $\mathbf{S}$. Then

$$v(\mathbf{S}) = \begin{bmatrix} \mathbf{S}_0 \\ \mathbf{S}_1 \\ \vdots \\ \mathbf{S}_{n-1} \end{bmatrix}$$

Stacking replaces $\mathbf{S}$ by a partitioned vector, each partition of which is a column of $\mathbf{S}$. (We note in passing that multidimensional arrays are stored in computer memory in stacked form.) It is easy to prove (Nissen, 1968) that for any conformable matrices $\mathbf{S}$, $\mathbf{Q}$, and $\mathbf{B}$,

$$v(\mathbf{SQB}) = (\mathbf{B}^T \times \mathbf{S})v(\mathbf{Q})$$

Let $\mathbf{Q}$ represent the working array as it is just before the $\mathbf{B}$ multiplications of Eq. (4-45).

$$\mathbf{Q} = \mathbf{S}'_s [\mathbf{S}'_r \mathbf{D}' \circ \mathbf{T}_r (\mathbf{T}_s z^T)^T]^T$$

The elements of $\mathbf{Q}$ are all real numbers. Then

$$\mathbf{X} = \mathbf{B}_r \mathbf{Q} \mathbf{B}^T_s$$

or

$$v(\mathbf{X}) = (\mathbf{B}_s \times \mathbf{X}\mathbf{B}_r)v(\mathbf{Q}) \qquad (4\text{-}46)$$

If we can find $\mathbf{B}_s \times \mathbf{XB}_r$ easily, the rest of the operation is simple, since as we have pointed out, the data will already be stacked in computer memory. The ingredients of the transform have already been computed, and it is necessary only to select the correct elements of $\mathbf{Q}$ for each element of $\mathbf{X}$; $(\mathbf{B}_s \times \mathbf{XB}_r)$ contains the instructions for selecting them.

This matrix $(\mathbf{B}_s \times \mathbf{XB}_r)$ is of order N, and we need a trick to avoid doing N^2 multiplications. These multiplications can easily be sidestepped by observing that the elements of any $\mathbf{B}$ matrix can only be 0, ± 1, or $\pm j$. Hence the magnitudes of the product elements will be either 0 or 1. As we loop through these matrices, we need only add or subtract the angles, and we can do this in units of $\pi/2$. If at any level of nesting a $\mathbf{B}$ element is zero, then the result is zero and all inner loops can be bypassed. Whenever the program makes it to the innermost level, it is only necessary to determine whether the corresponding element of $\mathbf{Q}$ is added to the real or imaginary part of $\mathbf{X}$.

In computing the inverse transform, we begin with a complex vector, $\mathbf{X}$, and must start by packing this vector into a real working array in such a way that the real inverse transform will eventually emerge. For a small-N transform, the inverse of $\mathbf{W}$ is $1/N_i$ times its conjugate; but since $\mathbf{W}$ (and $\mathbf{W}^{-1}$) are symmetric, we may also write

$$\mathbf{W}^{-1} = \mathbf{W}^* / N_i$$

where $*$ indicates the conjugate transpose. Then for any small-N transform,

$$\mathbf{W}^{-1} = \frac{1}{N}\mathbf{T}^T \mathbf{C}^T \mathbf{S}'^T \mathbf{B}^* \tag{4-47}$$

We can again derive a nested form for composite N; for two dimensions,

$$\mathbf{x} = -\frac{1}{N}\mathbf{T}_r^T \{\mathbf{T}_s^T D^T \circ \mathbf{S}_s'^T [\mathbf{S}_r'^T \mathbf{B}_r^* (\mathbf{B}_s^* Z^T)^T]^T\}^T \tag{4-48}$$

Here the multiplications by $\mathbf{B}^*$ come first, and these give us the instructions for unpacking the elements of $\mathbf{X}$. The same shortcuts used in finding $(\mathbf{B}_s \times \mathbf{XB}_r)$ apply here.

4.4 NUMBER-THEORETIC TRANSFORMS

We have repeatedly described the use of the DFT for finding convolutions. For applications such as digital filtering, the DFT presents some drawbacks: the data to be convolved are normally real integers, while the DFT requires complex floating-point computations; further, these operations inevitably result in some round-off error in the results. Hence it would be helpful if we could find some alternative to the DFT.

It is of interest to determine just what a transform must be like for it to possess the convolution property. The following exposition is based on Agarwal and Burrus (1974). Let $\{x(n)\}$ and $\{y(n)\}$ be two sequences of length N which are to be convolved, and let $\{z(n)\}$ be the result:

$$z(n) = x(n) * y(n) = \sum_{k=0}^{N-1} x(k)y(n-k) \tag{4-49}$$

We will assume that the convolution is cyclical, that is, that either the indices are evaluated modulo N or that $\{x\}$ and $\{y\}$ can be extended as needed with period N; then $\{z(n)\}$ is likewise of length N.

We will also assume that the transform can be represented as a matrix multiplication: if $\mathbf{X}$ is the transform of $\mathbf{x}$, then $\mathbf{X} = \mathbf{Tx}$, where $\mathbf{T}$ is a nonsingular $N \times N$ matrix: $\mathbf{T} = [t_{ij}]$. ($\mathbf{T}$ must be nonsingular since the transform must have an inverse.) Then the transform has the convolution property if $z(n) = x(n) * y(n)$ implies $\mathbf{Z} = \mathbf{X} \circ \mathbf{Y}$ (where $\circ$ denotes element-by-element multiplication), and conversely.

Therefore we wish to see what form $\mathbf{T}$ must take for us to be able to find $\mathbf{z}$ from

$$\mathbf{Z} = [\mathbf{Tx} \circ \mathbf{Ty}]$$

$$\mathbf{z} = \mathbf{T}^{-1}\mathbf{Z}$$

We will do this by finding two expressions for $Z(k)$, one from $\mathbf{Z} = \mathbf{Tz}$ and the other from $\mathbf{Z} = \mathbf{X} \circ \mathbf{Y}$, and using them to establish a relation among the elements of $\mathbf{T}$.

If $\mathbf{Z} = \mathbf{Tz}$, then any element of $\mathbf{Z}$ is given by

$$Z(k) = \sum_{n=0}^{N-1} t_{kn} z(n)$$

$$= \sum_{n=0}^{N-1} t_{kn} \sum_{p=0}^{N-1} x(p)y(n-p)$$

Let $q = n - p$ so that $n = p + q$; then

$$Z(k) = \sum_{q=0}^{N-1} \sum_{p=0}^{N-1} x(p)y(q)t_{k,p+q} \tag{4-50}$$

If $\mathbf{Z} = \mathbf{X} \circ \mathbf{Y}$, then any element of $\mathbf{Z}$ is given by

$$Z(k) = X(k)Y(k)$$

$$= \sum_{p=0}^{N-1} t_{kp} x(p) \sum_{q=0}^{N-1} t_{kq} y(q)$$

$$= \sum_{p=0}^{N-1} \sum_{q=0}^{N-1} x(p)y(q)t_{kp}t_{kq} \qquad \text{(4-51)}$$

For the cyclical convolution property to hold, these two expressions for $Z(k)$ must be equivalent; hence we require that

$$t_{k,p+q} = t_{kp}t_{kq} \qquad \forall \; k,p,q \qquad \text{(4-52)}$$

where the indices are added modulo N. If we start out with $p = q = 1$, we see that $t_{k2} = t_{k1}^2$ and in general $t_{km} = t_{k1}^m$. Now note that because the indices are added modulo N, we can write

$$t_{kp} = t_{k,p+N}$$
$$= t_{kp}t_{k1}^N$$

Therefore, $t_{k1}^N = 1$, or more to the point, t_{k1} is an Nth root of unity. In fact, since all the elements are seen to be integer powers of t_{k1}, they are all Nth roots of unity. Furthermore since **T** is nonsingular, the t_{k1} must all be different, and since there are N distinct Nth roots of unity, these must be the N different t_{k1} values. Without loss of generality, we can assume that t_{11} is the principal Nth root.

We have obtained this result without any assumptions about the nature of **T** or of the numbers constituting $\{x(n)\}$ and $\{y(n)\}$. For $\{x\}$ and $\{y\}$ real or complex, this leads directly to the defining equation of the DFT; W is the principal Nth root of unity, as we have seen, and the **W** matrix is composed of powers of W.

The same principle is applicable to integer arithmetic, however, provided we can find a suitable candidate for an Nth root. Consider the following example, in which we work with the integers modulo 17. The powers of 4 modulo 17 are as follows:

n:	0	1	2	3	4	5	6	7	8	9	$\cdots$
4^n:	1	4	16	13	1	4	16	13	1	4	$\cdots$

Since $4^4 = 1$ modulo 17, we say that 4 is the principal fourth root of unity modulo 17. It is left to the reader to verify that the other roots are 16, 13, and, trivially, 1. Hence we can form the matrix **T** as follows:

$$\mathbf{T} = \begin{bmatrix} 4^0 & 4^0 & 4^0 & 4^0 \\ 4^0 & 4^1 & 4^2 & 4^3 \\ 4^0 & 4^2 & 4^4 & 4^6 \\ 4^0 & 4^3 & 4^6 & 4^9 \end{bmatrix}$$

$$= \begin{bmatrix} 1 & 1 & 1 & 1 \\ 1 & 4 & 16 & 13 \\ 1 & 16 & 1 & 16 \\ 1 & 13 & 16 & 4 \end{bmatrix} \quad \text{(mod 17)}$$

The inverse of **T** can be found by conventional means, doing all arithmetic modulo 17 and noting that since $13 \cdot 4 = 52 \equiv 1 (\text{mod } 17)$, we can represent 4^{-1} as $13 (\text{mod } 17)$. This yields

$$\mathbf{T}^{-1} = 13 \begin{bmatrix} 1 & 1 & 1 & 1 \\ 1 & 13 & 16 & 4 \\ 1 & 16 & 1 & 16 \\ 1 & 4 & 16 & 13 \end{bmatrix}$$

Suppose we wish to convolve the sequences $\mathbf{x}^T = [2, 1, 0, 0]$ and $\mathbf{y}^T = [1, 2, 3, 4]$. The transforms are obtained in the usual way, except that all arithmetic must be done modulo 17. We obtain

$$\mathbf{X}^T = [3, 6, 1, 15] \ (\text{mod } 17)$$

$$\mathbf{Y}^T = [10, 7, 15, 6] \ (\text{mod } 17)$$

Then we should be able to find **z** by multiplying the transforms element by element, again modulo 17; this gives

$$\mathbf{Z}^T = [13, 8, 15, 5] \ (\text{mod } 17)$$

The convolution should be the inverse transform of **Z**; **T**-1**Z** gives us

$$\mathbf{z}^T = [6, 5, 8, 11] \ (\text{mod } 17)$$

which as the reader can verify, is $x(n) * y(n)$.

There are several things to be said about this process. First, we see clearly that as long as we follow the rules of modular arithmetic and select a suitable modulus, we can find an integer that is the Nth root of unity and build our transforms around this. The key is finding a suitable modulus; it is necessary that both the Nth root of unity and N^{-1} exist modulo this number. We will discuss the selection of a suitable modulus below.

Second, we must make sure that the starting values and the results all lie within the range 0 to $M - 1$, where M is the modulus. (We took care to choose an example which met this requirement.) If signed arithmetic is to be used, the values must lie within the range $-M/2$ to $M/2$. Overflowing these limits will cause the values to wrap around modulo M, exactly as happens in the case of integer arithmetic in a digital computer. This wraparound is similar to aliasing in sampled data, in that it arises from exceeding an inherent range limitation and in that it takes a similar form; it is sometimes referred to as amplitude aliasing.

Finally, the transform itself has no "meaning." The elements of the DFT can be interpreted as the amplitudes and phases of a sinusoidal decomposition of the input sequence. No such interpretation can be put on the number-theoretic transform. Indeed, in modular arithmetic no number is greater than or less than any other number, and the concept of magnitude has no meaning.

The notion of greater-than arises ultimately from the linear ordering imposed on the natural numbers by the successor operation. Among the natural numbers we know that 13 is greater than 8, because we can start from 8 and reach 13 by repeatedly finding successors, but not the other way around. But because of the cyclical nature of modular numbers, we can start with either number in a modulo-17 system and reach the other by going forward; hence the concept of greater-than breaks down.

It is disconcerting to engineers with a long experience of associating meanings with Fourier and Laplace transforms to find a meaningless transform; by way of reassurance we point out that first, we do not ordinarily work in the transform domain with these transforms and second, because all operations are defined in terms of integer arithmetic, there is never any round-off error. These transforms are thus uniquely suited to integer digital arithmetic.

We now put what we have seen on a somewhat more rigorous basis. In the following all arithmetic is modulo some integer M. Notice that these properties are the same *mutatis mutandis* as the corresponding properties of the DFT.

1. Defining equation:

$$X(k) = \sum_{n=0}^{N-1} x(n)a^{nk} \tag{4-53}$$

where a is the principal Nth root of unity: $a^N \equiv 1(\mod M)$.

2. Orthogonality property:

$$\sum_{k=0}^{N-1} a^{mk}a^{-nk} = N\delta_{mn} \tag{4-54}$$

where δ_{mn} is the Kronecker delta. To prove this, let $p = m - n$; then we can rewrite this sum as

$$\sum_{k=0}^{N-1} a^{pk}$$

Clearly if $m = n$, $p = 0$, and this is $\sum a^0 = N$; for $m \neq n$ we may write

$$\sum_{k=0}^{N-1} a^{pk} = \frac{a^{Np} - 1}{a^p - 1}$$

Multiplying by $a^p - 1$, we have

$$(a^p - 1) \sum_{k=0}^{N-1} a^{pk} = a^{Np} - 1$$

Since $a^{Np} = 1$, this product is zero. But in that case either $(a^p - 1)$ or the sum must be zero; since $(a^p - 1)$ cannot be zero, our result follows.

This proof, from Agarwal and Burrus (1975), depends on the well-known fact that if $xy = 0$, then x or y (or both) must be zero. In systems of numbers modulo M, however, this may not be so, in which case we say that x and y are *zero divisors*. [For example, in a modulo-36 system, 9 and 4 are zero divisors, since $9 \cdot 4 \equiv 0 (\text{mod } 36)$.] If $(a^p - 1)$ could be a zero divisor, our proof would fail. Vanwormhoudt (1977) addresses this possiblity and shows that $(a^p - 1)$ will never be a zero divisor.

3. Inverse transform:

$$x(n) = N^{-1} \sum_{k=0}^{N-1} a^{-nk} X(k) \tag{4-55}$$

which can be seen by substituting Eq. (4-53) for $X(k)$ and applying the orthogonality property.

4. Even and odd functions: If $f(n) = f(-n)$, then $F(k) = F(-k)$. If $f(n) = -f(-n)$, then $F(k) = -F(-k)$.

5. Symmetry: If $T\{x\}$ is the transform of x, then $T\{T\{f(n)\}\} = Nf(-n)$. This can be seen by straightforward evaluation of the repeated transforms and the application of the orthogonality property.

6. Shifting: If $X(k) = T\{x(n)\}$, then $T\{x(n-p)\} = a^{kp} X(k)$.

4.4.1 CHOICE OF A MODULUS

There are both theoretical and practical constraints on the choice of the modulus M. To start with the theoretical constraints, we have seen that the cyclical convolution property requires that both the Nth root of unity and N^{-1} exist. We wish to find what limitations these requirements impose on the modulus M. We will assume that $M > N$. From number theory, if N is the smallest positive integer for which $a^N = 1(\text{mod } M)$, then N divides $\phi(M)$, where $\phi(M)$ is Euler's *totient function*, the number of integers less than M and coprime to M. If M is prime, $\phi(M) = M - 1$; hence if M is prime, the Nth root of unity exists if N divides $M - 1$.

Furthermore if N and M are coprime, the inverse $N^{-1}(\text{mod } M)$ also exists. By Euler's theorem,

$$N^{\phi(M)} \equiv 1 \,(\mathrm{mod}\ M)$$

hence

$$N^{\phi(M)-1} \equiv 1/N \,(\mathrm{mod}\ M)$$

In our example above, M was 17, $\phi(M) = 16$, and N was 4, which divides 16; hence the fourth root of unity exists (mod 17). Furthermore $N^{\phi(M)-1}$ is $4^{15} \equiv 13(\mathrm{mod}\ 17)$, and we saw that this was, indeed, the inverse of N.

    If M is composite, the picture is more complicated. The number theory involved is elementary, but the reasoning is laborious. Suppose M is factored as

$$M = \prod_i p_i^{r_i}$$

then
$$\phi(M) = M \prod_i (p_i - 1)/p_i \qquad (4\text{-}56)$$

We know we need an a such that $a^N \equiv 1(\mathrm{mod}\ M)$. If M is composite, we also need

$$a^N \equiv 1(\mathrm{mod}\ p_i^{r_i}) \qquad \forall\, i \qquad (4\text{-}57)$$

The matrix of the transform is

$$\mathbf{T} = \begin{bmatrix} a^0 & a^0 & a^0 & a^0 & \dots \\ a^0 & a^1 & a^2 & a^3 & \dots \\ a^0 & a^2 & a^4 & a^6 & \dots \\ a^0 & a^3 & a^6 & a^9 & \dots \end{bmatrix}$$

Now suppose that $a^2 \equiv a^5 \,(\mathrm{mod}\ M)$. Then

$$a^{2k} \equiv a^{5k} \,(\mathrm{mod}\ M) \qquad k = 0, ..., N-1$$

and so row 2 and row 5 would be identical and the matrix would be singular. Hence we also require that

$$a^p \neq a^q \,(\mathrm{mod}\ M) \qquad \forall\, p \neq q \qquad (4\text{-}58)$$

and in fact

$$a^p \neq a^q \,(\mathrm{mod}\ p_r^i) \qquad \forall\, p \neq q, \forall\, i \qquad (4\text{-}59)$$

For this to be true, we require that

$$a^q \neq 1 \,(\mathrm{mod}\ p_i^r) \qquad q = 1, ..., N-1 \qquad (4\text{-}60)$$

because otherwise $a^{q+1} = a(\mathrm{mod}\ p_i^r)$ and the condition of Eq. (4-59) would be violated. This narrows down our choices, for now not only must

$$a^N \equiv 1\ (\mathrm{mod}\ p_i^r)$$

but N must be the smallest integer for which this is so.

Finally, the requirement of an inverse means there must be some integer z for which $z^N = 1\ (\mathrm{mod}\ M)$. This is possible only if $(N, M) = 1$, or equivalently, if $p_i \nmid N$.[†]

By Euler's theorem if a and M are coprime, then $a^{\phi(m)} = 1\ (\mathrm{mod}\ M)$; hence $a\phi(p_i^{r_i}) = 1\ (\mathrm{mod}\ p_i^{r_i})$. Therefore for $a^N = 1\ (\mathrm{mod}\ p_i^{r_i})$, N must divide $\phi(p_i^{r_i})$. (Note that this is a necessary condition for a suitable a to exist; it is not a sufficient condition for any a, N, and M.) Since $\phi(p_i^{r_i})$ is given by

$$\phi(p_i^{r_i}) = p_i^{r_i - 1}(p_i - 1)$$

N must divide either $p_i^{r_i - 1}$ or $(p_i - 1)$. We already require $(N, p_i) = 1$, so N must divide $p_i - 1$. Since this must apply simultaneously for all i, we require

$$N | (p_1 - 1)\ \text{and}\ N | (p_2 - 1)\ \text{and}\ N | (p_3 - 1)\ \text{and}\ \cdots$$

or equivalently,

$$N | \gcd(p_1 - 1, p_2 - 1, ..., p_m - 1) \tag{4-61}$$

This is a necessary condition on N and the factors of M for $N - 1$ and a suitable a to exist.

It is also a sufficient condition. For if $N | \gcd(p_1 - 1, p_2 - 1, ..., p_m - 1)$, then for any i, $N | (p_i - 1)$ and in fact $N | \phi(p_i)$. We can then find some number b_i such that

$$b_i^N = 1\ (\mathrm{mod}\ p_i^{r_i}) \tag{4-62}$$

and, by the CRT, some number a such that

$$a = b_i\ (\mathrm{mod}\ p_i^{r_i}) \tag{4-63}$$

and that is our desired base.

[†]$a|b$ means a divides b; $a \nmid b$ means a does not divide b; $\gcd(a, b)$ is greatest common divisor of a and b. See App. F.

4.4.2 PRACTICAL CONSIDERATIONS

For these transforms to be competitive with alternative methods, the arithmetic must be cheap. Agarwal and Burrus give three requirements which arise from this. First, for fast transforms along the lines of the FFT to exist, N should be composite, ideally a power of 2. Second, it would be desirable to eliminate as many multiplications as possible. Multiplications can always be replaced by shift-and-add operations; these will go faster only if the binary representations of the powers of a can be cast in forms having few 1 bits. Finally, since all arithmetic will be done modulo M, it is also desirable for M to be representable in a form containing few 1 bits. Hence in the search for suitable M, N, and a, interest has focussed on values that meet these requirements.

The ideal M would be a power of 2, since its binary form has only one 1; but by Eq. (4-61), the longest transform possible with such a modulus is $N = 1$. The next best alternatives are numbers of the form $2^p - 1$ or $2^p + 1$. In the case of $2^p - 1$, we need consider only cases where p is prime, since for a composite exponent pq, $2^p - 1$ divides $2^{pq} - 1$ and, by Eq. (4-61), the size of $2^p - 1$ will determine the length of the transform. Numbers of the form $2^p - 1$ where p is prime are Mersenne numbers; the use of Mersenne numbers was discussed by Rader (1972). For these numbers, transforms of length $2p$ can be found; however, since $2p$ has only two factors, it does not lend itself to FFT-type transforms.

In the case of $M = 2^p + 1$, p must be even to be of interest. For $2 = -1 \pmod 3$ and hence for odd p, $2^p = -1^p \pmod 3$ and hence $3 | 2^p + 1$ if p is odd; by Eq. (4-61) this means the maximum transform length is 2. Furthermore, we will require p to be a power of 2. The practical reason is that such a number has the binary form $1000 \cdots 0001$ and thus contains only two nonzero bits; the theoretical reason is that if p has an odd factor z, so that $p = z2^k$, then

$$2^{2^k} + 1 \text{ divides } 2^{z2^k} + 1$$

and hence, by Eq. (4-61) again, the factor z adds nothing to the length of the transform. Numbers of the form

$$2^{2^k} + 1$$

are known as Fermat numbers:

$$F_n = 2^{2^n} + 1,$$

and Agarwal and Burrus term transforms based on these numbers Fermat-number transforms (FNTs).

Fermat speculated that all such numbers were prime; however, the only known Fermat primes are $F_0, ..., F_4$. For these numbers, we may choose any convenient $N < R$. Numbers F_5 through F_8 are known to be composite; it is also known that the factors of any composite Fermat number F_n are all of the form, $k2^{n+2} + 1$. This means that by Eq. (4-61), the maximum transform length is 2^{n+2}. Agarwal and Burrus show that the a for which this maximum length is obtained is given by

$$a = 2^{2^{n-2}}(2^{2^{n-1}} - 1)$$

They denote this value of a as $\sqrt{2}$, because $a^2 = 2(\text{mod } F_n)$. Notice that for all these moduli, transform lengths that are powers of 2 are readily available. In this case, radix-2 FFT-type algorithms can be obtained directly from FNTs simply by substituting a for W.

Arithmetic for a FNT modulo $F_n = 2^b + 1, b = 2^n$, requires a word length of $b + 1$ bits. This is usually an awkward size: for F_4, we need 17 bits and for F_5, 33 bits. Agarwal and Burrus propose simply omitting the number 2^b and working only in the range $(0, 2^b - 1)$, on the grounds that the probability that 2^b will occur in the course of computation is very small. McClellan (1975, 1976) uses a mod number system in which the value 2^b is reserved for zero. Whenever this bit pattern is found in computation, operations using this value as an operand are modified to simulate an operand of zero; since this is the only number which requires $2^b + 1$ bits, the extra bit can be omitted without error. Leibowitz (1976) proposed a diminished-1 representation in which the number n is represented by the bit pattern normally used to represent $n - 1$; thus for example in a 4 bit system the number 1 is represented as 0000. The number zero is again mapped to 2^b; that is, in our 4-bit example zero is represented as 10000. In both McClellan's and Leibowitz's systems, the resulting arithmetic cost is roughly comparable to that of 1's-complement arithmetic, and the multiplications can be implemented as inexpensive shift-and-add operations.

4.5 SUMMARY

In this chapter, we have surveyed a number of fast algorithms for Fourier transforms and for convolution. These algorithms range from those that are of purely theoretical interest to those that are used on a day-to-day basis in many signal processing and analysis applications.

One point should be repeated for emphasis. These algorithms are fast mainly in the sense that they minimize the number of multiplications. Very little is known about minimizing the number of additions and subtractions. As we pointed out early in the chapter, minimizing multiplications may or may not have practical significance, depending on the hardware being used; many special-purpose DSP chips may do multiplications as fast as they do additions. On the other hand, a great deal of DSP is done in general-purpose digital computers, in which the difference between addition and multiplication times may be great. In evaluating any of these fast algorithms, a great deal of consideration must be given to where they will be used.

Notice the large part played by number theory in the more advanced techniques. Number theory has traditionally been regarded as one of the least useful areas of mathematics; for example, this is implicit in much of Hardy's (1945) essay. However, we are coming to find more and more applications of number theory in engineering and particularly in signal processing; see Schroeder (1979) for an excellent and readable survey.

PROBLEMS

4.1. Rearrange the decimation-in-frequency FFT work with a permuted input vector and yield an in-order vector.

4.2. The DFTs of two real sequences $\{a(n)\}$ and $\{b(n)\}$ of length N can be computed simultaneously by placing $\{a(n)\}$ in the real part of a complex sequence $x(n)$ of length N and placing $\{b(n)\}$ in the imaginary part. Find the method for recovering the transforms $\{A(k)\}$ and $\{B(k)\}$ from the complex transform $X(k)$.

4.3. Derive a decimation-in-frequency version of the radix-3 FFT.

4.4. (a) Use the remainder theorem of algebra to show that

$$X(k) = X(z) \bmod (z - W_N^{-k})$$

(b) Use the fact that $(z - W_N^{-k})$ divides $(z^N - 1)$ to show that

$$X(k) = [X(z) \bmod (z^N - 1)] \bmod (z - W_N^{-k})$$

(c) Use the fact that $z^{2^p} - 1$, can be factored as follows:

$$z^{2^p} - 1 = (z^{2^{p-1}} - 1)(z^{2^{p-1}} + 1)$$

$$= (z^{2^{p-1}} - 1)(z^{2^{p-2}} + j)(z^{2^{p-2}} - j)$$

to find an alternative representation of the split-radix transform.

4.5. In speeding up the 5-point convolution, we decomposed $z(ad+bc)+(bd-ac)$ in order to obtain the product with only three multiplications. There are four such decompositions, of which the form given there is one; find the other three.

4.6. It is a major project to find a Winograd small-N transform for $N = 25$, unless one has a symbolic-manipulation program available; but it is not difficult to derive the decomposition of the 25-point DFT by separating the elements whose subscripts are multiples of 5. Use powers of 2(mod 25) to form the Rader mapping of the reduced transform, and find a way to permute the $(0, 5, ..., 20)$ elements.

4.7. Prove the formula for the inverse of the number-theoretic transform, Eq. (4-55).

4.8. Let **A** be the matrix of an N-point number-theoretic transform with modulus M. Let the subscripts of the upper-left-hand corner of **A** be $[0, 0]$. Use Eq. (4-55) to show that the inverse of **A** is given by

$$a^{-1}[i, j] = (1/N)a[i, N - j]$$

where $1/N$ is computed modulo M and $(N - j)$ is computed modulo N.

4.9. Prove the symmetry property of the number-theoretic transform.

4.10. Let M be 61; find a principal sixth root of 1 modulo M. Using this root, construct the **A** matrix for a six-point number-theoretic transform and find its inverse. Use these matrices to convolve the sequences

$$\mathbf{x} = (3, 2, 1, 0, 0, 0)$$

$$\mathbf{y} = (2, 1, 3, 2, 4, 1)$$

by multiplying their transforms. Compute the cyclical convolution of **x** and **y** by hand and verify that the results are the same as those obtained by the number-theoretic transform.

4.11. Write a computer program for computing number-theoretic transforms with any specified modulus M and any transform size N. You will need procedures for searching for an Nth root, for constructing the **A** matrix and its inverse, for performing integer and matrix-vector multiplications modulo M, and for finding the reciprocal of an integer modulo M. (In doing integer multiplications modulo M, make sure the intermediate product is put in a data type long enough to contain it without overflow.) Use the method of Prob. 4.8 to construct the inverse of the **A** matrix.

REFERENCES

R. C. Agarwal and C. S. Burrus. "Fast convolution using Fermat number transforms with applications to digital filtering." *IEEE Trans. ASSP*, vol. ASSP-22 #2, p. 87, Apr. 1974.

————, "Number theoretic transforms to implement fast digital convolution." *Proc. IEEE*, vol. 63, no.4, p. 550, Apr. 1975.

R. C. Agarwal and J. W. Cooley. "New algorithms for digital convolution." *IEEE Trans. ASSP*, vol. ASSP-25, no. 5, pp. 392–410, Oct. 1977.

B. Aramabepola and P. J. W. Rayner. "Discrete transforms over residue class polynomial rings with applications in computing multidimensional convolutions." *IEEE Trans. ASSP*, vol. ASSP-28, no.4, p 407, Aug. 1980.

P. R. Chevillat. "Transform-domain digital filtering with number theoretic transforms and limited word lengths." *IEEE. Trans. ASSP*, vol. ASSP-26, pp. 284–290, Aug. 1978.

S. A. Cook, "On the minimum computation time of functions." Thesis. Cambridge, Mass., Harvard University, 1966.

J. W. Cooley and J. W. Tukey. "An algorithm for the machine calculation of complex Fourier series." *Math. Comp.*, vol. 19, no. 90, pp. 297–301, April 1965.

J. W. Cooley et al.. "The fast Fourier transform algorithm: programming considerations in the calculation of sine, cosine, and Laplace transforms." *J. Sound Vib.*, vol. 12, pp. 315–337, July 1970.

P. Duhamel. "Implementation of 'split-radix' FFT algorithms for complex, real, and real-symmetric data." *IEEE Trans. ASSP*, vol. ASSP-34, no. 2, pp. 285–295, Apr. 1986.

I. J. Good. "The interaction algorithm and practical Fourier series." *J. Royal Stat. Soc.*, ser. B, vol. 20, pp. 361–372 1958; addendum, vol. 22, pp. 372–375 1960.

G. H. Hardy, *A Mathematican's Apology*.

D. E. Knuth, *Seminumerical Algorithms*, vol. 2. *The Art of Computer Programming*. Reading, Mass.: Addison-Wesley 1969.

D. P. Kolba and T. W. Parks. "A prime factor FFT algorithm using high-speed convolution." *IEEE Trans. ASSP*, vol. ASSP-25, no. 4, pp. 281–294, Aug., 1977.

L. M. Leibowitz. "A simplified binary arithmetic for the Fermat number transform." *IEEE Trans. ASSP*, vol. ASSP-24 #5, pp. 356–359, Oct 1976.

J. B. Martens. "Recursive cyclotomic factorization—a new algorithm for calculating the discrete Fourier transform." *IEEE Trans. ASSP*, vol. ASSP-32, no. 4, pp. 750–761, Aug. 1984.

J. H. McClellan. "Hardware realization of a Fermat number transform." *IEEE Trans. ASSP*, vol. ASSP-24, no. 3, June 1976.

J. McClellan and C. M. Rader. "There is something much faster than the fast Fourier transform." seminar notes, Oct., 21 1976.

D. H. Nissen. "A note on the variance of a matrix." *Econometrica*, vol. 36, no. 3-4, pp. 603–604, July-Oct. 1968.

H. J. Nussbaumer, *Fast Fourier Transform and Convolution Algorithms*, Berlin: Springer-Verlag 1981.

T. W. Parsons. "A Winograd-Fourier transform algorithm for real-valued data." *IEEE Trans. ASSP*, vol. ASSP-27, no. 4, pp. 398–402, Aug. 1979.

C. M. Rader. "Discrete Fourier transforms when the number of data samples is prime." *Proc. IEEE*, vol. 56, pp. 1107–1108, June 1968.

———, "Discrete convolutions via Mersenne transforms." *IEEE Trans. Computers*, vol. C-21, pp. 1269–1273, Dec. 1972.

M. R. Schroeder, *Number Theory in Science and Communication*. Berlin: Springer-Verlag 1984.

H. F. Silverman. "An introduction to programming the Winograd Fourier transform algorithm (WFTA)." *IEEE Trans. ASSP*, vol. ASSP-25, no. 2, pp. 152–165, Apr. 1977; corrections and addendum, vol. ASSP-26, no. 3, p. 268, June 1978(a); further corrections, vol. ASSP-26, no. 5, p. 482, Oct. 1978(b).

L. H. Thomas. "Using a computer to solve problems in physics." *Applications of Digital Computers*, Boston: Ginn and Co. 1963.

A. L. Toom, *Doklady Academiia Nauk SSR*. vol. 150, pp. 496–498, 1963. English Tr. *Soviet Math*. vol. 3, pp. 714–763, 1963.

M. C. Vanwormhoudt. "On number theoretic Fourier transforms in residue class rings," *IEEE Trans. ASSP*, vol. ASSP-25, p. 585, Dec. 1977.

E. Vegh and L. M. Leibowitz. "Fast complex convolution in finite rings." *IEEE Trans. ASSP*, vol. ASSP-24, no. 4, p. 343, Aug. 1976.

M. Vetterli and H. J. Nussbaumer. "Simple FFT and DCT algorithms with reduced number of operations." *Sig. Proc.*, vol. 6, pp. 267–278, July 1984.

S. Winograd. "Some bilinear forms whose multiplicative complexity depends on the field of constants." *Math. Sys. Theor.*, vol. 10, pp. 169–180, 1977.

———, "On computing the discrete Fourier transform." *Proc. Nat. Acad. Sci. USA*, vol. 73, no. 4, pp. 1005–1006, Apr. 1976.

CHAPTER 5

■ ■

DESIGN OF DSP ALGORITHMS

Filtering is a fundamental application of digital signal processing. We will start with a review of conventional IIR and FIR digital filters and their design. Then we will consider applications of digital filters to multirate signal processing and, finally, special types of filter, including filters without multipliers and nonlinear filters.

5.1 TRADITIONAL FREQUENCY-DOMAIN FILTERS

The starting point for digital filter design is the linear system difference equation, together with the system function obtained by means of the Z transform:

$$y(n) = a_1 y(n-1) + a_2 y(n-2) + \cdots + a_N y(n-N)$$
$$+ b_0 x(n) + b_1 x(n-1) + b_2 x(n-2) + \cdots + b_M x(n-M) \quad \text{(5-1)}$$

$$Y(z) = \frac{b_0 + b_1 z^{-1} + b_2 z^{-2} + \cdots + b_M z^{-M}}{1 - (a_1 z^{-1} + a_2 z^{-2} + \cdots + a_N z^{-N})} X(z) \quad \text{(5-2)}$$

The filter design problem consists of determining the coefficients $\{a_i\}$ and $\{bi\}$ to obtain the desired behavior. Implementation of the filter consists of finding a structure to realize the corresponding difference equation economically and in a form that minimizes such problems as error propagation and coefficient-quantization sensitivity.

Digital filters fall into two general categories. If the $\{a_i\}$ coefficients are all zero, then the denominator polynomial in Eq. (5-2) is unity and we have a finite-duration-impulse-response (FIR) filter, as explained in

Chap. 1. Poles in the system function correspond to decaying exponentials or sinusoids which, in principle, continue forever; hence if the denominator polynomial is not a constant, we have an infinite-duration-impulse-response (IIR) filter. Each of these types has its own advantages and disadvantages and its own design techniques. Notice that there is no equivalent in the lumped-parameter analog world to the FIR filter. A filter that contains a feedback path is termed recursive, and one that does not, nonrecursive. The natural FIR filter structure is nonrecursive, but there are recursive implementations as well.

5.1.1 STRUCTURES

5.1.1.1 IIR Filter Structures Probably the obvious way to implement a digital filter is by direct implementation of the difference equation. This can be done most simply by using the canonical form of Fig. 5-1.

Its simplicity makes it an attractive choice, and it is probably the easiest form to implement in software, but when the order of the filter is high, it is more sensitive to coefficient-quantization errors than the other forms to be described, and, in many cases, more sensitive to round-off errors as well.

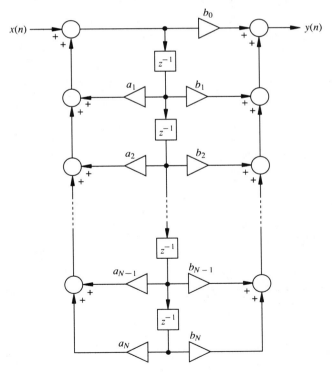

Figure 5-1 *Direct-form digital filter structure.*

If we factor the numerator and denominator polynomials of Eq. (5-2) and move b_0 outside the product to make the numerator monic, we can write

$$H(z) = b_0 \prod_{i=1}^{K} \frac{N_i(z)}{D_i(z)}$$

where $N_i(z) = 1 + a_i z^{-1}$ for factors corresponding to real zeroes or

$$N_i(z) = 1 + a_i z^{-1} + b_i z^{-2}$$

for factors corresponding to conjugate complex zero pairs; and similarly

$$D_i(z) = 1 + c_i z^{-1}$$

for factors corresponding to real poles or

$$D_i(z) = 1 + c_i z^{-1} + d_i z^{-2}$$

for factors corresponding to conjugate complex pole pairs. Then the corresponding filter structure is the cascade form, Fig. 5-2. Each section is an order-2 direct-form filter, except when real poles and zeroes are to be realized. The cascade form offers the convenience that individual poles and zeroes can be separately determined by the coefficients of each section, and it is also less sensitive to coefficient-quantization errors.

The system function can be written in partial-fraction form:

$$H(z) = A + \sum_{i=1}^{P} H_i(z)$$

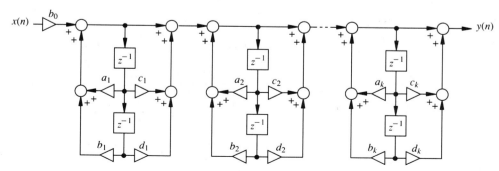

Figure 5-2 *Cascade-form digital filter structure.*

where $H_i(z)$ is either

$$H_i(z) = \frac{a_i}{1 + c_i z^{-1}}$$

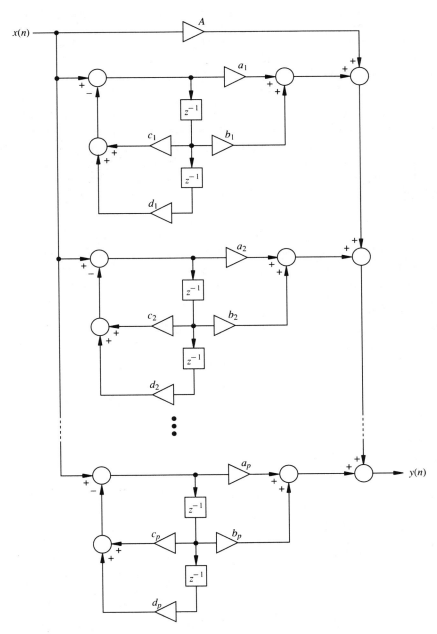

Figure 5-3   *Partial-fraction form digital filter structure.*

for real poles or

$$H_i(z) = \frac{a_i + b_i z^{-1}}{1 + c_i z^{-1} + d_i z^{-2}}$$

for complex pole pairs. Then each partial fraction can be implemented by an order-1 or order-2 direct form and the outputs summed, as shown in Fig. 5-3. As with the cascade form, the poles are readily determined by the denominator coefficients of each section, but since the zeroes result from cancellation among the outputs, their locations are not evident from the filter parameters.

5.1.1.2 FIR Filter Structures Direct-form structures can be used with FIR filters, and usually the coefficient sensitivities and round-off error problems are less severe than with IIR filters, since there is no feedback involved. A direct-form FIR structure is shown in Fig. 5-4(a). An equivalent structure, the transposed direct form, is also occasionally of importance. Here the delays occur in the summation path, as shown in

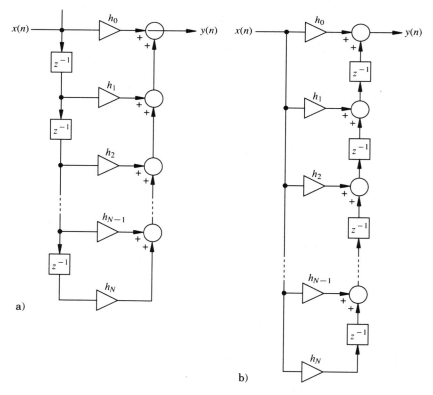

Figure 5-4 *(a) Direct-form FIR digital filter structure; (b) transposed direct-form structure.*

Fig. 5-4(b). The $\{h_i\}$ coefficients in Fig. 5-4 correspond to the $\{b_i\}$ in Eq. 5.2.

Many FIR designs are symmetric: that is for a filter of length N,

$$h(n) = h(N - 1 - n) \qquad 0 \le n < N$$

We can take advantage of this symmetry to reduce the number of multiplications required, using the structure shown in Fig. 5-5 for N odd. Here the delay path is bent into a U, so that the oldest and newest samples can be summed and multiplied by $h(0)$, and so on, as shown.

5.1.1.3 Lattice Filter Structures The lattice filter is most commonly associated with system models derived from the Levinson recursion, described

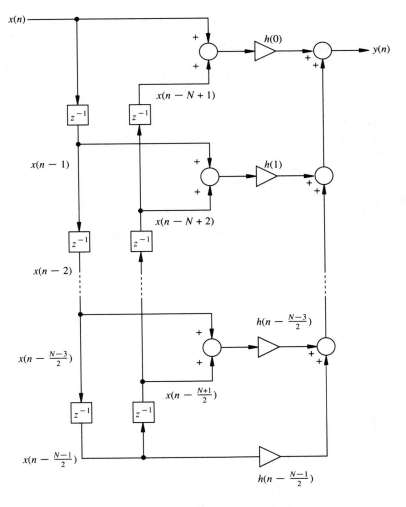

Figure 5-5 *Symmetric direct-form FIR digital filter structure.*

in App. E. The recurrence relations between successive prediction errors are given there by

$$e_p^+(n) = e_{p-1}^+(n) + k_p e_{p-1}^-(n)$$

$$e_p^-(n) = e_{p-1}^-(n-1) + k_p e_{p-1}^+(n-1)$$

(5-3)

or, in the Z-transform domain,

$$E_p^+(z) = E_{p-1}^+(z) + k_p E_{p-1}^-(z)$$

$$E_p^-(z) = z^{-1}[E_{p-1}^-(z) + k_p E_{p-1}^+(z)]$$

(5-4)

These error sequences can be made to provide the basis for a lattice filter; a single section of the filter corresponds to Eq. (5-4) applied to a particular value of p, as shown in Fig. 5-6(a). Here the inputs to the section are the forward and reverse prediction errors for an order-$(p-1)$ filter and the outputs are the corresponding errors for an order-p filter. Figure 5-6(b) shows a series of these sections cascaded to produce the entire filter.

Any conventional all-pole or all-zero filter can be converted to an equivalent lattice filter by means of the transformation given in Eq. (E-20) provided none of the resulting PARCOR coefficients $\{k_i\}$ have magnitudes ≥ 1.

An inverse filter can be similarly obtained. Again, from Eq. (5-4) we may construct the filter section shown in Fig. 5-7(a); the entire filter produced by cascading these sections is shown in Fig. 5-7(b). As is shown in App. E, the stability of the lattice filter is guaranteed if $|k_i| < 1$ for all i.

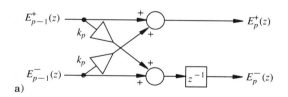

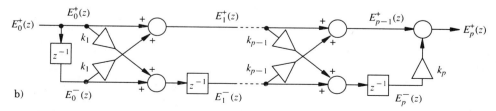

Figure 5-6 *All-zero lattice filter structures. (a) single section; (b) entire filter.*

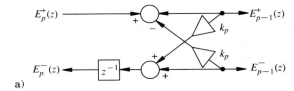

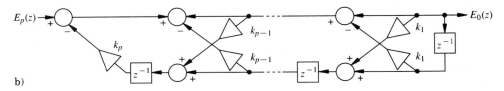

Figure 5-7   *All-pole lattice filter structures. (a) single section; (b) entire filter.*

5.1.2 DESIGN

5.1.2.1 Design of IIR Filters Most design techniques for IIR filters are based on corresponding analog designs, and the techniques center on finding a suitable way of modifying the analog designs for digital use. There are two such techniques of interest to us here, *impulse invariance* and *bilinear transformation*.

In the impulse-invariance design, the impulse response of the digital filter is a sampled version of the impulse response of the analog proto-type. The design process is relatively straightforward, except that care is needed to avoid aliasing. In a high-pass analog filter, for example, the frequency response is as sketched in Fig. 5-8. But the frequency response is the Fourier transform of the impulse response; when the impulse re-sponse is sampled, that part of its transform above f_s will be aliased back onto the part below f_s. Hence the impulse-invariance technique can be applied only to filters whose frequency response cuts off (becomes negligible) at a frequency well below $f_s/2$. In such a case, the digital filter's response will closely match that of the analog prototype; in any

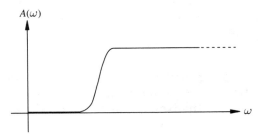

Figure 5-8   *Illustrating high-pass frequency response not suitable for impulse-invariance design.*

other case, aliasing will cause the digital version's frequency response to be different from the prototype's.

To see the design procedure for impulse-invariant filters, consider the impulse response of the analog prototype. This can be written as a sum of decaying exponentials or sinusoids by means of a partial-fraction expansion:

$$H(s) = \sum_{i=1}^{N} H_i(s) \tag{5-5}$$

For real poles,

$$H_i(s) = \frac{a_i}{s - c_i}$$

Each such term corresponds to an exponential,

$$h_i(t) = a_i \exp(c_i t) \qquad t \geq 0 \tag{5-6}$$

For complex pole pairs, it is convenient to write

$$H_i(s) = \frac{a_i s + b_i}{(s - c_i)^2 + d_i^2}$$

Each term of this type corresponds to

$$h_i(t) = \exp(c_i t)(a_i \cos d_i t + e_i \sin d_i t) \tag{5-7}$$

where $e_i = (b_i + a_i c_i)/d_i$

According to the superposition property, the impulse response is given by

$$h(t) = \sum_{i=1}^{N} h_i(t)$$

We wish to create a digital filter whose impulse response is a sampled version of the analog prototype's impulse response:

$$h_d(n) = h(nT)$$

where T is the sampling interval. Since superposition applies in both domains, we can transform the individual components of the analog impulse response and sum the transforms to obtain the desired filter.

The Z transform of Eq. (5-6) is

$$H_i(z) = \frac{a_i}{1 - \exp(c_i T)z^{-1}} \tag{5-8}$$

and the Z transform of Eq. (5-7) is

$$H_i(z) = \frac{\exp(c_i T)(a_i \cos d_i T + e_i \sin d_i T)z^{-1}}{1 - 2\exp(c_i T)\cos(d_i T)z^{-1} + \exp(2c_i T)z^{-2}} \tag{5-9}$$

Hence the desired digital filter is given by

$$H(z) = \sum_{i=1}^{N} H_i(z) \tag{5-10}$$

where the individual terms are obtained from Eq. (5-8) or (5-9) as needed.

In the bilinear transformation, we introduce the following transformation, which maps the entire $j\Omega$ axis onto the unit circle in the Z plane:

$$s = 2f_s \frac{1 - z^{-1}}{1 + z^{-1}} \tag{5-11}$$

The $j\Omega$ axis is infinite in extent, while the unit circle is finite; naturally something has to give. In fact, the mapping is nonlinear, as can be seen by evaluating Eq. (5-11) on the unit circle $z = e^{j\omega T}$:

$$s = 2f_s \frac{1 - e^{-j\omega T}}{1 + e^{-j\omega T}} = j\Omega$$

Multiplying numerator and denominator by $e^{j\omega T/2}$, we obtain

$$s = 2f_s \frac{j \sin \omega T/2}{\cos \omega T/2}$$

Hence we have the frequency distortion function

$$\Omega = 2f_s \tan \omega/2f_s$$

or more conveniently,

$$f_a = \frac{f_s}{\pi} \tan\left(\pi \frac{f_d}{f_s}\right) \tag{5-12}$$

where $f_a = \Omega/2\pi$, $f_d = \omega/2\pi$, and the a and d subscripts refer to the analog and digital frequencies, respectively. The consequence of this frequency distortion is that the frequencies of interest in the specification of the analog prototype must be predistorted using Eq. (5-12) in order to make the corresponding frequencies in the digital filter fall at the desired points.

Aside from this initial complication, the process is straightforward. We start with the system function of the predistorted analog filter, substitute for s throughout using Eq. (5-11), and clean up the result. Note,

194 CHAPTER 5   Design of DSP Algorithms

however, that the warping of the frequency scale means that this technique is useful only in realizing filters with a piecewise-constant frequency response. A differentiator, for example, must have a gain which is directly proportional to frequency; if we start with an analog prototype which meets this specification and predistort the frequency scale, the resultant digital filter will not have the required linear gain characteristic.

5.1.2.2 Design of FIR Filters

These designs are best performed using computer-assisted optimization procedures, although for noncritical cases a process based on Fourier transform techniques is often adequate.

FIR filters are possibly harder to design than IIR filters, but they offer some advantages over the latter. Their stability is guaranteed, since there are no finite poles, they are generally less sensitive to quantization and round-off errors, and they can be designed so as to introduce no phase distortion in the signal.

Linear-Phase Filters. In many applications, phase is of critical importance, and it will be desirable to filter the signal in such a way that phase is undisturbed. Ideally, we would like to have zero phase shift over the frequency range of interest; in practice, we will settle for a filter whose phase shift is directly proportional to frequency; such a filter is said to be *linear phase*. FIR filters are ideally adapted to this. It was shown in Chap. 1 that a system with a real, even impulse response will have no phase shift and that if this impulse is shifted to make it causal, the system will have a linear phase response. Such a shifted response is said to be symmetric: if the duration of the impulse response is N, then

$$h(n) = h(N - 1 - n) \qquad 0 \leq n < N \qquad (5\text{-}14)$$

where N can be either odd or even.

It should be clear that no physically realizable IIR filter can be exactly linear-phase, because the symmetry requirement would make its impulse response extend to minus infinity as well as to plus infinity. No finite delay will make this filter realizable, and no computer memory can contain the samples needed to implement this filter in non-real time. It is shown in App. B that if $h(n) = h(-n)$, then any poles of $H(z)$ must lie on the unit circle or occur in reciprocal pairs, one within the unit circle and a corresponding one outside. Hence, for stability, the filter must have no poles at all, that is be FIR.

[In some cases if the signal to be processed is of finite duration, it may be feasible to pass the signal through an IIR filter, reverse the output in time, and pass the reversed output through the same filter a second time. It can be shown (Prob. 5.2) that this procedure results in a linear-phase output. The output of the first pass will be lengthened by the impulse response of the filter and hence is, in principle, no longer finite; in practice, if the impulse response dies out quickly enough, the output can be truncated without undue error. This technique has occasionally been used in the analog domain; for example, running a tape recording backward through the recorder a second time to remove phase errors introduced by the equalization circuitry.]

It is occasionally of interest to have only a constant derivative of phase with respect to frequency. Such a filter is said to have constant group delay. It can be shown (Rabiner and Gold, 1975) that such a filter will have odd symmetry–that is

$$h(n) = -h(N - 1 - n) \qquad 0 \le n < N \tag{5-15}$$

Fourier Transform Design.   Since the impulse response of a filter is the inverse Fourier transform of its frequency response, it would seem natural to use Fourier techniques in the design of a digital filter. As a first try, suppose we take the desired frequency response and do a Fourier series expansion in the time domain as follows. We will normalize the frequency scale so that f_s is mapped to 2π; then

$$h(n) = \frac{1}{2\pi} \int_{-\pi}^{\pi} H(e^{j\omega}) e^{j\omega n} \, d\omega$$

If we apply this to a low-pass filter as in Fig. 5-9(a), we have

$$h(n) = \int_{-\omega_c}^{\omega_c} e^{j\omega n} \, d\omega = \frac{\sin \omega_c n}{\pi n}$$

which gives us the impulse response shown in Fig. 5-9(b). This impulse response is not finite: the Fourier series expansion of a rectangular pulse has an infinite number of terms.

In order to make this filter practical, we must truncate the impulse response at some point. (It is also necessary to shift it in time to make it causal if processing is to be done in real time.) Truncating the impulse response multiplies it by a rectangular window, however, and in the frequency domain this convolves the desired frequency response with a $\sin x/x$ curve whose width depends on the point at which the truncation was done. The truncation and its consequences are shown in Fig. 5-9(c) and (d). The overshoot and the ripples result from the well-known Gibbs phenomenon. The width of the main lobe of the $\sin x/x$ function changes the sharp cutoff of the desired filter to a gradual cutoff, and the side lobes give rise to the ripple. If the time window is made wider, the $\sin x/x$ function becomes narrower and thus the transition region becomes narrower, but the amplitude of the overshoot does not change, since the height of the side lobes of $\sin x/x$ does not change: widening the window merely scales the $\sin x/x$ curve in frequency.

The solution to this difficulty is to apply a time-domain weighting function to the impulse response. The weighting function multiplies the impulse response by some nonrectangular window function $w(n)$, and hence the frequency response is convolved with its transform $W(\omega)$. We have here all the problems associated with spectral estimation by means

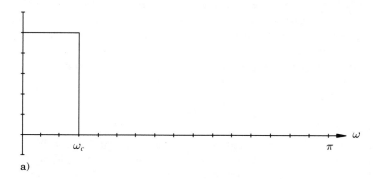

a)

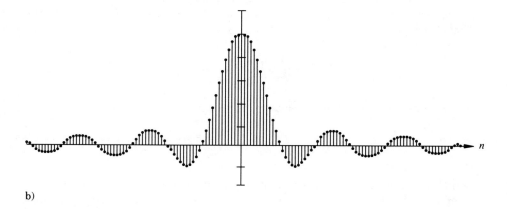

b)

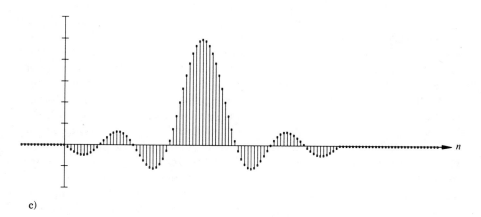

c)

Figure 5-9 *Fourier transform design of FIR filters: (a) ideal low-pass characteristic; (b) corresponding impulse response; (c) truncated and shifted impulse response.*

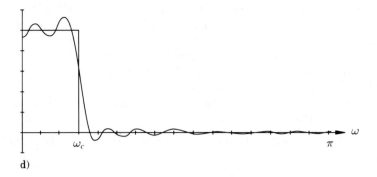

d)

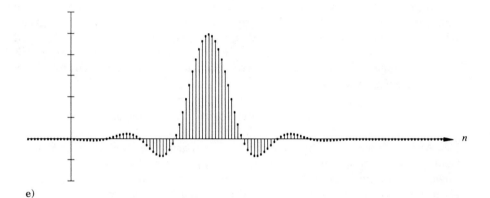

e)

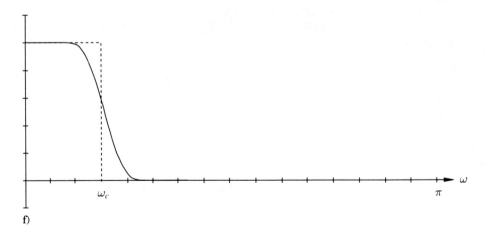

f)

Figure 5-9 *Fourier transform design of FIR filters: (d) resultant frequency response showing overshoot and ripples; (e) Hamming-weighted impulse response; (f) low-pass characteristic from Hamming-weighted filter.*

of Fourier transforms that are described in detail in elementary DSP courses: we wish to find a windowing function that will give the smallest side lobes (and hence the smallest ripple in the frequency response) with the narrowest main lobe (and hence the narrowest transition band). We also have all the repertoire of Fourier analysis windowing functions (the most useful of which are described in App. D) to draw upon. Figure 5-9(e) shows the impulse response of Fig. 5-9(c) with Hamming weighting applied, and Fig. 5-9(f) shows the resultant low-pass filter characteristic.

Duration of the Impulse Response. Note that as with the unweighted characteristic, the magnitude of the ripple depends on the side lobes of $W(\omega)$, and if the width of the window is changed, $W(\omega)$ is only scaled in frequency without changing the side lobes. Hence widening the time window narrows the transition band but does not change the magnitude of the ripples. From this we may conclude that for a given window shape and hence for a given ripple, the time-window width and the transition bandwidth are inversely proportional. Indeed if the width of the transition band is Δf and the duration of the time window is N samples, we may write

$$N\Delta f = \text{(constant depending on ripple tolerances)} \qquad (5\text{-}16)$$

Although we have obtained this result from a consideration of Fourier transform designs, it is applicable as an approximation to all FIR filters however designed. There is no tidy way of expressing the constant, and all published formulae are obtained empirically. In filter design the parameters of interest, as shown in Fig. 5-10, are normally Δf, δ_1, and δ_2; the latter two are the passband and stopband ripples, respectively.

Herrmann et al. (1973) give the following improvement on our rule of thumb:

$$N = 1 + \frac{D_\infty(\delta_1, \delta_2)}{\Delta F} - f(\delta_1, \delta_2)\Delta F \qquad (5\text{-}17)$$

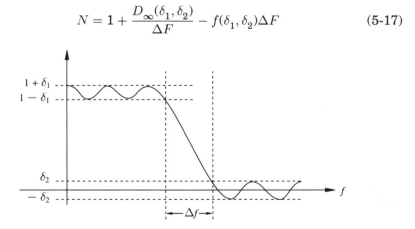

Figure 5-10 *Illustrating Δf (transition bandwidth), δ_1 (pass-band ripple), δ_2 (stop-band ripple).*

where ΔF is normalized to the sampling frequency,

$$\Delta F = \Delta f/f_s$$

$$D_\infty(\delta_1, \delta_2) = [0.005309(\log \delta_1)^2 + 0.07114 \log \delta_1 - 0.4761] \log \delta_2$$
$$- [0.00266(\log \delta_1)^2 + 0.5941 \log \delta_1 + 0.4278]$$

and $f(\delta_1, \delta_2) = 0.51244 \log(\delta_1/\delta_2) + 11.01$

A simpler, less unwieldy approximation is given by Kaiser (1974):

$$N = \frac{-10 \log(\delta_1 \delta_2) - 13}{14.6 \Delta F} + 1 \tag{5-18}$$

where again $\Delta F = \Delta f/f_s$. In these approximations, all logs are to the base 10.

It is well known that abrupt changes in frequency response are associated with a lengthy impulse response. In analog filters, such changes are associated with complex poles close to the $j\Omega$ axis; these poles contribute slowly decaying sinusoidal components. The same condition occurs with sharp-cutoff IIR filters, where the poles lie close to the unit circle. We now see that a similar property applies to FIR filters. FIR filters in general tend to be very long; filters with upwards of 100 taps are not unusual. Modern fast DSP integrated circuits, such as those described in Chap. 10, make such large filters practical; furthermore, the symmetry inherent in linear-phase FIR filters can be used to halve the number of multiplications.

Frequency Sampling.   It is impossible to achieve the step discontinuity of Fig. 5-9(a) in any realizable filter. An alternative design strategy (Gold and Jordan, 1969) is to accept this fact, specify the performance of the filter in the passband and stopband, and leave the transition gains unspecified; then we vary these gains to obtain a filter with the minimum ripple. This is known as the frequency-sampling technique.

This technique requires repeated iterations and is best implemented as a computer-aided design. The procedure is treated in exhaustive detail in Rabiner et al. (1970); we will only summarize briefly here. The gains are usually set at uniformly spaced frequencies; the passband gains would normally be set to 1 and the stopband gains to 0. Suppose that three transition gains are available for adjustment; let these gains be T_1, T_2, and T_3. The optimization proceeds as follows: initially the transition values are all set to 1. Then the first transition value T_1 is varied to minimize the largest side lobe. After that, the first two transition values are adjusted until the largest side lobe is minimized. Rabiner et al. show that the path of steepest descent is a straight line, so T_1 and T_2 can be varied together along this straight line. If the location of the largest side lobe changes, the paths of steepest descent will change; hence after the best combination of T_1 and T_2 is found, the process returns to perturbing

T_1 alone and then adjusts T_1 and T_2 together, as before. This process is repeated until the improvement in side lobe suppression is less than some specified threshold.

Then T_1, T_2, and T_3 are varied together, again along a straight line, until the best performance has been obtained. The procedure next returns to adjusting T_1 alone, then T_1 and T_2, and then all three; again this entire procedure is repeated until the improvement between successive iterations is below the threshold.

The frequency-sampling filter has its own structure. The Z transform of the impulse response will be

$$H(z) = \sum_{n=0}^{N-1} h(n)z^{-n} \tag{5-19}$$

We wish to express $H(z)$ in terms of the discrete Fourier transform (DFT) of $h(n)$. Let H_k be the kth Fourier coefficient in the DFT of $h(n)$. Then $h(n)$ is given by the inverse transform

$$h(n) = \frac{1}{N} \sum_{k=0}^{N-1} H_k W^{nk} \tag{5-20}$$

where W is $e^{j2\pi/N}$. If we substitute this into Eq. (5-19), we have

$$H(z) = \frac{1}{N} \sum_{n=0}^{N-1} z^{-n} \sum_{k=0}^{N-1} H_k W^{nk}$$

$$= \frac{1}{N} \sum_{k=0}^{N-1} H_k \sum_{n=0}^{N-1} z^{-n} W^{nk}$$

The second sum is $(1 - z^{-N} W^{Nk})/(1 - z^{-1}W^k)$. But $W^{Nk} = 1^k = 1$; hence the numerator can be factored to give

$$H(z) = \frac{1 - z^{-N}}{N} \sum_{k=0}^{N-1} \frac{H_k}{1 - z^{-1}W^k} \tag{5-21}$$

This leads to the structure shown in Fig. 5-11. Notice that each denominator term in the sum of Eq. (5-21) places a pole on the unit circle; hence the filter as shown is not strictly stable. In practice, each denominator term can be made to be $1 - az^{-1}W^k$, where a is a constant slightly less than unity, and the multiplier $1 - z^{-N}$ is adjusted accordingly. Since all but one of these denominators are complex, they are combined in conjugate pairs to obtain real coefficients.

Computer-Aided Design Techniques. The most widely used of these is due to McClellan et al. (1973). The program is very long, and it

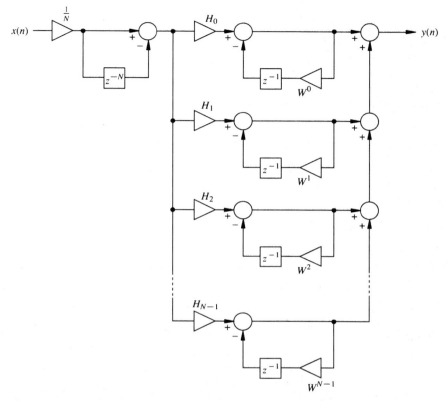

Figure 5-11   *Structure of frequency-sampling digital filter.*

is not feasible to present it here; in any case it is available in McClellan et al. and in IEEE (1981). We will sketch the logic of the procedure here.

Suppose we have a filter of length N with prescribed passband and stopband ripple δ_1 and δ_2, respectively. We will specify the filter response at N + 1 points, as in the frequency-sampling technique. Here, however, we will take the ripple into account. That is, we will specify that at a certain number of points the passband gain shall be $1 + \delta_1$ and at certain points in between the passband gain shall be $1 - \delta_1$. Similarly, in the stopband, the gain shall be alternately δ_2 and $-\delta_2$. Initially we do not know where these extrema will occur in the optimized filter, so we guess: typically we choose equally spaced points in the passband and stopband, as shown in Fig. 5-12.

We have now specified the gain function at $N + 1$ points, and we can pass a polynomial of degree N through these points, using Lagrangian interpolation. We then inspect the polynomial and determine where the extrema actually fell. These extrema will generally lie near our specified frequencies, but not at them, and the deviations will be greater than

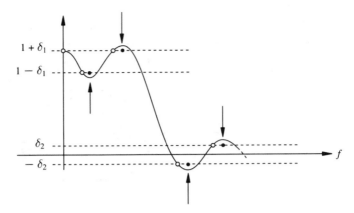

Figure 5-12 *Illustrating one iteration in McClellan et al, (1973) FIR design procedure. Open circles indicate specified extrema, arrows indicate locations of actual extrema, and solid circles show new specified extrema.*

specified. In Fig. 5-12, the specified extrema are shown as circles, and the interpolated polynomial can be seen passing through them, although the actual extrema, marked by arrows in the figure, do not coincide with the circles. We now assume that the actual locations of the extrema are correct, and we revise our specification so that the values we want will occur at the locations of the actual extrema: that is, we move our points to the locations indicated by the dots in Fig. 5-12. (The points at 0 and π, and those marking the edges of passbands and stopbands, remain fixed, however: we must not move those.)

This procedure constitutes one iteration of the process. We then construct a new polynomial passing through the new points, and once more we compare the locations of the actual extrema. Typically, they will have shifted slightly again, and again we move our specified points to the positions of the new extrema. This process will eventually converge; after a sufficient number of iterations, the actual extrema will differ from our specifications by an amount less than some predetermined threshold, and the algorithm will terminate. It is then necessary only to convert the polynomial coefficients to filter coefficients.

5.2 SAMPLING-RATE ALTERATION

It is occasionally necessary to change the rate at which data are sampled. The simplest way to do this would presumably be to convert the signal back to analog and resample at the new rate. This is usually unacceptable because of noise and errors introduced by the conversions. Hence it is best if the change in the sampling rate can be done in the digital domain. This has been described in detail in Schafer and Rabiner (1973); the material which follows is based in part on this paper.

5.2.1 DECIMATION AND INTERPOLATION

The simplest cases are those where one of the rates is an integer multiple of the other. If the new rate is L times the old rate, we can interpolate an extra $L-1$ samples after each of the original ones. If the new rate is $1/M$ times the old, then we can save every Mth sample and discard the rest. This latter process is known as decimation.

In decimation if the original signal was sampled at or near the Nyquist rate, then the decimated sequence will be under-sampled, and the signal will be corrupted by aliasing. Hence before decimation, it is generally necessary to low-pass filter the signal to remove those frequency components that would be aliased. Since we never leave the sampled domain, we can do this filtering digitally.

In drawing block diagrams for sample-rate conversions, we will want a symbol to represent the decimation process. It is customary to use a box with a down-pointing arrow, followed by the decimation factor, as shown in Fig. 5-13(a). This symbol represents the operation of retaining every Mth sample and throwing away the rest, as shown in Fig. 5-13(b) for $M=4$.

To model the decimation process, it is convenient to separate the process into two steps: first we set all samples whose indices are not integer multiples of M to zero, and second, we discard the samples that were set to zero. This permits us to consider the effect of under-sampling and the change in time scale separately. We can analyze the first step in two ways, one immediately clear and the other useful later on. Let the original sequence be $x(n)$ and let the result of the first step be $x'(n)$, defined by[†]

$$x'(n) = \begin{cases} x(n) & M|n \\ 0 & \text{otherwise} \end{cases}$$

[This sequence is shown in Fig. 5-13(b).] We observe that this first step is equivalent to multiplying $\{x(n)\}$ by a train of impulses spaced M samples apart:

$$x'(n) = x(n) \sum_{i=-\infty}^{\infty} \delta(n - Mi) \qquad (5\text{-}22)$$

It is easily established that the transform of such a Kronecker impulse train is another (Dirac) impulse train spaced $2\pi/M$ samples apart. Since multiplication in time is equivalent to convolution in frequency, we conclude that $X'(\omega)$, the transform of the decimated sequence, consists of M replications of $X(\omega)$ spaced at intervals of $2\pi/M$.

[†]$M|n$ means M divides n; that is, n is an integer multiple of M. See App. F.

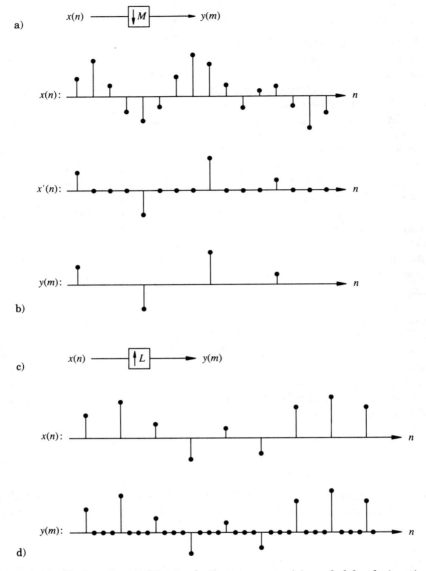

Figure 5-13 *Decimation and interpolation processes: (a) symbol for decimation by factor of M; (b) sampled sequence before and after decimation; (c) symbol for interpolation by factor of L; (d) sampled sequence before and after interpolation.*

The second way of looking at this step is to represent the impulse train by a Fourier series expansion: since the impulse train has a period M, we have

$$\sum_{i=-\infty}^{\infty} \delta(n - Mi) = \frac{1}{M} \sum_{k=0}^{M-1} e^{j2\pi nk/M}$$

Hence
$$x'(n) = x(n)\frac{1}{M}\sum_{k=0}^{M-1} e^{j2\pi nk/M} \qquad (5\text{-}23)$$

The Z transform is

$$X'(z) = \frac{1}{M}\sum_{n=-\infty}^{\infty} x(n)\sum_{k=0}^{M-1} e^{j2\pi nk/M}z^{-n}$$

Reversing the order of the sums and simplifying, this is

$$X'(z) = \frac{1}{M}\sum_{k=0}^{M-1}\sum_{n=-\infty}^{\infty} x(n)(ze^{-j2\pi k/M})^{-n}$$

The second sum is now recognizable as the Z transform of $x(n)$ evaluated at $ze^{-j2\pi k/M}$; hence

$$X'(z) = \frac{1}{M}\sum_{k=0}^{M-1} X(ze^{-j2\pi k/M}) \qquad (5\text{-}24)$$

If we adopt the FFT abbreviation $W_M = e^{-j2\pi/M}$, this may be written still more concisely as

$$X'(z) = \frac{1}{M}\sum_{k=0}^{M-1} X(W_M^k z) \qquad (5\text{-}25)$$

We can relate this form to our more intuitive analysis by evaluating Eq. (5-25) on the unit circle:

$$X'(e^{j\omega}) = \frac{1}{M}\sum_{k=0}^{M-1} X(e^{j(\omega-2\pi k/M)}) \qquad (5\text{-}26)$$

in which case the Fourier transform is seen to be M replications of $X(e^{j\omega})$ spaced at intervals of $2\pi/M$, as stated previously.

Finally, suppose the input sequence is filtered before decimation:

$$y(n) = \sum_{\ell=-\infty}^{\infty} h(\ell)x(n-\ell)$$

Then if $y'(n)$ is the sequence after multiplication by $\sum \delta(n-iM)$, the Z transform of $y'(n)$ is simply

$$Y'(z) = \frac{1}{M}\sum_{k=0}^{M-1} X(W_M^k z)H(W_M^k z) \qquad (5\text{-}27)$$

This is illustrated in Fig. 5-14, where the new sampling rate is one half the old. Figure 5-14(a) shows the Fourier transform of the original analog signal $X_a(\Omega)$. Before decimation the transform of the sampled signal is as shown in Fig. 5-14(b). If the sequence is decimated without prior filtering, the surviving samples represent the original time function sampled at $f_s/2$; the result, shown in Fig. 5-14(c), is as we would expect: the new, more closely spaced replications of $X(e^{j\omega})$ overlap, and portions of the adjacent images are aliased onto the baseband. In Fig. 5-14(d), the original sequence has been filtered to reduce its bandwidth, and in Fig. 5-15(e), the transform after decimation is free of aliasing.

In all this development, we still have not actually decimated the signal by throwing away those samples whose indices are not multiples of M. It is important to take note of the initial step separately, however, because first it throws the aliasing problem into relief and second, $X'(e^{j\omega})$ in Eq. (5-26) is precisely what we would obtain if a decimator and an interpolator were placed back to back with no intervening system (since the interpolator can be viewed as simply putting back the zeroes which the decimator removed) as in Fig. 5.13(e). This back-to-back structure will be of importance to us when we consider maximally decimated filter banks below.

The second step consists simply of a change in the time scale. If the input to the decimator is taken as $x'(n)$, then let the output be $y(m)$, where m is now used as the index of the decimated sequence; $m = Mn$. Then we have $y(m) = x'(Mn) = x(Mn)$, as shown in Fig. 5-13(b). This compression in time corresponds to an expansion in frequency, and from the time-scaling property of the Fourier transform,

$$Y(e^{j\omega}) = X'(e^{j\omega/M}) \tag{5-28}$$

and in fact

$$Y(z) = X'(z^{1/M}) \tag{5-29}$$

Combining this with Eq. (5-25), we have

$$Y(z) = \frac{1}{M} \sum_{k=0}^{M-1} X(W_M^k z^{1/M}) \tag{5-30}$$

Hence the single replication from $\omega = -\pi/M$ to $\omega = \pi/M$ is stretched to cover the entire range from $-\pi$ to π, as shown in Fig. 5-14(f). But note that this single replication will include any overlaps from other replications that may have resulted from aliasing; this is why we analyzed the first step separately.

In interpolation, the added points are initially zeroes. We symbolize this operation by a box with an upward-pointing arrow, followed by the interpolation factor, as shown in Fig. 5-13(c); this symbol represents the

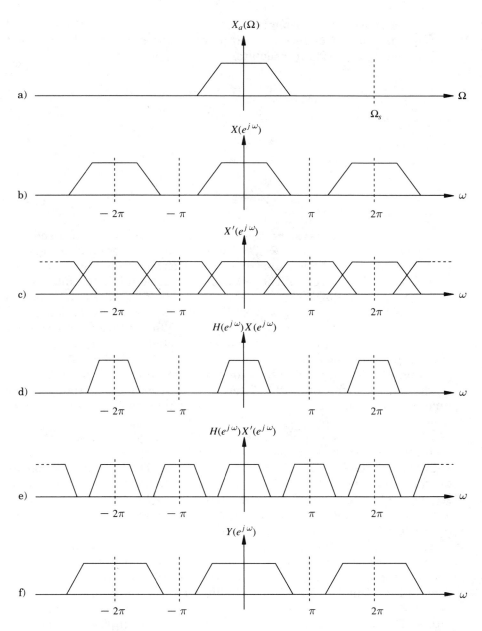

Figure 5-14 *Digital decimation by a factor of 2. (a) Fourier transform of analog signal; (b) transform of sampled signal; (c) transform of sequence with every other sample set to 0; (d) transform of sampled sequence as filtered before decimation; (e) transform of filtered sequence with every other sample set to 0; (f) transform of decimated sequence.*

insertion of $(L - 1)$ zero samples after each input sample, as shown in Fig. 5-13(d) for $L = 4$.

The analysis of interpolation is significantly simpler than that of decimation, since it can be treated as only another scale change. Let $x(n)$ be the input to the interpolator and y the output; then

$$y(n) = \begin{cases} x(n/L) & L|n \\ 0 & \text{otherwise} \end{cases}$$

Hence

$$Y(e^{j\omega}) = X(e^{jL\omega}) \tag{5-31}$$

or

$$Y(z) = X(z^L) \tag{5-32}$$

Thus the entire transform from $\omega = 0$ to $\omega = 2L\pi$ is compressed into the region from $\omega = 0$ to $\omega = 2\pi$. We must state it in these terms, because while it is true that the portion from 0 to 2π is compressed into the range from 0 to $2\pi/L$, the rest of the transform is not empty: it contains the periodic repetitions of $X(e^{j\omega})$, as shown in Fig. 5-15. Figure 5-15(a) shows the transform of the analog signal; the transform of the sampled signal is shown before interpolation in Fig. 5-15(b). When the zero samples are added, the baseband of the new sequence includes one or more images of the input sequence's transform, as shown in Fig. 5-15(c). Again, low-pass digital filtering is necessary to remove these images; the transform of the interpolated signal after filtering is shown in Fig. 5-15(d). In the time domain, this filtering has the effect of replacing the zero samples by interpolated values.

If the new sampling rate is a rational multiple of the original rate, then the change can be managed by interpolation followed by decimation. Let the new rate $f' = (L/M)f$, L and M integers. Then we can interpolate to an intermediate rate equal to Lf and then decimate this intermediate sequence by a factor of M to obtain the rate $(L/M)f$. Notice that in order to preserve the full bandwidth of the input sequence, we must interpolate first and decimate second. This is illustrated in the block diagram in Fig. 5-16(a), which illustrates the process of reducing the sampling rate by a factor of 3/4. Notice also that in this structure, the two filtering operations can profitably be consolidated and replaced by a single filter. The bandwidth of this filter must be the minimum of the bandwidths that would be required by the interpolation and decimation filters. The filtered output has the transform shown in Fig. 5-16(c); after decimation, the transform appears as shown in Fig. 5-16(d).

5.2.1.1 Band-Pass Signals Interpolation and decimation can be applied to band-pass signals by heterodyning them down to the baseband and decimating or interpolating the baseband signal. There are three ways in which a real band-pass signal can be heterodyned: by synchronous

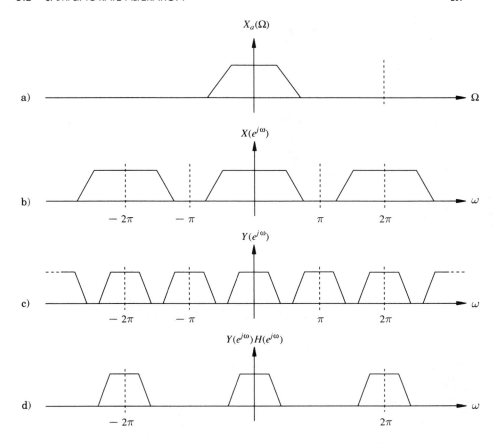

Figure 5-15 *Digital interpolation: (a) Fourier transform of analog signal; (b) transform of sampled signal; (c) transform of interpolated sequence before filtering; (d) transform of interpolated sequence after filtering.*

quadrature detection (the Weaver modulator), by real detection, and by integer-band sampling.

Consider the band-pass signal shown in Fig. 5-17(a). The signal is centered about a carrier frequency ω_c, as shown and has a bandwidth of ω_b. If we multiply the time sequence by $\exp(-j\omega_c n)$, the Fourier transform is convolved with $\delta(\omega - \omega_c)$; this shifts the transform so that the positive-frequency portion is centered about $\omega = 0$, as shown in Fig. 5-17(b). The negative-frequency part is now centered about $-2\omega_c$. The baseband signal can now be low-pass filtered and decimated, as before. Since the negative-frequency part of the original band-pass signal is the complex conjugate of the positive-frequency part, there will be enough information in the baseband signal to reconstruct both the positive- and negative-frequency parts of the original signal after decimation; hence it is not necessary to do a second heterodyning operation for the negative-frequency part.

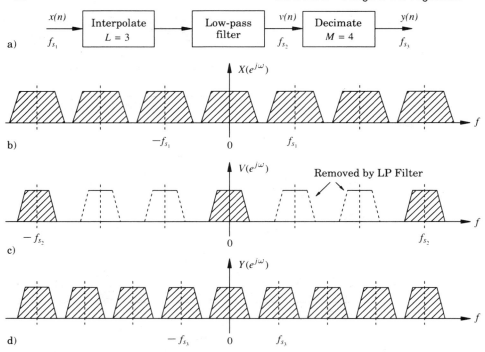

Figure 5-16 *Change in sampling rate by rational (noninteger) ratio: (a) block diagram; (b) transform of original sequence; (c) transform of interpolated and filtered sequence; (d) transform of decimated sequence.*

There is, in general, no guarantee that the transform of the original signal is conjugate-symmetric about ω_c, however; hence when the heterodyning operation is done, the baseband signal will not, in general, be conjugate-symmetric. This means that the corresponding baseband time sequence will not be real; then each sample requires a pair of numbers, one for the real part and one for the imaginary part, and we have, in effect two sequences.

If we rewrite $\exp(-j\omega_c n)$ as $\cos(\omega_c n) - j\sin(\omega_c n)$, then we can obtain the real and imaginary parts of the heterodyned sequence explicitly. If the sequence is $s(n)$, then the output of the heterodyner is

$$s'(n) = s(n)\exp(-j\omega_c n)$$
$$= s(n)\cos(\omega_c n) - js(n)\sin(\omega_c n)$$
$$= x(n) + jy(n)$$

Hence the baseband sequence can be obtained by applying the input to two multipliers, as shown in Fig. 5-18(a). The outputs give us two sequences, one of which, $x(n)$, is the real part of the baseband sequence and

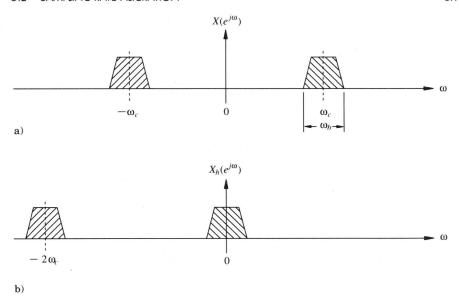

Figure 5-17 *Simple shift of band-pass signal to baseband by heterodyning.*

the other, $y(n)$, the imaginary part. It can be shown (Prob. 5.1) that to normalize the baseband signal amplitudes, the sine and cosine functions should have amplitude $\sqrt{2}$. If we need to reconstruct the original band-pass signal, we can apply the two sequences $x(n)$ and $y(n)$ to a matching pair of multipliers, as shown in Fig. 5-18(b).

If it is undesirable to have a complex baseband sequence, the hetero-dyning procedure described above can be modified to produce a real base-band sequence. We know from the properties of the Fourier transform that a sequence is real if and only if its Fourier transform is conjugate-symmetric. Hence if we heterodyne the positive part of $S(e^{j\omega})$ down to a center frequency of $\omega_b/2$ and the negative part of $S(e^{j\omega})$ up to $-\omega_b/2$, as shown in Fig. 5-19, the new transform is still conjugate-symmetric and the baseband sequence will be real.

5.2.1.2 Integer-band sampling These heterodyning processes are compli-cated and expensive to implement. In some cases, there is a simpler way of shifting a band-pass sequence down to the baseband. If the band-width of the sequence begins and ends at consecutive integer multiples of f_s/M, as shown in Fig. 5-20, then we can accomplish the frequency shift by under-sampling by a factor M; as a result of aliasing, we will obtain a complete set of images, one pair of which will be at the baseband. Low-pass filtering will now select the baseband images and remove all the others. This process is so much cheaper and simpler than heterodyning that it is frequently worth going to some trouble in the system design phase to ensure its feasibility.

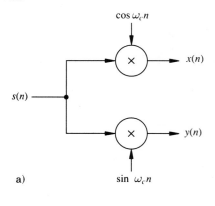

a)

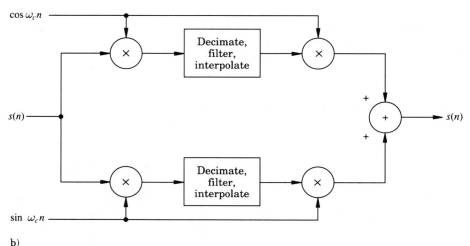

b)

Figure 5-18   *(a) Generating conjugate-symmetric baseband signal by means of two multipliers; (b) reconstruction of original band-pass signal after changing sampling rate.*

5.2.2 FILTERS FOR DECIMATION AND INTERPOLATION

The processes of decimation by dropping excess samples and of interpolation by inserting zeroes are, or ought to be, conceptually simple. The complexity arises from the problem of filtering, which is always necessary to avoid aliasing or to remove undesired images. For this filtering to be feasible in real-time or near-real-time applications, it must be done quickly—that is, cheaply—and the bulk of the research effort in multirate digital signal processing has concentrated on efficient filtering techniques.

The filter of choice is usually a FIR filter. In decimation or interpolation, it is generally desirable to preserve the phase relations, and we

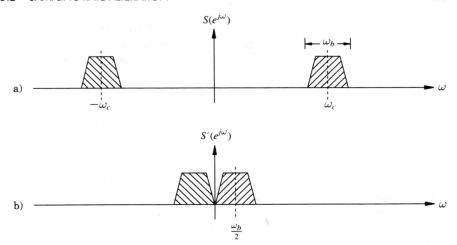

Figure 5-19   *Generation of conjugate-symmetric baseband signal by heterodyning to $\pm\omega_b/2$.*

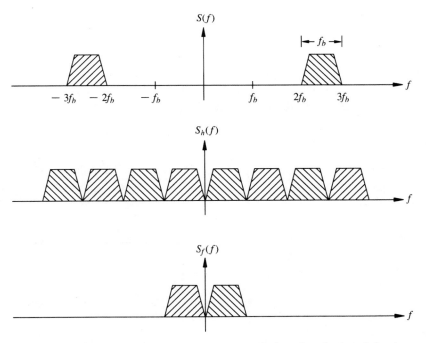

Figure 5-20   *Generation of conjugate-symmetric baseband signal by integer-band sampling. (a) Fourier transform of original sequence; (b) transform of undersampled signal; (c) transform of baseband sequence obtained by filtering.*

have seen that it is a simple matter to construct FIR filters that have no phase distortion.

It should also be clear that the length of a FIR filter for decimation or interpolation should preferably be odd; otherwise $(N-1)/2$ is not an integer and the output samples correspond to values midway between sampling points. This can most easily be seen by considering an impulse response of even length, as in Fig. 5-21(a). The output point will correspond to the midpoint of this response; but because there are an even number of points, the midpoint occurs halfway between samples, as can be seen by shifting it back to make it symmetric about $n = 0$, as in Fig. 5-21(b), and so such a filter includes a delay of one-half sample. This is usually undesirable, particularly in the case of an interpolating filter, where we expect the filter's output to be the nonzero input samples filled in with interpolated values, that is, where the filtered values of the nonzero samples are the same as the nonfiltered values.

The nature of the decimation and interpolation processes provides our first simplification. A digital filter normally produces one output sample for every input sample. In the case of decimation, we can speed up the filtering process by computing only every Mth output point, since these are the only ones that will be required after decimation. Hence where the normal convolution for a FIR filter would be

$$y(n) = \sum_{i=0}^{N-1} x(n-i)h(i)$$

the decimation filter takes the form

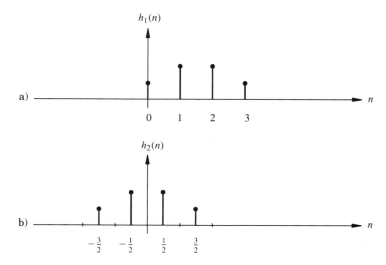

Figure 5-21 *Illustrating one-half-sample shift resulting from FIR filter of even length. (a) symmetric impulse response; (b) equivalent noncausal impulse response.*

$$y(n) = \sum_{i=0}^{N-1} x(Mn - i)h(i)$$

We show the practical significance of this in Fig. 5-22. Figure 5-22(a) depicts the basic decimation system, consisting of a filter followed by a decimator. In Fig. 5-22(b), we show the individual h(i) terms as separate branches in the diagram, and here we see that after all the output samples of the filter are computed, $M - 1$ out of every M values are thrown away by the decimator. There is clearly no reason why the points that are thrown away need be computed at all; hence in Fig. 5-22(c), we move the decimation step back to the input of each filter tap. The number of multiplications is thus reduced by a factor of M. Throwing samples away costs nothing, so the replication of the decimators does not materially increase the complexity of the process.

A similar saving is possible in the interpolation process. In this case, $L - 1$ out of every L samples is zero, and for these samples no multiplication need be done. This is illustrated in Fig. 5-23. Again we show the original form of the system in Fig. 5-23(a): an interpolator inserts $(L - 1)$ zero samples after each input sample. The filter is shown in detail in Fig. 5-23(b). Here we use the transposed direct-form structure of Fig. 5-4(b). Because of the interpolator, each tap in the filter is multiplying a meaningful value only $1/L$ of the time; in between times, it is multiplying a zero input. Just as there is no reason for the decimator to compute values that are to be thrown away, so there is also no reason for the interpolator to multiply zeroes; hence in Fig. 5-23(c), we move the interpolation step to the output of each filter tap. The multiplications now proceed at the lower rate, and each multiplication has a meaningful operand; the zeroes are inserted afterward.

We can obtain a further reduction from the symmetry requirement. If

$$y(n) = \sum_{k=0}^{N-1} h(k)x(n - k)$$

and $h(N - 1 - k) = h(k)$, then

$$y(n) = h\left(\frac{N-1}{2}\right)x\left(n - \frac{N-1}{2}\right) + \sum_{k=0}^{(N-3)/2} h(k)[x(n - k) + x(n - N - k + 1)]$$

where N is odd. The structure corresponding to this expression is the same as that shown in Fig. 5-5; as can be seen, for large N this approximately halves the number of multiplications required. If we include the effects of both decimation and symmetry, the number of multiplications per second is given by

$$R = \tfrac{1}{2}Nf_s/L$$

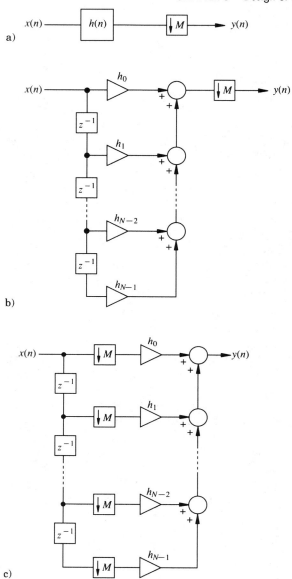

Figure 5-22   *Filtering a decimated sequence: (a) block diagram of filter followed by decimator; (b) direct-form FIR filter followed by decimator.   (c) decimation process moved into filter to decrease number of multiplications required.*

Two issues arise in the interpolation process that do not occur in decimation. First, it is clearly desirable that the interpolated sequence generated by the filter pass through the unfiltered samples, as shown in Fig. 5-24. Suppose we are interpolating by a factor of L and let the sequence after insertion of zeros be $x'(n)$. That is,

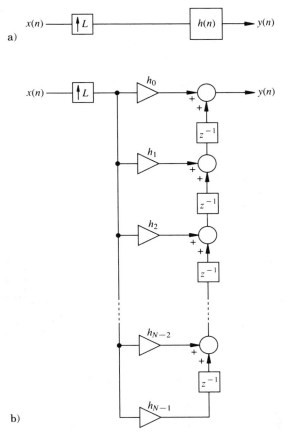

Figure 5-23 *Filtering an interpolated sequence: (a) block diagram of interpolator followed by filter; (b) interpolator followed by transposed direct-form FIR filter.*

$$x'(n) = \begin{cases} x(n/L) & L|n \\ 0 & \text{otherwise} \end{cases}$$

Then if $x'(n)$ is applied to the filter and the interpolated output is $y(n)$, we require

$$y(n) = \begin{cases} x'(n) & L|n \\ (\text{some interpolated value}) & \text{otherwise} \end{cases}$$

The filter output is given by

$$y(n) = \sum_{k=-p}^{p} x'(n-k)h(k)$$

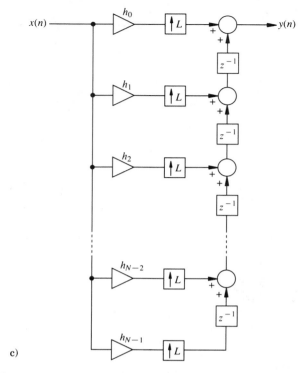

Figure 5-23 *Filtering an interpolated sequence: (c) interpolation moved into filter to decrease number of multiplications.*

When $L|n$, the original samples of $\{x\}$ are multiplied by $h(Lq)$, q an integer, as shown in Fig. 5-24(b). Since all other input samples are zero,

$$y(n) = \sum_{q=-p/L}^{p/L} x[(n-q)/L]h(Lq)$$

Clearly the only way we can guarantee that $y(n) = x(n/L)$ is to constrain $h(n)$ to be unity for $n = 0$ and zero for all other integer multiples of L.

Second, it is generally desirable that the same number of nonzero input points contribute to each output sample. But the interpolation filter is looking at an input sequence which is mostly zeroes, and if the length of the filter is not chosen with care, it may find fewer nonzero samples in some positions than in others. This is illustrated in Fig. 5-25(a), where $L = 5$ and the length of the filter is N. For the filter always to span the same number P of input samples, its length must be $N = LP$; since N must be odd, we will make $N = LP - 1$ if LP is even.

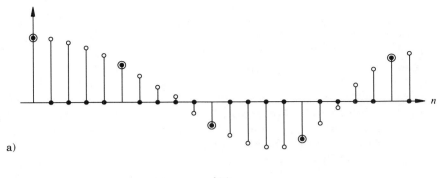

a)

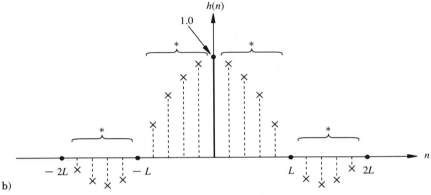

b)

Figure 5-24   *Illustrating requirement that filtered sequence pass through nonzero samples of unfiltered sequence: (a) sequences superimposed, dots indicate unfiltered values [x'(n)] and circles are filtered values [y(n)]; (b) Required form of impulse response; asterisks indicate other values as required.*

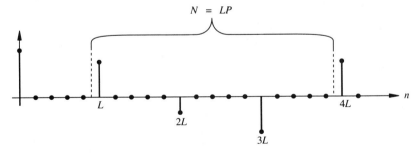

Figure 5-25   *Illustrating samples spanned by interpolating filter, L = 4 and length of filter= N.*

5.2.3 MULTISTAGE DECIMATION AND INTERPOLATION

We have seen that filtering accounts for most of the expense in decimation and interpolation. Bellanger et al. (1974) and Shively (1975) observed that for large M and N, the computational cost could be reduced significantly if the decimation were done in stages rather than all at once. Suppose we wish to decimate by a factor of 64. Rather than decimate immediately by $M = 64$, we will decimate by a factor of $M_1 = 8$, then by a factor of $M_2 = 4$, and finally by a factor of $M_3 = 2$. This necessitates three filtering operations, and it is not immediately clear what that buys us. But from Fig. 5-26, we can see that the filters for the first two stages can have a lot of slop; that is, they can have very wide transition bands. We know (Sec. 5.1.2.2) that the length of a FIR filter is inversely proportional to the width of the transition band. In the first filtering stage, the transition band has a width of nearly $f_s/8$, and hence a relatively short filter can be used to remove the images. In subsequent stages, the transition band is narrower and the filter longer, but this is partially offset by the fact that the sampling rate has been reduced.

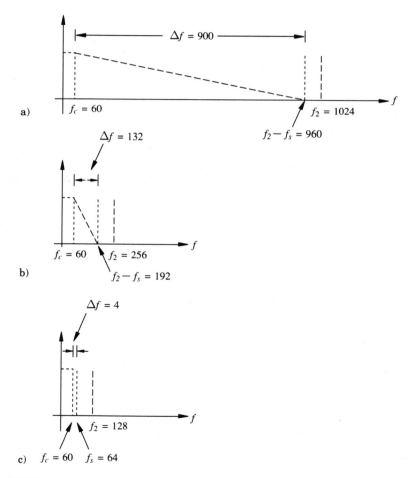

Figure 5-26   *Transition-band requirements in multistage decimation filters: (a) first stage; (b) second stage; (c) final stage.*

Let us examine these alternatives in detail. From Eq. (5-18), we have an estimate for the filter length, which we can abbreviate here as follows:

$$N = Cf_s/\Delta f + 1 \qquad\qquad (5\text{-}33)$$

where C is a constant depending on the pass-band and stop-band ripple tolerances and Δf is the transition bandwidth. The number of multiplications is given by $R = \frac{1}{2}Nf_s/M$, as shown above.

Suppose the initial sampling frequency is $f_s = 8,192 Hz$. Let the corner frequency be 60 Hz. Our final sampling rate will be 128 Hz, so a single-stage decimation must be preceded by a filter with a transition band going from 60 to 64 Hz; this is a transition bandwidth Δf of 4 Hz. Suppose $C = 3$, a realistic figure; then we have

$$N = (3)(8,192)/4 + 1 = 6,145$$

Since the output sampling rate is 128 Hz, we need perform these multiplications only for every eighth output point; hence

$$R = (0.5)(6,145)(128) = 393.28 \times 10^3$$

Hence we require approximately 3.9×10^5 multiplications per second. This is our basis for comparison.

In our three-stage decimator, the first stage will divide the sampling rate by $M_1 = 8$. The input rate is $f_1 = 8,192$ Hz; the output rate is $f_2 = 1,024$ Hz. To determine the transition bandwidth, reflect that the final filter will have a corner frequency of $f_c = 60$ Hz and a stopband beginning at $f_s = 64$ Hz. Frequency components in the range from $f_2 - f_s$ to f_2 will be aliased into the range from 0 to f_s; hence the low-pass filter for this stage must remove these. The transition band goes from f_c to $f_2 - f_s$. Thus

$$\Delta f = f_2 - f_s - f_c$$

In our first stage, this is 900 Hz. Then the filter must have a length of

$$N_1 = 3f_1/\Delta f = 3 \cdot (8,192)/900 + 1 = 28.31$$

Since we cannot have a fractional number of filter points, this will round up to 29. Then the required multiplications per second are

$$R_1 = \frac{1}{2}Nf_2 = (0.5)(29)(1,024) = 14.848 \times 10^3$$

The second stage will divide the sampling rate by 4. Now $f_1 = 1{,}024$ Hz and $f_2 = 256$ Hz. The transition bandwidth is

$$\Delta f = 256 - 64 - 60 = 132 Hz$$

Then
$$N_2 = 3(1{,}024/132) + 1 = 24.27$$

which rounds up to 25; this requires

$$R_2 = (0.5)(25)(256) = 3.2 \times 10^3$$

multiplications per second.

The third stage halves the rate; this brings it to 128 Hz. Now the transition bandwidth has its final value of $\Delta f_3 = 4$ Hz. We have

$$N_3 = 3(256/4) + 1 = 193$$

This requires

$$R_3 = (0.5)(193)(128) = 12.35 \times 10^3$$

multiplications per second. The total number of multiplications over all three stages is approximately 30,400 per second; this is less than a tenth the number required for the single-stage decimator. The total number of filter taps is 247 as compared with the single-stage requirement of 6,145.

The advantage of multistage decimation is greatest when the overall decimation factor M is large. There is unfortunately no easy way to decide how many stages to use or how to select the optimum decimation factors for the individual stages. We will follow Shively's analysis (1975) for the general case. Let the corner frequency be f_c, let the sampling rate at the input to the ith stage be f_{s_i}, and let the decimation factor for that stage be M_i. Then the filter length must be

$$N_i = C M_i / [1 - 2 M_i (f_c / f_{s_i})] + 1$$

where C is the constant used above.

In order to reduce this to manageable form, let

$$r = f_c / F_{s_1}$$

and note that for $i > 1$,

$$F_{s_i} = f_s / (M_1 M_2 \cdots M_{i-1})$$

Then

$$N_i = C M_i / (1 - r M_1 M_2 \cdots M_i)$$

We assume that the filter is linear-phase, and hence symmetric; then the number of multiplications per output point can be reduced to approximately $N_i/2$. Therefore

$$R = \frac{C}{2} \left[\frac{1}{1 - 2rM_1} + \frac{1}{M_1(1 - 2rM_1M_2)} + \frac{1}{M_1M_2(1 - 2rM_1M_2M_3)} + \cdots \right]$$
$$+ \frac{1}{2} \left[\frac{1}{M_1} + \frac{1}{M_1M_2} + \cdots \right] \tag{5-34}$$

Selecting M_1, M_2, etc., and the number of decimation stages, to minimize R is clearly not a trivial problem; Shively suggests optimization procedures which will yield approximate solutions. In any case, it is not necessary to carry these analyses to extremes, since in general they will yield non-integer filter lengths, which have to be increased to the next integer value, and usually to the next odd integer value.

Curves presented in Crochiere and Rabiner (1975) show that the greatest reduction is obtained by going from a single stage to two stages. It generally pays to put the largest decimation factor into the first stage. It should be noted, however, that multistage decimation by successive halving of the sampling rate permits the use of half-band filters; as we will see below, these permit another reduction by a factor of nearly 2 in the computational cost.

5.2.4 POLYPHASE FILTERS

Let us return to the decimation filter of Fig. 5-22(c). We can put this filter through a series of transformations which will lead us to a new form which is of considerable interest and utility. We will illustrate this with a specific example: Suppose we are decimating by a factor of $M = 3$, and suppose the filter has nine taps, as shown in Fig. 5-27(a). Then next to each filter tap, we can write its input sequence. The values shown reflect the fact that each decimator is throwing away two out of every three of its input samples.

But note that there is a considerable amount of redundancy among these input sequences. For example, the input to h_0 is the sequence, $\{x(n), x(n + 3), x(n + 6), ...\}$, while the input to h_3 is $\{x(n - 3), x(n), x(n + 3),...\}$, which is simply a delayed version of the input to h_0. The input to h_6 is similarly a delayed version of the input to h_0. Corresponding conditions apply to h_1, h_4, and h_7, and likewise to h_2, h_5, and h_8.

These observations indicate that we need no more than M delays at the high-rate end of the filter. Suppose we make do with only $M - 1 = 2$ delays, as in Fig. 5-27(b), and consolidate the three groups of filter taps, as shown. We may write this as follows: for a decimation factor of M, the original filter has

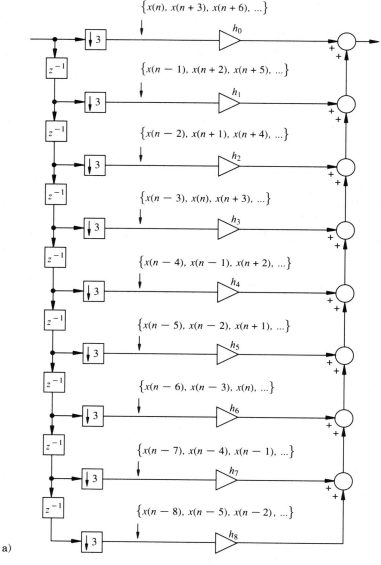

Figure 5-27 *Polyphase decimation filter. (a) filter with embedded decimation, showing redundancy among samples.*

$$y(n) = \sum_{k=0}^{N-1} h_k x(Mn - k) \qquad (5\text{-}35)$$

Because of the overlap between taps M points apart, let $k = Mq + i$, where $i = 0, 1, ..., N - 1$ and $q = 0, 1, ..., \lfloor N/M \rfloor$. Then we may rewrite Eq. (5-35) as

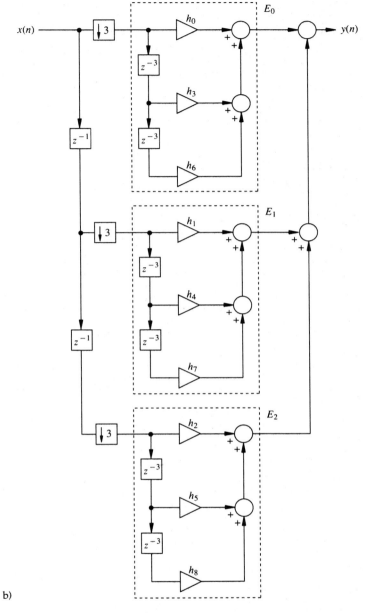

Figure 5-27   *Polyphase decimation filter. (b) filter after grouping and consolidating filter taps.*

$$y(n) = \sum_{i=0}^{M-1} \sum_{q} h_{Mq+i} x[M(n-q) - i] \qquad (5\text{-}36)$$

Each of the filters in Fig. 5-27(b) corresponds to one term in the sum on i in Eq. (5-36). Next we move the decimators back to the left of the M-sample delays, as shown in Fig. 5-27(b). Since the filters are now operating at the lower sampling rate, this means that their delays are one-sample delays, as shown in the figure. The individual filters in Fig. 5-27(b) are called polyphase filters.

Note that with this input structure, consisting of the chain of $M - 1$ delays followed by M decimators, the top filter receives as its input $\{x(n), x(n + 3), x(n + 6), ...\}$; the middle filter receives the sequence $\{x(n - 1), x(n + 2), x(n + 5), ...\}$; and the bottom filter receives $\{x(n - 2), x(n+1), x(n+4), ...\}$. Thus the delay/decimator chain has the effect of distributing consecutive samples among the three filters. Because of this, some people find it helpful to represent this structure as a commutator, as shown in Fig. 5-28. The wiper of the commutator, moving from input to input in synchronism with the sampling times, distributes the incoming samples among the filters in the same way.

We may sum up what we have done so far by writing

$$H(z) = \sum_{k=0}^{M-1} z^{-k} E_k(z^M) \qquad (5\text{-}37)$$

The z^{-k} factors represent the delays preceding the decimators; the E_k are the polyphase filters, and the argument is written as z^M to reflect the fact that a one-sample delay at the low rate is equivalent to an M-

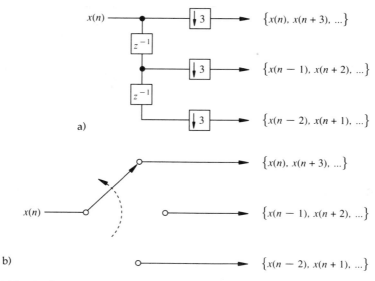

a)

b)

Figure 5-28   (a) Input delay/decimator chain of polyphase filter; (b) equivalent commutator.

sample delay at the high rate. The equivalence given by Eq. (5-37) is the basis of the polyphase transformation.

Each of these filters is a decimated version of the original filter, as can be seen from Fig. 5-29. In Fig. 5-29(a) we have the impulse response of the original filter. In Fig. 5-29(b to d) we have the impulse responses of the new filters, as reconstructed from Fig. 5-27(b). The offset of the first point of the impulse response from the origin represents a fractional-sample time delay at the lower rate. Such a delay, we know, corresponds to a linear phase-shift term. When these structures were first proposed (Bellanger et al., 1976), the authors took these delays as their starting point, which is why they named this structure a *polyphase* filter.

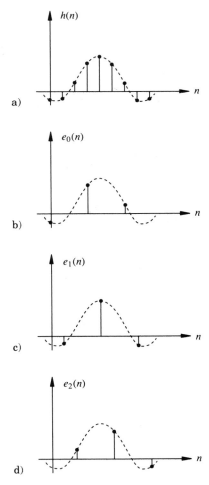

Figure 5-29   (a) Overall impulse response of polyphase filter. (b), (c), and (d) impulse responses of new filters in Fig. 5-28(b).

It is not yet clear what this is going to do for us, although the relatively simple structure of the new filters, which typically have approximately N/M delays each, lends itself well to the fast convolution methods of Chap. 4. Viewed as conventional filters, however, they are not generally linear-phase, as can be seen from Fig. 5-29, and therefore cannot

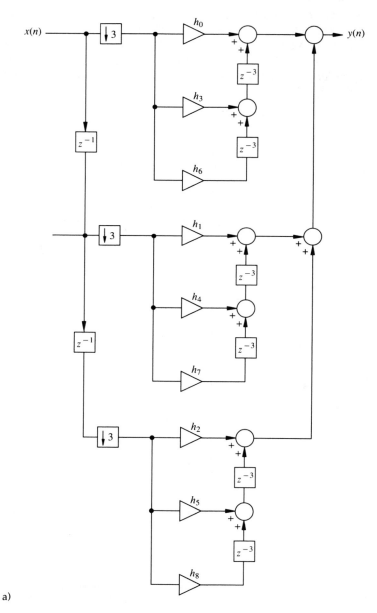

a)

Figure 5-30   *(a) Polyphase decimation filter with new filters in transposed direct form.*

be implemented with the symmetrical linear-phase structure. There are other things we can do, however.

In Fig. 5-30(a), we restructure the new filters in transposed direct form. It should be clear that concurrent outputs from the three filters can be added directly; this permits us to consolidate the delay elements among the three filters, as shown in Fig. 5-30(b). This results in a structure of appealing simplicity—simpler, indeed, than a casual glance at Fig. 5-30(b) would suggest. For a practical decimator, N may be of the order of 50 or more. In the polyphase structure shown here, the number of delays, and hence the number of storage elements required, is approximately $M + N/M$, in most cases a significantly smaller number than N.

The polyphase structure has another interesting feature. The ideal frequency response of our original filter is the usual rectangular low-pass function of Fig. 5-31(a), cutting off at some value ω_c. In the polyphase version, the new filters have been moved over to the low-rate domain and decimated accordingly, as we have seen. That must mean that if we redraw the ideal frequency response of any of these new filters, scaled to the new sampling rate, we will get an all-pass filter, since the gain

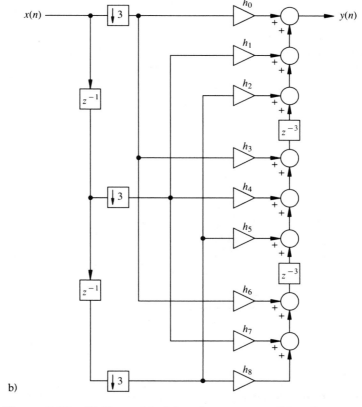

Figure 5-30   *(b) filter with delay elements consolidated.*

is now unity over the entire range from 0 to π and the discontinuity, which gave us so much trouble, has disappeared. All-pass filters are well understood, and relatively easy to implement; Bellanger et al. give design methods for these all-pass filters.

We can carry out a similar set of transformations, starting with the interpolation filter of Fig. 5-23(c) (see Prob. 5.3). The resultant polyphase filter is shown in Fig. 5-32. In all cases, the characteristic polyphase structure is the delay/decimator chain at the input of the decimation filter and the interpolator/delay chain at the output of the interpolation filter.

The polyphase structure has other uses beside sampling-rate changes. One important application is in the realization of digital filter banks. We shall have more to say about this later; at the moment, we will summarize the approach proposed by Bellanger et al. A uniform filter bank consists of a set of M band-pass filters of uniform width. The frequency response of the mth filter is a frequency-shifted version of a baseband prototype, where the shift in frequency is $m f_s/M$. We can represent this shift by replacing z with $z e^{j2\pi m/M}$:

$$H_m(z) = H_0(z e^{-j2\pi m/M})$$

where H_0 is the transfer function of the baseband prototype. If we apply the polyphase transformation of Eq. (5-37) to H_0, we obtain

$$H_m(z) = \sum_{k=0}^{M-1} z^{-k} \exp(-j2\pi mk/M) P_k(z^M)$$

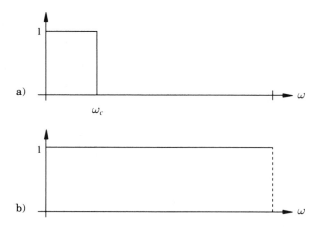

Figure 5-31   (a) Ideal frequency response of original filter: (b) frequency response of new filters, scaled to lower sampling rate.

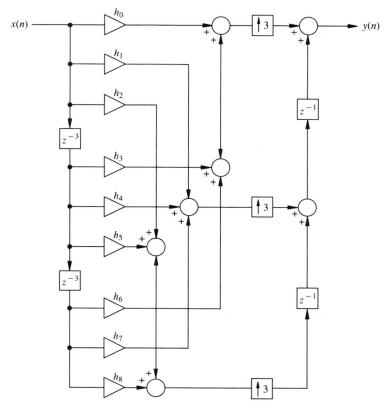

Figure 5-32 *Polyphase interpolating filter.*

where $\{P_k\}$ are the new filters. Notice that the form of P_k is independent of m. Next, suppose we let $W = \exp(-j2\pi/M)$; then we may write an expression for the entire filter bank as follows:

$$
\begin{bmatrix}
H_0(z) \\
H_1(z) \\
\vdots \\
H_{M-1}(z)
\end{bmatrix}
=
\begin{bmatrix}
W^0 & W^0 & \cdots & W^0 \\
W^0 & W^1 & \cdots & W^{M-1} \\
\vdots & \vdots & & \vdots \\
W^0 & W^{M-1} & \cdots & W^{(M-1)^2}
\end{bmatrix}
\begin{bmatrix}
P_0(z^M) \\
z^{-1}P_1(z^M) \\
\vdots \\
z^{-(M-1)}P_{M-1}(z^M)
\end{bmatrix}
$$

The matrix of powers of W will be recognized as a DFT matrix, and the inputs to the DFT are the several outputs of the polyphase filter taps. Hence Bellanger et al. note that if we find the polyphase form of the baseband filter, the only additional computations necessary to produce the entire filter bank are those necessary for the DFT. The resulting structure is shown in Fig. 5-33.

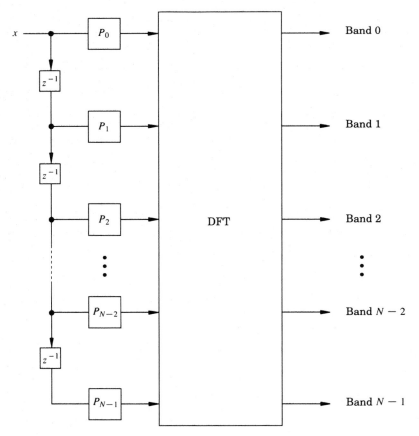

Figure 5-33 *Filter bank realized by cascading polyphase filter with DFT.*

5.2.5 HALF-BAND FILTERS

A half-band filter is a FIR filter whose frequency response has the following characteristics:

1. $H(e^{-j\omega}) = H(e^{j\omega})$

2. $H(e^{j\omega}) + H(e^{j(\pi-\omega)}) = H(1).$

That is, H is an even function of ω, and its response has odd symmetry about the point $(\pi/2, H(1)/2)$.† These characteristics are shown in Fig. 5-34(a). To see what the impulse response of such a filter must be like, it is simpler to consider a modified version of the filter $H'(e^{j\omega})$, defined by

---

†Note $H(1)$ is $H(e^{j0})$, the d-c gain.

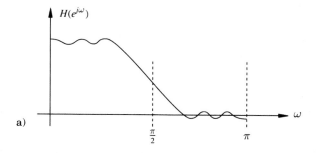

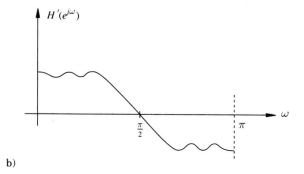

Figure 5-34   *(a) Half-band filter frequency response; (b) frequency response with $\frac{1}{2}H(0)$ subtracted, showing odd symmetry about $\pi/2$.*

$$H'(e^{jw}) = H(e^{jw}) - \tfrac{1}{2}H(1)$$

This subtraction makes $H(e^{j\pi/2}) = 0$, as in Fig. 5-34(b). We may now use well-known properties of the DFT to obtain the modified impulse response. We have

$$H'(e^{-j\omega}) = H'(e^{j\omega})$$

$$H'(e^{j(\omega-\pi)}) = -H'(e^{j\omega})$$

From the shifting property of the DFT, the latter property gives us

$$e^{-jn\pi}h'(n) = -h'(n)$$

Since $e^{-j\pi} = -1$, this is

$$h'(n)(-1)^n = -h'(n)$$

This can be satisfied only if $h'(n) = 0$ for all even n. If we return to the original filter of Fig. 5-34(a), we have

$$H(e^{j\omega}) = H'(e^{j\omega}) + \tfrac{1}{2}H(1)$$

hence $$h(n) = h'(n) + \tfrac{1}{2}H(1)\delta(n)$$

$$= h'(n) + H'(1)\delta(n)$$

Since $$H'(1) = \sum_{n=0}^{N-1} h'(n)$$

the impulse response $h(n)$ has the following characteristics:

1. $h(-n) = h(n)$.

2. $h(n) = 0$ for n even, $n \neq 0$.

3. $h(0) = \sum_{n \neq 0} h(n)$.

The principal advantage of the half-band filter is that the zeroes in its impulse response reduce the number of multiplies required by almost half. The symmetry about $\pi/2$ means that the ripple in the passband is equal to that in the stopband; this is not normally a disadvantage. Indeed, we will use this equality to good advantage in discussing perfect-reconstruction filter banks below.

5.2.6 MIRROR FILTERS

Let $h_1(n)$ be a FIR filter with real coefficients. We define the corresponding *mirror filter* as follows:

$$h_2(n) = (-1)^n h_1(n) \tag{5-38}$$

That is, the mirror filter is obtained from h_1 by changing the signs of the odd-numbered samples in its impulse response. The reason for the name can be seen from the following consideration. Let $H_1(z)$ and $H_2(z)$ be the corresponding Z transforms. Then

$$H_2(z) = \sum_{n=-\infty}^{\infty} z^{-n} h_2(n)$$

$$= \sum_{n=-\infty}^{\infty} z^{-n}(-1)^n h_1(n)$$

$$= H_1(-z) \tag{5-39}$$

Thus if $H_1(z)$ looks like Fig. 5-35(a), for example, then $H_2(z)$ must look like Fig. 5-35(b), the mirror image of Fig. 5-35(a).

The Fourier transform of $h_1(n)$ and $h_2(n)$ can be found by evaluating $H_1(z)$ and $H_2(z)$ on the unit circle, as usual; from this it is clear that the magnitude of the frequency response of h_2 is likewise the mirror image

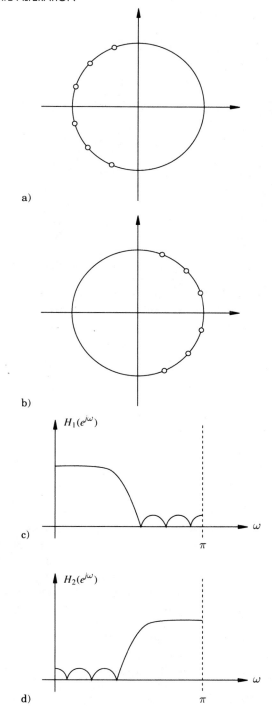

Figure 5-35 *(a) and (b) Zeros of FIR filter and corresponding mirror filter; (c) and (d) frequency responses of (a) and (b).*

of that of h_1, as shown in Fig. 5-35(c) and (d). Since $h_2(n) = e^{j\pi n}h_1(n)$, we can invoke the shifting property of the Fourier transform and write

$$H_2(e^{j\omega}) = H_1(e^{j(\omega+\pi)}) \tag{5-40}$$

A frequency of π corresponds to one half the sampling rate. The frequency responses of filters H_1 and H_2 are reflected about $\pi/2$, which is one-quarter of the sampling rate; hence these filters are commonly called quadrature mirror filters (QMFs).

QMFs are of particular interest in subdividing the frequency range of a signal for economical coding (Esteban and Galand, 1977). Many signals do not have a uniform frequency spectrum. If the spectrum amplitude tends to be small in a particular frequency band, and if that band can be encoded separately, then it can be encoded with fewer bits. Furthermore, since the subbands resulting from the division are narrower than the original bandwidth, it should be possible to sample each one at a lower rate. In that case, the increase in the number of bands will be offset by a reduction in the sampling rate per band. In particular if there are M subbands, each down-sampled by a factor M, the system is said to be *maximally decimated*.

In such applications, it is highly desirable that the bandwidth of the signal be subdivided in such a way that the pieces can be reassembled later with no error other than that introduced by the coding. This is termed *perfect reconstruction*; formally, $Y(z)$ is a perfect reconstruction of $X(z)$ if

$$Y(z) = cz^{-k}X(z)$$

that is, if the recovered signal equals the original signal to within a constant gain factor and possibly a delay. The obvious way to split a frequency range into subbands is to use extremely high-quality, sharp-cutoff filters, so that there will be no possibility of aliasing when the subbands are down-sampled. The inevitable imperfection of even the highest-quality filters entangles us in two conflicting requirements, however: First, the transition bands of the filters *must not* overlap, so there will be no aliasing. Second, the transition bands *have* to overlap; otherwise there will be gaps in the spectral coverage of the signal. The QMF technique of Esteban and Galand provides the first of a number of ingenious ways out of this dilemma.

Let us consider a system that divides the frequency range into two equal parts, decimates the two subbands by a factor of 2, and after transmission, recovers these bands by interpolation. The filters at the transmission end are termed *analysis filters* and those at the receiving end *synthesis filters*. Both analysis and synthesis filters are FIR, linear-phase filters. Such a system is shown in Fig. 5-36. (In most of what follows, we will not require that the filters be causal, for notational simplicity; causality can always be obtained by inserting delays as required.) We

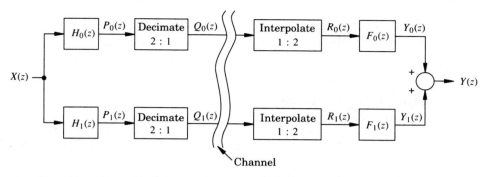

Figure 5-36 *Subband coding and transmission system with two equal subbands.*

wish to determine what constraints are required on the filters in order that there be perfect reconstruction of the signal. We have seen Eq. (5-25) that at the output of a decimator and interpolator placed back to back,

$$R_i(z) = \frac{1}{M} \sum_{k=0}^{M-1} P_i(W_M^k z) \qquad (5\text{-}41)$$

In the present example, $M = 2$ and we have

$$\begin{aligned} R_0(z) &= \tfrac{1}{2}[P_0(z) + P_0(-z)] \\ &= \tfrac{1}{2}[H_0(z)X(z) + H_0(-z)X(-z)] \end{aligned} \qquad (5\text{-}42)$$

and similarly for R_1. Then

$$Y(z) = R_0(z)F_0(z) + R_1(z)F_1(z) \qquad (5\text{-}43)$$

In order to prepare for generalization later on, it is expedient to write this in matrix form. Combining Eqs. (5-42) and (5-43),

$$\begin{aligned} Y(z) &= [R_0(z) \quad R_1(z)] \begin{bmatrix} F_0(z) \\ F_1(z) \end{bmatrix} \\ &= \tfrac{1}{2}[X(z) \quad X(-z)] \begin{bmatrix} H_0(z) & H_1(z) \\ H_0(-z) & H_1(-z) \end{bmatrix} \begin{bmatrix} F_0(z) \\ F_1(z) \end{bmatrix} \end{aligned} \qquad (5\text{-}44)$$

The matrix

$$\begin{bmatrix} H_0(z) & H_1(z) \\ H_0(-z) & H_1(-z) \end{bmatrix}$$

is known as the *alias-component* (AC) matrix. For general M, the AC matrix takes the form

$$[A_{mn}] = [H_n(W_M^{-m} z)]$$

Multiplying in Eq. (5-44), we obtain

$$Y(z) = \tfrac{1}{2}[X(z) \quad X(-z)] \begin{bmatrix} H_0(z)F_0(z) & + H_1(z)F_1(z) \\ H_0(-z)F_0(z) + H_1(-z)F_1(z) \end{bmatrix}$$

The first row gives us the desired response and the second row the aliased components. We can eliminate the aliased components if

$$H_0(-z)F_0(z) + H_1(-z)F_1(z) = 0 \tag{5-45}$$

If H_0 and H_1 are QMFs, however, we have $H_1(z) = H_0(-z)$. In that case, we can satisfy Eq. (5-45) by making

$$F_0(z) = H_0(z) \tag{5-46}$$

$$F_1(z) = -H_0(-z)$$

$$\text{Then} \quad Y(z) = \tfrac{1}{2}X(z)[H_0(z)H_0(z) - H_0(-z)H_0(-z)]$$

$$+ \tfrac{1}{2}X(-z)[H_0(-z)H_0(z) - H_0(z)H_0(-z)]$$

$$= \tfrac{1}{2}X(z)[H_0^2(z) - H_0^2(-z)] \tag{5-47}$$

Hence we will obtain perfect reconstruction provided $H_0^2(z) - H_0^2(-z) = 2$. Note that this technique does not eliminate aliasing by sharp-cutoff filtering, but by *cancellation*; note also that the aliased components are present in the subbands and are not cancelled until the recombination step at the output.

To see what constraints this approach puts on H_0, we will rewrite Eq. (5-47) using Fourier transforms:

$$Y(e^{j\omega}) = \tfrac{1}{2}X(e^{j\omega})[H_0^2(e^{j\omega}) - H_0^2(e^{j(\omega+\pi)})]$$

We must now separate the magnitude and phase of $H_0(e^{j\omega})$. If $h_0(n)$ is symmetric, causal, and of length N, then

$$H_0(e^{j\omega}) = H(\omega)e^{-j\omega(N-1)/2}$$

where $H(\omega)$ is real and the exponential accounts for the shift needed to make the filter realizable. Then

$$Y(e^{j\omega}) = \tfrac{1}{2}X(e^{j\omega})[H^2(\omega)e^{-j\omega(N-1)} - H^2(\omega + \pi)e^{-j(\omega+\pi)(N-1)}]$$

$$= \tfrac{1}{2}X(e^{j\omega})[H^2(\omega) - e^{-j\pi(N-1)}H^2(\omega + \pi)]e^{-j\omega(N-1)}$$

Since $e^{j\pi(N-1)} = (-1)^{N-1}$, the magnitude of $Y(e^{j\omega})$ is

$$|Y(e^{j\omega})| = \tfrac{1}{2}|X(e^{j\omega})|[H^2(\omega) - (-1)^{N-1}H^2(\omega + \pi)]$$

It follows that for perfect reconstruction, $N - 1$ must be odd, so that the factor of (-1) will disappear; otherwise $|Y(\pi/2)|$ will always be zero, no matter what $H(\omega)$ is. Hence our requirements are

$$N \text{ is even} \qquad\qquad (5\text{-}48)$$
$$\text{and } H^2(\omega) + H^2(\omega + \pi) = 2$$

This second requirement cannot be met exactly by linear-phase band-splitting filters; however, good approximations are possible; these are discussed in Johnson and Crochiere (1979), Cheung and Winslow (1980), Galand and Esteban (1980), Esteban and Galand (1981), and Foo and Turner (1982).

Notice that we are dealing with a completely different kind of constraint from the kind usually associated with band splitting. There is, in particular, no inherent requirement that the cutoff be sharp. Indeed, $H(\omega) = \sqrt{2}\cos(\omega/2)$ satisfies Eq. (5-48) perfectly with strikingly poor rolloff characteristics. In practice, a reasonably good approximation to an ideal low-pass filter is desirable, since otherwise the aliased components present in the subbands interfere with efficient coding. The virtue of the QMF approach is that these alternative constraints are frequently easier to satisfy with good signal reconstruction at the other end. Johnson and Crochiere (1979) give a particularly good example of this; they approximate $H(\omega)$ with a 32-point Hamming-window filter and obtain a reconstructed signal with a maximum deviation from unity gain of only 0.25 dB.

Clearly a single FIR filter can be used to realize both H_0 and H_1, since their coefficients are identical except for signs. This fact leads to the configuration shown in Fig. 5-37. Similar economies can be realized in implementing F_0 and F_1. If more than two subbands are required, one popular implementation is the tree structure shown in Fig. 5-38, obtained by applying the basic two-subband technique recursively. Here each level of the tree splits the preceding subband into two equal parts; this permits any desired decomposition into 2^p subbands. Designs of IIR mirror filters are discussed in Barnwell (1982) and Millar (1985).

5.2.7 PERFECT-RECONSTRUCTION SUBBAND FILTER BANKS

If we relax the requirement that the filters be linear-phase, perfect reconstruction is possible. This was first shown by Smith and Barnwell (1984); a general solution, which permits multichannel filter banks without resorting to the tree structure of Fig. 5-38, was presented by Vaidyanathan (1987a), on whose paper we will draw for the following material.

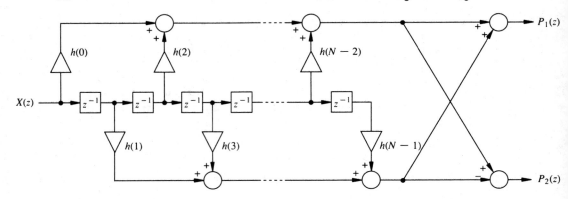

Figure 5-37   *Use of single FIR filter to implement H_1 and H_2.*

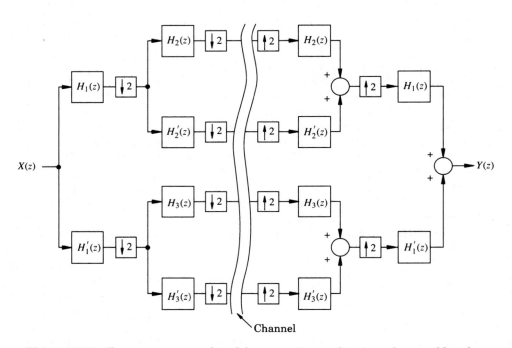

Figure 5-38   *Tree structure produced by recursive application of two-subband technique.*

Let us take the two-channel quadrature mirror filter bank as our starting point. Perfect reconstruction requires that $Y(z) = X(z)$. This means that from Eq. (5-44),

$$\begin{bmatrix} H_0(z) & H_1(z) \\ H_0(-z) & H_1(-z) \end{bmatrix} \begin{bmatrix} F_0(z) \\ F_1(z) \end{bmatrix} = \begin{bmatrix} 1 \\ 0 \end{bmatrix} \qquad (5\text{-}49)$$

Smith and Barnwell (1984) obtained a perfect-reconstruction filter bank from this same general starting place by requiring that

$$H_0(z) \text{ be FIR but not linear-phase}$$

$$H_1(z) = zH_0(-z^{-1}) \qquad (5\text{-}50a)$$

$$F_0(z) = H_0(z^{-1}) \qquad (5\text{-}50b)$$

$$F_1(z) = z^{-1}H_0(-z) \qquad (5\text{-}50c)$$

Then Eq. (5-49) becomes

$$\begin{bmatrix} H_0(z) & zH_0(-z^{-1}) \\ H_0(z-1) & zH_0(z^{-1}) \end{bmatrix} \begin{bmatrix} H_0(-z) \\ z^{-1}H_0(-z) \end{bmatrix}$$

$$= \begin{bmatrix} H_0(z)H_0(z^{-1}) + H_0(-z)H_0(-z^{-1}) \\ H_0(-z)H_0(z^{-1}) - H_0(-z)H_0(z^{-1}) \end{bmatrix} = \begin{bmatrix} P_0(z) + P_1(z) \\ 0 \end{bmatrix}$$

where $P_0(z)$ and $P_1(z)$ are the "product filters":

$$P_0(z) = H_0(z)H_0(z^{-1})$$

$$\text{and } P_1(z) = H_1(z)H_1(z^{-1}) \qquad (5\text{-}51)$$

$$= H_0(-z^{-1})H_0(-z)$$

$$= P_0(-z)$$

If $P_0(z)$ and $P_1(z)$ can be found such that they are factorable as in Eq. (5-51) and such that

$$P_0(e^{j\omega}) + P_1(e^{j\omega}) = 1 \qquad (5\text{-}52)$$

then we have solved the perfect reconstruction problem.

Smith and Barnwell obtained such a pair by starting with conventional symmetric odd-order half-band filters as in Eq. (5-48). To make them factorable, they obtained a modified pair of filters,

$$p_0'(n) = ap_0(n) + b\delta(n)$$

$$p_1'(n) = ap_1(n) + b\delta(n)$$

Since $F\{\delta(n)\} = 1$, this has the effect of raising $P_0(e^{j\omega})$ and $P_1(e^{j\omega})$ by b; a suitably chosen b will make the minima of $P_0(e^{j\omega})$ and $P_1(e^{j\omega})$ equal to zero. By suitably adjusting a and b together, $P_0(e^{j\omega}) + P_1(e^{j\omega})$ can also be made to sum to 1: This will satisfy Eq. (5-52). Then, by virtue of the symmetry of $p_0'(n)$ and $p_1'(n)$, all the zeroes of $P_0'(z)$ and $P_1'(z)$ will occur in reciprocal pairs, except those on the unit circle, which will be of even multiplicity. In that case, $H_0(z)$ and $H_0(z^{-1})$ can be obtained by selecting reciprocal factors of $P_0'(z)$, and similarly for $H_1(z)$ and $H_1(z^{-1})$.

5.2.7.1 M-Channel Filter Banks We would now like to find a general solution for M-channel filter banks in which the decimation factor is also M, that is, for maximally decimated filter banks. To see how Vaidyanathan found such a solution, we will start with a polyphase implementation of our two-channel filter bank. This takes the form shown in Fig. 5-39(a), with E_0 and E_1 the polyphase analysis filters and D_0 and D_1 the polyphase synthesis filters.

The corresponding subband filter bank can be obtained by moving the filters E_0, E_1, D_0, and D_1 outside the decimators and interpolators so that the latter are back to back, as shown in Fig. 5-39(b). Initially, our interest will be in the form of Fig. 5-39(a), but it is well to establish the correspondence between these two forms right away, since the subband form is our ultimate goal.

In moving the filters, we must correct for the difference in sampling rate, since (for example) every one-sample delay to the right of the decimator becomes a two-sample delay when shifted to the left. Indeed, we may say in general that if we move any linear system $H(z)$ from the right to the left of an M-fold decimator, the equivalent system becomes $H(z^M)$, as shown in Fig. 5-40(a). (We saw an instance of this in deriving

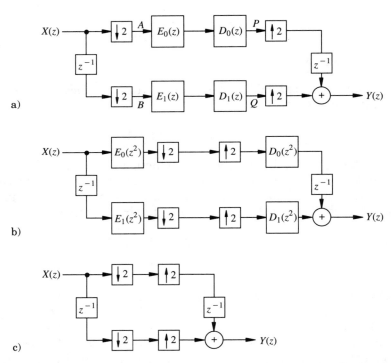

Figure 5-39 (a) Polyphase representation of two-channel maximally decimated filter bank; (b) filter bank obtained from (a); (c) trivial system resulting from cancellation of matrices in (a).

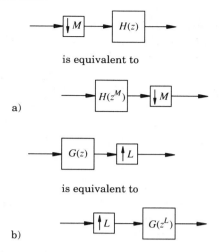

a)

is equivalent to

b)

Figure 5-40 *Illustrating transformations in moving systems past decimators and interpolators.*

our first polyphase structure. In Fig. 5-27(c), we moved the *decimators* to the left of the second string of delays and, to compensate, had to change the delays from z^{-3} to z^{-1}.) Similarly if we move any linear system $G(z)$ past an L-fold interpolator to its right, the equivalent system likewise becomes $G(z^L)$, as shown in Fig. 5-40(b). Moving these filters outside, as described, gives us the structure of Fig. 5-39(b).

The primary virtue of the polyphase structure in Fig. 5-39(a) is that it permits us to generalize to a matrix form. From the figure we have

$$\begin{bmatrix} P(z) \\ Q(z) \end{bmatrix} = \begin{bmatrix} D_0(z) & 0 \\ 0 & D_1(z) \end{bmatrix} \begin{bmatrix} E_0(z) & 0 \\ 0 & E_1(z) \end{bmatrix} \begin{bmatrix} A(z) \\ B(z) \end{bmatrix}$$

For perfect reconstruction, a sufficient condition is that the product of the two matrices be **I** for $z = e^{j\omega}$, since in that case as far as the overall input-output relation is concerned, the two matrices cancel and the system is equivalent to the trivial structure shown in Fig. 5-39(c), which certainly provides perfect reconstruction. In between the two matrices, the sequence in each channel is filtered by its own analysis filter. There is no reason why these matrices cannot have off-diagonal elements, however. The off-diagonal elements correspond to interchannel connections and offer us a solution of greater generality and flexibility. Hence we contemplate a polyphase-like system such as that shown in Fig. 5-41(a). In that case,

$$\begin{bmatrix} P(z) \\ Q(z) \end{bmatrix} = \begin{bmatrix} D_{00}(z) & D_{01}(z) \\ D_{10}(z) & D_{11}(z) \end{bmatrix} \begin{bmatrix} E_{00}(z) & E_{01}(z) \\ E_{10}(z) & E_{11}(z) \end{bmatrix} \begin{bmatrix} A(z) \\ B(z) \end{bmatrix}$$

$$= \mathbf{D}(z)\mathbf{E}(z) \begin{bmatrix} A(z) \\ B(z) \end{bmatrix}$$

And again we will have perfect reconstruction if $\mathbf{D}(z)\mathbf{E}(z) = \mathbf{I}$. If the matrix $\mathbf{E}(z)$ is unitary on the unit circle, then the corresponding $\mathbf{D}(z)$ matrix is easy to find, because for $\mathbf{E}(z)$ unitary on the unit circle,

$$\mathbf{E}^T(z^{-1})\mathbf{E}(z)\big|_{z=e^{j\omega}} = c\mathbf{I} \qquad (5\text{-}53)$$

By the principle of analytic continuation, if Eq. (5-53) holds on the unit circle, then it holds for all z. (Such a matrix is known variously as a paraunitary or lossless matrix.) The corresponding subband structure is shown in Fig. 5-41(b).

We may immediately generalize to M channels with the structure shown in Fig. 5-42. We find a lossless matrix $\mathbf{E}(z^M)$ for the analysis filters; we immediately know that the corresponding synthesis matrix is given by $\mathbf{E}^T(z^{-M})$. This structure can be shown to provide perfect reconstruction by a series of steps parallelling those in Fig. 5-39.

This structure has several appealing properties. First, it solves the perfect reconstruction problem; second, the order of the synthesis filters is the same as that of the analysis filters; and finally, it is not difficult to show a structure which automatically guarantees that $\mathbf{E}(z)$ will be lossless. This structure is shown in Fig. 5-43; it takes the form of a cascade of alternating constant matrices $\mathbf{K}_i$ and delay matrices $\mathbf{\Delta}_i$. The constant matrices are unitary or, for real coefficients, orthogonal matrices; these are clearly lossless. The delay matrices are diagonal matrices consisting only of powers of z^{-1}; being diagonal, they are trivially lossless. Since the product of lossless matrices is clearly lossless, it follows that the entire cascade of Fig. 5-43 will be lossless. The corresponding analysis cascade is constructed by reversing the sequence of Fig. 5-43, transposing the

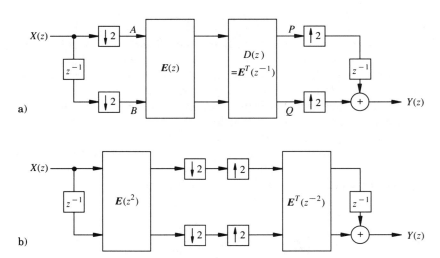

Figure 5-41 (a) Polyphase system with general matrices substituted for individual channel filters of Fig. 5-40; (b) filter bank obtained from (a).

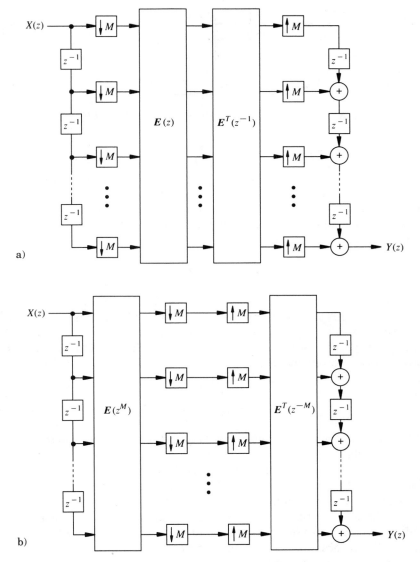

Figure 5-42 *M-channel maximally-decimated filter bank: (a) polyphase form;*
(b) filter-bank form.

constant matrices, and substituting z for z^{-1} in the delay matrices. For
example, for $M = 2$ we have at the analysis end

$$\mathbf{E}(z) = \begin{bmatrix} k_{00} & k_{01} \\ -k_{01} & k_{00} \end{bmatrix} \begin{bmatrix} 1 & 0 \\ 0 & z^{-2} \end{bmatrix} \begin{bmatrix} k_{10} & k_{11} \\ -k_{11} & k_{10} \end{bmatrix} \dots \begin{bmatrix} k_{p0} & k_{p1} \\ -k_{p1} & k_{p0} \end{bmatrix}$$

where $k_{i0}^2 + k_{i1}^2 = 1$; then at the synthesis end we have

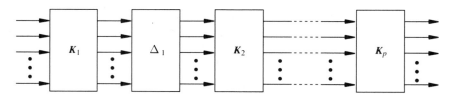

Figure 5-43   *Cascaded-lattice implementation of* $\mathbf{E}(z)$.

$$\mathbf{E}^T(-z) = \begin{bmatrix} k_{p0} & -k_{p1} \\ k_{p1} & k_{p0} \end{bmatrix} \begin{bmatrix} 1 & 0 \\ 0 & z^2 \end{bmatrix} \cdots \begin{bmatrix} 1 & 0 \\ 0 & z^2 \end{bmatrix} \begin{bmatrix} k_{00} & -k_{01} \\ k_{01} & k_{00} \end{bmatrix}$$

(We have z^{-2} and z^2 because these structures are outside of the decimators and interpolators.) In a causal system, the delay matrices take the form,

$$\begin{bmatrix} z^{-2} & 0 \\ 0 & 1 \end{bmatrix}$$

Notice that with this structure, losslessness and hence the perfect-reconstruction property is in a sense *built in*, inherent in the structure of the filter. It is worth observing also that for a two-channel bank, the constant matrices can be rewritten as

$$k_{i1} \begin{bmatrix} 1 & k_{i2}/k_{i1} \\ -k_{i2}/k_{i1} & 1 \end{bmatrix}$$

in which case the losslessness property is immune to coefficient quantization effects.

The design of such a filter bank, as described by Vaidyanathan, consists of selecting a suitable form for the Δ matrices and computing the elements of the constant matrices by minimizing the gain in the filter stopbands. (Since perfect reconstruction applies in the passbands and the passbands are, of course, nonoverlapping, making the stopband gain of the filters small is sufficient to make the passband gain close to unity.) Computing the orthogonal matrix coefficients is best done with an optimization program; Vaidyanathan used the ZXMWD routine (*IMSL*, 1980).

5.3 FILTERS WITHOUT MULTIPLIERS

Digital filters are usually designed to provide satisfactory or optimal performance under the assumption of infinite precision for the coefficient values. These filters are then customarily implemented in hardware by rounding the coefficients to the allowable number of bits. Munson (1981) has shown that FIR filters using rounded coefficient values are optimal for both a minimum time-domain norm, max $|y_n - \tilde{y}_n|$, and a minimum mean-squared error norm $E\{(y_n - \tilde{y}_n)^2\}$, where y_n and $\tilde{y}_n$ are the outputs of the infinite-precision and the rounded-coefficient filters, respectively. This, however, is not the case for weighted frequency-domain criteria of the form

$$\max M(\omega)|H(\omega) - \hat{H}(\omega)|$$

or
$$\int_0^\pi W(\omega)|H(\omega) - \hat{H}(\omega)|^2 \, d\omega$$

where $M(\omega)$ and $W(\omega)$ are suitable weighting functions. In the first instance, integer linear programming (Lim et al., 1982 and Kodek, 1980) can be used to optimize the filter performance with the finite-bit constraint. A detailed account of these integer programming methods is available in the indicated references.

Our intent in this section, however, is to describe certain FIR filters whose coefficients are quantized to the extreme, that is, to the ternary values $\{-1, 0, 1\}$. Such filters, which can be constructed without multipliers, offer obvious advantages in speed of operation and simple hardware realizations. All problems accruing from finite-precision multiplication and coefficient quantization are circumvented. Thus there is no roundoff noise, and there are no hidden limit cycles. The only noise appearing at the filter output is due to that in the filter input—for example, from a/d signal quantization.

Two types of filters without multipliers are discussed here. The first type, the orthogonal binomial filters (Haddad, 1971) has inherently quantized ternary-valued coefficients. The second type consists of a tandem connection of nonrecursive structures followed by a recursive integrator. The coefficients in the nonrecursive structure are constrained to be 0 and ± 1 and must be such as to cancel the pole or poles in the recursive part.

5.3.1 FILTERS WITH DISCRETE COEFFICIENTS

Finite word-length effects are always a source of error in digital filters. Conventional filter design techniques assume coefficients drawn from the real number field and approximate these coefficients in the computer with floating-point numbers. The design is reduced to practice by suitably rounding the computed values, and this rounding inevitably causes the performance of the practical filter to be inferior to the computed filter.

In view of this fact, some investigators have explored the possibility of finding optimum FIR digital filters subject to the constraint that the coefficient values be finite-precision fractions or, with suitable normalization, finite-length integers. In principle, the optimum coefficient values can be found by an exhaustive search; in practice, the combinatorics make this impossible, and various standard integer-programming packages have been used to mount an efficient search.

Kodek (1980) used such a package, called MPOS (multipurpose optimization system) to design FIR filters, mostly low-pass. The improvements over filters obtained by rounding floating-point coefficient values are considerable.

For a given tolerance, there is a trade-off between filter length and coefficient word length b, in the sense that a coarser set of coefficients may yield acceptable performance if the length of the filter is increased. Kodek found that if the total number of bits Nb in the filter is taken as a parameter, then there is an optimum word length for any given value of Nb. In a later paper, Kodek and Steiglitz (1980) showed that first, for any finite word length, there is a nonzero lower bound on the approximation error, regardless of how large N is made, and second, such trade-off as there is comes to an abrupt end: for a word length less than some minimum number of bits, there is no filter length that will yield the desired performance.

These integer-programming packages are faster than an exhaustive search, but they are still slow. Lim and Parker (1983) observe that run time increases exponentially with the length of the filter, and the time required for filters of length greater than approximately 40 taps is prohibitive. Hence a lot of effort has gone into finding faster ways of searching the coefficient space. Lim and Parker investigated a modified integer-programming technique in which a branch-and-bound technique was combined with a depth-first search. Goodman and Carey (1977) present eight half-band filters whose coefficients are all integer values.

5.3.2   THE BINOMIAL FILTERS

The binomial-Hermite family of orthogonal sequences were introduced in Chap. 2. The transforms of these sequences, Eqs. (2-81) and (2-82), are repeated as

$$X_0(z) = (1 + z^{-1})^N \tag{5-54}$$

$$X_r(z) = \left(\frac{1 - z^{-1}}{1 + z^{-1}}\right)^r (1 + z^{-1})^N \tag{5-55}$$

$$= (1 - z^{-1})^r (1 + z^{-1})^{N-r} \tag{5-56}$$

Digital filters based on Eqs (5-55) and (5-56) can be realized without multipliers as shown in Fig. 5-44. These filters exhibit low-pass (for $r = 0$), band-pass ($0 < r < N$) and high-pass ($r = N$) behavior with linear-phase and magnitude responses that are almost Gaussian. The frequency response of these filters may be written

$$X_r(e^{j\omega}) = A_r(\omega)e^{j\phi_r(\omega)} \tag{5-57}$$

where the magnitude and phase responses are given by

$$A_r(\omega) = 2^N (\sin \omega/2)^r (\cos \omega/2)^{N-r} \tag{5-58}$$

$$\phi_r(\omega) = (r\pi - N\omega)/2 \tag{5-59}$$

These magnitude responses are shown in Fig. 5-45 for $N = 6$. The maximum response occurs at

$$\omega_m = 2 \sin^{-1} \sqrt{r/N} \tag{5-60}$$

with value $\qquad A_r(\omega_m) = 2^N (r/N)^{r/2} (1 - r/N)^{(N-r)/2} \tag{5-61}$

In Haddad (1971), these responses are shown to be almost Gaussian with half-power bandwidths

$$BW \approx \begin{cases} 2.34/\sqrt{N} & r > 0 \\ 1.66/\sqrt{N} & r = 0 \end{cases} \tag{5-62}$$

It is well known that the binomial sequence $\binom{N}{k}$ approaches a Gaussian for large N (Papoulis, 1984). It is not surprising, then, that the magnitude of the Fourier transform [Eq. (5-58) with $r = 0$] is approximately Gaussian.

By suitable choice of r and N, a band-pass filter with any bandwidth and center frequency can be realized without multipliers. For a sampling frequency of 10 kHz, center frequency 1 kHz, and half-power bandwidth 0.15kHz, Eqs. (5-55) and (5-62) yield $r = 59$ and $N = 617$.

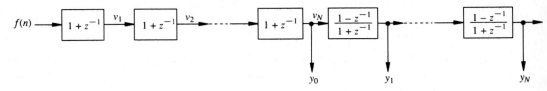

a)

$$v_j(n) = v_{j-1}(n) + v_{j-1}(n - 1)$$

$$y_r(n) = -y_r(n - 1) + y_{r-1}(n) - y_{r-1}(n - 1)$$

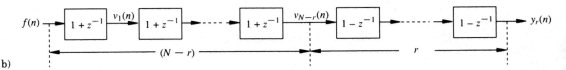

b)

Figure 5-44   *Canonic processors for binomial filter: (a) sequential form for filter bank, data available as serial streams; (b) batch processing, single band-pass filter.*

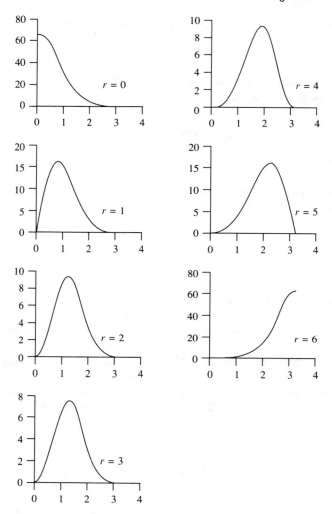

Figure 5-45 *Magnitude frequency responses of the binomial family of filters, N = 6.*

While the value of N in this example appears large, the computations are trivially simple. As shown in Fig. 5-44(a), the sequential canonic form processes the data stream $\{f(0), f(1), \cdots\}$ sequentially. Note that only additions and subtractions are performed and an entire bank of low-pass and band-pass filters can be realized and operated simultaneously. The maximum signal at the low-pass filter output is

$$|y_o(n)|_{\max} = 2^N$$

when $f(n) = 1$ for all n. This bound permits sizing the registers which store the data. Alternatively, we may scale each of the low-pass add blocks by one half and thereby bound the output at each constituent block in the low-pass section by unity.

A batch-processing configuration is shown in Fig. 5-44(b), where $(N - r)$ stages of the add operator $(1 + z^{-1})$ are followed by r stages of the differencing operator $(1 - z^{-1})$. These operations can be performed rapidly and easily by batch processing. To see this, suppose $(L + 1)$ successive data samples are stored in the linear array

$$\mathbf{f}^T = [f(0), f(1), ..., f(L)] \qquad f(-1) = 0 \qquad (5\text{-}63)$$

Let the successive outputs of the first add stage be designated as

$$\mathbf{v}_1 = [v_1(0), v_1(1), ..., v_1(L)] \qquad (5\text{-}64)$$

Then
$$v_1(n) = f(n) + f(n - 1) \qquad (5\text{-}65)$$

implies
$$\mathbf{v}_1 = \mathbf{Sf} \qquad (5\text{-}66)$$

where $\mathbf{S}$ is the transmission matrix for a single add stage:

$$\mathbf{S} = \begin{bmatrix} 1 & 0 & 0 & \cdots & 0 & 0 \\ 1 & 1 & 0 & \cdots & 0 & 0 \\ 0 & 1 & 1 & \cdots & 0 & 0 \\ . & . & . & . & . & \\ 0 & 0 & 0 & \cdots & 1 & 1 \end{bmatrix} \qquad (5\text{-}67)$$

Similarly, a single-stage batch differencing operation can be derived from

$$g(k) = h(k) - h(k - 1) \qquad k = 0, 1, 2, ... \qquad (5\text{-}68)$$
$$g(-1) = 0$$

and expressed by

$$\mathbf{g} = \mathbf{Dh} \qquad (5\text{-}69)$$

where
$$\mathbf{h}^T = [h(0), h(1), ..., h(L)]$$
$$\mathbf{g}^T = [g(0), g(1), ..., g(L)]$$

and
$$\mathbf{D} = \begin{bmatrix} 1 & 0 & 0 & \cdots & 0 & 0 \\ -1 & 1 & 0 & \cdots & 0 & 0 \\ 0 & -1 & 1 & \cdots & 0 & 0 \\ . & . & . & . & . & . \\ 0 & 0 & 0 & \cdots & -1 & 1 \end{bmatrix} \qquad (5\text{-}70)$$

Combining the add and difference operators implicit in Eq. (5-56), we obtain for the band-pass filters

$$\mathbf{y}_r = \mathbf{D}^r \mathbf{S}^{N-r} \mathbf{f} \qquad (5\text{-}71)$$

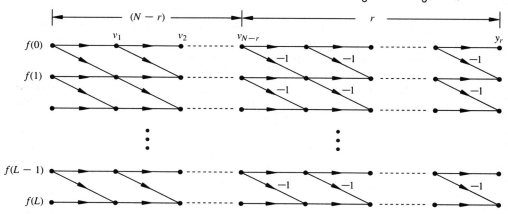

Figure 5-46   *Band-pass filter structure for batch-canonic processor. (Unmarked branches have unity gain.)*

This batch-canonic structure is shown in Fig. 5-46. The output vector $\mathbf{y}_r$ is available in the computation time required to execute the **D** and **S** operations. These array operations resemble the FFT butterfly, except that they are much simpler. All operations are sums or differences performed in place and with constant geometry.

5.3.3   THE DELTA MODULATION CLASS OF FIR FILTERS

The ideas behind this class of multiplier-free filters can be traced to concepts used in delta modulation signal coding. In a pulse-code modulation (PCM) system, the signal is sampled at the Nyquist rate to produce the sequence $\{x_n = X(nT_N)\}$. These samples are then quantized to b bits, encoded, and transmitted, as shown in Fig. 5-47(a).

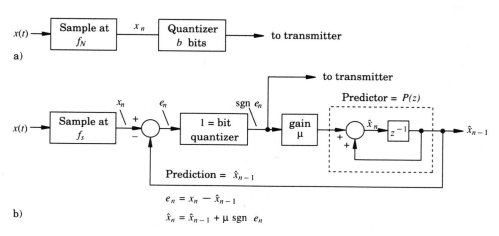

Figure 5-47   *(a) PCM system; (b) delta-modulation system.*

In the delta modulation system of Fig. 5-47(b), the analog signal is sampled at a rate f_s much greater than the Nyquist rate f_N. This produces a sequence $\{x_n\}$ of highly correlated samples. Next the *difference* between x_n and $\hat{x}_{n-1}$, our prediction of it, is formed, quantized to 1 bit, and transmitted. Note that since a single bit is transmitted for each difference and we must indicate negative differences as well as positive ones, there is no way of signalling a slope of zero. The 1-bit signal is then scaled by a gain μ and used as the input to the predictor. The process relies on over-sampling the signal, so that the differences are small and can be encoded in a 1-bit quantizer.

The trade-off is between relatively slow sampling with several bits per sample for the PCM system as compared with high-frequency sampling and only 1 bit per sample for the delta modulator. The delta modulator suffers from two conflicting error sources. First, slope-overload distortion occurs when μ is too small to allow $\hat{x}_n$ to follow a steep rise in $x(t)$. Second, the staircase approximation $\hat{x}_n$ hunts around slowly-varying parts of $x(t)$ because there is no way to represent a difference of zero. This error is known as granularity, and the culprit here is too large a step size μ. Finding a suitable compromise depends on over-sampling at a high enough rate; Abate (1967) gives the rule of thumb

$$\frac{\text{Maximum slew rate of system}}{\text{RMS slew rate of signal}} = \ln \frac{\text{sampling frequency}}{\text{signal bandwidth}}$$

Let us assume that an FIR filter is to be used to approximate some desired analog impulse response $g(t)$, as shown in Fig. 5-48(a). We choose to do this by first truncating $g(t)$ to the length T shown and then sampling $g(t)$ at some approximate Nyquist rate $f_N = 2B$, as in Fig. 5-48(a). A PCM-inspired FIR filter is illustrated in Fig. 5-49. The infinite-precision impulse response obtained by the foregoing (or some similar) procedure is

$$g(n) = \sum_{k=0}^{N-1} g_k \delta_{n-k} \qquad (5\text{-}72)$$

The infinite-precision coefficients (or samples) g_k are rounded off to b bits to give the quantized-coefficient filter as

$$g_Q(n) = \sum_{k=0}^{N-1} Q[g_k]\delta_{n-k} \qquad (5\text{-}73)$$

where $Q[g_k]$ is the b-bit representation of g_k.

Now suppose that the desired impulse response is sampled at a high rate $f_s = k f_N$ to produce the closely spaced sequence $g_n = g(nT_s) = g(n/f_s)$ shown in Fig. 5-48(b). The *difference* $g_n - g_{n-1} = \nabla g_n$ could be quantized to the ternary set $\{-1, 0, 1\}$ in keeping with the delta modulation methodology.

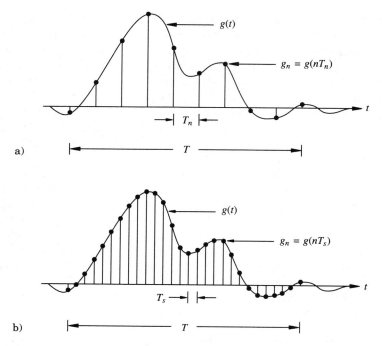

Figure 5-48 *(a) Nyquist-sampled impulse response; (b) impulse response sampled at $f_s = k f_N$.*

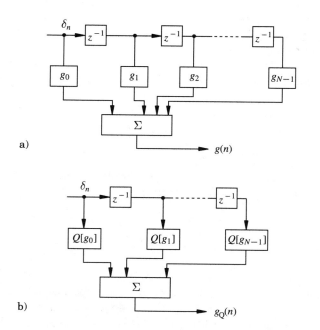

Figure 5-49 *(a) FIR filter with infinite-precision coefficients; (b) FIR filter with rounded coefficients.*

The steps involved are shown in Fig. 5-50. The infinite-precision filter is shown in Fig. 5-50(a), followed by the cancelling operations of differencing $(1 - z^{-1})$ and digital integrating $1/(1 - z^{-1})$. Next a ternary-valued quantizer is inserted following the difference operator in Fig. 5-50(b). Hence

$$g(n) = g(n - 1) + \nabla g(n) \qquad \text{[Fig. 5-50(a)]}$$

$$\tilde{g}(n) = \tilde{g}(n - 1) + Q[\nabla g(n)] \qquad \text{[Fig. 5-50(b)]}$$

where $Q[\nabla g(n)]$ is a ternary-valued, finite-duration sequence of length L. In the last stage, the difference operator is absorbed into the tap coefficients, which are now individually quantized to ternary values and constrained to cancel the accumulator pole. Thus

$$\hat{g}(n) = \hat{g}(n - 1) + d(n) \qquad (5\text{-}74)$$

where $\qquad d(n) = 0, 1, \text{ or } -1 \qquad n = 0, 1, ..., L - 1 \qquad (5\text{-}75)$

The last stage in Fig. 5-50 is the purely quantized structure with no multipliers. For this structure we can now choose $\{d(n)\}$ to minimize the error measure

$$\sum_{n=0}^{L-1} [g(nT_s) - \mu \hat{g}(n)]^2 \qquad (5\text{-}76)$$

subject to these constraints:

1. $d(n) \in \{-1, 0, 1\} \qquad n = 0, 1, ..., L - 1.$ $\qquad (5\text{-}77)$

2. $d(n)$ be selected such that

$$D(z) = \sum_{n=0}^{L-1} d(n) z^{-n} \qquad (5\text{-}78)$$

have a zero at $z = 1$ to cancel the integrator pole of $C(z)$ at $z = 1$. This constraint is imposed to ensure that the composite transfer function

$$\hat{H}(z) = C(z)D(z) = \frac{1}{1 - z^{-1}} D(z) \qquad (5\text{-}78)$$

will have a finite-duration impulse response.

Figure 5-50(c) is the basic structure studied by Benvenuto et al. (1986), but with a broader class of "resonators" $C(z)$. They used a dynamic programming algorithm to find the $\{d(n)\}$ that minimize the error measure Eq. (5-76) subject to its associated constraints 1 and 2. Their algorithm searched over combinations of filter size L and step size μ for different resonators $C(z)$. They found that the double integrator $C(z) = 1/(1 - z^{-1})^2$ and the double integrator followed by $1/(1 - z^{-2})$

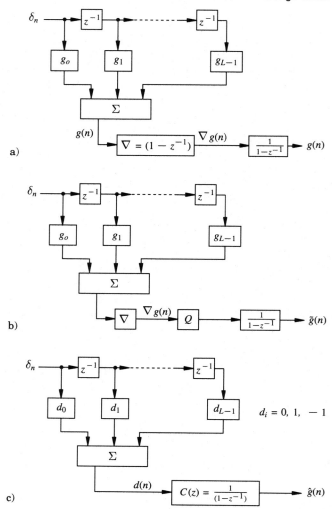

Figure 5-50 *(a) Infinite-precision FIR filter: (b) infinite-precision FIR filter with ternary quantization; (c) final FIR filter with no multipliers.*

are both better than a single integrator from the standpoint of trackability [ability to follow steep changes in $g(n)$] and accuracy (as determined by the granularity μ). (Recall that these same conflicting demands on μ arose in the original delta modulation signal coding scheme.) The collection of resonators tested were

$$C_1(z) = \frac{1}{1 - z^{-1}} \qquad C_3(z) = \frac{1}{(1 - z^{-1})^3}$$

$$C_2(z) = \frac{1}{(1 - z^{-1})^2} \qquad C_4(z) = \frac{1}{(1 - z^{-1})^2(1 - z^{-2})}$$

$$C_5(z) = \frac{1 - z^{-2}}{1 - z^{-1}}$$

In an earlier paper, Van Gerwen et al. (1975) postulated the structure of Fig. 5-50(c), allowing the $\{d(n)\}$ coefficients to be powers of 2 (which can be implemented by simple shifts of the data) with $C(z)$ a class of resonator with poles on the unit circle and coefficients in $\{-1, 0, 1\}$. Specifically, they considered $C(z)$ structures made up of tandem combinations of one or more of the following:

$$\frac{1}{1 - z^{-1}}, \quad \frac{1}{1 - z^{-1} + z^{-2}}, \quad \frac{1}{1 + z^{-2}}, \quad \frac{1}{1 + z^{-1} + z^{-2}}, \quad \frac{1}{1 + z^{-1}}$$

Benvenuto et al. concluded with the observation that the C(z) resonator chosen depended on the type of filter desired. Some typical cases are

Low-pass: $C(z) = 1/(1 - z^{-1})$
Band-pass
    (with center frequency near $\pi/2$): $C(z) = 1/(1 + z^{-2})$
High-pass: $C(z) = 1/(1 + z^{-1})$

5.4 OTHER TIME-DOMAIN FILTERS

5.4.1 MEDIAN FILTERS

A median filter is a nonlinear process which replaces the input sequence with the running median over a window of some specified length. The window always spans an odd number of points so that the median will always be one of the points. Let the window width be $2N + 1$, and let med $[x(m), x(n)]$ represent the median of the x values from $x(m)$ through $x(n)$. Then for an input sequence $\{x(i)\}$,

$$y(i) = \text{med}(x(i - N), x(i + N)) \tag{5-79}$$

A brute-force implementation of a median filter simply copies the values $x(i - N), \cdots, x(i + N)$, sorts the copied values, and outputs the central point in the sorted sequence. [For short windows (roughly $N < 4$), the brute-force procedure is usually optimal; for longer windows, there are a number of possible shortcuts.]

Note that if the central point in the window is already the median, it will be unchanged by the filter. Note also that the output of a median filter is always one of the input points: if the central point is not the median, one of the neighboring values, moved to the center by the sorting process, will be selected instead.

The significance of this kind of filter can be seen by examining Fig. 5-51. The dots represent the input data; there is a small amount of noise in the data, with two noise spikes at $i = 16$ and at $i = 40$ and 41.

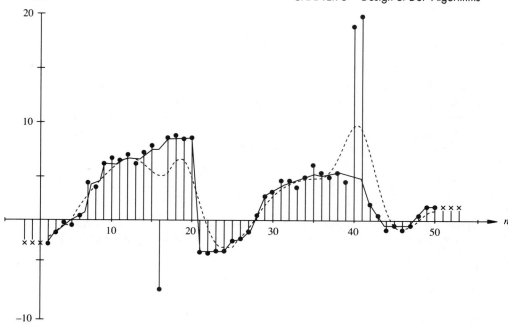

Figure 5-51   *Noisy sequence filtered by linear smoothing filter and by seven-point median filter: dotted line, linear filter output; solid line, median filter output.*

There is also an abrupt transition from $x(20) = 9.1$ to $x(21) = -4.1$. The dotted line shows the output of a seven-point raised-cosine linear smoothing filter; the solid line shows the output of a seven-point median filter.

The linear-filter output is somewhat smoother than that of the median filter; it reduces the impulse noise but does not get rid of it entirely, and it smoothes over the transition. The median filter completely removes the impulses, since the median of every seven-point window containing the impulse will always be one of the neighboring points. The sharp transition is preserved almost perfectly. In some cases transitions of this sort are important; perhaps the most obvious example is in the preservation of sharp edges in image processing. A sharp edge will be an abrupt transition of the sort shown here, and smoothing it will have the effect of blurring the edge. It is also desirable to remove impulse noise from images entirely; the median filter does this by substituting one of the neighboring points for the outlier.

The beginning and end of a sequence always present problems; for the median filter, these problems are most commonly solved by appending N points at each end of the given sequence. The points appended to each end have values equal to the respective endpoints. In Fig. 5-51, the points marked with x represent the appended values. Some authors have investigated other alternatives; for example, if the sequence is periodic with period p, it may be extended by repeating the first p samples at the beginning and the last p samples at the end. The way in which a signal is extended can have profound effects on the behavior of the median filter, particularly under repeated filtering operations. We will assume throughout that the sequence is extended as shown in Fig. 5-51.

Since median filters are nonlinear devices, all the traditional analysis techniques based on superposition and decomposition into orthogonal functions are of no use to us here. Nevertheless, a certain amount of analysis has been done, most notably by Gallagher and various collaborators (Gallagher, 1981; Nodes, 1982; Arce, 1982; and Fitch, 1984, 1985); we will draw heavily on these papers in the following discussion.

Gallagher starts by defining certain characteristic signal types. Let the window width be $2N + 1$, as before. Then

A *constant neighborhood* is a sequence of at least $(N + 1)$ consecutive identical points.

An *edge* is a monotonically increasing or decreasing sequence bounded by constant neighborhoods.

An *impulse* is a set of at most N points bounded by identically valued constant neighborhoods and different from these neighborhoods.

An *oscillation* is a sequence not part of a constant neighborhood, an edge, or an impulse.

A *root* is a sequence which is not modified by median filtering.

It is clear at once that a constant neighborhood will never be changed by median filtering, since the median of a set of points all equal to x_0 will be x_0. At either endpoint of a constant neighborhood, we have $N + 1$ points equal to x_0 and N points of other values; the median value of such a set will still be x_0. It should also be clear that median filtering will not change an edge. At any point along the edge, including its starting and ending points, the values are by definition already sorted in either ascending or descending order, and the middle point of the window is already the median and hence will not be changed. It follows from these two considerations that a sufficient condition for a sequence to be a root is that it consist only of constant neighborhoods and edges.

This is also a necessary condition. Consider a window moving from left to right across a sequence which is a root. As a result of the appended points, the first point in the signal is one of $N + 1$ identical values and hence is unchanged. If the next sample is identical to the preceding ones, the same reasoning applies. Suppose the next point is greater. Then if it is unchanged, none of the following N points can be less than the current point; otherwise one of them would be the median and the point would be changed. Hence the window contains a monotone sequence of length at least $N + 2$. A similar argument applies if the new point is less than the preceding points. The same reasoning can be applied as the window moves in its traversal of the root to any new point, including the final point, since the final point is followed by N identical appended points. Hence for a sequence to be a root, every set of $N + 2$ consecutive points must be monotone. Such a signal can consist only of edges separated by constant neighborhoods.

An impulse will always be removed by a median filter. When the window is centered on any point in the impulse, the window contains at most N impulse points and at least $N + 1$ points equal to some value x_0. Hence the median is again x_0 and this will replace the impulse point.

The picture is less simple when the filter encounters an oscillation. Let the last unchanged point be p, and let the first point to be changed be $q = p + 1$. Since p and its predecessors are unchanged, they must be monotone; suppose they are monotone nondecreasing. Since point p was not changed, it must have been the median; since none of the N points prior to point p were greater than point p, the remaining N points in the window must be as least as great as p. With this as background, let us turn our attention to q. We have seen that point q must be at least equal to p. Then the N predecessors to point q are also monotone. Then at least one of the N points following q must be less than q; otherwise q would be the median and would not be changed. But in that case, the median is less than q.

Hence we see that the first point to be changed following a monotone nondecreasing subsequence (1) must be at least equal to the last unchanged point and (2) will be lowered in value. A similar line of reasoning covers the case where p follows a monotone nonincreasing subsequence. Since the direction in which the filter moves is immaterial, the same property must hold for the last point to be changed, *mutatis mutandis*. The consequence of this is that median filters tend to extend root subsequences at the expense of adjoining oscillations.

In a median filter, the only parameters available for manipulation by the designer are the length of the filter and the number of passes made over the signal. From the foregoing discussion, it is clear that the only effect of changing the length of the filter is to change the length of the longest impulse that the filter can remove, to change the minimum length of a constant neighborhood, and hence to change the form of the root. If repeated passes are made over the signal, then the output tends toward a root. It can be shown (Prob. 5.5) that if point q in our discussion above is changed by a filtering pass, it will be invariant under all subsequent passes. Hence each pass reduces the length of all oscillations by 2, and after a maximum of $\frac{1}{2}(L-2)$ passes the sequence will be reduced to a root. Note that a root is not necessarily the right answer; it will be only if the noise-free signal was a root and if the added noise does not transform the signal into the ancestor of a different root.

It should be observed that these analyses apply only to a one-dimensional median filter. Many of the concepts, such as monotonicity, do not generalize readily to two-dimensional data; two-dimensional windows can have a shape parameter as well as a size parameter, and some of the properties, most notably the monotonicity of a root, break down when applied to two-dimensional data. Narendra (1978) introduced the concept of a separable median filter, which processes two-dimensional data by filtering first the rows and then the columns. Characteristics of root signals under separable filtering have been investigated in detail by Nodes and Gallagher (1983).

Fitch et al. (1984) developed a decomposition to take the place that orthogonal-function decomposition has in linear filtering. The primitive signals are binary signals obtained by threshold decomposition, and these primitive signals can be superposed. The threshold decomposition is defined as follows: let the threshold value be t; then

$$y_t(m) = \begin{cases} 1 & \text{if } x(m) \geq t \\ 0 & \text{otherwise} \end{cases} \tag{5-80}$$

If the signal is quantized and the range of the signal comprises k values, for example from 0 to $k-1$, then there are $k-1$ thresholds $1, 2, ..., k-1$. (It should be noted that these primitive functions are not, in general, orthogonal.) The original signal can be reconstructed from the decomposition as follows:

$$x(m) = \sum_{t=1}^{k-1} y_t(m) \tag{5-81}$$

The importance of this decomposition arises from the fact that the result of median filtering the original signal can also be reconstructed from the results of median filtering the binary signals:

$$M[x(m)] = \sum_{t=1}^{k-1} M[y_t(m)] \tag{5-82}$$

where $M[\cdot]$ represents the result of median filtering.

It follows from this property that anything which can be proved concerning median filtering of binary signals can readily be extended to multilevel signals. It also follows that a median filter for multilevel signals can be constructed by thresholding the input signal, median filtering the binary signals, and summing the filter outputs. A binary signal can be median filtered by counting the number of ones in the window. This means that no sorting operation is necessary and the filter is simple to implement in hardware, if the number of levels is not excessive.

5.4.2 MODIFIED MEDIAN FILTERS

Since the length of the filter is the principal design parameter available, some thought has been given to ways of making such filters more flexible. The modifications that have been considered include (1) using an order statistic other than the median, (2) filtering recursively, (3) using a linear combination of order statistics, and (4) using a linear combination of points lying within some neighborhood of the median.

Nodes and Gallagher (1982) considered order statistics other than the median. Suppose we select the second-largest point in the window. Then the filter will tend to favor high values over low values. If the filter length is suitably selected, this provides a particularly robust AM detector, as shown in Fig. 5-52. Here the sampling rate has been chosen so that one cycle of the carrier is approximately 17 samples long, and the filter window is 17 samples long. The input sequence has been extended by N points at the beginning and at the end, as was done in the case of the median filter. Figure 5-52(a) shows detection of a noise-free signal; the output is approximately a boxcar signal which follows the peaks closely.

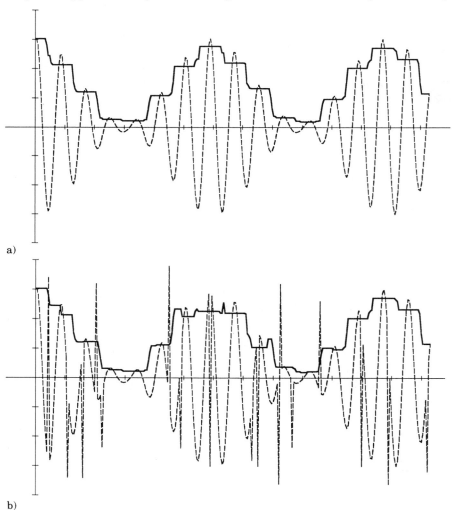

a)

b)

Figure 5-52 *Detection of amplitude-modulated signal with ranked- order filter: (a) noise-free signal; (b) noisy signal.*

In Fig. 5-52(b), the signal is contaminated with impulse noise, but the detected output is substantially the same as in 5-52(a), with only minor perturbations from the noise. The detector thus has impulse-suppression capabilities similar to those of the median filter.

These filters, which we will call ranked-order filters (ROFs), have the following properties. Proofs of these properties are simple, but long-winded, and they are omitted here; they are given in Nodes and Gallagher (1982). Let ROF(N, n) stand for an nth ranked-order filter with window length $2N + 1$; such a filter outputs the nth point from the sorted input. The filter used in Fig. 5-52 is ROF(8, 16).

1. A constant neighborhood must have a minimum length of $N + 1 + |N + 1 - n|$ points.

2. An impulse higher than its surroundings will be eliminated if its width is less than $2N + 2 - n$; an impulse lower than its surroundings will be eliminated if its width is less than n.

3. A rising edge will be shifted to the left by $n - N - 1$ points; a falling edge will be shifted to the right by $n - N - 1$ points.

4. Any constant region of at least $2N + 2 - n$ points will be lengthened by $2(n - N - 1)$ points if it is surrounded by constant neighborhoods of lesser values and will be shortened by the same amount if it is surrounded by constant neighborhoods of greater values. (This is a corollary of property 3.)

5. If n is not the median, only constant signals are invariant under ROF(N, n) filtering.

6. If n is not the median, any finite-length signal will be reduced to a constant by repeated passes of an ROF(N, n).

Nodes and Gallagher also give expressions for the probability density function and probability distribution of the output of an ROF(N, n) filter whose input consists of independent, identically distributed samples with a known density and distribution.

Bovik et al. (1983) considered filters whose outputs are linear combinations of order statistics of the points in the window, for example, 0.6 times the median plus 0.4 times the second-highest value. Such filters are clearly more flexible than simple ROFs, and in some applications they perform better than either median or linear filters, but they are expensive to design. The design requires computing the correlation matrix of order statistics of the noise, a laborious procedure. When noise is highly impulsive (that is, more so than the Laplacian case), the weighted order-statistic filter tends toward a median filter.

Lee and Kassam (1985) considered three generalizations of the median filter, called the L filter, the M filter, and the modified trimmed-mean (MTM) filter. Let the window length be $2N + 1$, as usual, and let $x_{(j)}^k$ be the jth-largest sample in the window centered about point k. Then the L filter is essentially the same as Bovik's:

$$y(k) = \sum_{h=1}^{2N+1} a_j x_{(j)}^k \qquad (5\text{-}83)$$

Lee and Kassam show that if the filter weights are symmetric, that is, if

$$a_j = a_{2N+2-j}$$

then the L filter is invariant under scaling and addition of a constant. If we represent the filtering operation by $S(\{x\})$, then

$$S(a\{x\} + b) = aS(\{x\}) + b$$

Lee and Kassam also show that the L filter preserves edges in a rather limited sense. Since a monotone sequence is already sorted, the output of the L filter is simply the output of an ordinary linear FIR filter with the same weights. Thus an edge will still be an edge, although the exact values of the input samples will not generally be preserved, and if the width of the edge is comparable to the window width, there will be some blurring of the edge.

If the operation of a median filter is observed in detail, it will be seen that impulses are removed by virtue of the fact that they will tend to fall at one end or the other of the sorted values. Since all values enter into the output of a general L filter, this filter will not reject impulses, although it will suppress them if the coefficients for $j = 1$ and $j = 2M+1$ are small. But we could remove them altogether if these coefficients were zero. This consideration leads to the α-trimmed mean filter. This filter is constructed as follows: the coefficients are zero for the first T points and the last T points; the remaining coefficients all have some constant value, normalized for unity gain. That is,

$$y(k) = \sum_{j=T+1}^{2N+1-T} \frac{1}{2(N-T)+1} x_{(j)}^k \qquad (5\text{-}84)$$

It is customary to specify T as a fraction of the window width:

$$T = \lfloor \alpha(2N + 1) \rfloor$$

Such a filter combines linear smoothing with rejection of all impulses of length $\leq T$. In particular, an edge of width 0 will be smoothed into a linear ramp of width $2(N - T) + 1$.

The M filter is the second generalization. Let $\Psi(x)$ be an odd, continuous, sign-preserving function. Then the M filter has an output which is the solution of the equation

$$\sum_{i=k-N}^{k+N} \Psi[x(i) - y(k)] = 0 \qquad (5\text{-}85)$$

This filter can be thought of as an estimator, where the estimate $y(k)$ is based on the $2N + 1$ values of x and $\Psi(\cdot)$ is similar to a cost function. The authors show that if Ψ is strictly increasing, Eq. (5-85) always has a solution, the M filter is invariant under addition of a constant (but not under scaling), and that the M filter is also edge-preserving in a limited sense. A special case of the M filter is the limiter type M (LTM) filter, for which if $g(x)$ is a strictly increasing, odd, continuous function,

$$\Psi(x) = \begin{cases} g(p) & x > p \\ g(x) & |x| \le p \\ -g(p) & x < -p \end{cases} \qquad (5\text{-}86)$$

where p is a positive constant that determines the point at which the cost function reaches its limit. A special case of the LTM filter is the standard type M (STM) filter, for which $g(x) = ax$. This filter tends to operate like a running-mean FIR filter in the absence of edges and impulses and like a median filter at edges.

The third generalization is the modified trimmed mean (MTM) filter, which, like the α-trimmed median filter, averages those samples in the window that lie close to the median. The difference is that where the α-trimmed median filter averages a constant fraction of the points in the window, the MTM filter averages as many samples as fall within a predetermined distance from the median.

We have seen that threshold decomposition provides one way of eliminating the need to sort values when doing median filtering. Algorithms for finding order statistics tend to be closely related to sorting algorithms; Ataman et al. (1980) provide an alternative method which is similar to the radix-exchange sort (Knuth, 1973). Let $k = (N + 1)/2$, so the median is the kth-largest sample in the window. Suppose the samples are represented in unsigned binary notation. Then we can locate the median by inspecting the bits of the samples, starting with the MSB and working downward until we find the first bit position where not all of the samples have zero bits in that position. If more than half the samples have a 1 in that position, then one of them is the median, and we need no longer consider the other samples. The median is now the kth largest sample in the remaining subset. We now move to the next bit position and note how many samples have a 1. If more than k 1s appear, we can again throw out the remaining samples and continue. If $p < k$ 1s are found, then the median is the $(k - p)$th-largest sample of the 0 subset. We continue in this way until the size of the subset under consideration is 1; this is the median. As in the radix-exchange sort, there is approximately a $\log_2 n : n$ reduction in the cost of finding the median.

5.5 SUMMARY

The beginning of this chapter was mostly a review of the most common types of digital filter and techniques for designing them. Of this material, probably only the relation between transition bandwidth and filter length will be completely new to most readers.

The operations of decimation and interpolation and their applications, either for alteration of the sampling rate or for filter banks, have come to be known as multirate DSP. Notice that the polyphase filter, which starts out as merely an interesting and handy implementation of the filters needed for interpolation and decimation, becomes increasingly important until in the section on perfect-reconstruction filter banks, it is at the heart of the process.

Of the other types discussed here, perhaps the median filters are of the greatest interest, because of their behavior and range of applications. As the capabilities of linear systems have been more and more thoroughly explored, interest has turned toward nonlinear techniques, and among these, median filters are some of the most promising. We will encounter these filters again in Chap. 9, where they are used to attenuate impulse noise in images while preserving sharp edges.

PROBLEMS

5.1.   Show that the sine and cosine functions in Fig. 5-18(a) must have amplitudes of $\sqrt{2}$.

5.2.   Phase distortion can be removed from an output of a linear, time-invariant system by reversing it in time and passing it through the system a second time. Let $x(t)$ be a finite-duration signal and let $h(t)$ be any linear time-invariant system. Let $y(t)$ be the output of the system when $x(t)$ is its input: $y(t) = x(t) * h(t)$. Let $y^R(t)$ be the time reversal of $y(t)$: that is, $y^R(t) = y(-t)$. Let $z^R(t)$ be the output of $h(t)$ when fed $y^R(t)$: $z^R(t) = y^R(t) * h(t)$. Show that in the total process from $x(t)$ to $z(t)$ no phase distortion is introduced.

5.3.   Apply the multistage principle to the design of interpolators.

5.4.   Starting with the interpolation filter of Fig. 5-23(c), consider the output sequences at each interpolator, including the inserted zeroes, and observe the sequences of nonzero samples at the various points along the delay chain. Use these observations to replace the delay chain with a shorter chain with only $L - 1$ delays. The new filters resulting from this consolidation will be in transposed direct form. Replace them with direct-form structures to obtain the interpolation filter shown in Fig. 5-32.

5.5.   Let q be the first point to be changed by a median filter. Show that the value of point q will not be changed if the resultant sequence is filtered a second time by the same median filter.

5.6.   Define the finite-length sequence $x(m) = m, m = 0, \cdots, M - 1$. Show that the primitive functions obtained by threshold decomposition of $x(m)$ as defined in Eq. (5-80) are not orthogonal.

5.7.   Prove Eq. (5-82).

5.8.   In the system of Fig. 5-38, show that if $H^2(\omega) + H^2(\omega + \pi) = 2$, then the same filter pair can be used throughout; that is, $H_1(z) = H_2(z) = H_3(z)$ and $H_1'(z) = H_2'(z) = H_3'(z)$ will give perfect reconstruction.

5.9.   Modify the design of Fig. 5-38 to produce an octave-band decomposition of the input signal—a decomposition in which the kth subband is twice as wide as the $(k - 1)$th band.

5.10.   Consider the binomial family, Eq. (5-56) with $N = 3$. Let

$$H_0(z) = X_0(z) + \sqrt{3}X_1(z)$$
$$H_1(z) = \sqrt{3}X_2(z) - X_3(z)$$

This pair, H_0 and H_1, constitute a two-channel perfect-reconstruction filter bank. With reference to Fig. 5-36, show that the conditions for perfect reconstruction are met if the synthesis filters are

$$F_0(z) = X_0(z) - \sqrt{3}X_1(z)$$
$$F_1(z) = \sqrt{3}X_2(z) + X_3(z)$$

Also show that $|H_0(e^{j\omega})|^2 + |H_1(e^{j\omega})|^2 = $ constant.

5.11.   Frequency sampling. Let $H(e^{j\omega}) \leftrightarrow h(n)$ be the transfer function and impulse response of an ideal filter, and let

$$H(e^{j\omega})\big|_{\omega = 2\pi k/N} = H(k)$$

where N is odd. Show that an interpolating polynomial passing through the N points $H(k)$ is given by

$$\hat{H}(e^{j\omega}) = \frac{1}{N}\sum_{k=0}^{N-1} H(k)\frac{\sin(N/2)(\omega - \ell 2\pi/N)}{\sin(1/2)(\omega - \ell 2\pi/N)}$$

and that the corresponding impulse response is

$$\hat{h}(n) = \begin{cases} \sum_{r=-\infty}^{\infty} h(n - rN) & |n| \le (N-1)/2 \\ 0 & \text{otherwise} \end{cases}$$

5.12. A symmetric band-pass filter can be represented as a frequency translation of a low-pass prototype. Let $H_L(e^{j\omega})$ be a low-pass filter, band-limited to $\pm\alpha$, with real impulse response $h_L(n)$. With $\omega_0 > \alpha$, define

$$H_{BP}(e^{j\omega}) = \tfrac{1}{2}[H_L(e^{j(\omega-\omega_0)}) + H_L(e^{j(\omega+\omega_0)})]$$

(a) Show that $h_{BP}(n) = h_L(n)\cos\omega_0 n$. (b) Consider a modulated signal,

$$x(n) = x_L(n)\cos\omega_0 n$$

as the input to the band-pass filter in (a), where $x_L(n)$ is a real, low-frequency signal band-limited to $\pm\beta$, with transform $X_L(e^{j\omega})$. Show the equivalence of the two structures in Fig. 5-P1.

5.13. Consider a complex-valued signal $f(n)$ with band-limited transform $F(e^{j\omega})$ such that $F(e^{j\omega}) = 0$ for $\omega < -\alpha_1$ and $\omega > \alpha_2$. Let $f(n) = x(n) + jy(n)$, where $x(n)$ and $y(n)$ are real.
(a) Show that

$$x(n) = \tfrac{1}{2}[f(n) + f^*(n)]$$
$$y(n) = \tfrac{1}{2}[f(n) - f^*(n)]$$

where * indicates the complex conjugate.
(b) Show that an asymmetric band-pass filter with real impulse response can be expressed as

$$H_{BP}(e^{j\omega}) = \tfrac{1}{2}[F(e^{j(\omega_0+\omega)}) + F^*(e^{j(\omega_0-\omega)})]$$
$$h_{BP}(n) = x(n)\cos\omega_0 n + y(n)\sin\omega_0 n$$

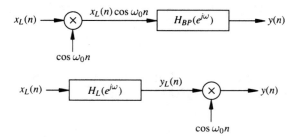

Figure 5-P1   *Problem 5.12.*

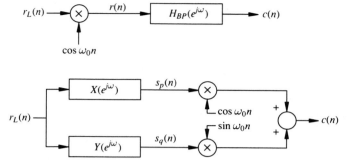

Figure 5-P2   *Problem 5.14.*

5.14.   A modulated signal

$$r(n) = r_L(n) \cos \omega_0 n$$

where $r_L(n) \leftrightarrow R_L(e^{j\omega})$ is a low-frequency signal band-limited to $\pm \alpha$, is applied to the nonsymmetrical band-pass filter $H_{BP}(e^{j\omega})$ of Prob. 5.13. Let the center frequency be ω_0.

(a) Show that the filter response is

$$c(n) = s_p(n) \cos \omega_0 n + s_q(n) \sin \omega_0 n$$

where $S_p(e^{j\omega}) = X(e^{j\omega}) R_L(e^{j\omega})$

$S_q(e^{j\omega}) = Y(e^{j\omega}) R_L(e^{j\omega})$

$$X(e^{j\omega}) = \begin{cases} \frac{1}{2}[H_{BP}(e^{j(\omega-\omega_0)}) + H_{BP}(e^{j(\omega+\omega_0)})] & |\omega| < \alpha \\ 0 & \text{otherwise} \end{cases}$$

$$Y(e^{j\omega}) = \begin{cases} \frac{1}{2}[H_{BP}(e^{j(\omega-\omega_0)}) - H_{BP}(e^{j(\omega+\omega_0)})] & |\omega| < \alpha \\ 0 & \text{otherwise} \end{cases}$$

(b) Show that the representations in Fig. 5-P2 are equivalent.

5.15.   (a) Show that the system (E-9) can be used to design an IIR filter whose output will have a specified autocorrelation when the input is white noise.

(b) Use the Levinson recursion to design an order-7 IIR filter whose output autocorrelation is $r(n) = \frac{1}{2}(1 - |n|/7 + 0.875^{|n|})$ when the input is white noise.

5.16.   Show that if the reflection coefficients of an order-p FIR lattice filter satisfy

$$|k_i| < 1 \quad i < p$$
$$|k_p| = 1$$

then all filter zeroes lie on the unit circle.

REFERENCES

J. E. Abate. "Linear and adaptive delta modulation." *Proc. IEEE*, vol. 55, pp. 298–308, Mar. 1967.

R. Ansari and B. Liu. "Efficient sample rate alteration using recursive (IIR) digital filters." *IEEE Trans. ASSP*, vol. ASSP-32, no. 6, pp. 1366–1373, Dec. 1983.

G. R. Arce and N. C. Gallagher. "State description for the root-signal set of median filters." *IEEE Trans. ASSP*, vol. ASSP-30, no. 6, pp. 894–902, Dec. 1982.

E. Ataman et al. "A fast method for real-time median filtering." *IEEE Trans. ASSP*, vol. ASSP-28, no. 4, pp. 415–421, Aug. 1980.

T. P. Barnwell III. "Subband coder design incorporating recursive quadrature filters and optimum ADPCM coders." *IEEE Trans. ASSP*, vol. ASSP-30, no. 4, pp. 751–765, Oct. 1982.

M. G. Bellanger. "Computation rate and storage estimation in multirate digital filtering with half-band filters." *IEEE Trans. ASSP*, vol. ASSP-25, no. 4, pp. 344–346, Aug. 1977.

M. G. Bellanger et al. "Interpolation, extrapolation, and reduction of computation speed in digital filters." *IEEE Trans. ASSP*, vol. ASSP-22, no. 4, pp. 231–235, Aug. 1974.

M. G. Bellanger et al. "Digital filtering by polyphase network: application to sample-rate alteration and filter banks." *IEEE Trans. ASSP*, vol. ASSP-24, no.2, pp. 109–114, Apr. 1976.

N. Benvenuto et al. "Realization of finite impulse response filters using coefficients +1, 0, and -1." *IEEE Trans. Comm.*, vol. COM-33, pp. 1117–1125, Oct. 1985.

———, "Dynamic programming methods for designing FIR filters using coefficients -1, 0, and +1." *IEEE Trans. ASSP*, vol. ASSP-34, no. 4, pp. 785–792, Aug. 1986.

A. C. Bovik et al. "A generalization of median filtering using linear combinations of order statistics." *IEEE Trans. ASSP*, vol. ASSP-31, no. 6, pp. 1342–1350, Dec. 1983.

C. S. Burrus. "Block realization of digital filters." *IEEE Trans. Audio and Electroacous.*, vol. AU-20, no. 4, pp. 230–235, Oct. 1972.

R. S. Cheung and R. L. Winslow. "High quality 16 kb/s voice transmission: the subband coder approach." *ICASSP-80*, pp. 319–322.

R. E. Crochiere and L. R. Rabiner. "Optimum FIR digital filter implementations for decimation, interpolation, and narrow-band filtering." *IEEE Trans. ASSP*, vol. ASSP-23, no. 5, pp. 444–456, Oct. 1975.

——, "Further considerations in the design of decimators and interpolators." *IEEE Trans. ASSP*, vol. ASSP-24, no. 4, pp. 296–311, Aug. 1976.

——, *Multirate Digital Signal Processing*. Englewood Cliffs, N. J.: Prentice-Hall, 1983.

A. W. Crooke and J. W. Craig. "Digital filters for sample-rate reduction." *IEEE Trans. Audio and Electroacous.*, vol. AU-20, no. 4, pp. 308–315, Oct. 1972.

S. Darlington. "On digital single-side-band modulators." *IEEE Trans. Circ. Theory*, vol. CT-17, pp. 409–414, Aug. 1970.

D. J. Esteban and C. Galand. "Application of quadrature mirror filters to split band voice coding schemes." *ICASSP-77*, pp. 191–195.

——, "HQMF: halfband quadrature mirror filters." *ICASSP-81*, pp. 220–223, Apr. 1981.

J. P. Fitch et al. "Median filtering by threshold decomposition." *IEEE Trans. ASSP*, vol. ASSP-32, no. 6, pp. 1183–1188, Dec. 1984.

J. P. Fitch et al. "Root properties and convergence rates of median filters." *IEEE Trans. ASSP*, vol. ASSP-33, no. 1, pp. 230–240, Feb. 1985.

S. W. Foo and L. F. Turner. "Design of nonrecursive quadrature mirror filters." *IEE Proc.*, vol. 129, part G, no. 3, pp. 61–67, June, 1982.

N. C. Gallagher and G. L. Wise. "A theoretical analysis of the properties of median filters." *IEEE Trans. ASSP*, vol. ASSP-29, no. 6, pp. 1136–1141, Dec. 1981.

C. Galand and D. J. Esteban. "16kbps real time QMF sub-band coding implementation." *ICASSP-80*, pp. 332–335.

D. J. Goodman and M. J. Carey. "Nine digital filters for decimation and interpolation." *IEEE Trans. ASSP*, vol. ASSP-25, no. 2, pp. 121–126, Apr. 1977.

R. A. Haddad. "A class of orthogonal nonrecursive binomial filters." *IEEE Trans. Audio and Electroacous.*, vol. AU-19, no. 4, pp. 296–304, Dec. 1971.

O. Herrmann et al. "Practical design rules for optimum finite impulse response lowpass digital filters," Bell Sys. Tech. J., vol. 52, no. 6, pp. 769–799, July-Aug. 1973.

E. B. Hogenauer. "An economical class of digital filters for decimation and interpolation." *IEEE Trans. ASSP*, vol. ASSP-29, no. 2, pp. 155–162, Apr. 1981.

N. S. Jayant. "Average- and median-based smoothing techniques for improving digital speech quality in the presence of transmission errors." *IEEE Trans. Comm.*, vol. COM-24, pp. 1043–1045, Sept. 1976.

J. D. Johnson and R. E. Crochiere. "An all-digital 'commentary grade' subband coder." *AES*, vol. 27, no. 11, pp. 855–865, Nov. 1979.

J. F. Kaiser. "Nonrecursive digital filter design using the $I_0 -$ sinh window function." *Proc. 1974 IEEE Int. Symp. Circ. and Sys.*, pp. 20–23, Apr. 22-25, 1974.

E. P. F. Kan and J. K. Aggarwal. "Multirate digital filtering." *IEEE Trans. Audio and Electroacous.*, vol. AU-20, no. 3, pp. 223–225, Aug. 1972.

D. E. Knuth, *The Art of Computer Programming*. Vol. 3, *Sorting and Searching*. Reading, Mass.: Addison-Wesley, 1973.

D. M. Kodek. "Design of optimal finite word length FIR filters using integer programming techniques." *IEEE Trans. ASSP*, vol. ASSP-29, no. 9, pp. 304–308, Jun. 1980.

D. M. Kodek and K. Steiglitz. "Filter-length word-length tradeoffs in FIR digital filter design." *IEEE Trans. ASSP*, vol. ASSP-28, no. 6, pp. 739–744, Dec. 1980.

Y. H. Lee and S. A. Kassam. "Generalized median filtering and related nonlinear filtering techniques." *IEEE Trans. ASSP*, vol. ASSP-33, no. 3, pp. 672–683, June, 1985.

Y. C. Lim et al. "Finite word length FIR filter design using integer programming over a discrete coefficient space." *IEEE Trans. ASSP*, vol. ASSP-30, no. 4, pp. 661–664, Aug. 1982.

Y. C. Lim and Constantinides. "Linear-phase FIR digital filter[s] without multipliers." *Proc. 1979 IEEE Int. Symp. Circ. Sys.*, pp. 185–188.

Y. C. Lim and S. R. Parker. "FIR filter design over a discrete powers-of-two coefficient space." *IEEE Trans. ASSP*, vol. ASSP-31, no. 3, pp. 583–591, Jun. 1983.

J. P. Marques de Sá. "A new design method of optimal finite wordlength linear phase FIR digital filters." *IEEE Trans. ASSP*, vol. ASSP-31, no. 4, pp. 1032–1034, Aug. 1983.

J. H. McClellan et al. "A computer program for designing optimum FIR linear-phase digital filters." *IEEE Trans. Audio and Electroacous.*, vol. AU-21, no. 6, pp. 506–526, Dec. 1973.

R. A. Meyer and C. S. Burrus. "Design and implementation of multirate digital filters." *IEEE Trans. ASSP*, vol. ASSP-24, no. 1, pp. 53–58, Feb. 1976.

P. C. Millar. "Recursive quadrature mirror filters–criteria specification and design method." *IEEE Trans. ASSP*, ASSP-33, no. 2, pp. 413–420, Apr. 1985.

F. Mintzer and B. Liu. "The design of optimal multirate bandpass and bandstop filters." *IEEE Trans. ASSP*, vol. ASSP-26, no. 6, pp. 534–543, Dec. 1978.

T. Miyawaki and C. W. Barnes. "Multirate recursive digital filters: a general approach and block structures." *IEEE Trans. ASSP*, vol. ASSP-31, no. 5, pp. 1148–1154, Oct. 1983.

D. C. Munson, Jr.. "On finite wordlength FIR filter design." *IEEE Trans. ASSP*, vol ASSP–29, p. 329, Apr. 1981.

P. M. Narendra. "A separable median filter for image noise." *Proc. IEEE Conf. Pattern Recog. and Image Proc.*, 1978.

T. A. Nodes and N. C. Gallagher. "Median filters: some modifications and their properties." *IEEE Trans. ASSP*, vol. ASSP-30, no. 5, pp. 739–746, Oct. 1982.

————, "Two-dimensional root structures and convergence properties of the separable median filter." *IEEE Trans. ASSP*, vol. ASSP-31, no. 6, pp. 1350–1365, Dec. 1983.

L. R. Rabiner and R. E. Crochiere. "A novel implementation for narrow-band FIR digital filters." *IEEE Trans. ASSP*, vol. ASSP-23, no. 5, pp. 457–464, Oct. 1975.

L. R. Rabiner and B. Gold, *Theory and Application of Digital Signal Processing*. Englewood Cliffs, N. J.: Prentice-Hall, 1975.

L. R. Rabiner et al. "Some comparisons between FIR and IIR digital filters." *Bell Sys. Tech. Jour.*, vol. 53, pp. 305–331, Feb. 1974.

L. R. Rabiner et al. "FIR digital filter design techniques using weighted Chebyshev approximation." *Proc. IEEE*, vol. 63, pp. 595–610, Apr. 1975.

T. A. Ramstad. "Digital methods for conversion between arbitrary sampling frequencies." *IEEE Trans. ASSP*, vol. ASSP-32, no. 3, pp. 577–591, Jun. 1984.

R. W. Schafer and L. R. Rabiner. "A digital signal processing approach to interpolation." *Proc. IEEE*, vol. 61, pp. 692–702, Jun. 1973.

R. R. Shively. "On multistage FIR filters with decimation." *IEEE Trans. ASSP*, vol. ASSP-23, no. 4, pp. 353–357, Aug. 1975.

M. J. T. Smith and T. P. Barnwell III. "A procedure for designing exact reconstruction filter banks for tree structured subband coders." *ICASSP-84*, pp. 27.1.1–27.1.4, Apr. 1984.

J. W. Tukey. *Exploratory Data Analysis*. Reading, Mass.: Addison-Wesley, 1977.

P. P. Vaidyanathan. "Theory and design of M-channel maximally decimated quadrature mirror filters with arbitrary M, having the perfect-reconstruction property." *IEEE Trans. ASSP*, vol. ASSP-35, no. 4, pp. 476–492, April, 1987(a).

————, "Quadrature mirror filter banks, M-band extensions, and perfect-reconstruction techniques." *IEEE ASSP Magazine*, vol. 4, no. 3, pp. 4–20, July, 1987(b).

P. J. Van Gerwen et al. "A new type of digital filter for data transmission." *IEEE Trans. Comm.*, vol. COM-23, no. 2, pp. 222–234, Feb. 1975.

CHAPTER 6

■ ■

LEAST-SQUARES OPTIMAL AND ADAPTIVE FILTERS

Chapter 5 presented techniques for the design of digital filters from frequency-domain requirements. Often the design tried to approximate certain ideal behavior, for example, an ideal low-pass or band-pass filter. The signal to be filtered and the process generating the signal were not prime considerations.

In this chapter we consider these and pose a twofold problem: how to model the signal generating process and how to design algorithms to filter these signals.

Both deterministic and random signals are treated here from the least-squares point of view. The least-squares approach provides a mechanism for designing fixed filters when the properties of the signal source are known. More importantly, it provides a vehicle for adaptive filter design that can operate in an environment of changing signal properties.

The source signal is modelled as the output of a linear discrete-time system with parameters that are either known (for a fixed algorithm) or unknown (in the adaptive case). Noise added to the observations completes the signal description. The least-squares algorithm is then required to do the "best" filtering of the signal, employing as much of the a priori signal and noise models as is known. If these a priori properties are unknown, then the least-squares algorithm is required to identify the changed conditions and to adapt its parameters to the new signal environment.

6.1 CLASSICAL LEAST SQUARES

This technique for estimating the parameters of a phenomenon from redundant measurements was developed by Gauss in 1795 and later, independently, by Legendre in 1806. Gauss dealt with the problem of estimating the parameters of the motion of astronomical bodies from a collection of observations and a foreknowledge of the laws governing the motion. Least squares then and now, therefore, provides a guide to the estimation of the parameters of a dynamic process from noisy measurements. Furthermore, least squares provides a mathematically tractable solution since it leads to a set of linear equations for the parameters.

Modern applications of least squares lead to optimal filter design on the one hand and to adaptive filters on the other. Arguably, it is the most useful technique for estimation and optimization today.

In this chapter we will use least squares to design fixed-parameter FIR filters, to provide a vehicle for optimal recursive filtering and its progeny, the Kalman filter, and thence to determine adaptive recursive algorithms for updating the coefficients in an optimized FIR filter.

6.1.1 ELEMENTARY DATA FITTING

To fix ideas, we begin with a simple example of fitting a straight line to observations, as indicated in Fig. 6-1. We are given a collection of data $\{z(t_1), ..., z(t_L)\}$, that are noisy measurements of a process we believe to be of the form

$$y = a + bt \tag{6-1}$$

where a and b are unknown. Our measurements are of

$$z_k = y_k + v_k \tag{6-2}$$

where v_k is the kth observation error, with variance σ_k^2.

We let the estimate be

$$\hat{y}_k = \hat{a} + \hat{b}t_k \tag{6-3}$$

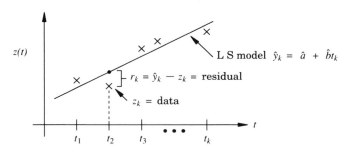

Figure 6-1   *Least-squares fit to straight line.*

We assume the observation errors are zero-mean and uncorrelated with each other and have variances σ_k^2, or a covariance matrix

$$\mathbf{R} = \text{diag}[\sigma_1^2, \sigma_2^2, ..., \sigma_L^2]$$

We define the residual to be

$$r_k = z_k - \hat{y}_k$$
$$= (y_k - \hat{y}_k) + v_k \tag{6-4}$$

(A weighted residual would be r_k/σ_k.) The model for the estimator is shown in Fig. 6-2.

The least-squares estimation problem may be stated: Determine $\hat{a}$ and $\hat{b}$ to minimize the sum of the squared weighted residuals

$$J = \sum_{k=1}^{L} (r_k/\sigma_k)^2 = J(\hat{a}, \hat{b}) \tag{6-5}$$

We can find $\hat{a}$ and $\hat{b}$ by setting

$$\frac{\partial J}{\partial \hat{a}} = 0 \qquad \frac{\partial J}{\partial \hat{b}} = 0 \tag{6-6}$$

This gives two equations in two unknowns and allows us to solve for $\hat{a}$ and $\hat{b}$ in terms of the observations. For the case where $\sigma_k = 1$ for all k, straightforward differentiation leads to

$$0 = \frac{\partial J}{\partial \hat{a}} = 2 \sum_k r_k \frac{\partial r_k}{\partial \hat{a}}$$

But

$$\frac{\partial r_k}{\partial \hat{a}} = \frac{\partial}{\partial \hat{a}} [z_k - (\hat{a} + \hat{b}t_k)] = -1$$

or

$$\sum_k (-1) r_k = 0$$

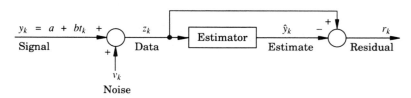

Figure 6-2   *Model of estimator.*

Expanding this last term gives

$$\sum_k z_k = \sum_k \hat{y}_k = \sum_k \left(\hat{a} + \hat{b}t_k \right)$$

$$= \hat{a} \sum_k 1 + \hat{b} \sum_k t_k \tag{6-7}$$

Similarly $\dfrac{\partial J}{\partial \hat{b}} = \sum_k 2r_k \dfrac{\partial r_k}{\partial \hat{b}} = 0$ and $\dfrac{\partial r_k}{\partial \hat{b}} = t_k$

Hence $\displaystyle\sum_k t_k r_k = 0$

Upon expansion this becomes

$$\sum_k t_k z_k = \hat{a} \sum_k t_k + \hat{b} \sum_k t_k^2 \tag{6-8}$$

Combining Eqs. (6-7) and (6-8) in matrix form, we have

$$\begin{bmatrix} \sum_{k=1}^{L} 1 & \sum_{k=1}^{L} t_k \\ \sum_{k=1}^{L} t_k & \sum_{k=1}^{L} t_k^2 \end{bmatrix} \begin{bmatrix} \hat{a} \\ \hat{b} \end{bmatrix} = \begin{bmatrix} \sum_{k=1}^{L} z_k \\ \sum_{k=1}^{L} t_k z_k \end{bmatrix} \tag{6-9}$$

These linear equations can be solved for $\hat{a}$ and $\hat{b}$.

As a prelude to the more general case, we will reformulate this straight-line data fit in vector-matrix form. Let

$$z_k = a + bt_k + v_k = \begin{bmatrix} 1 & t_k \end{bmatrix} \begin{bmatrix} a \\ b \end{bmatrix} + v_k \tag{6-10}$$

and with $\mathbf{H}_k = \begin{bmatrix} 1 & t_k \end{bmatrix}$ a row vector

$$\mathbf{x} = \begin{bmatrix} a \\ b \end{bmatrix} \qquad \text{the parameter vector} \tag{6-11}$$

we have $z_k = \mathbf{H}_k \mathbf{x} + v_k,$ the observation model   (6-12)

Now the estimate becomes

$$\hat{y}_k = \hat{z}_k = \hat{a} + \hat{b}t_k = \begin{bmatrix} 1 & t_k \end{bmatrix} \begin{bmatrix} \hat{a} \\ \hat{b} \end{bmatrix}$$

$$= \mathbf{H}_k \hat{\mathbf{x}} \tag{6-13}$$

If we stack the observations,

$$\begin{bmatrix} z_1 \\ z_2 \\ \vdots \\ z_L \end{bmatrix} = \begin{bmatrix} 1 & t_1 \\ 1 & t_2 \\ \vdots & \vdots \\ 1 & t_L \end{bmatrix} \mathbf{x} + \begin{bmatrix} v_1 \\ v_2 \\ \vdots \\ v_L \end{bmatrix}$$

or $$\mathbf{z} = \mathbf{Hx} + \mathbf{v} \qquad\qquad (6\text{-}14)$$

is now the observation model. Next let

$$\hat{\mathbf{z}} = \mathbf{H\hat{x}} \qquad\qquad (6\text{-}15)$$

and let the residual vector be

$$\mathbf{r} = \mathbf{z} - \hat{\mathbf{z}} = (\mathbf{z} - \mathbf{H\hat{x}}) \qquad\qquad (6\text{-}16)$$

The least-squares estimation problem is to choose $\hat{\mathbf{x}}$ so as to minimize the sum of the squared residuals

$$J = (\mathbf{z} - \mathbf{H\hat{x}})^T(\mathbf{z} - \mathbf{H\hat{x}}) \qquad\qquad (6\text{-}17)$$

Note that J as defined in Eq. (6-17) is the same as in Eq. (6-5) when $\sigma_k = 1$. We already know that the solution to this problem can be obtained by solving Eq. (6-9); our intent is to generalize the problem, and the solution, in the following sections.

6.1.2 GENERAL LEAST SQUARES

In this section, we extend and generalize the preceding problem formulation and solution, and the purely deterministic result obtained is given a statistical interpretation.

1. We are given a set of L observations $\{z_i\}$ at times $\{t_i\}$:

$$\mathbf{z}^T = [z_1, z_2, ..., z_L] \qquad\qquad (6\text{-}18)$$

2. There is an N-dimensional vector parameter $\mathbf{x}$ related to $\mathbf{z}$ by

$$\mathbf{z} = \mathbf{Hx} + \mathbf{v} \qquad\qquad (6\text{-}19)$$

In Eq. (6-19), $\mathbf{z}$ is the data vector, $\mathbf{H}$ is a known $L \times N$ matrix ($L \geq N$) describing the theoretical relationship between $\mathbf{x}$ and $\mathbf{z}$, and $\mathbf{v}$ represents measurement errors or noise. The specific nature of $\mathbf{H}$ depends on the application. It has been called an observation matrix in some instances and a data matrix (as in Sect. 6.5) in others.

3. Determine a linear estimator

$$\hat{\mathbf{x}} = \mathbf{Az} \qquad\qquad (6\text{-}20)$$

and define

$$\hat{\mathbf{z}} = \mathbf{H\hat{x}} \qquad\qquad (6\text{-}21)$$

such that J, the sum of the squared residuals is minimized, where

$$J = (\mathbf{z} - \mathbf{H}\hat{\mathbf{x}})^T(\mathbf{z} - \mathbf{H}\hat{\mathbf{x}}) \tag{6-22}$$

The problem formulation is depicted in Fig. 6-3. Setting $\partial J/\partial \hat{\mathbf{x}} = \mathbf{0}$ yields the set of linear equations

$$(\mathbf{H}^T\mathbf{H})\hat{\mathbf{x}} = \mathbf{H}^T\mathbf{z} \tag{6-23}$$

and the least-squares estimate is

$$\hat{\mathbf{x}} = [(\mathbf{H}^T\mathbf{H})^{-1}\mathbf{H}^T]\mathbf{z} \tag{6-24}$$

The details are as follows: First, we establish some inter-matrix differentiation rules. The gradient is a vector, defined by

$$\frac{\partial f}{\partial \mathbf{x}} = \left[\begin{array}{cccc} \dfrac{\partial f}{\partial x_1} & \dfrac{\partial f}{\partial x_2} & \cdots & \dfrac{\partial f}{\partial x_N} \end{array}\right]^T \tag{6-25}$$

It may be shown that

$$\frac{d}{d\mathbf{x}}(\mathbf{b}^T\mathbf{x}) = \mathbf{b} \tag{6-26}$$

and

$$\frac{d}{d\mathbf{x}}(\mathbf{x}^T\mathbf{A}\mathbf{x}) = 2\mathbf{A}\mathbf{x}$$

Expanding Eq. (6-22), we have

$$\begin{aligned} J &= \mathbf{z}^T\mathbf{z} - \mathbf{z}^T\mathbf{H}\hat{\mathbf{x}} - \hat{\mathbf{x}}^T\mathbf{H}^T\mathbf{z} + \hat{\mathbf{x}}^T(\mathbf{H}^T\mathbf{H})\hat{\mathbf{x}} \\ &= \mathbf{z}^T\mathbf{z} - 2\mathbf{z}^T\mathbf{H}\hat{\mathbf{x}} + \hat{\mathbf{x}}^T(\mathbf{H}^T\mathbf{H})\hat{\mathbf{x}} \end{aligned} \tag{6-27}$$

Then by Eq. (6-26),

$$\frac{\partial J}{\partial \hat{\mathbf{x}}} = -2\mathbf{H}^T\mathbf{z} + 2(\mathbf{H}^T\mathbf{H})\hat{\mathbf{x}} \tag{6-28}$$

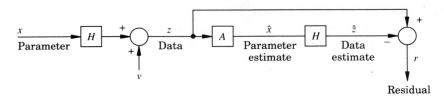

Figure 6-3   *Least-squares estimator problem formulation.*

Setting the gradient in Eq. (6-28) to zero gives the set of N equations in Eq. (6-23) and the least-squares solution in Eq. (6-24), repeated here for convenience:

$$(\mathbf{H}^T\mathbf{H})\hat{\mathbf{x}} = \mathbf{H}^T\mathbf{z} \tag{6-23}$$

$$\hat{\mathbf{x}} = [(\mathbf{H}^T\mathbf{H})^{-1}\mathbf{H}^T]\mathbf{z} \tag{6-24}$$

Eq. (6-23) is called the deterministic normal equation. It has a unique solution provided $\mathbf{H}$, the $L \times N$ observation matrix, $L \geq N$, has at least N linearly independent columns. Then $\mathbf{H}^T\mathbf{H}$ is nonsingular; in fact, $\mathbf{H}^T\mathbf{H}$ is real, symmetric, and positive definite, that is, a covariance-type matrix.

Some insight into the least-squares estimation may be garnered by rewriting Eq. (6-28) in terms of the residuals $\mathbf{r}$:

$$\frac{\partial J}{\partial \hat{\mathbf{x}}} = -2\mathbf{H}^T(\mathbf{z} - \mathbf{H}\hat{\mathbf{x}}) = -2\mathbf{H}^T\mathbf{r} = 0$$

or $\qquad \mathbf{H}^T\mathbf{r} = 0 \tag{6-29}$

This equation asserts that the minimum error residual is orthogonal to each column of the observation matrix $\mathbf{H}$. Another way of interpreting this result is to multiply Eq. (6-29) by $\hat{\mathbf{x}}^T$ to obtain

$$\hat{\mathbf{x}}^T\mathbf{H}^T\mathbf{r} = (\mathbf{H}\hat{\mathbf{x}})^T\mathbf{r} = \hat{\mathbf{z}}^T\mathbf{r} = 0$$

or $\qquad \hat{\mathbf{z}}^T\mathbf{r} = 0 \tag{6-30}$

That is, the optimal least-squares estimator makes the residual error $\mathbf{r}$ orthogonal to the data estimate $\hat{\mathbf{z}}$, as shown in Fig. 6-4.

Equations (6-29) and (6-30) constitute the deterministic orthogonality principle, which is central to least-squares estimation. This version is the deterministic counterpart of the stochastic orthogonality principle used in linear mean-squared estimation. The latter asserts that the

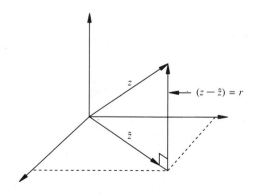

Figure 6-4 *Orthogonality principle displayed.*

best linear estimate, in the mean-squared sense, is one which makes the estimation error $(\mathbf{x} - \hat{\mathbf{x}})$ orthogonal to the data $\mathbf{z}$; that is,

$$E\{(\mathbf{x} - \hat{\mathbf{x}})\mathbf{z}^T\} = 0.$$

Finally, the minimum of J is obtained by substituting Eq. (6-24) into Eq. (6-27). The last term in Eq. (6-27) is

$$\hat{\mathbf{x}}^T(\mathbf{H}^T\mathbf{H})\hat{\mathbf{x}} = \hat{\mathbf{x}}^T(\mathbf{H}^T\hat{\mathbf{z}})$$

This cancels one of the cross-product terms in Eq. (6-27) and leaves

$$J_{\min} = \mathbf{z}^T\mathbf{z} - \mathbf{z}^T\mathbf{H}\hat{\mathbf{x}} \geq 0$$

or
$$J_{\min} = \mathbf{z}^T\mathbf{z} - \mathbf{z}^T[\mathbf{H}(\mathbf{H}^T\mathbf{H})^{-1}\mathbf{H}^T]\mathbf{z}$$

$$= \mathbf{z}^T[\mathbf{I} - \mathbf{H}(\mathbf{H}^T\mathbf{H})^{-1}\mathbf{H}^T]\mathbf{z} \geq 0 \tag{6-31}$$

The second term in Eq. (6-31) is nonnegative; this indicates the reduction in the error measure obtained by the least-squares estimator.

6.1.2.1 A Probabilistic Interpretation Suppose that the measurements are not equally accurate. In that event, we should weight the residuals to reflect our confidence in the measurements. We achieve this by an as yet unspecified weight matrix $\mathbf{W}$. Now the error measure is

$$J = (\mathbf{z} - \mathbf{H}\hat{\mathbf{x}})^T\mathbf{W}(\mathbf{z} - \mathbf{H}\hat{\mathbf{x}}) \tag{6-32}$$

where $\mathbf{W}$ is a real, symmetric, positive-definite matrix. The resulting weighted least-squares equations are:

Estimate: $\hat{\mathbf{x}} = [(\mathbf{H}^T\mathbf{W}\mathbf{H})^{-1}\mathbf{H}^T\mathbf{W}]\mathbf{z}$ \hfill (6-33)

Error measure: $J_{\min} = \mathbf{z}^T\mathbf{W}\mathbf{z} - \mathbf{z}^T[\mathbf{W}\mathbf{H}(\mathbf{H}^T\mathbf{W}\mathbf{H})^{-1}\mathbf{H}^T\mathbf{W}]\mathbf{z} \geq 0$ \hfill (6-34)

Orthogonality: $\hat{\mathbf{z}}^T(\mathbf{W}\mathbf{r}) = 0$ \hfill (6-35)

Note that the weighted residual $\mathbf{W}\mathbf{r}$ is now orthogonal to the data estimate $\hat{\mathbf{z}}$.

Suppose that the measurement errors are modelled as zero-mean random variables with a covariance matrix

$$\mathbf{R} = E\{\mathbf{v}\mathbf{v}^T\} \tag{6-36}$$

In this case, we can take $\mathbf{R}^{-1}$ as the weight matrix in Eqs. (6-32) to (6-35). This resulting weighted least-squares estimator has the same form as the linear maximum-likelihood estimator which minimizes the mean-squared error

$$E\{(\mathbf{x} - \hat{\mathbf{x}})^T(\mathbf{x} - \hat{\mathbf{x}})\}$$

In the maximum-likelihood case, the measurement errors are modelled via $\mathbf{R}$, but the parameter vector $\mathbf{x}$ has unknown statistics.

We now pose the question, how good is the weighted least-squares estimator? With $\hat{\mathbf{x}}$ as the parameter or "state" vector estimate, we define the state estimation error as

$$\tilde{\mathbf{x}} = \mathbf{x} - \hat{\mathbf{x}} \tag{6-37}$$

This vector, $\tilde{\mathbf{x}}$, is sometimes called the state-estimate residual, to distinguish it from $\mathbf{r} = \mathbf{z} - \hat{\mathbf{z}}$, the measurement residual. For convenience let

$$\mathbf{A} = \begin{cases} (\mathbf{H}^T\mathbf{H})^{-1}\mathbf{H}^T & \text{(unweighted least squares)} \\ (\mathbf{H}^T\mathbf{R}^{-1}\mathbf{H})^{-1}\mathbf{H}^T\mathbf{R}^{-1} & \text{(weighted least squares)} \end{cases} \tag{6-38}$$

Then $\hat{\mathbf{x}} = \mathbf{A}(\mathbf{Hx} + \mathbf{v})$ $\tag{6-39}$

and $\tilde{\mathbf{x}} = \mathbf{x} - \hat{\mathbf{x}} = (\mathbf{I} - \mathbf{AH})\mathbf{x} - \mathbf{Av}$ $\tag{6-40}$

In both unweighted and weighted least squares, we have selected $\mathbf{A}$ to make $\mathbf{AH} = \mathbf{I}$ and thus to get

$$\tilde{\mathbf{x}} = -\mathbf{Av} \tag{6-41}$$

and $\tilde{\mathbf{x}}\tilde{\mathbf{x}}^T = \mathbf{A}(\mathbf{vv}^T)\mathbf{A}^T$ $\tag{6-42}$

Now with $\mathbf{P}$ defined as the covariance matrix of $\mathbf{x}$, we get

$$\mathbf{P} = E\{\tilde{\mathbf{x}}\tilde{\mathbf{x}}^T\} = \mathbf{A}E\{\mathbf{vv}^T\}\mathbf{A}^T = \mathbf{ARA}^T \tag{6-43}$$

By direct substitution for $\mathbf{A}$ in Eq. (6-43), we find

$$\mathbf{P} = (\mathbf{H}^T\mathbf{R}^{-1}\mathbf{H})^{-1} \tag{6-44}$$

When $\mathbf{R} = \mathbf{I}$, that is, when the measurement errors are uncorrelated and have unit variances, Eq. (6-44) simplifies to the standard least-squares formula

$$\mathbf{P} = (\mathbf{H}^T\mathbf{H})^{-1} \tag{6-45}$$

6.2 RECURSIVE LEAST SQUARES AND SEQUENTIAL ESTIMATION

The technique of least squares was developed in Sect. 6.1 in a form suitable for batch processing of data. As more data become available, it would be prohibitive to recompute a new parameter estimate on an expanding batch basis. The recursive least squares (RLS) approach avoids

Table 6-1. Least Squares Estimation

Data Model:
$$\mathbf{z} = \mathbf{H}\mathbf{x} + \mathbf{v}$$
where

$\mathbf{x}$ is N-vector to be estimated
$\mathbf{H}$ is $L \times N$ measurement matrix
$\mathbf{v}$ is additive measurement noise, and zero mean, with covariance matrix $\mathbf{R}$

$\mathbf{r} = \mathbf{z} - \mathbf{H}\hat{\mathbf{x}}$	the measurement residual
$J = (\mathbf{z} - \mathbf{H}\hat{\mathbf{x}})^T \mathbf{R}^{-1}(\mathbf{z} - \mathbf{H}\hat{\mathbf{x}})$	the error measure
$\hat{\mathbf{x}} = [(\mathbf{H}^T\mathbf{R}^{-1}\mathbf{H})^{-1}\mathbf{H}^T\mathbf{R}^{-1}]\mathbf{z}$	the weighted LS estimate
$\mathbf{P} = (\mathbf{H}^T\mathbf{R}^{-1}\mathbf{H})^{-1}$	covariance of $\tilde{\mathbf{x}} = \mathbf{x} - \hat{\mathbf{x}}$
$\mathbf{H}^T\mathbf{R}^{-1}\mathbf{r} = 0$	the orthogonality property

---

this by combining the old parameter estimate with the new measurement in a one-step recursive update.

We will extend the RLS estimation principle to obtain the Kalman filter in Sec. 6.2.2. This recursive structure also provides the framework for adaptively updating the coefficients of an FIR filter in Sec. 6.4.

6.2.1 THE RECURSIVE ALGORITHM

Let us assume we are given $\mathbf{z}_1$ a set of noisy measurements and

$$\mathbf{z}_1 = \mathbf{H}_1\mathbf{x} + \mathbf{v}_1 \tag{6-46}$$

Initially, we batch process these following the probabilistic interpretation of Table 6.1:

$$\hat{\mathbf{x}}_1 = [(\mathbf{H}_1^T\mathbf{R}_1^{-1}\mathbf{H}_1)^{-1}\mathbf{H}_1^T\mathbf{R}_1^{-1}]\mathbf{z}_1 \tag{6-47}$$

and $\quad \mathbf{P}_1 = (\mathbf{H}_1^T\mathbf{R}_1^{-1}\mathbf{H}_1)^{-1}$ is the covariance of $\{\tilde{\mathbf{x}}_1\}$.

We next accept a new measurement vector $\mathbf{z}_2$ of dimension k_2, of the form

$$\mathbf{z}_2 = \mathbf{H}_2\mathbf{x} + \mathbf{v}_2 \tag{6-48}$$

and $\mathbf{R}_2$ is the measurement noise covariance for $\mathbf{v}_2$.

The composite measurement is obtained by stacking Eqs. (6-46) and (6-48) as follows:

$$\mathbf{z} = \begin{bmatrix} \mathbf{z}_1 \\ \mathbf{z}_2 \end{bmatrix} = \begin{bmatrix} \mathbf{H}_1 \\ \mathbf{H}_2 \end{bmatrix} \mathbf{x} + \begin{bmatrix} \mathbf{v}_1 \\ \mathbf{v}_2 \end{bmatrix} \qquad \mathbf{R} = \begin{bmatrix} \mathbf{R}_1 & 0 \\ 0 & \mathbf{R}_2 \end{bmatrix} \tag{6-49}$$

Let $\hat{\mathbf{x}}_2$ be the least-squares estimate of $\mathbf{x}$ after incorporating the new observation $\mathbf{z}_2$, and let $\mathbf{P}_2$ be the associated estimation-error covariance. The dimension of the parameter-vector estimate, of course, does not change, although the data vector on which this estimate is based is now enlarged. Following the least squares procedure of Sec. 6.1, we form

$$J(\hat{\mathbf{x}}_2) = (\mathbf{z} - \mathbf{H}\hat{\mathbf{x}}_2)^T \mathbf{R}^{-1}(\mathbf{z} - \mathbf{H}\hat{\mathbf{x}}_2) \tag{6-50}$$

where
$$\mathbf{z} - \mathbf{H}\hat{\mathbf{x}}_2 = \begin{bmatrix} \mathbf{z}_1 \\ \mathbf{z}_2 \end{bmatrix} - \begin{bmatrix} \mathbf{H}_1 \\ \mathbf{H}_2 \end{bmatrix} \mathbf{x}_2 = \begin{bmatrix} (\mathbf{z}_1 - \mathbf{H}_1\hat{\mathbf{x}}_2) \\ (\mathbf{z}_2 - \mathbf{H}_2\hat{\mathbf{x}}_2) \end{bmatrix} \tag{6-51}$$

and
$$\mathbf{R}^{-1} = \begin{bmatrix} \mathbf{R}_1^{-1} & 0 \\ 0 & \mathbf{R}_2^{-1} \end{bmatrix}$$

The usual minimization gives

$$\hat{\mathbf{x}}_2 = [(\mathbf{H}^T\mathbf{R}^{-1}\mathbf{H})^{-1}\mathbf{H}^T\mathbf{R}^{-1}]\mathbf{z} \tag{6-52}$$

and
$$\mathbf{P}_2 = (\mathbf{H}^T\mathbf{R}^{-1}\mathbf{H})^{-1}$$

Expanding each of the terms in Eq. (6-52),

$$\mathbf{P}_2^{-1} = \mathbf{H}^T\mathbf{R}^{-1}\mathbf{H} = \begin{bmatrix} \mathbf{H}_1 \\ \mathbf{H}_2 \end{bmatrix}^T \begin{bmatrix} \mathbf{R}_1^{-1} & 0 \\ 0 & \mathbf{R}_2^{-1} \end{bmatrix} \begin{bmatrix} \mathbf{H}_1 \\ \mathbf{H}_2 \end{bmatrix}$$

$$= \mathbf{H}_1^T\mathbf{R}_1^{-1}\mathbf{H}_1 + \mathbf{H}_2^T\mathbf{R}_2^{-1}\mathbf{H}_2$$

$$= \mathbf{P}_1^{-1} + \mathbf{H}_2^T\mathbf{R}_2^{-1}\mathbf{H}_2 \tag{6-53}$$

Similarly,
$$\mathbf{H}^T\mathbf{R}^{-1}\mathbf{z} = [\mathbf{H}_1^T \ \ \mathbf{H}_2^T] \begin{bmatrix} \mathbf{R}_1^{-1} & 0 \\ 0 & \mathbf{R}_2^{-1} \end{bmatrix} \begin{bmatrix} \mathbf{z}_1 \\ \mathbf{z}_2 \end{bmatrix} \tag{6-54}$$

$$= \mathbf{H}_1^T\mathbf{R}_1^{-1}\mathbf{z}_1 + \mathbf{H}_2^T\mathbf{R}_2^{-1}\mathbf{z}_2$$

Combining Eqs. (6-53) and (6-54) with Eq. (6-52),

$$\hat{\mathbf{x}}_2 = (\mathbf{P}_1^{-1} + \mathbf{H}_2^T\mathbf{R}_2^{-1}\mathbf{H}_2)^{-1}(\mathbf{H}_1^T\mathbf{R}_1^{-1}\mathbf{z}_1 + \mathbf{H}_2^T\mathbf{R}_2^{-1}\mathbf{z}_2) \tag{6-55}$$

Using the matrix identity (the matrix inversion lemma)

$$(\mathbf{A} + \mathbf{BCD})^{-1} = \mathbf{A}^{-1} - \mathbf{A}^{-1}\mathbf{B}(\mathbf{C}^{-1} + \mathbf{DA}^{-1}\mathbf{B})^{-1}\mathbf{DA}^{-1} \tag{6-56}$$

in Eq. (6-55) and reorganizing $\hat{\mathbf{x}}_1$ in the form of Eq. (6-47) eventually leads to

$$\hat{\mathbf{x}}_2 = \hat{\mathbf{x}}_1 + \mathbf{P}_1\mathbf{H}_2^T(\mathbf{H}_2\mathbf{P}_1\mathbf{H}_2^T + \mathbf{R}_2)^{-1}(\mathbf{z}_2 - \mathbf{H}_2\hat{\mathbf{x}}_1)$$

$$= \hat{\mathbf{x}}_1 + \mathbf{K}_2(\mathbf{z}_2 - \mathbf{H}_2\hat{\mathbf{x}}_1) \tag{6-57}$$

where $\mathbf{K}_2$ is the recursive weighted LS gain matrix

$$\mathbf{K}_2 = \mathbf{P}_1 \mathbf{H}_2^T (\mathbf{H}_2 \mathbf{P}_1 \mathbf{H}_2^T + \mathbf{R}_2)^{-1} \tag{6-58}$$

Equations (6-53), (6-57), and (6-58) provide the basis for the recursive update of the parameter estimate. The starting point is a prior estimate $\hat{\mathbf{x}}_1$ and covariance $\mathbf{P}_1$. At the next time instant, the measurement $\mathbf{z}_2$ is taken. We pretend to know $\mathbf{H}_2$, the observation matrix, and $\mathbf{R}_2$, the noise covariance associated with this measurement.

The gain $\mathbf{K}_2$ is calculated in Eq. (6-58) and used in the updated Eq. (6-57) to give $\hat{\mathbf{x}}_2$ as the optimum least-squares combination of $\hat{\mathbf{x}}_1$ and $\mathbf{z}_2$. Next, $\mathbf{P}_2$, the covariance for the estimation error ($\tilde{\mathbf{x}}_2 = \mathbf{x} - \hat{\mathbf{x}}_2$) is calculated in Eq. (6-53). Now we are ready to accept the next measurement at the next time index and repeat the recursion.

To generalize these equations, let n be the current time index and let $\mathbf{z}_n$ be the L-dimensional measurement vector at this time. Then the RLS algorithms are

$$\hat{\mathbf{x}}_n = \hat{\mathbf{x}}_{n-1} + \mathbf{K}_n (\mathbf{z}_n - \mathbf{H}_n \hat{\mathbf{x}}_{n-1}) \tag{6-59}$$

$$\mathbf{K}_n = \mathbf{P}_{n-1} \mathbf{H}_n^T (\mathbf{H}_n \mathbf{P}_{n-1} \mathbf{H}_n^T + \mathbf{R}_n)^{-1} \tag{6-60}$$

$$\mathbf{P}_n = (\mathbf{P}_{n-1}^{-1} + \mathbf{H}_n^T \mathbf{R}_n^{-1} \mathbf{H}_n)^{-1} \tag{6-61}$$

$\mathbf{K}_n$ is the optimized $N \times L$ gain matrix and $\mathbf{P}_n$ is the $N \times N$ covariance matrix for the estimation error $\tilde{\mathbf{x}}_n$ at time n.

There is yet another form for the $\mathbf{P}_n$ recursion. It is left as an exercise for the reader to show that

$$\begin{aligned} \mathbf{P}_n &= \mathbf{P}_{n-1} - \mathbf{K}_n \mathbf{H}_n \mathbf{P}_{n-1} \\ &= (\mathbf{I} - \mathbf{K}_n \mathbf{H}_n) \mathbf{P}_{n-1} \end{aligned} \tag{6-62}$$

The recursive algorithm of Eq. (6-59) has the predictor-corrector structure shown in Fig. 6-5, a notable feature of which is the feedback loop that creates the residual $\mathbf{r}_n = (\mathbf{z}_n - \hat{\mathbf{z}}_n)$. If the estimator is well-behaved, $\hat{\mathbf{x}}_n$ will be close to $\mathbf{x}$, and we would expect $\hat{\mathbf{z}}_n$ to be close to $\mathbf{z}_n$. The residual is small, and the filter tends to extrapolate along the inner prediction loop. On the other hand, a divergence between $\hat{\mathbf{x}}_n$ and $\mathbf{x}$ would be sensed at $\mathbf{r}_n$, and the corrective signal applied through $\mathbf{K}_n$ would tend to null out the incipient divergence. Predictor-corrector filters of this type tend to be robust; that is, they can successfully ride out such divergences.

Another interpretation of the RLS filter can be found by writing Eq. (6-59) as

$$\hat{\mathbf{x}}_n = (\mathbf{I} - \mathbf{K}_n \mathbf{H}_n) \hat{\mathbf{x}}_{n-1} + \mathbf{K}_n \mathbf{z}_n \tag{6-63}$$

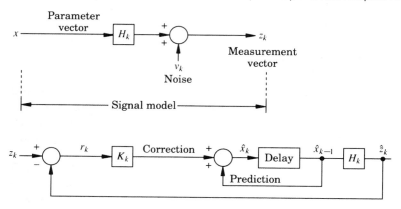

Figure 6-5   *Recursive least-squares estimator.*

This form shows that the updated estimate is a weighted sum of the old estimate $\hat{\mathbf{x}}_{n-1}$ and the new data $\mathbf{z}_n$. The gain $\mathbf{K}_n$ is chosen to allocate these weights optimally.

The RLS algorithm of Eqs. (6-59) to (6-61) requires an initialization $\hat{\mathbf{x}}_0$, along with $\mathbf{P}_0$. These could be assigned from a priori information, or they could be obtained by a first-pass batch estimate via Eq. (6-33) and (6-44) with $\mathbf{W} = \mathbf{R}^{-1}$.

6.2.2 THE KALMAN FILTER

In this section, we provide a brief sketch of the Kalman filter, viewed as an extension and generalization of the RLS filter of the preceding section. Our intent here is to indicate the broader class of signals the Kalman filter can accommodate. Our approach is purely heuristic, so no proofs are offered. The reader can find detailed derivations in many references, for example, Kalman (1960), Kalman and Bucy (1961), and Balakrishnan (1984).

The recursive least-squares estimator of Eqs. (6-59) to (6-62) is optimal for the estimation of a constant from noisy data. A generalization of this signal model allows $\mathbf{x}$ to change with time and in fact to be a random process generated by

$$\mathbf{x}_{n+1} = \mathbf{\Phi}\mathbf{x}_n + \mathbf{w}_n \tag{6-64}$$

where $\mathbf{x}_n$ is the N-dimensional state vector

$\mathbf{\Phi}$ is the time-invariant state transition matrix

$\mathbf{w}_n$ is a white-noise process, the "state" noise

In the least-squares case, $\mathbf{w}_n$ was zero and $\mathbf{\Phi} = \mathbf{I}$. The measurement model on the other hand is the same:

$$\mathbf{z}_n = \mathbf{H}\mathbf{x}_n + \mathbf{v}_n \tag{6-65}$$

As before, $\mathbf{H}$ is the observation matrix, and $\mathbf{v}_n$ is now a zero-mean white-noise process representing measurement errors. The measurement noise and the state noise are uncorrelated. We can summarize these noise models as follows:

$$
\begin{aligned}
E\{\mathbf{w}_n\} &= 0 && \text{zero mean state noise} \\
E\{\mathbf{w}_n\mathbf{w}_j^T\} &= Q\delta_{n-j} && \text{white state process} \\
E\{\mathbf{v}_n\} &= 0 && \text{zero mean measurement noise} && \text{(6-66)} \\
E\{\mathbf{v}_n\mathbf{v}_j^T\} &= R\delta_{n-j} && \text{white process} \\
E\{\mathbf{v}_n\mathbf{w}_j^T\} &= 0 \text{ for all } n, j && \text{uncorrelated state and measurement noise}
\end{aligned}
$$

We have assumed that $\mathbf{w}_n$ and $\mathbf{v}_n$ are stationary and that Φ and $\mathbf{H}$ are constant. These assumptions are not necessary for the Kalman filter but rather are simplifications for our purposes here. This signal model is shown in Fig. 6-6 and should be compared with the least-squares parameter model of Fig. 6-5.

The Kalman filter generates the optimal linear estimate $\hat{\mathbf{x}}_n$ from the data $\{\mathbf{z}_k, k \leq n\}$; this filter minimizes the mean-squared error

$$
E\{(\mathbf{x}_n - \hat{\mathbf{x}}_n)^T(\mathbf{x}_n - \hat{\mathbf{x}}_n)\}
$$

Not surprisingly, it can be shown that the Kalman filter has the recursive predictor-corrector structure shown in Fig. 6-7. It has the same form as the RLS filter, except that now the predicted state is $\Phi\hat{\mathbf{x}}_{k-1}$ rather than $\hat{\mathbf{x}}_{k-1}$. The major computational difference is the formula for calculating the gain $\mathbf{K}_n$. The Kalman gain is more complex and involves all the process parameters, $\mathbf{Q}$ as well as $\mathbf{R}$ and Φ as well as $\mathbf{H}$.

We may summarize as follows:

Let $\qquad \hat{\mathbf{x}}_n^- =$ estimate of $\mathbf{x}_n$ using $\{\mathbf{z}_k, k \leq n-1\}$,

the predicted estimate;

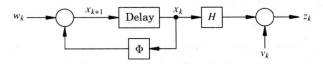

Figure 6-6 *Signal model for Kalman filter.*

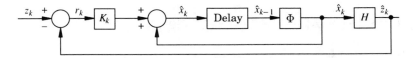

Figure 6-7 *Recursive Kalman filter structure.*

$$\hat{\mathbf{x}}_n = \text{estimate of } \mathbf{x}_n \text{ using } \{\mathbf{z}_k, k \leq n\}$$

the updated estimate;

and let $\mathbf{P}_n^-$ and $\mathbf{P}_n$ be the prediction-error and update-error covariance matrices, respectively:

$$\mathbf{P}_n^- = \text{cov}\{\mathbf{x}_n - \hat{\mathbf{x}}_n^-\}$$

$$\mathbf{P}_n = \text{cov}\{\mathbf{x}_n - \hat{\mathbf{x}}_n\} \tag{6-67}$$

Then the Kalman filter equations are

Prediction: $\hat{\mathbf{x}}_n^- = \boldsymbol{\Phi}\hat{\mathbf{x}}_{n-1}$ (6-68)

$$\mathbf{P}_n^- = \boldsymbol{\Phi}\mathbf{P}_{n-1}\boldsymbol{\Phi}^T + \mathbf{Q} \tag{6-69}$$

Updating: $\mathbf{K}_n = \mathbf{P}_n^-\mathbf{H}^T[\mathbf{H}\mathbf{P}_n^-\mathbf{H}^T + \mathbf{R}]^{-1}$ (6-70)

$$\hat{\mathbf{x}}_n = \hat{\mathbf{x}}_n^- + \mathbf{K}_n[\mathbf{z}_n - \mathbf{H}\hat{\mathbf{x}}_n^-] \tag{6-71}$$

$$\mathbf{P}_n = [\mathbf{P}_n^{-1} + \mathbf{H}^T\mathbf{R}^{-1}\mathbf{H}]^{-1} \tag{6-72}$$

As before, the algorithm is initialized with an initial state estimate $\hat{\mathbf{x}}_0$, along with its covariance matrix $\mathbf{P}_0$.

These equations are generalizations of the RLS estimator, Eqs. (6-59) to (6-61). The major differences are:

1. The parameter vector $\mathbf{x}$, which is not constant in time, but a dynamic process driven by state noise. This is manifested by the matrices $\boldsymbol{\Phi}$ and $\mathbf{Q}$ in the Kalman formulation.

2. The steady-state value of the gain matrix $\mathbf{K}_n$. In the RLS filter, $\mathbf{K}_n \to 0$ as $n \to \infty$, and the corrective signal in the predictor-corrector structure is driven to zero. This effectively opens the outer loop, causing the estimate and state parameters to diverge under mismatched conditions. In the Kalman filter, on the other hand, the $\mathbf{Q}$ term in Eq. (6-69) serves as a forcing function in the covariance recursions. This term prevents $\mathbf{P}_n$ from going to zero as n $\to \infty$, which in turn keeps $\mathbf{K}_n$ nonzero.

3. When certain conditions are satisfied (Anderson and Moore, 1979), the steady-state Kalman gain $\mathbf{K}_\infty$ takes on the value of the optimal Wiener filter. In fact, a suboptimal Kalman filter could use this steady-state value for all n.

6.3 LEAST SQUARES DESIGN OF FIR FILTERS

The least-squares philosophy espoused in this chapter finds application in two broad areas: 1. Modelling and design of fixed FIR filters. An example is that of a channel equalizer or whitening filter. 2. Design of adaptive least-squares filters that are capable of modifying the parameters in a changing environment.

We first develop the fixed LS design of the FIR structure shown in Fig. 6-8. Then we adapt the design to the signal environment.

6.3.1 LEAST-SQUARES INVERSE MODELING: THE EQUALIZER PROBLEM

As shown in Fig. 6-8, the process or channel has a transfer function $H(z)$ and an associated impulse response $h(n)$. If $H(z)$ were autoregressive $[H(z) = 1/A(z)$ all-pole and known] then we could choose a FIR filter $W(z) = A(z)$ that would exactly cancel $H(z)$ and achieve

$$H(z)A(z) = 1$$

or perfect equalization or deconvolution.

If the input ξ_n is a deterministic impulse δ_n, the output under these idealized conditions would also be an impulse $y_n = \delta_n$. On the other hand, if ξ_n is a white-noise process with autocorrelation equal to δ_n, then the equalization $W(z)H(z) = 1$ makes y_n likewise a white-noise process with autocorrelation δ_n. Thus the inverse filter is said to be a whitening filter.

The real problem at hand, however, is that $H(z)$ may not be purely autoregressive. It could be of the form

$$H(z) = N(z)/A(z)$$

Cancellation compensation would then require

$$W(z) = A(z)/N(z)$$

an IIR structure. Furthermore, if $N(z)$ were nonminimum phase (that is, with zeroes outside the unit circle), then the equalizer would be unstable. We also look to a design approach that is extendable to the adaptive-filter case—that is, when $h(n)$ is not known a priori, but has to be inferred from measurements and subsequently tracked. The adaptive case will be developed in Sect. 6.4.

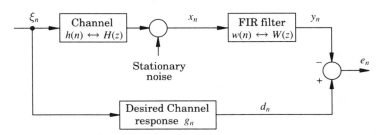

Figure 6-8 *Inverse modelling.*

We may state the problem as follows: For the structure of Fig. 6-8, find the coefficients $w(n), 0 \leq n \leq N$, of the FIR filter $W(z)$ that minimizes the sum of the squared residuals,

$$J = \sum e^2(n) \qquad (6\text{-}73)$$

We assume that $\{h(n), n \geq 0\}$ is known and that the order of the filter N is selected in advance.

The limits on the summation in Eq. (6-73) lead to differing error measures, solution methods, and results. There are two cases of general interest,

$$J_L = \sum_{n=0}^{L} e^2(n) \qquad (6\text{-}74\text{a})$$

$$J_\infty = \sum_{n=0}^{\infty} e^2(n) \qquad (6\text{-}74\text{b})$$

We develop the method using the error measure of Eq. (6-74a) first. As we shall see, this leads to the the covariance method of solution; while the second error measure leads to the correlation method.

From Fig. 6-8, the error residual is

$$e_n = d_n - \sum_{k=0}^{n} w_{n-k} x_k \qquad n = 0, 1, ..., L; \; L \geq N \qquad (6\text{-}75)$$

Equation (6-75) can be expressed in vector-matrix form:

$$\begin{bmatrix} x_0 & 0 & \cdots & 0 \\ x_1 & x_0 & \cdots & 0 \\ \vdots & \vdots & & \\ x_N & x_{N-1} & \cdots & x_0 \\ \hdashline x_{N+1} & x_N & \cdots & x_1 \\ \vdots & \vdots & & \\ x_L & x_{L-1} & \cdots & x_{L-N} \end{bmatrix} \begin{bmatrix} w_0 \\ w_1 \\ \vdots \\ w_N \end{bmatrix} + \begin{bmatrix} e_0 \\ e_1 \\ \vdots \\ e_N \\ \hdashline \vdots \\ e_L \end{bmatrix} = \begin{bmatrix} d_0 \\ d_1 \\ \vdots \\ d_N \\ \hdashline \vdots \\ d_L \end{bmatrix} \qquad (6\text{-}76)$$

or
$$\mathbf{Xw} + \mathbf{e} = \mathbf{d} \qquad (6\text{-}77)$$

where $\mathbf{X}$ is the $(L+1) \times (N+1)$ data matrix
$\quad \mathbf{w}$ is the $(N+1)$ vector of FIR weights
$\quad \mathbf{e}$ is the $(L+1)$ error vector
$\quad \mathbf{d}$ is the $(L+1)$ desired response vector

The least-squares filter design procedure is to choose $\mathbf{w} = \hat{\mathbf{w}}$ to minimize

$$J_L = \sum_{n=0}^{L} e^2(n) = \mathbf{e}^T \mathbf{e}$$

But $$\mathbf{e} = \mathbf{d} - \mathbf{Xw}$$

therefore $$J_L = (\mathbf{d} - \mathbf{Xw})^T(\mathbf{d} - \mathbf{Xw}) \tag{6-78}$$

This is the same problem statement as the one encountered in the earlier least-squares problems. Equations (6-77) and (6-78) suggest the block diagram of Fig. 6-9. Comparing this with Fig. 6-3 and making the identifications

$$\mathbf{x} \to \mathbf{w} \qquad \mathbf{z} \to \mathbf{d}$$

$$\mathbf{v} \to \mathbf{e} \qquad \mathbf{H} \to \mathbf{X}$$

we can invoke the least-squares results previously obtained to get

$$\hat{\mathbf{w}} = [(\mathbf{X}^T\mathbf{X})^{-1}\mathbf{X}^T]\mathbf{d} \tag{6-79}$$

The matrix to be inverted in Eq. (6-79) is the $(N + 1) \times (N + 1)$ covariance matrix,

$$\mathbf{R}_L = \begin{bmatrix} \sum_{i=0}^{L} x_i^2 & \cdots & \sum_{i=0}^{L-N} x_i x_{i+N} \\ \sum_{i=0}^{L-1} x_i x_{i+1} & \sum_{i=0}^{L-1} x_i^2 & \cdots \\ \cdots & & \cdots \\ \sum_{i=0}^{L-N} x_i x_{i+N} & \cdots & \sum_{i=0}^{L-N} x_i^2 \end{bmatrix} = \mathbf{X}^T\mathbf{X} \tag{6-80}$$

Note that because of the limits used in the error measure of Eq. (6-74a), the terms along the diagonals of $\mathbf{R}_L$ are not equal. Thus $\mathbf{R}_L$ in Eq. (6-80) is what is known as a covariance matrix (real, symmetric, positive-semidefinite) rather than a correlation matrix. Inversion of $\mathbf{R}_L$ is achieved efficiently by a Cholesky decomposition similar to that discussed in App. E.

Now if $L \to \infty$, all diagonal terms in $\mathbf{R}_L$ are equal, and $\mathbf{R}_L$ becomes $\mathbf{R}$, a correlation matrix:

$$\mathbf{R} = \begin{bmatrix} r_0 & r_1 & \cdots & r_N \\ r_1 & r_0 & \cdots & \\ \hdashline & & & \\ r_N & \cdots & \cdots & r_0 \end{bmatrix} \tag{6-81}$$

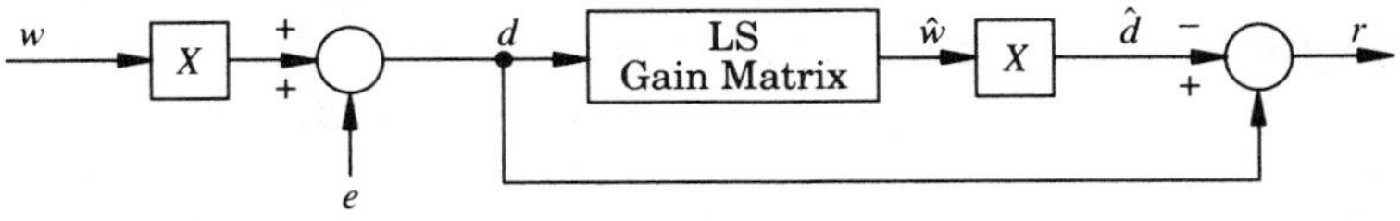

Figure 6-9   *FIR inverse modelling as LS estimation.*

where
$$r_m = \sum_{i=0}^{\infty} x_i x_{i+m} = r_{-m}$$

This is a Toeplitz matrix, the inversion of which can be done iteratively by means of a Durbin-Levinson recursion along the lines of that developed in App. E.

A second way of obtaining a correlation matrix is to window the data $x(n)$. If we take $x(n)$ to be zero for $n > L - N$, then the data matrix $\mathbf{X}$ for the truncated data is

$$\mathbf{X} = \begin{bmatrix}
x_0 & 0 & 0 & \cdots & 0 \\
x_1 & x_0 & 0 & \cdots & 0 \\
\cdots & \cdots & \cdots & \cdots & \cdots \\
x_N & x_{N-1} & x_{N-2} & \cdots & x_0 \\
\hdashline
x_{N+1} & x_N & x_{N-1} & \cdots & x_1 \\
\cdots & \cdots & \cdots & \cdots & \cdots \\
x_{L-N} & x_{L-N-1} & x_{L-N-2} & \cdots & x_{L-2N} \\
0 & x_{L-N} & x_{L-N-1} & \cdots & x_{L+1-2N} \\
0 & 0 & x_{L-N} & \cdots & x_{L+2-2N} \\
\cdots & \cdots & \cdots & \cdots & \cdots \\
0 & 0 & 0 & \cdots & x_{L-N}
\end{bmatrix} \qquad (6\text{-}82)$$

In this case $\mathbf{X}^T\mathbf{X}$ is also Toeplitz:

$$\mathbf{X}^T\mathbf{X} = \begin{bmatrix}
\rho_0 & \rho_1 & \cdots & \rho_N \\
\rho_1 & \rho_0 & \cdots & \\
\cdots & \cdots & \cdots & \cdots \\
\rho_N & \cdots & \cdots & \rho_0
\end{bmatrix} \qquad (6\text{-}83)$$

where
$$\rho_0 = \sum_{i=0}^{L-N} x_i^2$$

$$\rho_m = \sum_{i=0}^{L-N} x_i x_{i+m} \qquad (6\text{-}84)$$

The inversion of this fudged matrix can also be accomplished with the aid of the Levinson recursion rather than the more complex Cholesky decomposition that would be required if $x(n)$ were not windowed. The resulting estimate is only an approximation to the least-squares solution of Eq. (6-79), however.

The optimal least-squares filter coefficients given by Eq. (6-79) and Eq. (6-80) depend only on the observed sequence $\{x_n\}$ and on the desired channel output signal $\{d_n\}$ This is the formulation needed for the adaptive FIR filter, described in the next section; the adaptive filter will be data-dependent, since the channel $\{h_n\}$ is unknown or varying or both.

For a fixed filter design, we can assume that $\{h_n\}$ is known and unchanging. In this case we take

$$\xi_n = \delta_n \tag{6-85}$$

then
$$d_n = g_n \tag{6-86}$$

where g_n is the desired channel impulse response. Now in the absence of noise,

$$x_n = h_n \tag{6-87}$$

where h_n is the given channel impulse response. Therefore for a fixed filter design, we take

$$\begin{aligned} d_n &\rightarrow g_n \\ x_n &\rightarrow h_n \\ X &\rightarrow H \end{aligned} \tag{6-88}$$

The optimal FIR filter is then

$$\hat{\mathbf{w}} = (\mathbf{H}^T\mathbf{H})^{-1}\mathbf{H}^T\mathbf{g} \tag{6-89}$$

where
$$\mathbf{g}^T = [g_0, g_1, ..., g_L] \tag{6-90}$$

and H is the $(L+1) \times (N+1)$ transmission matrix (Freeman, 1966). It has the same form as $\mathbf{X}$ in Eq. (6-76), with the substitution of h_i for x_i. Similarly, the covariance matrices $\mathbf{R}_L$ and $\mathbf{R}$ are the same as before, again with the substitution of h_i for x_i in Eqs. (6-80) and (6-81).

For a fixed filter, the correlation matrix approach makes more sense. In this case with $L \rightarrow \infty, (\mathbf{H}^T\mathbf{H}) = \mathbf{R}$ has the form of Eq. (6-81), and

$$\mathbf{H}^T\mathbf{g} = \begin{bmatrix} r_{gh}(0) \\ \vdots \\ r_{gh}(N) \end{bmatrix} = \mathbf{r}_{gh} \tag{6-91}$$

where
$$r_{gh}(k) = \sum_{i=0}^{\infty} h_i g_{i+k},$$

This gives us

$$\hat{\mathbf{w}} = \mathbf{R}^{-1}\mathbf{r}_{gh} \tag{6-92}$$

In the foregoing, $r_{gh}(k)$ is a deterministic cross correlation sequence.

Finally for the whitening filter, the desired response is $g_n = \delta_n$. For this case,

$$\mathbf{g}^T = [1, 0, ..., 0]$$

and
$$\hat{\mathbf{w}} = h_0 \mathbf{R}^{-1} \begin{bmatrix} 1 \\ 0 \\ 0 \\ \vdots \\ 0 \end{bmatrix} \tag{6-93}$$

This last equation has the form of the Yule-Walker equation.

6.3.2 THE LATTICE FILTER AS LEAST-SQUARES FIR

The least-squares inverse equation (6-93) can be solved using the Levinson iteration (App. E). Our interest here is to show that the lattice filter equations of Sec. 5.1.1.3 can be viewed as the LS FIR filter of Eq. (6-93).

To put the two on the same footing, we normalize the FIR coefficient weights via

$$\mathbf{a} = \frac{1}{w_0}\mathbf{w} = \begin{bmatrix} 1 \\ a_1 \\ \vdots \\ a_N \end{bmatrix} \tag{6-94}$$

and define
$$\frac{h_0}{w_0} \triangleq E_N \tag{6-95}$$

Substitution of these into Eq. (6-93) gives

$$\hat{\mathbf{a}} = \mathbf{R}^{-1} \begin{bmatrix} E_N \\ 0 \\ \vdots \\ 0 \end{bmatrix} \tag{6-96}$$

where $\mathbf{R}$ is a symmetric Toeplitz matrix of the form of Eq. (6-81), repeated here for convenience:

$$\mathbf{R} = \begin{bmatrix} r_0 & r_1 & \cdots & r_N \\ r_1 & r_0 & \cdots & \\ \cdots & \cdots & \cdots & \cdots \\ r_N & \cdots & \cdots & r_0 \end{bmatrix} \tag{6-81}$$

The derivation in App. E can be used iteratively to calculate the $\{a_i\}$ coefficients, with the initialization starting at

$$E_0 = r_0$$

After N iterations, we determine $E_N = h_0/w_{0,N}$, and from this $w_{0,N}$ is found. Those same iterations led us to the lattice filter structure shown in Fig. 5-6, which can now be interpreted as an alternate realization of the least-squares FIR inverse filter. The iterations are spelled out in App. E.

6.4   LEAST-SQUARES ADAPTIVE FILTERS

In the FIR design of Sec. 6.3, the inverse filter was selected on the basis of a known channel characterization $h(n)$. The filter weights were de-termined off-line, and once the filter is implemented, they remain fixed. Adaptive filters, on the other hand, are time-varying signal processors. These digital filters have time-varying coefficients whose values are ad-justed at regular intervals in real time to minimize an ongoing error measure. That is, the filters *adapt* their parameters to a sensed error measure.

In this section, we will describe the inverse modelling process as an adaptive filter whose coefficients are changed in accordance with the Widrow (1985) least-mean-square (LMS) adaptive algorithm.

6.4.1   ADAPTIVE LEAST-SQUARES PROBLEM FORMULATION

The adaptive version of the least-squares inverse filter is shown in Fig. 6-10. The adaptive filter consists of two parts: the FIR filter structure that processes the data stream and the adaptive algorithm that adjusts the parameters in that filter. The input signal (or *test signal*) ξ_k is known, as is the desired response d_k. The response of the unknown channel to ξ_k is corrupted with noise to generate x_k the available channel output signal. For the deterministic problem, we assume that the additive noise is zero, so that x_k represents only the channel response to ξ_k. Following the steps in Sec. 6.3, we set up the error e_k as

$$e_k = d_k - y_k$$

$$= d_k - \sum_j w_{k-j} x_j \tag{6-97}$$

where
$$w_k = 0, k < 0, k > N$$

We will not specify the test signal ξ_k, but rather formulate the prob-lem in terms of x_k, the observed channel output, and d_k, the desired output of an ideal channel. Then Eq. (6-97) is the same as Eq. (6-75). Following Eqs. (6-75) to (6-78), we have

$$\mathbf{Xw} + \mathbf{e} = \mathbf{d} \tag{6-98}$$

and
$$J = \mathbf{e}^T \mathbf{e} = (\mathbf{d} - \mathbf{Xw})^T (\mathbf{d} - \mathbf{Xw}) \tag{6-99}$$

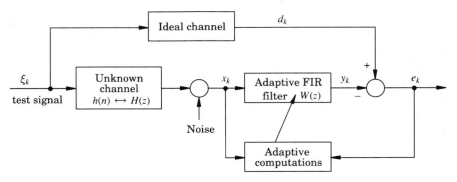

Figure 6-10 *Adaptive inverse filter structure.*

As before, we choose **w** to minimize the performance measure (also termed the mean squared error or MSE when a stochastic signal model is used).

With

$$\mathbf{R} = \mathbf{X}^T\mathbf{X} \qquad \text{the data covariance matrix} \tag{6-100}$$

$$\mathbf{p}^T = \mathbf{d}^T\mathbf{X} \qquad \text{the cross covariance vector}$$

the quadratic performance measure can be written as

$$J = \mathbf{d}^T\mathbf{d} - 2\mathbf{p}^T\mathbf{w} + \mathbf{w}^T\mathbf{R}\mathbf{w} \tag{6-101}$$

The gradient (see Eq. 6-26) is then

$$\frac{\partial J}{\partial \mathbf{w}} = \nabla = 2\mathbf{R}\mathbf{w} - 2\mathbf{X}^T\mathbf{d} = 2(\mathbf{R}\mathbf{w} - \mathbf{p}) \tag{6-102}$$

The optimum weight $\mathbf{w}^*$ is obtained when $\nabla = 0$:

$$\mathbf{w}^* = \mathbf{R}^{-1}\mathbf{X}^T\mathbf{d} = \mathbf{R}^{-1}\mathbf{p} \tag{6-103}$$

At this setting, the minimum MSE is

$$J^* = \mathbf{d}^T\mathbf{d} - \mathbf{p}^T\mathbf{w}^* \tag{6-104}$$

For any arbitrary weight **w**, ∇ will be nonzero. If we premultiply Eq. (6-102) by $\frac{1}{2}\mathbf{R}^{-1}$,

$$\tfrac{1}{2}\mathbf{R}^{-1}\nabla = \mathbf{w} - \mathbf{R}^{-1}\mathbf{p} \tag{6-105}$$

Noting that $\mathbf{w}^* = \mathbf{R}^{-1}\mathbf{p}$, Eq. (6-105) becomes

$$\mathbf{w}^* = \mathbf{w} - \tfrac{1}{2}\mathbf{R}^{-1}\nabla \tag{6-106}$$

Equation Eq. (6-106) asserts that if $\mathbf{R}^{-1}$ is known and ∇ is the (non-zero) gradient corresponding to $\mathbf{w}$, then we can calculate $\mathbf{w}^*$ in just one iteration. The difficulty is that $\mathbf{R}^{-1}$ is not known precisely, nor for that matter can ∇ be calculated accurately. These reservations are particularly of consequence when the processes are nonstationary, that is, when the channel parameters are slowly changing with time. Thus we are led to a less ambitious iteration, taking small bites rather than attempting to correct the filter in one fell swoop. Thus we try

$$
\boxed{
\begin{aligned}
\mathbf{w}_{k+1} &= \mathbf{w}_k - \gamma \mathbf{R}^{-1}\nabla_k \\
\nabla_k &= 2(\mathbf{R}\mathbf{w}_k - \mathbf{p})
\end{aligned}
}
\tag{6-107}
$$

To assess the convergence properties of Eq. (6-107), let

$$
\tilde{\mathbf{w}}_k = \mathbf{w}_k - \mathbf{w}^*
\tag{6-108}
$$

Subtracting $\mathbf{w}^*$ from both sides of Eq. (6-107) and using Eq. (6-106) for ∇_k leads to

$$
\tilde{\mathbf{w}}_{k+1} = (1 - 2\gamma)\tilde{\mathbf{w}}_k
\tag{6-109}
$$

with solution
$$
\tilde{\mathbf{w}}_k = (1 - 2\gamma)^k \tilde{\mathbf{w}}_0
\tag{6-110}
$$

Hence the iteration of Eq. (6-107) is stable for $|1-2\gamma| < 1$ and monotonic for $0 < 1 - 2\gamma < 1$. The limits on γ are thus

$$
\begin{aligned}
&0 < \gamma < 1 \qquad &&\text{stable} \\
&0 < \gamma < 1/2 \qquad &&\text{stable and nonoscillatory}
\end{aligned}
\tag{6-111}
$$

We can similarly develop a recursion formula for the performance measure:

$$
J_k = \mathbf{d}^T\mathbf{d} - 2\mathbf{p}^T\mathbf{w}_k + \mathbf{w}_k^T\mathbf{R}\mathbf{w}_k
\tag{6-112}
$$

After substituting Eqs. (6-108), (6-103), and (6-110) in this last equation, we obtain

$$
J_k = J^* + (1 - 2\gamma)^{2k} E_0
\tag{6-113}
$$

$$
E_0 = \tilde{\mathbf{w}}_0^T \tilde{\mathbf{R}}\mathbf{w}_0
$$

We observe that $J_k \rightarrow J^*$ monotonically from above at a rate determined by $(1 - 2\gamma)^{2k}$.

The foregoing iterations require a knowledge of $\mathbf{R}$ at each iteration of ∇_k. Often $\mathbf{R}$ is not known with precision, or it may vary. In those cases, we need an alternate formulation which does not depend on an a priori covariance matrix. Such a formulation is developed in the next section as the LMS algorithm.

6.4.2 THE LMS ALGORITHM

In the absence of an explicit model for the data covariance $\mathbf{R}$, we may replace the iterative algorithm of Eq. (6-107) by the steepest-descent version,

$$\mathbf{w}_{k+1} = \mathbf{w}_k - \mu\nabla_k \tag{6-114}$$

This iteration starts with an initial guess $\mathbf{w}_0$. At each step, the estimate is incremented by a term proportional to the negative gradient. The proportionality factor μ controls the convergence rate of the algorithm.

To get some foundation, let us again define $\tilde{\mathbf{w}}$ by Eq. (6-108). Subtracting $\mathbf{w}^*$ from both sides of Eq. (6-114) and using Eq. (6-106) gives

$$\tilde{\mathbf{w}}_{k+1} = (\mathbf{I} - 2\mu\mathbf{R})\tilde{\mathbf{w}}_k \tag{6-115}$$

Now if we pretend that $\mathbf{R}$ is a Toeplitz correlation matrix, we can factor it into

$$\mathbf{R} = \mathbf{U}\mathbf{D}\mathbf{U}^T \tag{6-116}$$

where $\mathbf{U}$ is an orthonormal matrix ($\mathbf{U}^T\mathbf{U} = \mathbf{U}\mathbf{U}^T = \mathbf{I}$) and $\mathbf{D}$ is a diagonal matrix whose entries are the (real) eigenvalues of $\mathbf{R}$:

$$\mathbf{D} = \text{diag}\,[\lambda_1, \lambda_2, ..., \lambda_N], \qquad \lambda_i > 0 \qquad \forall i \tag{6-117}$$

Premultiplying both sides of Eq. (6-115) by $\mathbf{U}^T$ and substituting Eq. (6-116) gives

$$\mathbf{U}^T\tilde{\mathbf{w}}_{k+1} = \mathbf{U}^T\tilde{\mathbf{w}}_k - 2\mu\mathbf{D}\mathbf{U}^T\tilde{\mathbf{w}}_k$$

Now let $\mathbf{v}_k = \mathbf{U}^T\tilde{\mathbf{w}}_k$ so that

$$\mathbf{v}_{k+1} = (\mathbf{I} - 2\mu\mathbf{D})\mathbf{v}_k \tag{6-118}$$

The matrix $(\mathbf{I} - 2\mu\mathbf{D})$ is diagonal, with ith diagonal entry equal to $(1 - 2\mu\lambda_i)$. For stability and monotone convergence, we must maintain

$$|1 - 2\mu\lambda_i| < 1 \qquad \forall i$$

This is ensured by requiring

$$0 < \mu < 1/\lambda_{\text{max}} \tag{6-119}$$

A good choice is to set μ equal to half the upper limit in Eq. (6-119). A further approximation is possible. Note that

$$\sum \lambda_i = \text{trace } (\mathbf{R}) = N\sigma^2 \qquad (6\text{-}120)$$

for $\mathbf{R}$ a correlation matrix. In this interpretation, σ^2 is the average power in the signal $\{x_k\}$; statistically

$$\sigma^2 = E\{x_k^2\}$$

Hence if we take $\sum \lambda_i$ as the upper bound on λ_{max}, we have the constraint

$$0 < \mu < 1/N\sigma^2 \qquad (6\text{-}121)$$

Therefore we can summarize the steepest-descent algorithm for a signal with average power σ^2 and an order-N FIR filter as follows:

$$
\boxed{
\begin{aligned}
\mathbf{w}_{k+1} &= \mathbf{w}_k - \mu\nabla_k \qquad 0 < \mu < 1/N\sigma^2 \\
\nabla_k &= 2(\mathbf{Rw}_k - \mathbf{p})
\end{aligned}
}
\qquad (6\text{-}122)
$$

This steepest-descent algorithm still requires computation ∇_k from the data covariance $\mathbf{R}$.

The LMS algorithm provides a means to determine ∇_k directly from the available data and hence *adapts* to the data. In effect, we replace ∇_k in Eq. (6-122) by an estimate $\hat{\nabla}_k$ which is computed from the data. Thus, with the data vector, $\mathbf{x}_k^T = [x_k, x_{k-1}, ..., x_{k-N}]$ and with the scalar instantaneous error

$$e_k = d_k - \mathbf{w}_k^T \mathbf{x}_k$$

we have
$$\hat{\nabla}_k = \frac{\partial}{\partial \mathbf{w}} e_k^2$$

$$= 2e_k \frac{\partial}{\partial \mathbf{w}} [d_k - \mathbf{w}_k^T \mathbf{x}_k]$$

$$= -2e_k \mathbf{x}_k \qquad (6\text{-}123)$$

The LMS algorithm becomes

$$\mathbf{w}_{k+1} = \mathbf{w}_k + \frac{2u}{N\sigma^2} e_k \mathbf{x}_k \qquad 0 < u < 1 \qquad (6\text{-}124a)$$

In the predictor-corrector form, the last equation becomes

$$\mathbf{w}_{k+1} = \mathbf{w}_k + \mathbf{k}_k (d_k - \mathbf{x}_k^T \mathbf{w}_k), \qquad \mathbf{k}_k = \frac{2u}{N\sigma^2} \mathbf{x}_k \qquad (6\text{-}124b)$$

The LMS algorithm uses the data e_k and $\mathbf{x}_k$ themselves. It does not require a calculation of $\mathbf{R}$, but it does require a value for the average signal power σ^2. This could be estimated on line by

$$\hat{\sigma}^2 = \frac{1}{N} \sum_1^N x_k^2 = \frac{1}{N} \mathbf{x}_k^T \mathbf{x}_k \tag{6-125}$$

or an a priori value could be used.

This LMS algorithm and variations on it have been successfully used in various applications (Cowan and Grant, 1985), particularly as telephone channel equalizers.

The choice of u in Eq. (6-124) involves a compromise between rate of convergence and steady-state mean squared error after convergence. Widrow and Stearns (1985) have shown that the excess mean squared error, defined as the difference between the actual MSE and the minimum MSE, can be approximated by

$$\frac{MSE_{\text{excess}}}{MSE_{\text{min}}} \approx uN \qquad 0 < u < 1 \tag{6-126}$$

where N is the number of coefficients in the FIR filter. This trade-off between dynamic response and steady-state error is a recurring theme in estimation. Recent attempts to improve the MSE have led to recursive, predictor-corrector types of filter structure that inherently provide some signal smoothing along with convergent parameter estimation for channel equalization. The adaptive recursive least-squares estimator is representative of this class of FIR filter.

The operation of the FIR filter with the adaptive LMS algorithm is shown in Fig. 6-11. The FIR filter is represented in state-variable form, x_k is the scalar input, $\mathbf{x}_k$ is the data or state vector, and y_k is the scalar output, related by

$$\mathbf{x}_{k+1} = \mathbf{Z}\mathbf{x}_k + \mathbf{b}x_k$$

where
$$\mathbf{Z} = \begin{bmatrix} 0 & 0 & 0 & \cdots & 0 & 0 & 0 \\ 1 & 0 & 0 & \cdots & 0 & 0 & 0 \\ 0 & 1 & 0 & \cdots & 0 & 0 & 0 \\ \hdashline 0 & 0 & 0 & \cdots & 0 & 1 & 0 \end{bmatrix} \qquad \mathbf{b} = \begin{bmatrix} 1 \\ 0 \\ 0 \\ \cdots \\ 0 \end{bmatrix} \tag{6-127}$$

and
$$y_k = \mathbf{w}_k^T \mathbf{x}_k$$

At each cycle, the data vector $\mathbf{x}_k$ and the error e_k are fed to the adaptive algorithm which calculates the updated $\mathbf{w}_{k+1}$ and adjusts the filter coefficients accordingly.

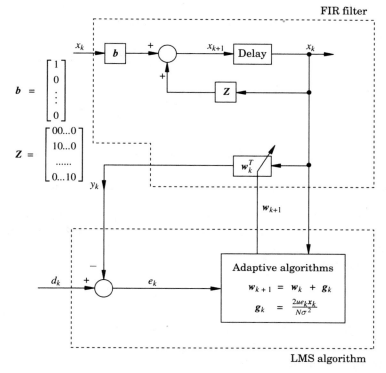

Figure 6-11   *Structure of the LMS adaptive FIR filter.*

6.4.3 RECURSIVE LEAST-SQUARES (RLS) ADAPTIVE FILTER

The RLS adaptive algorithm for updating the coefficients of the FIR filter is superior to the LMS (steepest-descent) algorithm of Sec. 6.4.2 in convergence properties, eigenvalue sensitivity, and excess MSE. The price paid for this improvement is additional computational complexity.

The RLS algorithm is readily obtained as a special case of the recursive least-squares estimators of Sec. 6.2. Equations Eq. (6-59), (6-60), and (6-61) are first particularized to the case where z_n is a scalar input at time n. This makes $\mathbf{H}_n$ in those equations a row vector and $\mathbf{R}_n$ a 1×1 matrix or scalar, which we scale to unity. $\mathbf{P}_n$ remains an $N \times N$ covariance matrix.

With the identifications

$$\mathbf{x} \to \mathbf{w} \quad \text{parameter vector}$$

$$\mathbf{H}_n^T \to \mathbf{x}_n \quad \text{data vector}$$

$$z_n \to d_n \quad \text{(scalar) reference signal}$$

$$\mathbf{K}_n \to \mathbf{k}_n \quad \text{gain vector}$$

the recursive least-squares estimator becomes

$$\hat{\mathbf{w}}_n = \hat{\mathbf{w}}_{n-1} + \mathbf{k}_n(d_n - \mathbf{x}_n^T \hat{\mathbf{w}}_{n-1}) \tag{6-128}$$

$$\mathbf{k}_n = \frac{\mathbf{P}_{n-1}\mathbf{x}_n}{1 + \mathbf{x}_n^T \mathbf{P}_{n-1}\mathbf{x}_n} \tag{6-129}$$

$$\mathbf{P}_n = (\mathbf{P}_{n-1}^{-1} + \mathbf{x}_n \mathbf{x}_n^T)^{-1} \tag{6-130}$$

or, alternatively, $$\mathbf{P}_n = \mathbf{P}_{n-1} - \mathbf{k}_n \mathbf{x}_n^T \mathbf{P}_{n-1} \tag{6-131}$$

As before, $\mathbf{P}_n$ is the covariance matrix for the error in the filter weights ($\hat{\mathbf{w}}_n - \mathbf{w}^*$).

The RLS algorithm, Eqs. (6-128) to (6-130) is seen to have a predictor-corrector form.† The gain $\mathbf{k}_n$ is calculated every cycle from the data vector $\mathbf{x}_n$ and the previous covariance $\mathbf{P}_{n-1}$. This gain then multiplies the *estimation residual*

$$d_n - \mathbf{x}_n^T \hat{\mathbf{w}}_{n-1}$$

and adds it to the old estimate to form the update.

Another interpretation of the predictor-corrector algorithm is afforded by writing Eq. (6-128) as

$$\hat{\mathbf{w}}_n = (\mathbf{I} - \mathbf{k}_n \mathbf{x}_n^T)\hat{\mathbf{w}}_{n-1} + \mathbf{k}_n d_n \tag{6-132}$$

This form shows the update as a weighted sum of the old estimate and the new reference signal d_n. Proper choice of $\mathbf{k}_n$ adjusts the weights in an optimal fashion.

In the foregoing RLS algorithm, all data and errors, however remote in time, are equally weighted. Thus the filter coefficients depend on all past inputs and the algorithm has a long, expanding memory. In order to stress more recent data, we employ an ageing factor on the data. This is accomplished by modifying the performance measure Eq. (6-78) to read

$$J_L = \sum_{k=0}^{L} \lambda^{L-k} e^2(k) = \mathbf{e}^T \mathbf{\Lambda} \mathbf{e} \tag{6-133}$$

where $$0 < \lambda \le 1$$

$$\mathbf{e}^T = [e_0, e_1, ..., e_L] \tag{6-134}$$

---

†This is similar to the LMS algorithm, Eq. (6-124b). In the LMS case, $\mathbf{k}_n$ was proportional to the signal amplitude vector $\mathbf{x}_n$ and inversely proportional to the signal power. In the RLS case of Eq. (6-129) $\mathbf{k}_n$ depends on $\mathbf{x}_n$ and $\mathbf{P}_n$.

and $$\mathbf{\Lambda} = \text{diag}\,[\lambda^L, \lambda^{L-1}, ..., 1] \tag{6-135}$$

This error measure can be recognized as a variant of the weighted least-squares form encountered in Eq. (6-32). In the present instance, the weighting is *exponential* in time, regulated by the parameter λ. We can use the results of Eq. (6-33) to write

$$\hat{\mathbf{w}} = (\mathbf{X}^T \mathbf{\Lambda} \mathbf{X})^{-1} (\mathbf{X}^T \mathbf{\Lambda}) \mathbf{d}$$

When this least-squares result is made recursive, it can be shown that the following equations result (see Prob. 6.13):

$$\boxed{\begin{aligned}
\hat{\mathbf{w}}_n &= \hat{\mathbf{w}}_{n-1} + \mathbf{k}_n [d_n - \mathbf{x}_n^T \hat{\mathbf{w}}_{n-1}] \\
\mathbf{k}_n &= \mathbf{P}_{n-1} \mathbf{x}_n (\lambda + \mathbf{x}_n^T \mathbf{P}_{n-1} \mathbf{x}_n)^{-1} \\
\lambda \mathbf{P}_n &= \mathbf{P}_{n-1} - \mathbf{k}_n \mathbf{x}_n^T \mathbf{P}_{n-1}
\end{aligned}} \tag{6-136}$$

This latter set of equations is the RLS adaptive algorithm with exponentially time-weighted data. The filter structure is shown in Fig. 6-12. Using $\hat{\mathbf{w}}_{n-1}$, the FIR filter generates the inner product

$$y_n = \mathbf{x}_n^T \hat{\mathbf{w}}_{n-1}$$

The error residual $(d_n - \mathbf{x}_n^T \hat{\mathbf{w}}_{n-1})$ is formed and used in the adaptive algorithm to generate $\hat{\mathbf{w}}_n$, which is used in the FIR filter in the next cycle.

The algorithm can be initialized in various ways. The simplest is

$$\hat{\mathbf{w}}_0 = \mathbf{0} \tag{6-137}$$
$$\mathbf{P}_0 = \sigma^2 \mathbf{I}$$

where σ^2 is a large positive number. The convergence can be accelerated by gathering an initial batch of data $\mathbf{x}_0$ and $\mathbf{d}$ as in Eq. (6-76) and using a standard least-squares fit via Eq. (6-79):

$$\mathbf{P}_0 = (\mathbf{X}^T \mathbf{X})^{-1}$$
$$\hat{\mathbf{w}}_0 = \mathbf{P}_0 \mathbf{X}^T \mathbf{d} \tag{6-138}$$

But the latter approach requires a waiting interval to collect the data; hence the advantage of Eq. (6-138) over (6-137) could be illusory.

Comparisons (Hayken et al., 1987) of the LMS and RLS algorithms show that:

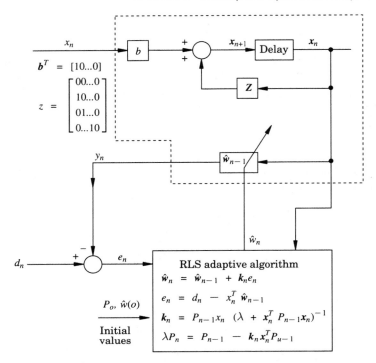

$$\mathbf{b}^T = [10...0]$$

$$z = \begin{bmatrix} 00...0 \\ 10...0 \\ 01...0 \\ 0...10 \end{bmatrix}$$

RLS adaptive algorithm

$$\hat{w}_n = \hat{w}_{n-1} + k_n e_n$$

$$e_n = d_n - x_n^T \hat{w}_{n-1}$$

$$k_n = P_{n-1} x_n (\lambda + x_n^T P_{n-1} x_n)^{-1}$$

$$\lambda P_n = P_{n-1} - k_n x_n^T P_{u-1}$$

$P_o, \hat{w}(o)$

Initial values

Figure 6-12 *Structure of the exponentially weighted RLS adaptive filter.*

1. The RLS algorithm converges much more rapidly than the LMS algorithm.

2. The RLS convergence is insensitive to the spread of the eigenvalues of the data covariance matrix.

3. The LMS convergence depends on the choice of u, whereas the RLS convergence is relatively insensitive to the value of λ.

4. Theoretically for n approaching infinity, the RLS algorithm produces zero excess MSE, whereas the LMS result is bounded below by Eq. (6-126).

The major drawback in the RLS scheme is the computational complexity in the calculation of the gain $\mathbf{k}_n$ and the covariance $\mathbf{P}_n$. In particular, round-off errors in updating $\mathbf{P}_n$ from Eq. (6-136) can lead to numerical instability, such as a negative-definite covariance matrix. Bierman (1975) recognized the problem in the context of the Kalman filter update and proposed the *square-root* factorization to resolve the numerical accuracy problem. (See also Giordano and Hsu, 1985.) However, this does not alleviate the computational burden in Eq. (6-136).

6.4.4 AN ADAPTIVE LATTICE FILTER

The lattice filter provides yet another structure for an adaptive FIR filter. In this case, the lattice adjusts the reflection (or PARCOR) coefficients $\{k_p\}$ in accordance with an adaptive criterion.

The lattice filter equations in Eq. (5-3) are repeated here as

$$e_p^+(n) = e_{p-1}^+(n) + k_p e_{p-1}^-(n)$$
$$e_p^-(n) = e_{p-1}^-(n-1) + k_p e_{p-1}^+(n-1) \qquad p = 1, 2, ..., L$$
$$e_0^+(n) = x(n) \tag{6-139}$$

where $e_p^+(n)$, $e_p^-(n)$ are the forward and backward prediction errors of the lattice filter in Fig. 5-6 and App. E. The lattice structure replaces the $[w(n) \leftrightarrow W(z)]$ block in the generic adaptive-filter configuration of Fig. 6-10.

The previous algorithms described here updated the coefficient vector $\mathbf{w}$ by the LMS algorithm Eq. (6-129), in which the gradient is approximated by a formulation using the observed instantaneous error Eq. (6-128).

For the lattice filter, the signal $e_p^+(n)$ at the pth stage is the one-step prediction error for the order-p filter. Hence we take the point of view that the least-squares error measure for the order-p filter is

$$J_p = [e_p^+(n)]^2 \tag{6-140}$$

This suggests that we update each reflection coefficient separately by using the error measure at that stage and choosing a step size μ_p commensurate with the signal level at that stage. Therefore, we take

$$k_p(n+1) = k_p(n) - \mu_p \frac{\partial J_p}{\partial k_p} \tag{6-141}$$

From Eqs. (6-139) and (6-140), we find

$$\frac{\partial J_p}{\partial k_p} = 2e_p^+(n) \frac{\partial e_p^+(n)}{\partial k_p} = 2e_p^+(n) e_{p-1}^-(n) \tag{6-142}$$

The adaptive lattice coefficient equation then becomes

$$k_p(n+1) = k_p(n) - 2\mu_p e_p^+(n) e_{p-1}^-(n) \tag{6-143}$$

Following the discussion in Sect. 6.4.2, we can take the step size

$$0 < \mu_p < 1/\sigma_p^2 \tag{6-144}$$

where σ_p^2 is the signal power at the pth stage. This σ_p^2 parameter must be provided a priori by a suitable model, or it can be estimated on line, for example, by

$$\sigma_p^2 \approx \frac{1}{M} \sum_{k=n-M}^{n-1} |e_p^+(k)|^2. \tag{6-145}$$

Some features of this adaptive lattice are:

1. The coefficients at any iteration must satisfy the bound,

$$0 < |k_p| < 1$$

2. After convergence, the filter becomes a standard lattice filter, one property of which is that the backward prediction errors $\{e_p^-(n), p = 1, 2, ..., L\}$ are mutually orthogonal. This means that the associated correlation matrix is diagonal, a property which lends itself to further adaptation (Widrow and Wallach, 1984).

More detailed descriptions of adaptive lattices can be found in the original papers by Makhoul (1978, 1981) and Griffith (1977) and in recent texts by Cowan and Grant (1985) and Widrow and Stearns (1985).

6.5 SUMMARY

In this chapter, we used classical least squares as a vehicle for introducing optimal and adaptive digital filters and to demonstrate the link between optimal estimation and digital filter design. The simple notions of elementary data fitting (least-squares line of regression) provided us with a springboard to least-squares FIR filter design with applications to deconvolution and fixed channel equalization using a batch or block-processing mode.

The recursive least-squares filter was shown to be a simple extension of the batch-mode processor to accommodate an expanding data base. This sequential estimator has a predictor-corrector structure, which when optimized becomes the Kalman filter.

The least-squares approach provided a natural framework for the adaptive-filter algorithm, particularly the adaptive FIR filter implemented by the LMS or the RLS update routine.

Interested readers can delve more deeply into this subject by studying the literature cited. For a description of the adaptive IIR filters, which we did not discuss, we refer them to the tutorial paper by Shynk (1989).

PROBLEMS

6.1. Prove the formulae for the differentiation of vector products and quadratic forms Eq. (6-26).

6.2. Show that the (r, k) entry in the covariance matrix Eq. (6-80) can be expressed as

$$\phi_{r,k} = \sum_{i=0}^{L-k} h_i h_{i+(k-r)}$$

Derive the recursion formula

$$\phi_{r,k} = \phi_{r+1,k+1} + h_{L-k} h_{L-r} \qquad r \le k$$

with starting value

$$\phi_{r,N} = \sum_{i=0}^{L-N} h_i h_{i+(N-r)}$$

Discuss how these recursions can be used to compute all the entries in the covariance matrix Eq. (6-80).

6.3. Given the system of Fig. 6-P1; let $\mathbf{r} = (\mathbf{z} - \mathbf{H}\hat{\mathbf{x}})$ be the residual. Find $\hat{\mathbf{x}}$ that minimizes $(\mathbf{C}\hat{\mathbf{x}})^T(\mathbf{C}\hat{\mathbf{x}})$ subject to the constraint that $\mathbf{r}^T\mathbf{r} \le \epsilon^2$. Hint: Set up

$$V = \lambda(\mathbf{r}^T\mathbf{r} - \epsilon^2) + \hat{\mathbf{x}}^T(\mathbf{C}^T\mathbf{C})\hat{\mathbf{x}}$$

(λ a Lagrangian multiplier), take

$$\frac{\partial V}{\partial \hat{\mathbf{x}}} = 0 \qquad \frac{\partial V}{\partial \lambda} = 0$$

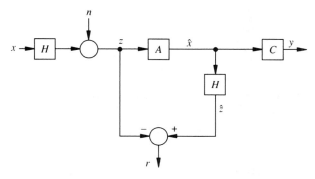

Figure 6-P1 *Problem 6.3.*

and show

$$\hat{\mathbf{x}} = [\mathbf{H}^T\mathbf{H} + 1/\lambda(\mathbf{C}^T\mathbf{C})]^{-1}\mathbf{H}^T\mathbf{z}$$

6.4. Show that both forms of the **A** matrix in Eq. (6-38) satisfy $(\mathbf{I} - \mathbf{AH}) = \mathbf{0}$.

6.5. Derive Eq. (6-62) from the recursive Eqs. (6-59) through (6-61).

6.6. Show that the minimum least-squares error for the whitening filter of Eq. (6-93) is

$$J = 1 - h_0^2(\mathbf{R})_{11}^{-1}$$

where **R** is the autocorrelation matrix Eq. (6-81).

6.7. Let the channel impulse response in Fig. 6-8 be

$$h(n) = \begin{cases} 1 & n = 0, 1, 2 \\ 0 & \text{otherwise} \end{cases}$$

and let $g(n) = \delta(n)$. Evaluate the optimal $\{w_n\}$ and the resulting least-squares error J.

6.8. Repeat Prob. 6.7 for the case that $G(z) = z^{-1}$ and compare the two cases.

6.9. Show that the least-squares error for the whitening filter Eq. (6-93) can be written as

$$J = \frac{1}{2\pi}\int_{-\pi}^{\pi}|1 - W(e^{j\theta})H(e^{j\theta})|^2 d\theta$$

6.10. Show that the minimized least-squares error J_∞ of Eq. (6-74b) can be expressed as

$$J_{min} = \rho - \mathbf{r}_{gh}^T\mathbf{w}$$

where $\rho = \sum_{n=0}^{\infty}|g(n)|^2$ and $\mathbf{r}_{gh}$ is given by Eq. (6-91).

6.11. Derive Eq. (6-120).

6.12. Show that the optimum least-squares coefficient vector $\mathbf{w}^*$ of Eq. (6-106) can be derived from the least-squares orthogonality condition Eq. (6-30) after an appropriate relabelling of variables that reconciles Fig. 6-10 with Fig. 6-3.

6.13. Derive the RLS adaptive algorithm with exponentially time-weighted data Eq. (6-136).

REFERENCES

A. V. Balakrishnan. *Kalman Filtering Theory*. New York: Optimization Software, Inc., 1984.

G. J. Bierman. "Measurement updating using the U-D factorization." *Proc. IEEE Conf. Decision and Control*, Houston, 1975.

———. *Factorization Methods for Discrete Sequential Smoothing*. New York: Academic Press, 1977.

C. F. N. Cowan and P. M. Grant, eds. *Adaptive Filters*. Englewood Cliffs, N. J.: Prentice-Hall, 1985.

H. Freeman. *Discrete-Time Systems*. New York: John Wiley, 1966.

A. A. Giordano and F. M. Hsu. *Least Squares Estimation*. New York: Wiley Interscience, 1985.

L. J. Griffiths. "A continuously adaptive filter implemented on a lattice structure." *Proc. IEEE Int. Conf. Acous., Speech, and Sig. Proc.*, pp. 683–686, Apr. 1977.

S. Haykin. *Adaptive Filter Theory*. Englewood Cliffs, N. J.: Prentice-Hall, 1986.

F. N. Hsu. "Square root Kalman filtering for high-speed data received over fading, dispersive high-frequency channels." *IEEE Trans. Inf. Theory*, vol. IT-28, Sept. 1982.

R. E. Kalman. "A new approach to linear filtering and prediction problems." *Trans. ASME J. Basic Eng.*, series D, vol. 82, pp. 35–46. 1960.

R. E. Kalman and R. Bucy. "New results in linear filtering and prediction theory." *Trans ASME J. Basic Eng.*, series D, vol. 83, pp. 95–108. 1961.

M. L. Honig and D. G. Messerschmitt. "Convergence properties of an adaptive lattice filter." *IEEE Trans. ASSP*, vol. ASSP-29, no. 3, pp. 642–653, June 1981.

J. Makhoul and L. J. Cosell. "Adaptive lattice analysis of speech." *IEEE Trans. ASSP*, vol. ASSP-29, pp. 654–659, June 1981.

J. Makhoul and R. Viswanathan. "Adaptive lattice methods for linear prediction." *Proc IEEE Int. Conf. Acous., Speech, and Sig. Proc.*, pp. 83–86, Apr. 1978.

J. S. Meditch. *Stochastic Optimal Linear Estimation and Control*. New York: McGraw-Hill, 1969.

J. Mendel. *Lessons in Digital Estimation Theory*. Englewood Cliffs, N. J.: Prentice-Hall, 1987.

J. J. Shynk. "Adaptive IIR filtering." *IEEE ASSP Magazine*, vol. 6, no. 2, pp. 4–21, Apr. 1989.

B. Widrow and S. D. Stearns. *Adaptive Signal Processing*. Englewood Cliffs, N. J.: Prentice-Hall, 1985.

B. Widrow and E. Wallach. "On the statistics and efficiency of the LMS algorithm with nonstationary inputs." *IEEE Trans. Inf. Theory*, vol. 30, pp. 211–221, Mar. 1984.

CHAPTER 7

■ ■

ESTIMATION METHODS

In this chapter, we will concentrate on estimation of models for Markov processes and on one-dimensional spectral estimation, with particular emphasis on those spectral techniques that are less widely known. The material on Markov estimation begins with a general discussion of Markov sequences and processes and of Markov chains. The section on spectral estimation similarly starts with the familiar subject of estimation by means of Fourier transforms and the use of windowing. The chapter closes with a description of the deconvolution problem, where we wish to estimate the input to a linear system from its output. If the parameters of the linear system are known, there are a number of analytic approaches to the solution; if the parameters are not known, the problem is known as *blind deconvolution* and the solutions involve a considerable amount of detective work and ad hoc approaches.

7.1 MARKOV SEQUENCES AND PROCESSES

A sequence of random variables $\{x_1, x_2, ..., x_n\}$ is said to be a *strict-sense* Markov (SSM) sequence if the probability density function of x_n conditioned on the entire past set $x_{n-1}, ..., x_1$ equals the density function conditioned only on the most recent sample, x_{n-1}. That is,

$$f(x_n | x_{n-1}, x_{n-2}, ..., x_1) = f(x_n | x_{n-1}) \tag{7-1}$$

As a consequence of this definition, it follows that the joint density function of a Markov sequence can be expressed as the product

$$f(x_n, x_{n-1}, ..., x_1) = f(x_n|x_{n-1}) \cdots f(x_2|x_1)f(x_1)$$

$$= f(x_1) \prod_{1=2}^{n} f(x_i|x_{i-1}) \tag{7-2}$$

where the elements of the set, $\{f(x_i|x_{i-1})\}$ are called transitional densities. Conversely Eq. (7-2) implies Eq. (7-1). From the defining equation for a Markov sequence, we can obtain an iterative relationship among the transitional densities:

$$f(x_r|x_{r-2}) = \int f(x_r, x_{r-1}|x_{r-2}) \, dx_{r-1}$$

$$= \int f(x_r|x_{r-1}, x_{r-2})f(x_{r-1}|x_{r-2}) \, dx_{r-1}$$

$$= \int f(x_r|x_{r-1})f(x_{r-1}|x_{r-2}) \, dx_{r-1}$$

More generally,

$$f(x_n|x_m) = \int f(x_n|x_r)f(x_r|x_m) \, dx_r \qquad n > r > m \tag{7-3}$$

By repeated use of this equation, known as the Chapman-Kolmogorov equation, we can, in principle, obtain $f(x_n|x_1)$, and hence the unconditional density $f(x_n)$, from

$$f(x_n) = f(x_n|x_1)f(x_1) \, dx_1 \tag{7-4}$$

The Markov sequence is *homogeneous* when the transitional densities are independent of the time index. The homogeneous sequence is said to be stationary when all the random variables $\{x_1, ..., x_n\}$ have the same probability density function. The latter is ensured for homogeneous sequences when the first two random variables x_1 and x_2 have the same distribution.

The definition of a Markov sequence via the conditional density of Eq. (7-1) suggests that the discrete-time Markov process can be modelled as the reponse of the first-order discrete-time filter,

$$x_n = \phi_n(x_{n-1}) + \xi_n \qquad n > 1 \tag{7-5}$$

where the input $\{\xi_n\}$ is a sequence of zero-mean, independent random variables and the initial state x_0 is a zero-mean random variable independent of ξ_n for $n \geq 1$. To simplify our notation, we take x_0 as ξ_0 and express Eq. (7-5) as the set

$$x_0 = \xi_0$$

$$x_1 = \phi_1(x_0) + \xi_1$$

$$x_2 = \phi_2(x_1) + \xi_2 \qquad\qquad (7\text{-}6)$$

$$\cdots$$

$$x_n = \phi_n(x_{n-1}) + \xi_n$$

To prove that the sequence defined by Eq. (7-6) is Markov, we must show that the joint density function for the sequence $\{x_n\}$ satisfies Eq. (7-1). The set of $n+1$ equations in Eq. (7-6) can be viewed as a transformation of the random variables $\{\xi_0, \xi_1, ..., \xi_n\}$ into the set $\{x_0, x_1, ..., x_n\}$. Hence the joint density function for the $\{x\}$ is

$$f(x_0, ..., x_n) = \frac{f_{\xi_0 \cdots \xi_n}(\xi_0, ..., \xi_n)}{|\mathbf{J}|}$$

where $|\mathbf{J}|$ is the Jacobian of the transformation. Invoking the independence of the $\{\xi\}$, we have

$$f(x_0, ..., x_n) = \frac{f_{\xi_0}(\xi_0) \cdots f_{\xi_n}(\xi_n)}{|\mathbf{J}|} \qquad\qquad (7\text{-}7)$$

The arguments ξ_i in this last equation are evaluated as the functional inverses of the transformation. From Eq. (7-7) we obtain

$$f_{\xi_i}(\xi_i) = f_{\xi_i}[x_i - \phi_i(x_{i-1})] \qquad\qquad (7\text{-}8)$$

which is recognized as the transitional density $f(x_i|x_{i-1})$. Finally, the Jacobian of this transformation is the determinant of a lower-triangular matrix with ones along the main diagonal:

$$\mathbf{J} = \begin{bmatrix} 1 & 0 & \cdots & 0 & 0 \\ \frac{\partial x_1}{\partial \xi_0} & 1 & \cdots & 0 & 0 \\ \cdot & \cdot & \cdots & & \\ \frac{\partial x_n}{\partial \xi_0} & \frac{\partial x_n}{\partial \xi_1} & \cdots & \frac{\partial x_n}{\partial \xi_{n-1}} & 1 \end{bmatrix}$$

so that $|\mathbf{J}| = 1$. Hence the joint density function for the sequence $\{x_n\}$ is

$$f(x_n, x_{n-1}, ..., x_0) = f_{\xi_0}(x_0) \prod_{i=1}^{n} f_{\xi_i}[x_i - \phi_i(x_{i-1})]$$

$$= f(x_0) \prod_{i=1}^{n} f(x_i|x_{i-1}) \qquad\qquad (7\text{-}9)$$

as required for a Markov sequence.

This sequence is homogeneous if the functions $\phi_i(\cdot)$ and the densities $f_\xi(\cdot)$ are independent of i. Thus the filter must be time-invariant (but not necessarily linear), and the independent random variables $\{\xi_i\}$ must be identically distributed.

It was stated in Chap. 6 that the optimal mean-squared estimate of a random variable x_n conditioned on its past values is the conditional mean

$$\hat{x}_n = E\{x_n | x_{n-1} \cdots\} \tag{7-10}$$

For Markov sequences, this reduces to

$$\hat{x}_n = E\{x_n | x_{n-1}\} = \int x_n f(x_n | x_{n-1}) \, dx_n \tag{7-11}$$

which is seen to depend only on the most recent observation, x_{n-1}. In this instance, the best estimator is a memoryless device—though possibly nonlinear and time-varying.

As an example of the foregoing, suppose $\{\xi_n\}$ in Eq. (7-5) is a sequence of zero-mean, independent, Gaussian random variables with a variance σ_n^2, and suppose $\phi_n(x) = x^2/(n+1)$. Then the transitional density is

$$f(x_n | x_{n-1}) = f[x_n - \phi_n(x_{n-1})] = \frac{1}{\sqrt{2\pi\sigma_n^2}} \exp\left[\frac{-1}{2\sigma_n^2}\left(x_n - \frac{x_{n-1}^2}{n+1}\right)\right]$$

and the conditional mean is $E\{x_n | x_{n-1}\} = x_{n-1}^2/(n+1) = \hat{x}_n$.

It is conceptually easy to extend the foregoing to second- and higher-order difference equations. For example,

$$x_n = \phi_n(x_{n-1}, x_{n-2}) + \xi_n \tag{7-12}$$

is characterized by

$$f(x_n | x_{n-1}, \ldots, x_0) = f(x_n | x_{n-1}, x_{n-2}) \tag{7-13}$$

whenever ξ_n is a zero-mean, independent sequence. The process generated by Eq. (7-12) would be called Markov 2, and so on for higher-order systems. But these can all be represented in a uniform manner by using state-vector notation. Thus the vector-valued sequence $\{\mathbf{x}_n\}$ is generated by the first-order vector difference equation,

$$\mathbf{x}_n = \phi_n(\mathbf{x}_{n-1}) + \xi_n \qquad \mathbf{x}_0 = \xi_0 \tag{7-14}$$

where $\{\boldsymbol{\xi}_n\}$ is a sequence of zero-mean, independent, vector random variables. All the components of $\boldsymbol{\xi}_n$ are independent of all the components of $\boldsymbol{\xi}_k$ for $n \neq k$, and the joint density function of the input sequence is

$$f(\boldsymbol{\xi}_0, \boldsymbol{\xi}_1, ..., \boldsymbol{\xi}_n) = f_0(\boldsymbol{\xi}_0) f_1(\boldsymbol{\xi}_1) \cdots f_n(\boldsymbol{\xi}_n) \qquad (7\text{-}15)$$

The conditional and joint densities take the form

$$f(\mathbf{x}_n | \mathbf{x}_{n-1}, \mathbf{x}_{n-2}, ...) = f(\mathbf{x}_n | \mathbf{x}_{n-1}) = f_n[\mathbf{x}_n - \boldsymbol{\phi}_n(\mathbf{x}_{n-1})] \qquad (7\text{-}16a)$$

and
$$f(\mathbf{x}_n, \mathbf{x}_{n-1}, ..., \mathbf{x}_0) = f_0(\mathbf{x}_0) \prod_{i=1}^{n} f(\mathbf{x}_i | \mathbf{x}_{i-1}) \qquad (7\text{-}16b)$$

The foregoing properties of SSM sequences provide the basis for optimal mean-squared estimation, and, as we have noted, involve conditional expectations. On the other hand, *linear* mean-squared estimation involves wide-sense properties of a process—means and covariances—rather than pdfs.

For the latter case, we define the more restricted class of *wide-sense Markov* (WSM) sequences $\{x_n\}$ as the response of a linear, causal, first-order, discrete-time system

$$x_n = a_n x_{n-1} + \xi_n \qquad n \geq 1 \qquad (7\text{-}17)$$

where $\{\xi_n\}$ is a sequence of zero-mean, orthogonal random variables and the initial state x_0 is zero-mean and orthogonal to ξ_n for $n \geq 1$. Thus[†]

$$\langle x_0 \rangle = \langle \xi_n \rangle = 0$$
$$\langle x_0 x_n \rangle = 0 \qquad n \geq 1 \qquad (7\text{-}18)$$
$$\langle \xi_n \xi_k \rangle = \sigma_k^2 \delta_{n-k}$$

We can compare the operational definition of a WSM sequence in Eq. (7-17) with that for the SSM in Eq. (7-5). The nonlinear difference equation becomes a linear one and the adjective "independent" is replaced by the weaker adjective "orthogonal" in the input signal description.

It is left to the reader to show that the autocorrelation $R(m,n) = \langle x_n x_m \rangle$ satisfies

$$R(m,n) = a_n R(m, n-1) \qquad m \leq n-1 \qquad (7\text{-}19a)$$

$$R(n,n) = a_n^2 R(n-1, n-1) + \sigma_n^2 \qquad (7\text{-}19b)$$

[†]We are using $\langle x \rangle$ to represent average or expectal value of x.

For the special case of a stationary (constant-variance) input and a time-invariant filter,

$$\sigma_n^2 = \sigma^2$$

$$a_k = a$$

Equation (7-19) gives the correlation sequence as

$$R(m, n) = \sum_{j=0}^{m} a^{m-j} a^{n-j} \sigma^2 = \frac{\sigma^2}{1 - a^2} a^{|n-m|} - a^{n+m+2}$$

Additionally, if the filter is stable, with $|a| < 1$ and if as $m \to \infty$ and $n \to \infty$, the difference $(m - n)$ is finite, then in the steady state

$$\lim_{m,n \to \infty} R(m, n) = \frac{\sigma^2}{1 - a^2} a^{|m-n|} = R(m - n) \qquad (7\text{-}20)$$

This form implies that $\{x_k\}$ is WSS in the steady state. Alternatively, we can allow the operation of the filter to commence as some time N_0 rather than 0; then as $N_0 \to -\infty$ under the conditions specified, $R(m - n) = R(n - m)$ as given above. Clearly, $R(m)$ satisfies

$$R(m + 1) = aR(m)$$

so that
$$R(m) = a^{|m|} R(0) \qquad (7\text{-}21)$$

(The absolute value is required in the exponent to make $R(m)$ an even function of m.) If $a = 1$ in the foregoing stationary filter, we get the unstable sequence known as the random walk:

$$x_n = x_{n-1} + \xi_n \qquad x_0 = \xi_0$$

which is equivalent to

$$x_n = \sum_{k=0}^{n} \xi_k \qquad (7\text{-}22)$$

Note that the random walk is just the response of a digital integrator (that is, an accumulator) to a zero-mean uncorrelated input.

7.1.1 MARKOV CHAINS

A Markov chain is a special case of the strict-sense Markov sequence defined by Eqs. (7-1) to (7-4) in which the random variable x_n is quantized— that is, $\{x_n\}$ in now a sequence of discrete random variables that can take on values $\alpha_1, \alpha_2, ..., \alpha_N$. A machine or system producing such a discrete-valued sequence is called a *finite-state machine*, and the set of values $\{\alpha_i\}$ that the output can assume at any time are called states.

When $x_n = \alpha_i$, we say that the machine is in the ith state at time n. The notation,

$$p_i(n) = P\{x_n = \alpha_i\} \tag{7-23}$$

is read as "the probability that the random variable x has the value α_i at time n" or "the probability that the chain is in the ith state at time n." At any time n, one of the states must be occupied, and consequently

$$\sum_{i=1}^{N} p_i(n) = 1 \tag{7-24}$$

By virtue of the Markov character of the sequence, the probability that $x_{n+1} = \alpha_j$, conditioned on the entire past $x_n = \alpha_i, x_{n-1} = \alpha_\ell, ...$, and so on, is equal to the probability of $x_{n+1} = \alpha_j$ conditioned only on the most recent value $x_n = \alpha_i$. That is,

$$P\{x_{n+1} = \alpha_j | x_n = \alpha_i, x_{n-1} = \alpha_\ell, ...\}$$
$$= P\{x_{n+1} = a_j | x_n = a_i\}$$
$$\equiv a_{ij}(n) \tag{7-25}$$

This conditional probability $a_{ij}(n)$, called the *transition probability*, is the probability that the chain will switch from state i at time n into state j at time $n + 1$. Since the chain can go into the jth state at time $n + 1$ by any one of N mutually exclusive transitions from states that may be occupied at time n, it follows that

$$p_j(n + 1) = \sum_{i=1}^{N} a_{ij}(n) p_i(n) \qquad j = 1, 2, ..., N \tag{7-26}$$

In vector form, this becomes the linear, first-order homogeneous vector difference equation

$$\mathbf{p}(n + 1) = \mathbf{A}^T(n)\mathbf{p}(n) \tag{7-27}$$

where $\mathbf{A}(n)$ is the $N \times N$ matrix $[a_{ij}(n)]$.

The probabilistic behavior of a Markov chain is sometimes represented by a *state-transition diagram* such as the one shown in Fig. 7-1 for a three-state machine. This is a directed graph in which the ith node represents the probability of being in state i and the edges represent the transitional probabilities of switching from one state to another. Thus

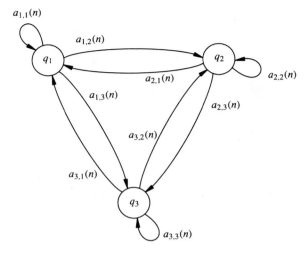

Figure 7-1   *Markov chain modelled as finite-state machine.*

the probability of occupying state i at time $n + 1$ is the sum of all transitions from the other nodes at time n, weighted by their transitional probabilities. Note that the elements in the ith row of $\mathbf{A}(n)$ correspond to the probabilities of transitions out of state i (including a transition out of state i back into itself). Clearly the sum of the weights on the edges leaving any node must sum to unity, or

$$\sum_{j=1}^{N} a_{ij}(n) = 1 \qquad \text{where } 0 \le a_{ij}(n) \le 1 \qquad (7\text{-}28)$$

Any matrix satisfying Eq. (7-28) is said to be a *stochastic matrix*.

As with Markov sequences, a chain is homogeneous if the transitional probabilities are independent of time. In this case, the difference equation reduces to the time-invariant form

$$\mathbf{p}(n + 1) = \mathbf{A}^T \mathbf{p}(n) \qquad (7\text{-}29)$$

with solution $\qquad\qquad \mathbf{p}(n) = (\mathbf{A}^T)^n \mathbf{p}(0) \qquad\qquad (7\text{-}30)$

The behavior of the chain is governed by the properties of $\mathbf{A}^n$, particularly as $n \to \infty$. These properties are determined by the initial "state," $\mathbf{p}(0)$ and by the eigenvalues of $\mathbf{A}$.

To see why this should be, let $\mathbf{P}(z)$ be the one-sided Z transform of $\mathbf{p}(n)$, and for economy of notation let $\mathbf{Q} = \mathbf{A}^T$. Then Eq. (7-29) may be written

$$z[\mathbf{P}(z) - \mathbf{p}(0)] = \mathbf{Q}\mathbf{P}(z) \qquad (7\text{-}31)$$

Solving for $\mathbf{P}(z)$, we have

$$\mathbf{P}(z) = z[z\mathbf{I} - \mathbf{Q}]^{-1}\mathbf{p}(0) \tag{7-32}$$

The natural frequencies of $\mathbf{P}(z)$ will be its poles. But the inverse of $z\mathbf{I}-\mathbf{Q}$ will be a fraction whose denominator is $\det(z\mathbf{I} - \mathbf{Q})$; hence the poles will be the solutions of

$$\det(z\mathbf{I} - \mathbf{Q}) = 0 \tag{7-33}$$

These are clearly the eigenvalues of $\mathbf{Q}$ (hence also of $\mathbf{A}$).
    Thus if $\{\lambda_i\}$, $1 \le i \le N$, are the roots of the characteristic polynomial

$$f(\lambda) = \det(\lambda\mathbf{I} - \mathbf{A}) = \prod_{i=1}^{N}(\lambda - \lambda_i) \tag{7-34}$$

all solutions are of the form

$$p_i(n) = \sum_{k=1}^{N} C_{ik}(\lambda_k)^n \tag{7-35}$$

where the $\{C_{ik}\}$ are constants and the roots assumed distinct.

7.1.2 ESTIMATION OF THE HOMOGENEOUS SYSTEM

Suppose the finite-state machine produces one of a finite number M of discrete outputs or messages $\{z_i\}$ in each state. Here the focus is not on the values of $\{x_i\}$, as previously, but on the outputs associated with the states; we will consider that $(\alpha_0, \alpha_1, \alpha_2, ...)$ are simply $(0, 1, 2, ...)$ and refer to the states as $\{q_i\}$. Then if these outputs are random, we can characterize them by the probability that the machine puts out z_j when it is in state q_i:

$$b_{ij} = P\{z = z_j | q = q_i\} \qquad 1 \le j \le M \tag{7-36}$$

For the entire machine, these probabilities can be represented by the matrix $\mathbf{B} = [b_{ij}]$, where the ith row vector $\mathbf{b}_i$ gives the output probabilities for state q_i. Clearly each row of $\mathbf{B}$ must sum to unity. The machine $\mathbf{M}$ is defined by the quadruple $\{N, \mathbf{q}(0), \mathbf{A}, \mathbf{B}\}$.
    This model has found extensive application in speech recognition, where the states correspond very roughly to the positions of the vocal organs. (See Chap. 8.) In such applications, it is essential to find a way of estimating $\mathbf{M}$ from an ensemble of observed output sequences $\{\mathbf{O}\}$. In speech recognition, this is known as training the model. The basics of training the model are given by Levinson et al. (1978), whom we follow here.

These estimation methods are iterative. They employ forward and backward recursive relations among the probabilities (Baum, 1972), which we will now derive. Suppose that at some time t the machine has passed through a series of states $(q_0, q_1, ..., q_i)$, producing a corresponding sequence of observations $\mathbf{O}(t) = (O_0, O_1, ..., O_t)$ where each observation is some output z_k. This event is termed a *partial sequence* of states and observations. We wish to find a recursive relation among the probabilities of such partial sequences.

Suppose we know the probability of this partial sequence:

$$\alpha_t(i) = P\{\mathbf{O}(t), q_i \text{ at time } t\}$$

Then we may find $\alpha_{t+1}(j)$ as follows: for any partial sequence ending in current state q_i,

$$\alpha_{t+1}(j) = \alpha_t(i)a_{ij}b_{jk} \qquad \text{for } q = q_i \text{ at time } t$$

The a_{ij} factor is the transition probability from q_i to q_j, b_{jk} is $P\{O_{t+1} = z_k | q = q_j\}$, and $\alpha_t(i)$ conditions the probability on the particular partial sequence ending in state q_i. Then summing over all states, we have

$$\alpha_{t+i}(j) = \sum_{i=0}^{N} \alpha_t(i)a_{ij}b_{jk} \tag{7-37}$$

To start this recursion, we set

$$\alpha_0(j) = p_j(0)b_{jk} \qquad j = 1, ..., M$$

The backward recursion is obtained similarly. Consider a partial sequence of states and observations starting from state q_i at time t and continuing to the end of the output at time T, producing the observations $(O_{t+1}, O_{t+2}, ..., O_T)$, and let $\beta_t(i)$ be the probability of this partial sequence. Then

$$\beta_t(i) = \sum_{j=1}^{N} \beta_{t+1}(i)a_{ij}b_{jk} \tag{7-38}$$

This recursion is started by setting $\beta_T = 1$ for all final states.

The iterative estimation procedure starts with initial guesses at $\mathbf{A}$, $\mathbf{B}$, and $\mathbf{p}(0)$. Levinson et al. show how these guesses can be used with the forward and backward recursions Eqs. (7-37) and (7-38) in conjunction with a large number of observation sequences $\{\mathbf{O}\}$ to derive new estimates of $\mathbf{A}$, $\mathbf{B}$, and $\mathbf{p}(0)$.

To update the $\mathbf{A}$ matrix, let $\gamma(i, j|\mathbf{O}, \mathbf{M})$ be the expected number of transitions from q_i to q_j, given the observed sequence $\mathbf{O}$ and our current estimate of the model $\mathbf{M}$, and let $\gamma(i|\mathbf{O}, \mathbf{M})$ be the expected number of transitions from state q_i to any other state under the same conditions. Then we can estimate a_{ij} as

$$\hat{a}_{ij} = \frac{\gamma(i, j|\mathbf{O}, \mathbf{M})}{\gamma(i|\mathbf{O}, \mathbf{M})} \tag{7-39}$$

To obtain these expected numbers, note that

$$\gamma(i|\mathbf{O}, \mathbf{M}) = \sum_{j=0}^{N} \gamma(i, j|\mathbf{O}, \mathbf{M}) \tag{7-40}$$

To find $\gamma(i, j|\mathbf{O}, \mathbf{M})$, let $\gamma(i, j, t|\mathbf{O}, \mathbf{M})$ be the expected number of transitions from q_i to q_j at time t. This will depend on the number of observations, which is a constant and can be ignored, and the probability of such a transition, $P\{i, j, t|\mathbf{O}, \mathbf{M}\}$. We may further write

$$P\{i, j, t|\mathbf{O}, \mathbf{M}\} = \frac{P\{i, j, t, \mathbf{O}|\mathbf{M}\}}{P\{\mathbf{O}|\mathbf{M}\}} \tag{7-41}$$

We will represent the denominator by P. The numerator can be found from the probabilities α and β: it must be the product of the probability of a partial sequence ending in q_i at time t, the probability of a transition to q_j with output $O_t = z_k$, and the probability of another partial sequence beginning in q_j at time $t+1$ and continuing to the end of the data. Hence

$$\gamma(i, j, t|\mathbf{O}, \mathbf{M}) = \frac{1}{P} \alpha_t(i) a_{ij} b_{jk} \beta_{t+1}(j) \tag{7-42}$$

Then the numerator of Eq. (7-39) is

$$\gamma(i, j|\mathbf{O}, \mathbf{M}) = \sum_{t=0}^{T-1} \gamma(i, j, t|\mathbf{O}, \mathbf{M}) \tag{7-43}$$

and the denominator is given by Eq. (7-40) and can be shown to be

$$\gamma(i|\mathbf{O}, \mathbf{M}) = \frac{1}{P} \sum_{t=0}^{T-1} \alpha_t(i) \beta_t(i) \tag{7-44}$$

To update the $\mathbf{B}$ matrix, let the expected number of times state q_j produces output z_k, conditioned on the observations $\mathbf{O}$ and the model $\mathbf{M}$, be $\delta(j, k|\mathbf{O}, \mathbf{M})$ and let $\delta(j|\mathbf{O}, \mathbf{M})$ be the expected number of times state q_j gives any output. Then our revised estimate of b_{jk} is

$$\hat{b}_{jk} = \frac{\delta(j, k|\mathbf{O}, \mathbf{M})}{\delta(j|\mathbf{O}, \mathbf{M})} \tag{7-45}$$

Clearly $\delta(j|\mathbf{O}, \mathbf{M}) = \gamma(j|\mathbf{O}, \mathbf{M})$; for the numerator we have

$$\delta(j, k|\mathbf{O}, \mathbf{M}) = \frac{1}{P} \sum_{t:O_t = z_k} \alpha_t(i)\beta_t(i) \qquad (7\text{-}46)$$

that is, the sum of Eq. (7-44) over only those times when the output $O_t = z_k$. The initial probabilities are updated with

$$p_i(0) = \frac{1}{P}\alpha_0(i)\beta_0(i) \qquad (7\text{-}47)$$

The chief practical difficulties with this estimation procedure are first, that the repeated multiplications by small probabilities tend to drive the α and β values toward zero, with the resultant possibility of underflow in the course of computation, and second, that for a model of any significant size a massive amount of data is required for reliable estimates. The first problem is usually avoided by periodically rescaling the values. There is no satisfactory solution to the second problem, but speech-recognition systems using this model show strikingly good performance even when trained on inadequate data (Rabiner, 1983). An additional problem is that the iterative procedure tends to home in on a local optimum rather than the global optimum; hence the training process must usually be repeated with a number of initial guesses.

7.2 SPECTRAL ESTIMATION

Estimation of the power spectral density of a sequence is a well-known procedure in data analysis. Traditionally, one found the Fourier transform of the autocorrelation function, using various windowing functions in order to reduce the effects of analyzing a finite-length sequence. These procedures are summarized in the classic work by Blackman and Tukey (1958). In recent years, various optimization approaches have been applied to the problem; of these, we have chosen the Capon (maximum-likelihood), Burg (maximum-entropy), and Pisarenko techniques. There are others as well; see Kay and Marple (1981) for an excellent summary and Childers (1978) for a good selection of reprints.

7.2.1 FOURIER TRANSFORM SPECTRAL ESTIMATION; WINDOWING

The traditional procedure for spectral estimation is by means of the DFT, using a time window for improved performance. We begin with a brief review of the windowing problem. Suppose we have a sinusoid to transform. If we select a finite-length sequence of N samples from the sinusoid, then this selection is equivalent to multiplying the original sequence by a rectangular window

$$w(n) = \begin{cases} 1, & n = 0, 1, ..., N - 1 \\ 0, & \text{otherwise} \end{cases}$$

The discrete Fourier transform of this window is given by

$$W(k) = \exp[-j\pi k(N-1)/N]\frac{\sin(\pi k)}{\sin(\pi k/N)} \qquad (7\text{-}48)$$

The exponential is a linear-phase term accounting for the fact that the rectangle is centered about $(N-1)/2$ instead of 0. The ratio of sines is known as the Dirichlet kernel; it plays much the same role in discrete Fourier transforms that $\sin x/x$ does in continuous Fourier transforms. For convenience, we will define

$$D_N(k) = \frac{\sin(\pi k)}{N \sin(\pi k/N)}$$

The factor of N in the denominator normalizes $D_N(0)$ to unity.

If the original signal was $x(n)$ and the windowed signal is $x'(n)$, then

$$x'(n) = x(n)w(n)$$

and, by the convolution property, their DFTs are related by

$$X'(k) = \frac{1}{N}X(k) * W(k)$$

Hence the impulse that would be expected from the sinusoid is broadened into a Dirichlet kernel. If the frequency of the sinusoid is an integer multiple of $1/N$, then its transform will be centered on one of the frequency samples. Hence $D_N(k)$ will be sampled at the middle of its main lobe, and at zero crossings elsewhere, as shown in Fig. 7-2(a). If the frequency is not an integer multiple of $1/N$, then the transform samples will fall elsewhere on $D_N(k)$, as shown in Fig. 7-2(b).

The two principal problems illustrated in Fig. 7-2(b) are known as leakage and the picket-fence effect. Leakage causes spectral components that ought to be concentrated at k_0 to spill over into other parts of the spectrum, where they may obscure smaller components at other frequencies. The picket-fence effect makes it difficult to estimate the amplitude of the sinusoid, since the main lobe of $D_N(k)$ is not sampled at its center. The picket-fence effect may be ameliorated by over-sampling the transform; this is done by extending the time function with zeroes. This zero padding does not increase the resolution of the transform, but it samples the spectrum, and hence the main lobe, at finer intervals.

Leakage is addressed by using a window function other than a rectangle. This is common practice wherever Fourier transforms arise; in antenna design it is called *shading* and in optics it is called *apodizing*. A window is chosen in which the side lobes are smaller than those of

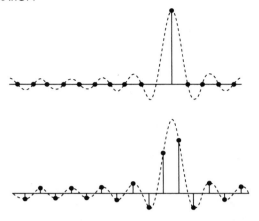

Figure 7-2 *Illustrating leakage and the picket-fence effect.*

$D_N(k)$. The reduction in side lobes is always purchased at the price of a widening of the main lobe. This widening tends to mitigate the picket-fence effect, but since the width of the main lobe affects the ability to resolve closely-spaced frequency components, it is desirable to minimize the widening and to attack the picket-fence effect with zero padding.

Much of the history of classical spectrum analysis centers around the search for windows that (1) have small side lobes, (2) have a narrow main lobe, and (3) are economical to compute. Most elementary DSP texts describe one or two of these windows; the most commonly used window functions are summarized in App. D, and Harris (1978) gives an exhaustive survey.

7.2.2 THE PERIODOGRAM

It is well known that the power spectrum of a signal is the Fourier transform of its autocorrelation function. Hence one starting point for estimating the power spectrum is to estimate the autocorrelation function and compute its Fourier transform. If no a priori model for the autocorrelation is assumed, we say the estimation is *nonparameteric*.

In all practical cases, we are dealing with a finite-length segment of data. We will assume that the data are uniformly sampled; hence we are given $\{x(k), k = 0, 1, ..., N - 1\}$ where N is the number of available points. The first practical problem arises when we try to estimate the autocorrelation of x. We wish to find

$$r(k) = \lim_{T \to \infty} \frac{1}{2T} \sum_{n=-T}^{T} x(n)\bar{x}(n + k) \qquad (7\text{-}49)$$

(where the overbar indicates the complex conjugate) and this is impossible, since we do not know $x(k)$ for k outside the interval $(0, N - 1)$. Indeed, we know the summand only in the interval in which $x(n)$ and

$x(n-k)$ overlap. There are two options open to us. First, we can restrict the limits on the sum to the range where the two functions overlap and correct for the changing overlap length:

$$\tilde{r}(k) = \frac{1}{N-k} \sum_{n=k}^{N-1} x(n)\bar{x}(n+k) \tag{7-50}$$

It can be shown (Prob. 7.1) that this estimate is unbiased, since $E\{\tilde{r}(k)\} = r(k)$. But notice that as k increases, the denominator decreases; hence if there is any noise or error in our measurement of x, the noise-to-signal ratio of $\tilde{r}(k)$ steadily increases. Furthermore, in practice, the use of $\tilde{r}(k)$ may result in a power-spectrum estimate which is not everywhere nonnegative.

The second alternative is the biased estimator

$$\hat{r}(k) = \frac{1}{N} \sum_{n-k}^{N-1} x(n)\bar{x}(n+k) \tag{7-51}$$

Effectively this assumes $x(n)$ is zero outside the interval $(0, N-1)$; this is equivalent to multiplying $x(k)$ by a rectangular window of length N. If we let $\hat{x}(n)$ be this windowed function, then $\hat{r}(k) = \hat{x}(k) * \bar{\hat{x}}(-k)$. Since the convolution of two rectangles of width N is a triangle of width $2N$, this has the effect of applying a Bartlett window (App. D) of width $2N$ to $r(k)$: $\hat{r}(k) = (|N-k|/N)r(k)$. The corresponding power-spectrum estimate is the Fourier transform of $\hat{r}(k)$

$$P_B(\omega) = \sum_{k=-\infty}^{\infty} \hat{r}(k)e^{-j\omega k} \tag{7-52}$$

and the convolution property of the Fourier transform gives us

$$P_B(\omega) = |\hat{X}(\omega)|^2/N \tag{7-53}$$

where $\hat{X}(\omega)$ is the Fourier transform of $\hat{x}(k)$. This process is known as the *periodogram* of x; because of the implied triangular weighting in $\hat{r}(k)$, it is also called the Bartlett estimate, as indicated by the B subscript in Eq. (7-53). Notice that $P_B(\omega)$ is guaranteed nonnegative. It can be shown (Prob. 7.4) that $P_B(\omega)$ can be calculated directly from the data using

$$P_B(\omega) = \frac{1}{N} \left| \sum_{k=0}^{N-1} x(k)e^{-jk\omega} \right|^2$$

7.2.3 MAXIMUM-LIKELIHOOD SPECTRAL ESTIMATION

All practical spectral analysis must deal with a finite amount of data. In some cases only a finite amount may be available, perhaps much less than we could wish for; in other cases we may be compelled to segment an essentially infinite data stream, either in order to make it computationally tractable or because we are interested in the spectral content of only a particular portion of the signal.

When the data segment is extremely short, the frequency resolution available using the Fourier transform is limited by the uncertainty principle (Papoulis, 1962). In recent years, researchers have explored the possibility of alternative methods of extracting spectral information, that are not limited by the uncertainty principle. We will consider three of the more interesting solutions that have been found, the maximum-likelihood estimator, the maximum-entropy estimator, and the Pisarenko estimator.

The maximum-likelihood (ML) spectral estimator was proposed by Capon (1969). The appropriateness of the name has been questioned; Kay (1988) prefers "minimum-variance" for reasons that will become clear below. The ML description has been used widely in the literature, however, and we will retain it here. Capon's paper deals with estimating two-dimensional spectra from data obtained from an array of sensors; his particular application was in the field of seismic analysis. Because the relation between this problem area and that of spectral analysis of a one-dimensional time series may not be immediately clear, we will follow the derivation given in Lacoss (1971).

We will approach Capon's method by modelling the spectrum analyzer as a tunable filter. If we wish to find the power in the input signal at some frequency ω, we can imagine tuning the filter to ω and measuring its output power. This filter must meet two criteria: First, if the input signal consists of nothing but a sinusoid at some frequency ω, we wish the output power of the filter to be exactly equal to that of the input. Second, any signal components at other frequencies must make the minimum possible contribution to the output power of the filter.

We will assume a filter of length N with coefficients $a_0, a_1, \cdots, a_{N-1}$. Then we know that the output of such a filter will be

$$y(n) = \sum_{k=0}^{N-1} a_k x(n - k) \tag{7-54}$$

and we will assume that the input sequence consists of a complex sinusoid of frequency ω, plus other, unrelated components:

$$x(n) = Ae^{j\omega n} + p(n) \tag{7-55}$$

Here we have singled out the desired component at ω and are using $p(n)$ to represent the rest of the signal; $p(n)$ is assumed to be zero mean, stationary, and uncorrelated with the sinusoid.

Our design criteria may be written as follows:

1. If $p(n) = 0$ (that is, if there are no other components), then $y(n) = x(n)$.

2. Otherwise, the output power of the filter is to be a minimum.

The reasoning behind the second criterion is as follows: We are going to use the output power as our estimate of A^2. That power will contain contributions from $Ae^{j\omega n}$ and from $p(n)$, and since these are uncorrelated, their powers will be additive. But because of the first criterion, the output power will not be less than A^2; hence if we minimize the variance, we will be minimizing the unwanted contributions from $p(n)$.
Consider requirement 1. This means

$$\sum_{k=0}^{N-1} a_k A e^{j\omega(n-k)} = A e^{j\omega n}$$

But we can factor $Ae^{j\omega n}$ out of the sum; hence we are left with

$$\sum_{k=0}^{N-1} a_k e^{-j\omega k} = 1$$

For what follows, it will be convenient to write our requirements in vector form. If we let

$$\mathbf{a} = [a_0, a_1, \ldots, a_{N-1},]^T$$
and $$\mathbf{e} = [1, e^{j\omega}, e^{j2\omega}, \cdots, e^{j(N-1)\omega}]^T \tag{7-56}$$

then requirement 1 is

$$\mathbf{e}^*\mathbf{a} = 1 \tag{7-57}$$

where the asterisk denotes the complex conjugate transpose.
Now consider requirement 2. The variance is

$$\sigma^2 = \langle y^2(n) \rangle$$

$$= \langle \sum_{i=0}^{N-1} \bar{a}_i \bar{x}(n-i) \sum_{k=0}^{N-1} a_k x(n-k) \rangle$$

where the overbars indicate complex conjugates. Interchanging the order of summations and expected values, this is

$$\sigma^2 = \sum_i \sum_k \bar{a}_i a_k \langle \bar{x}(n-i) x(n-k) \rangle$$

$$= \sum_i \sum_k \bar{a}_i a_k r_{i-k}$$

where r_n is the autocorrelation of the input x at lag n. This is a quadratic form in $\mathbf{a}$ and may be written

$$\sigma^2 = \mathbf{a}^*\mathbf{Ra} \tag{7-58}$$

where $\mathbf{R}$ is the $N \times N$ autocorrelation matrix $[r_{i-k}]$.

We now wish to minimize Eq. (7-58) subject to the constraint of Eq. (7-57). Using Lagrange's method, we have

$$F(\mathbf{a}) = \mathbf{a}^*\mathbf{Ra} + \lambda\mathbf{e}^*\mathbf{a}$$

where λ is the Lagrangian multiplier. Note that the derivative of a quadratic form $\mathbf{a}^*\mathbf{Ra}$ with respect to $\mathbf{a}$ is $[2\mathbf{a}^*\mathbf{R}]^T$ (see Sec. 6.2.2). Hence setting the derivative of $F(\mathbf{a})$ with respect to $\mathbf{a}$ to zero, we have

$$\left[2\mathbf{a}^*\mathbf{R} + \lambda\mathbf{e}^*\right]^T = 0$$

from which
$$\mathbf{a} = \frac{1}{2}\lambda\mathbf{R}^{*-1}\mathbf{e}$$

where the minus sign has been subsumed into the lambda. We use the constraint to find λ:

$$\mathbf{e}^*\mathbf{a} = \frac{1}{2}\lambda\mathbf{e}^*\mathbf{R}^{*-1}\mathbf{e} = 1$$

from which
$$\lambda = 2/\mathbf{e}^*\mathbf{R}^{*-1}\mathbf{e}$$

and our filter coefficients are

$$\mathbf{a} = \frac{\mathbf{R}^{*-1}\mathbf{e}}{\mathbf{e}^*\mathbf{R}^{*-1}\mathbf{e}}$$

$$= \frac{\mathbf{R}^{*-1}\mathbf{e}}{\mathbf{e}^*\mathbf{R}^{-1}\mathbf{e}} \tag{7-59}$$

Note that the denominator is a scalar. Notice also that the filter parameters are now expressed in terms of the data and frequency (that is, in terms of $\mathbf{R}$ and $\mathbf{e}$) alone. This should not be surprising, since they were optimized for the data and ω. It means, however, that the final spectral estimate will likewise be in terms of the data and ω alone; the filter was simply a convenient model for starting the derivation.

We are now in a position to find out what our spectral estimate will be. Using the formula for variance Eq. (7-58) and substituting Eq. (7-59), we have

$$\sigma^2 = \mathbf{a}^*\mathbf{R}\mathbf{a}$$

$$= \frac{1}{(\mathbf{e}^*\mathbf{R}^{-1}\mathbf{e})^2}\mathbf{e}^*\mathbf{R}^{-1}\mathbf{R}\mathbf{R}^{-1}\mathbf{e}$$

$$= \frac{1}{\mathbf{e}^*\mathbf{R}^{-1}\mathbf{e}}$$

But this minimum variance is our estimate of the spectral power at the frequency ω. Hence we may write

$$P_L(\omega) = \frac{1}{\mathbf{e}^*\mathbf{R}^{-1}\mathbf{e}} \tag{7-60}$$

where the L subscript identifies the maximum-likelihood estimate and $\mathbf{e}$ is determined by the frequency ω, as in Eq. (7-56), while $\mathbf{R}^{-1}$ is independent of frequency and need be computed only once. Furthermore,

$$\mathbf{e}^*\mathbf{R}^{-1}\mathbf{e} = \sum_{i=0}^{N-1}\sum_{k=0}^{N-1} e^{-ji\omega}e^{jk\omega}r_{i-k}$$

In evaluating this sum, we multiply all elements along any diagonal of $\mathbf{R}^{-1}$ by the same complex factor $e^{-ji\omega}e^{jk\omega} = e^{j(k-i)\omega}$. Thus if we let $n = k - i$, then

$$\mathbf{e}^*\mathbf{R}^{-1}\mathbf{e} = \sum_{n=-(N-1)}^{N-1} e^{jx\omega}\rho_n \tag{7-61}$$

where ρ_n is the sum of all elements of $\mathbf{R}^{-1}$ along the diagonal $k - i = n$. This form can be evaluated with the aid of a DFT; hence we do not have to go through the labor of computing $1/\mathbf{e}^*\mathbf{R}^{-1}\mathbf{e}$ for every frequency of interest.

7.2.4 MAXIMUM-ENTROPY SPECTRAL ESTIMATION

We can conveniently start our consideration of maximum-entropy (ME) spectral estimation from the standpoint of windowing. Traditional methods of spectral analysis are not well adapted to handling finite pieces of data, as we have seen. The domain of the continuous Fourier transform is infinite; the domain of the discrete Fourier transform is cyclical and hence imposes a spurious periodicity on the data. If the power spectrum is to be estimated from the autocorrelation function, the domain of the autocorrelation function is also infinite.

In analysis based on Fourier transforms we use some kind of time weighting function that mitigates the consequences of treating a finite segment of data as if it were infinite. These windows inevitably assume, in the case of the autocorrelation, that the data autocorrelations outside the window are all equal to zero. In the case of the discrete Fourier transform, we not only window the time data but also pad them with zeroes; we are thus assuming that the repetitions are far enough apart that they will do no particular harm and that the data points between the repetitions are zero. Burg (1975) denounces these assumptions as "affronts to common sense," and indeed, while we may not know what the values outside our window are, we can usually be almost certain that they are not all zero.

Burg attacks this problem in an entirely different way. Suppose we estimate the power spectrum from the available autocorrelations, but refuse to make any assumption about the nature of the signal outside of the window. Specifically, suppose we know the autocorrelation function of the signal up to a certain maximum lag. Depending on the unknown autocorrelations outside this limit, there are an infinite number of power spectra that are consistent with what we know. Burg addresses the problem of finding that power spectrum which (1) is consistent with the known autocorrelations and (2) makes the fewest assumptions about the remaining autocorrelations, or about the signal itself outside the available data.

There remains the question of how we got the few autocorrelations that we know. At first glance, it would appear that the windowing problem arises at the very first nonzero lag. We will find, however, that Burg's method actually manages to sidestep the computation of the autocorrelations altogether, while if the autocorrelations are important to us for other reasons, his method gives us a way of estimating them, and these estimates are likewise independent of any assumptions about the behavior of the signal outside the window. Burg starts with the known segment of data and obtains everything that is desired from that.

7.2.4.1 Maximum Entropy In information theory, *entropy* is the measure of our uncertainty about any set of events. Since we wish to leave the signal free to do as it will outside the window, we are attempting to maximize our uncertainty—to be "maximally noncommittal" (Ulrych and Bishop, 1975)—about this part of the signal. We do this specifically by maximizing the entropy of the power spectrum. Burg's method, briefly, starts with an expression for the entropy of a power spectrum and then finds the spectrum that maximizes this entropy subject to the condition that the spectrum be consistent with the known autocorrelations. The ME philosophy is not restricted to spectral analysis; it is a general approach, powerful and theoretically attractive, to a large class of problems; for a general discussion, see Jaynes (1963) or Papoulis (1984).

In order to develop Burg's method, we will start with a brief summary of the concept of entropy as it is used in information theory. Let x be a discrete random variable and let $p(x_i)$ be the probability that the variable takes on a specific value x_i. Then the entropy of x is defined as

$$H(x) = -\sum_{X_i} p(x_i) \log p(x_i) \tag{7-62}$$

where the sum is taken over all possible values of x_i. For example, if x can take on only two values, x_0 and x_1, then

$$H(x) = -p \log p - (1-p) \log(1-p)$$

where $p = P\{x = x_0\}$. $H(x)$ is plotted as a function of p for this example in Fig. 7-3.

This measure has the following characteristics:

1. $H(x)$ is a continuous function of the probabilities $p(x_i)$.

2. $H(x)$ is nonnegative and is zero only for cases where $p(x_i) = 0$ for all values of x_i but one. This is in accord with intuition, since in such cases there is no uncertainty about what value x will assume.

3. $H(x)$ is a maximum when all values are equally likely. This is the situation with which we would intuitively associate the greatest uncertainty.

4. If x can take on N different values, all equally likely, then $H(x)$ increases with N. Again this is in accord with our expectations; we would expect the uncertainty to increase with the number of possibilities.

5. $H(x)$ is additive in the following sense: Suppose we partition the possible values of x into two sets A and B. Then finding the value of x can be divided into two steps, that of determining whether x is in A

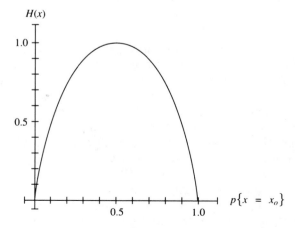

Figure 7-3 *Entropy function for two-valued random variable.*

or in B and that of determining the actual value of x given that it is known to be in A (or B). The entropies of these two steps will sum to the original entropy associated with x itself. This is also intuitively reasonable: if we learn which set contains x, we have removed a certain amount of uncertainty and if we then find the specific value of x, we must have removed the rest of the uncertainty. Shannon (1948) proves that Eq. (7-62) is the only measure of uncertainty that has these five characteristics.

If x is a continuous random variable with a probability density function (pdf) $f(x)$, then its entropy is defined as

$$H(x) = -\int_{-\infty}^{\infty} f(x) \log f(x) \, dx \qquad (7\text{-}63)$$

The base of the logarithm in Eq. (7-62) and (7-63) determines the unit of entropy. For discrete random variables, the base is usually 2 and the unit of entropy is the bit. This is also the name for a single binary storage unit in a digital computer, and that is no coincidence: it is clear from Eq. (7-62) that the entropy associated with a 16-bit word in a computer is exactly 16 bits provided all bit patterns are equally likely. For continuous random variables, it is usually convenient to use natural logarithms; in that case the unit is the nat. Because we will be integrating and differentiating logs, we will use nats.

For a continuous random variable with a fixed standard deviation σ, the pdf with the greatest entropy is Gaussian. This can be determined by a simple application of Lagrangian multipliers; see Prob. 7-6. This is important for us, since our basic problem is one of maximizing entropy, and this result indicates that we are now at liberty to make the Gaussian assumption. For a zero-mean Gaussian pdf with standard deviation σ,

$$f(x) = \frac{1}{\sqrt{2\pi}\sigma} e^{-(x^2/2\sigma^2)} \qquad (7\text{-}64)$$

and Eq. (7-63) gives us

$$H(x) = \ln \sqrt{2\pi e}\sigma \qquad \text{nats} \qquad (7\text{-}65)$$

For an n-dimensional, zero-mean, multivariate Gaussian density function with covariance matrix $\mathbf{C}$,

$$f(\mathbf{x}) = [(2\pi)^{n/2}|\mathbf{C}|]^{-1/2} \exp[\tfrac{1}{2}(-\mathbf{x}^T\mathbf{C}^{-1}\mathbf{x})] \qquad (7\text{-}66)$$

and after some labor,

$$H(\mathbf{x}) = \ln \sqrt{(2\pi e)^n|\mathbf{C}|} \qquad \text{nats} \qquad (7\text{-}67)$$

With this as background, we must now find an expression for the entropy associated with a power spectrum. We will assume that the time series is stationary and treat it as a multivariate random variable. As the portion of signal we consider becomes longer, the dimensionality of the random variable similarly increases. As the length of the sequence under consideration becomes infinite, Eq. (7-67) fails to converge. (Clearly an infinitely long message can convey an infinite amount of information.) We avoid this problem by using the entropy rate; this is the average entropy per sample point:

$$H'(\mathbf{x}) = \lim_{n \to \infty} \frac{1}{n} \ln \sqrt{(2\pi e)^n |\mathbf{C}_n|}$$

$$= \ln \sqrt{2\pi e} + \tfrac{1}{2} \lim_{n \to \infty} \frac{\ln |\mathbf{C}_n|}{n} \tag{7-68}$$

where $\mathbf{C}_n$ is the $(n + 1) \times (n + 1)$ covariance matrix of $\mathbf{x}$. If the signal is stationary, its autocorrelation depends only on the lag and hence its covariance matrix is Toeplitz and Hermitian. That is, $c_{ij} = r_{i-j}$. This allows us to use the fact that if $\mathbf{C}$ is Toeplitz, then

$$\lim_{n \to \infty} \frac{1}{n} \ln |\mathbf{C}_n| = \frac{1}{2W} \int_{-W}^{W} \ln S(f)\, df \tag{7-69}$$

where W is one half the sampling frequency and $S(f)$ is the power spectrum obtained from the $\{r_i\}$. This relation is not easy to prove without going rather far afield; proofs may be found in Smylie (1973) and in Papoulis (1984); see also Papoulis (1981). Hence we have the result

$$H'(\mathbf{x}) = \ln \sqrt{2\pi e} + \frac{1}{4W} \int_{-W}^{W} \ln S(f)\, df \tag{7-70}$$

7.2.4.2 Application to Spectral Analysis We can now return to our initial maximum-entropy spectral analysis problem. We assume that we know all the autocorrelations out to some maximum lag p and none beyond this maximum. If we knew the complete autocorrelation function, then we could compute the power spectrum in the usual way:

$$S(f) = \frac{1}{2W} \sum_{n=-\infty}^{\infty} r_n e^{-j2\pi f n T} \tag{7-71}$$

where $T = 1/f_s$ is the spacing between samples. But the relatively few autocorrelations known to us are not enough to go on; depending on the unknown values, there are an infinite number of spectra which could be computed from Eq. (7-71).

The maximum-entropy spectrum analysis problem may then be stated as follows: Find $S(f)$ which maximizes $H'(x)$, subject to the constraint that $S(f)$ yield the known autocorrelations, that is, that

$$r_n = \frac{1}{2W} \int_{-W}^{W} S(f) e^{j2\pi fnT} \, df \qquad |n| \le p \qquad (7\text{-}72)$$

This means we wish to maximize

$$H'(\mathbf{x}) = \ln \sqrt{2\pi e} + \frac{1}{4W} \int_{-W}^{W} \ln S(f) \, df$$

$$= \ln \sqrt{2\pi e} + \frac{1}{4W} \int_{-W}^{W} \ln \left[\frac{1}{2W} \sum_{n=-\infty}^{\infty} r_n e^{-j2\pi fnT} \right] df \qquad (7\text{-}73)$$

If we take the derivative of $H'(x)$ with respect to r_n, we have

$$\frac{\partial H'(\mathbf{x})}{\partial r_n} = \frac{1}{4W} \int_{-W}^{W} \left[\frac{1}{2W} \sum_{k=-\infty}^{\infty} r_k e^{-j2\pi fkT} \right]^{-1} e^{-j2\pi fnT} \, df$$

$$= \frac{1}{4W} \int_{-W}^{W} \frac{1}{S(f)} e^{-j2\pi fnT} \, df \qquad (7\text{-}74)$$

since the terms in the square brackets are the right-hand side of Eq. (7-71). We would naturally maximize the entropy by setting these derivatives to zero. The effect of the constraints Eq. (7-72), however, is that we may set the derivatives to zero only with respect to those autocorrelations r_n for $n > p$. But note that these derivatives are simply the coefficients of the Fourier series expansion of $1/S(f)$. Hence the rather surprising result of the maximum-entropy requirement is that the Fourier expansion of $1/S(f)$ must have a finite number of terms, specifically that $S(f)$ must be of the form

$$S(f) = \left[\sum_{n=-p}^{p} q_n e^{-j2\pi fnT} \right]^{-1} \qquad (7\text{-}75)$$

We must now determine how to find the coefficients $\{q_n\}$. This requires some ingenuity. The easiest way is probably to start by rewriting Eq. (7-75) in the Z-transform domain:

$$S(f) = S_D(z) \Big|_{z=e^{j2\pi fT}}$$

where $$S_D(z) = 1/Q(z)$$

and $Q(z)$ is the Z transform of $\{q_n\}$. Recall that a power spectrum is real, even, and nonnegative; then $S_D(z)$ must be factorable as

$$S_D(z) = P/[A(z)\bar{A}(z^{-1})] \qquad (7\text{-}76)$$

where the overbar indicates the complex conjugate, $P = 1/q_0$, and $A(z)$ has the form

$$A(z) = \sum_{i=0}^{p} a_i z^{-i}$$

Since P takes care of q_0, $a_0 = 1$.

If we let $R(z)$ be the Z transform of the autocorrelation function, then the requirement of Eq. (7-72) is equivalent to

$$R(z) = P/[A(z)\bar{A}(z^{-1})]$$

Then
$$R(z)A(z) = P/\bar{A}(z^{-1}) \tag{7-77}$$

The left-hand side of Eq. (7-77) must be the Z transform of a convolution. To write out this convolution explicitly, we need a way of expressing the inverse transform of $P/\bar{A}(z^{-1})$. We handle this by observing that $P/A(z)$ is the Z transform of the impulse response of a causal linear system. Let that impulse response be $h(n)$; then the right-hand side of Eq. (7-77) is the Z transform of $\bar{h}(-n)$, where the overbar indicates the complex conjugate. In that case, we have

$$r_n * a_n = \bar{h}(-n)$$

(where $*$ indicates convolution) or

$$\sum_{i=0}^{p} a_i r_{n-i} = \bar{h}(-n) \tag{7-78}$$

Since we must match all autocorrelations out to r_p, Eq. (7-78) must be satisfied for $n = 0, \cdots, p$. But if $h(n)$ is causal, then $\bar{h}(-n)$ is zero for all positive n. Hence we have

$$\sum_{i=0}^{p} a_i r_{n-i} = \begin{cases} \bar{h}(0) & n = 0 \\ 0 & 1 \le n \le p \end{cases} \tag{7-79}$$

This is a system of simultaneous equations; it may be recognized as having the same form as that of the normal equations for a linear predictor of order p (see App. E). This similarity is important to us, and we shall follow it up shortly.

Once we have found $\{a\}$, the coefficients $\{q\}$ can readily be computed from them, but as we shall see, the $\{a\}$ coefficients are actually the more generally useful. For example, if we wish to display the power spectrum, we can use the fact that

$$S(f) = \left. \frac{P}{|A(z)|^2} \right|_{z=e^{j2\pi fT}} \tag{7-80}$$

and obtain $S(f)$ from the reciprocal of the absolute-value-squared Fourier transform of the sequence $\{a_0, a_1, \cdots, a_p, 0, 0, \cdots\}$.

7.2.4.3 Linear Prediction and Burg's Method

Much of the remaining development is best understood if we take advantage of the formal similarity of Eq. (7-79) to the normal equations for a linear predictor, as given in Eq. (E-7). In particular, the recursive solution and recurrence relations developed in App. E are of central importance to us. That solution is repeated here for convenience:

$$a_0(0) = 1$$

$$E_0 = r_0$$

$$a_{n+1}(j) = a_n(j) + k_{n+1} a_n(n + 1 - j) \qquad (7\text{-}81\text{a})$$

$$E_{n+1} = E_n(1 - k_{n+1}^2) \qquad (7\text{-}81\text{b})$$

where
$$k_{n+1} = \frac{1}{E_n} \sum_{i=0}^{n} a_n(i) r_{n+1-i} \qquad (7\text{-}81\text{c})$$

This solution, and the recurrence relations, all hinge on the partial-correlation (PARCOR) coefficients $\{k_i\}$ defined by Eq. (7-81c), and indeed the recursive solution can be viewed as a procedure for finding the PARCOR coefficients.

So far we have assumed that we know the real autocorrelations and have used them, in conjunction with Eq. (7-81) to find the $\{a\}$ that will provide us with the power spectrum with the aid of Eq. (7-80). But we entered upon this whole analysis because we did not know the real autocorrelations. We have conveniently ignored this problem so far in our discussion, but now we must ask where we got these autocorrelations.

Burg's answer turns this question on its head. We know that $\mathbf{a}_p$ gives us the ME power spectrum; in the recursion, each new $\mathbf{a}$ vector depended only on the newest PARCOR coefficient k_n. Suppose we had an alternative way of minimizing the prediction error that would (1) be consistent with the recursive method we have found, (2) give us the $\{k_n\}$ coefficients directly, and (3) not rely on autocorrelation estimates. Then we could use these coefficients to compute the power spectrum directly with the aid of Eqs. (7-80) and (7-81a).

There is such a way. It is based on the recurrence relations among the individual forward and reverse prediction errors, given in App. E and repeated here for convenience. For economy in notation, let the forward prediction-error sequence of an order-p linear predictor, as defined in the appendix, be $\{f_p\}$ and let the error of the backward predictor be $\{b_p\}$. Then

$$f_p(n) = f_{p-1}(n) + k_p b_{p-1}(n)$$

$$b_p(n) = b_{p-1}(n - 1) + k_p f_{p-1}(n - 1)$$

$$(7\text{-}82)$$

Burg uses these recurrence relations to sidestep the computation of the autocorrelations, as follows: On each recursion step n, we use Eq. (7-82) to find the value of k_n that minimizes the sum of the mean-squared forward and backward prediction errors over the sequence. If the length of the sequence is N, this sum is

$$E = \frac{1}{2(N-n)} \sum_{j=n}^{N-1} f_n^2(j) + b_n^2(j+1) \tag{7-83}$$

$$= \frac{1}{2(N-n)} \sum_{j=n}^{N-1} [f_{n-1}(j) + k_n b_{n-1}(j)]^2 + [b_{n-1}(j) + k_n f_{n-1}(j)]^2$$

The limits of the sum must be n and $N-1$ to prevent the predictor from trying to use points outside the available sequence. (This will mean that fewer samples contribute to each new k_n.) If we differentiate this expression with respect to k_n, we get

$$\frac{\partial E}{\partial k_n} = \frac{1}{N-n} \sum_{j=n}^{N-1} k_n [f_{n-1}^2(j) + b_{n-1}^2(j)] + 2 f_{n-1}(j) b_{n-1}(j)$$

Setting this derivative to zero yields

$$k_n = -2P/Q \tag{7-84a}$$

where

$$P = \sum_{j=n}^{N-1} f_{n-1}(j) b_{n-1}(k) \tag{7-84b}$$

$$Q = \sum_{j=n}^{N-1} f_{n-1}^2(j) + b_{n-1}^2(j) \tag{7-84c}$$

Equation Eq. (7-84) gives us an expression for the PARCOR coefficient at step n which uses errors that we already know from the previous step and which requires no knowledge of the autocorrelations. Having found the new PARCOR coefficient, we can find the new **a** vector by Eq. (7-81a), and we can find the new prediction errors by Eq. (7-82). To start the process, we observe that for $n = 0$, the predicted values are all zero, and so the errors are simply the values of the sequence. The Burg recursion then proceeds as follows:

1. For $n = 0$,

$$f_0(j) = y(j) \qquad b_0(j) = y(j-1) \qquad j = 1 \text{ to } N \qquad a_0(0) = 1$$

2. For $n = 1$ to p do
   a. $k_n = -2P/Q$, from Eq. (7-84).
   b. $a_n(n) = k_n$ and $a_n(0) = 1$.
   c. From Eq. (7-81a), for $i = 1$ to $n - 1$ do

$$a_n(i) = a_{n-1}(i) + k_n a_{n-1}(n - i)$$

   d. From Eq. (7-82), for $j = n$ to $N - 1$ do

$$f_n(j) = f_{n-1}(j) + k_n b_{n-1}(j)$$

$$b_n(j) = b_{n-1}(j - 1) + k_n f_{n-1}(j - 1)$$

3. Obtain $S(f)$ using Eq. (7-80).

Step 3 is carried out by finding the reciprocal of the squared magnitude of the DFT of the sequence $\{a_0, a_1, \cdots, a_p, 0, 0, 0, \cdots\}$, as described previously. The zeroes are appended in order to provide adequate resolution; we will discuss this again below.

If it is desired to obtain ME estimates of the autocorrelations as well, these can be obtained as a by-product of the Burg recursion. To see how this can be done, we refer to Eq. (7-81c), rewritten here for convenience.

$$k_n = -\frac{1}{E_{n-1}} \sum_{i=0}^{n-1} a_{n-1}(i) r_{n-i} \qquad (7\text{-}85)$$

When we derive this equation in App. E, we assume we know r_n and want to find k_n. With the Burg recursion, however, we know k_n, and we can use this to find r_n:

$$r_n = -E_{n-1} k_n - \sum_{i=1}^{n-1} a_{n-1}(i) r_{n-i}$$

$$= -k_n \sum_{i=1}^{n} r_{n-i} a_{n-1}(n - i) - \sum_{i=1}^{n-1} r_{n-i} a_{n-1}(i)$$

$$= -\sum_{i=1}^{n} a_n(i) r_{n-i} \qquad (7\text{-}86)$$

The last line follows from Eq. (7-81a) and the fact that $a_{n-1}(n) = 0$; this result can also be seen from the last row of Eq. (E-9). We can incorporate the computation of the autocorrelations into the Burg recursion by appending the following initialization to step 1,

$$r_0 = \sum_{j=1}^{N} y^2(j)$$

and adding the following recursion at the end of step 2:

$$r_n = -\sum_{i=1}^{n} a_n(i)r_{n-i}$$

7.2.4.4 Choice of Order Following the parallel with linear prediction, we observe that the ME method models the signal as the output of an autoregressive process; in that case, the order of the model is given by p. Up to this point, nothing has been said about how this order is to be set. Experimentation with Burg's method will quickly show that this order is critical. Figure 7-4 shows the estimates obtained for a signal consisting of three sinusoids; the frequencies are approximately 764, 955, and 1,019 Hz. As the order of the model increases, the components show up with increasing clarity. Order 2 can only show their general location. Order 6 can resolve two of the components, but the estimates are off; the upper spectral line is roughly in between the two actual frequencies. Order 12 shows all frequencies clearly. When the order is increased to 24, however, a curious thing happens: the lower peak is split, with one maximum apparently on the correct component and the second slightly above it. (The highest-frequency peak may be split as well, but the resolution of this plot is insufficient to show it clearly.)

This *line splitting* is a characteristic of the Burg method when the order is too high. In our example, it seems obvious that an order of 24 is unnecessarily high for three sinusoids, but in a general case when the number of components is not known for sure, how can we tell?

Akaike (1969) has proposed two criteria. His *final prediction error* is the sum of the prediction error power and an estimate of the errors in the predictor coefficients. It is given by

$$F(p) = E_p(N + p + 1)/(N - p - 1) \qquad (7\text{-}87)$$

where D_p is the prediction-error power. As p increases, the decrease in prediction error is gradually offset by the fraction; the preferred order is the value of p for which $F(p)$ is a minimum.

Akaike's information criterion is an estimate of the maximum likelihood, given by

$$I(p) = 2[p - \ln(\text{maximum likelihood})] \qquad (7\text{-}88)$$

If the signal is assumed to have a Gaussian probability density, then Eq. (7-88) is approximately

$$I(p) = \ln E_p + 2p/N$$

At the beginning of the Burg iteration, the prediction error decreases rapidly and the effect of $2p/N$ is negligible, but ultimately the error is

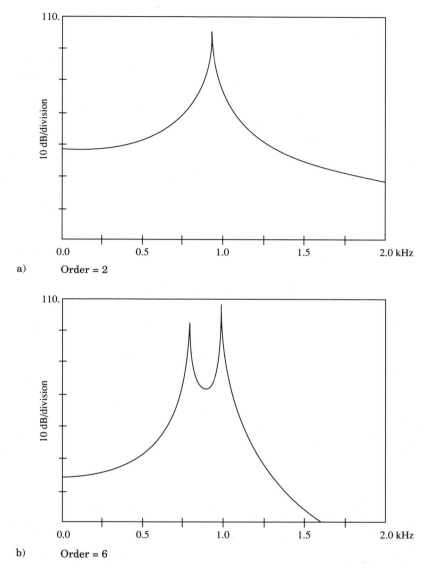

Figure 7-4 *Maximum entropy power spectral estimates: (a) order 2; (b) order 6.*

reduced to a point where further improvement is small, and the $2p/N$ term becomes significant. The point at which $I(p)$ reaches a minimum represents the optimum value of p.

$I(p)$ is plotted in Fig. 7-5 for our three-sinusoid example. In our example, the Gaussian assumption is not valid, and indeed in Fig. 7-5, $I(p)$ shows no minimum out through $p = 25$. Close inspection of the actual values shows a small local minimum at $p = 12$, however, which corresponds to what we have seen in Fig. 7-4.

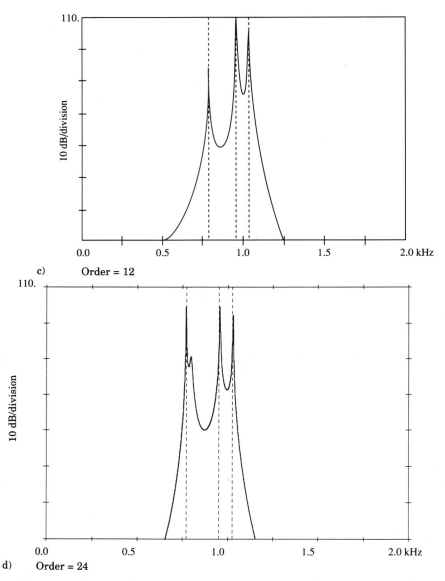

Figure 7-4 *Maximum entropy power spectral estimates: (c) order 12, (d) order 24.*

Notice also that the peaks in Fig. 7-4 are very sharp. It is therefore essential to evaluate $S(f)$ at very closely spaced frequencies; otherwise a peak may be missed completely, and the remarkable resolution of the Burg method will be lost. This is the reason for appending the zeroes to the predictor coefficients before taking the DFT: an order-10 predictor may well be padded out to 1,024 points before transforming.

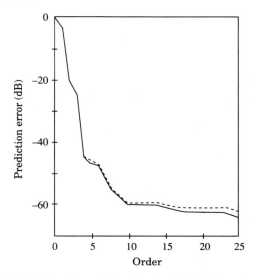

Figure 7-5 *Akaike's information criterion for the signals of Fig. 7.4.*

7.2.5 PISARENKO'S METHOD

The goal of spectrum analysis is the resolution of a finite-length sequence into a sum of sinusoidal components. None of the methods we have considered so far make explicit use of the assumption that the components being sought are, in fact, sinusoids. Pisarenko (1973) found a way of relating the autocorrelation matrix of a wide-sense stationary, zero-mean process directly to its sinusoidal components.

The following development is based on Kay and Marple (1981), whose treatment is simpler than Pisarenko's. Let $x(n)$ be a sequence of equally spaced samples of a sum of a finite number of sinusoidal components. Then $x(n)$ satisfies an order-$2p$ recurrence which we may write as

$$x(n) = -\sum_{i=1}^{2p} a_i x(n-i) \qquad (7\text{-}89)$$

where p is the number of sinusoidal components. For example, if $p = 1$ so that $x(n) = \sin(n\omega T + \theta)$, where T is the sampling interval, then $a_1 = -2\cos\omega T$ and $a_2 = 1$. If we define $a_0 = 1$, we may rewrite Eq. (7-89) more concisely as

$$\sum_{i=0}^{2p} a_i x(n-i) = 0 \qquad (7\text{-}90)$$

If Eq. (7-90) is written as a Z transform, we obtain

$$X(z)A(z) = X(z)\sum_{i=0}^{2p} a_i z^{-i} = 0 \tag{7-91}$$

Note that the roots of $A(z)$ will all lie on the unit circle.

Next, suppose we observe a signal consisting of a finite number of sinusoids plus white noise. We would like to pull these recurrence coefficients a_i out of this noisy signal. If we can do that, we will be able to find the frequencies of the sinusoids from the roots of $A(z)$. We can represent the observed signal by

$$y(n) = x(n) + w(n) \tag{7-92}$$

where $x(n)$ is the set of sinusoids, as before, and $w(n)$ is the white noise. If we substitute $y(n) - w(n)$ for $x(n)$ in Eq. (7-90), we obtain

$$\sum_{i=0}^{2p} a_i y(n-i) = \sum_{i=0}^{2p} a_i w(n-i) \tag{7-93}$$

Let $\mathbf{a} = [a_0, a_1, ..., a_{2p}]$ and let $\mathbf{y}^T = [y(n), y(n-1), ..., y(n-2p)]$ and similarly for $\mathbf{w}$ and $\mathbf{x}$. Then we may write Eq. (7-93) as

$$\mathbf{y}^T \mathbf{a} = \mathbf{w}^T \mathbf{a}$$

Premultiplying by $\mathbf{y}$ and taking expected values, we get

$$\langle \mathbf{y}\mathbf{y}^T \rangle \mathbf{a} = \langle \mathbf{y}\mathbf{w}^T \rangle \mathbf{a} \tag{7-94}$$

The expected value on the left is recognizable as the $(2p+1) \times (2p+1)$ autocorrelation matrix of the observations $\{y(i)\}$, which we will write as $\mathbf{R}_y$. On the right, since the noise is assumed white and hence uncorrelated with the sinusoids, we have

$$\langle \mathbf{y}\mathbf{w}^T \rangle = \langle \mathbf{x}\mathbf{w}^T + \mathbf{w}\mathbf{w}^T \rangle = \langle \mathbf{w}\mathbf{w}^T \rangle$$

and because the noise is assumed white, $\langle \mathbf{w}\mathbf{w}^T \rangle$ is a diagonal matrix all of whose diagonal elements are the noise power σ_w^2. Hence Eq. (7-94) reduces to

$$\mathbf{R}_y \mathbf{a} = \sigma_w^2 \mathbf{a}$$

or

$$(\mathbf{R}_y - \sigma_w^2 \mathbf{I})\mathbf{a} = \mathbf{0} \tag{7-95}$$

We recognize this as an eigenvalue problem; σ_w^2 will be one of the roots of

$$\det(\mathbf{R}_y - \lambda \mathbf{I}) = 0 \tag{7-96}$$

In Eq. (7-95), $\mathbf{a}$ will be the characteristic vector associated with σ_w^2. Marple (1976) has shown that if $y(n)$ meets our assumptions–that is, p sinusoids plus white noise–then σ_w^2 will be the smallest root of $\mathbf{R}_y$. For if λ_i is any root of Eq. (7-96), then

$$\mathbf{R}_y \mathbf{a}_i = \lambda_i \mathbf{a}_i$$

But
$$\mathbf{R}_y = \langle \mathbf{xx}^T + \mathbf{ww}^T \rangle$$

$$= \mathbf{R}_x + \sigma_w^2 \mathbf{I}$$

Substituting for $\mathbf{R}_y$, we have

$$\mathbf{R}_x \mathbf{a}_i = (\lambda_i - \sigma_w^2) \mathbf{a}_i$$

which means that $(\lambda_i - \sigma_w^2)$ is a characteristic root of $\mathbf{R}_x$ and hence must be positive. This means no λ_i can be less than σ_w^2; since σ_w^2 is one of the roots of Eq. (7-96), it must be the smallest one.

Hence the vector $\mathbf{a}$ is the characteristic vector associated with the smallest root of Eq. (7-96). This vector, scaled so that $a_0 = 1$, gives us the desired polynomial in z, as in Eq. (7-91).

This leads to the following procedure: Assume the existence of p sinusoids; find $\mathbf{R}_y$, and solve the eigenvalue problem. Select the smallest eigenvalue and find its associated vector $\mathbf{a}$, scaled to $a_0 = 1$. The roots of $A(z)$, as defined by Eq. (7-91), give the frequencies of the sinusoids. In theory, if the number of sinusoids is unknown, p can safely be overestimated. If p is too large, we simply get repeated values of the smallest eigenvalue; the extra roots of the polynomial will fall well short of the unit circle. In practice, overestimating p may give rise to spurious peaks.

The positions of the roots provide the frequencies of the sinusoids. To find the amplitudes, note that the autocorrelation of $\{y(i)\}$ is given by

$$r_k = \begin{cases} \sigma_w^2 + \sum_{i=0}^{p} p_i & k = 0 \\ \\ \sum_{i=0}^{p} p_i \cos \omega_i kT & k = 1, 2, ..., p \end{cases}$$

$$(7\text{-}97)$$
$$(7\text{-}98)$$

where T is the sampling interval. Knowing the frequencies $\{\omega_i\}$, we can solve Eq. (7-98) for the powers $\{p_i\}$.

7.2.6 COMPARISONS

It is instructive to compare the spectral estimation techniques we have considered. They all use a form of the autocorrelation matrix $\mathbf{R}$, and all but the Pisarenko estimate may be expressed in terms of the quadratic form $\mathbf{e}^* \mathbf{R} \mathbf{e}$. If we use the periodogram, our estimate of the autocorrelation is

$$\hat{r}_k = \sum_{n=k}^{N-1} x(n)\bar{x}(n-k)$$

We have seen that this imposes a triangular weighting on the estimate, since $(N-k)$ terms contribute to r_k. This weighting allows us to express the Bartlett estimate $P_B(\omega)$ in a format similar to Eq. (7-60). Following the same diagonal argument, we may write, with Lacoss (1971)

$$P_B(\omega) = \mathbf{e}^*\mathbf{R}\mathbf{e} \tag{7-99}$$

As we have seen, the maximum-likelihood estimate is

$$P_L(\omega) = 1/\mathbf{e}^*\mathbf{R}^{-1}\mathbf{e} \tag{7-100}$$

Burg (1972) provides a relation between the ME and ML estimates. The Burg estimate may be written (Lacoss, 1971)

$$P_E(\omega) = E_n/\mathbf{e}^*\mathbf{a}\mathbf{a}^*\mathbf{e} \tag{7-101}$$

where n is the order of the predictor, E_n is the prediction error, $\mathbf{a}_n$ is the vector of predictor coefficients, and $\mathbf{e}$ is as defined in Eq. (7-56). [This follows directly from Eq. (7-80).] Let $\mathbf{L}$ be an $n \times n$ lower-triangular matrix whose ith column is the order-i linear predictor $\mathbf{a}_i$, extended at the top with zeroes to a height of n. It can be shown (see Prob. 7.15) that

$$\mathbf{L}^*\mathbf{R}\mathbf{L} = \mathbf{P}$$

where $\mathbf{P}$ is the diagonal matrix,

$$\mathbf{P} = \text{diag}[E_n, E_{n-1}, ..., E_1]$$

In that case,

$$\mathbf{R}^{-1} = \mathbf{L}\mathbf{P}^{-1}\mathbf{L}^*$$

We return now to the ML estimate. We may write,

$$P_L^{-1}(\omega) = \mathbf{e}^*\mathbf{R}^{-1}\mathbf{e}$$

Substituting for $\mathbf{R}$, we have

$$P_L^{-1}(\omega) = \mathbf{e}^*\mathbf{L}\mathbf{P}^{-1}\mathbf{L}^*\mathbf{e}$$

Since $\mathbf{P}^{-1} = \text{diag}[1/E_n, 1/E_{n-1}, ...]$, this may be written as

$$P_L^{-1}(\omega) = \sum_{i=1}^{n} \mathbf{e}^*\mathbf{L}_i\mathbf{L}_i^*\mathbf{e}/E_i$$

where $\mathbf{L}_i$ is the ith column of $\mathbf{L}$. But each term in this sum is then the reciprocal of the order-i Burg estimate:

$$P_L^{-1}(\omega) = \sum_{i=1}^{n} P_E^{-1}(\omega) \qquad (7\text{-}102)$$

A Burg spectrum consists of a number of peaks which become sharper and more numerous as the order of the estimate increases, as we saw in Fig. 7-3; hence its reciprocal consists of a number of valleys which do the same. Equation (7-102) tells us that the reciprocal of the order-n ML estimate is the sum of the reciprocals of the Burg estimates of all orders through n; hence these different sets of valleys are averaged together. This means that the resolution of the ML estimate is, in general, lower than that of the ME estimate.

   Burg (1972) points out that although the resolution of the ML estimate is less fine, there are applications, most notably in processing spatial data from a nonuniformly spaced array of sensors, in which the ML estimate is much more tractable computationally than the ME estimate.

   The Pisarenko estimate is inherently a line spectrum; hence if the spectrum of the data is inherently continuous, this method will not provide a faithful representation of the spectrum. It is also expensive to compute, since it requires solving the eigenvalue problem and finding complex roots of a polynomial. Hence its applicability is generally limited to no more than a few dozen sinusoidal components.

   These four estimators are compared in Fig. 7-6. The data are autocorrelations for two sinusoids in white noise:

$$r(n) = \delta_n + 5.62\cos(0.375\pi n) + 10\cos(0.5\pi n), \qquad n = 0, ..., 9$$

The lower-frequency sinusoid is set to be 5 dB below the higher-frequency one. The Bartlett estimate is unable to resolve the two components; the ML estimate resolves them and shows amplitudes corresponding to those of the two sinusoids. The ME spectrum shows two sharp peaks; the amplitudes in this case correspond to the areas under these peaks. The Pisarenko estimate is shown as a line spectrum; the positions of the lines are determined from the roots of $A(z)$ and the heights of the lines are found by solving Eq. (7-98), as described previously; the heights are in agreement with the known amplitudes. Figure 7-6, however, is, in a sense, obtained from "cooked" data: it was obtained from a precomputed autocorrelation function of the form shown above. Figure 7-7 shows similar plots obtained from the time function itself, here a pair of sinusoids in white noise, with amplitudes and frequencies set to match those of Fig. 7-6. The autocorrelations were estimated from 63 samples of the signal. Notice (1) the poorer resolution of the Capon estimate, (2) the bias in the Burg estimate, and (3) the bias and the spurious peak in the Pisarenko estimate.

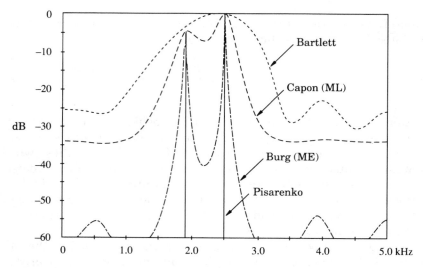

Figure 7-6   *Comparison of Bartlett, Capon, Burg, and Pisarenko estimates based on computed autocorrelations.*

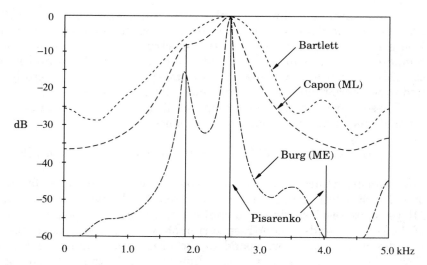

Figure 7-7   *Comparison of Bartlett, Capon, Burg, and Pisarenko estimates based on autocorrelations from raw data.*

7.3 DECONVOLUTION

When a signal passes through a shift-invariant linear system, the output, as we know, is the convolution of the signal and the impulse response of the system. We have generally viewed the effect of the linear system as something that we want to do to the signal. In many cases however, the signal, as it is available to us, has been degraded by the transmission medium through which it has passed. If the corrupting effect of the medium can be modelled as a linear system, then the original signal should be recoverable if we can undo the convolution that occurred when the signal passed through the medium. This process is known as deconvolution. (See also the discussion of the least-squares equalizer in Sec. 6.3.1.)

An example of this occurs in image processing. The image of an object is ideally a replica of the object itself, except possibly for a scale factor corresponding to the magnification of the system and the effects of perspective. In practical optical systems, however, the image of a point is not another point, but a finite, two-dimensional function. This can happen for a number of reasons, such as aberrations or other imperfections of the optical system, inaccuracies in focussing, or atmospheric turbulence. We can consider the point to be an impulse function, and its image is the two-dimensional impulse response of the optical system. As we observed in Chap. 3, this impulse response is commonly called the point-spread function. Every point in the ideal image is replaced in the actual image by the point-spread function.

Hence for any object, the image obtained is the convolution of the ideal image $f(x, y)$ and the point-spread function of the optical system. Hence we may write,

$$b(x, y) = f(x, y) ** h(x, y) \tag{7-103}$$

where $h(x, y)$ is the point-spread function. In addition, the image may be noisy; in photographic images, the graininess of the photographic emulsion is an important source of noise. This noise is usually modelled as additive and Gaussian; it is frequently not Gaussian and may be multiplicative, but the additive, Gaussian assumption is mathematically tractable and in practice yields satisfactory results. Hence if we include additive noise, Eq. (7-103) becomes

$$s(x, y) = b(x, y) + n(x, y)$$
$$= f(x, y) ** h(x, y) + n(x, y) \tag{7-104}$$

7.3.1 FOURIER TRANSFORMS

To estimate the ideal image, we must remove the effect of the convolution. There are several important ways of doing this. Probably the obvious approach is to try to undo the convolution by means of Fourier transforms. This method presents some practical problems, but because of its relative simplicity, we will consider it first. If we take Fourier transforms of both sides of Eq. (7-104), we have

$$S(u, v) = B(u, v) + N(u, v)$$
$$= F(u, v)H(u, v) + N(u, v)$$

where u and v are spatial frequencies, in which case we can estimate $f(x, y)$ by dividing by $H(u, v)$ and computing the inverse transform:

$$\hat{f}(x, y) = \mathfrak{F}^{-1} \frac{S(u, v)}{H(u, v)} \tag{7-105}$$

The principal difficulty in this process lies in the fact that $|H(u, v)|$ generally tends to zero for large u and v; that is, it is band-limited. Hence $B(u, v)$ likewise tends to zero, and if the optical system has circular symmetry, as it usually has, then $B(u, v)$ contains no useful information outside some circle $u^2 + v^2 = \rho_0^2$. Outside this region, the noise dominates. But if $|H(u, v)|$ falls off for large u and v, then its reciprocal will be very large in precisely the region where the signal-to-noise ratio is smallest, and the deconvolved image will be dominated by the high-frequency noise, which has been emphasized by $1/H$. This problem can be alleviated (Harris, 1966) by multiplying $S(u, v)$ by a window function $W(u, v)$ that is zero in those regions where $N(u, v)$ is the larger term. This has the effect of replacing the blur caused by $h(x, y)$ with one caused by $w(x, y)$, but in many cases this is acceptable.

An example is shown in Fig. 7-8. The point-spread function is Gaussian, $h(x, y) = \exp[-(x^2 + y^2)/r_0^2]$, and the signal-to-noise ratio is approximately 30 dB. The restoring filter's transfer function has been multiplied by a circular window. To minimize the effect of the discontinuity at the edge of the window, the window has a cosine taper at its edges:

$$W(\rho) = \begin{cases} 1 & \rho \le \rho_1 \\ \frac{1}{2}[1 + \cos \pi(\rho - \rho_1)/(\rho_2 - \rho_1)] & \rho_1 < \rho \le \rho_2 \\ 0 & \rho > \rho_2 \end{cases}$$

where ρ is the radial distance from the origin. In Fig. 7-8(f), the deconvolved data match the original image reasonably well.

7.3.2 HOMOMORPHIC DECONVOLUTION

Homomorphic deconvolution (Oppenheim, 1968, and Stockham, 1975) is a variant of this method. In this case, the logarithms of the power spectra are used. Since $S(\omega) = H(\omega)F(\omega)$, we have

$$\ln S(\omega) = \ln H(\omega) + \ln F(\omega) \tag{7-106}$$

in which case the original signal can be estimated by subtracting the log spectra:

$$\ln F(\omega) = \ln S(\omega) - \ln H(\omega) \tag{7-107}$$

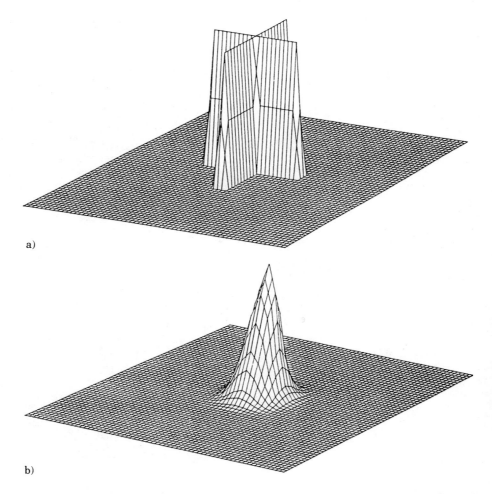

Figure 7-8   *Example of image restoration using Fourier transforms: (a) original image; (b) point-spread function.*

   This method is considerably more powerful than it may appear to be at first glance; much more can be done with additive components than with multiplicative ones. In some cases, the frequency behavior of $H(\omega)$ may differ greatly from that of $F(\omega)$, so that $H(\omega)$ may be removed from $S(\omega)$ by a second transformation.
   A specific example may make this clear. In voice processing, it is often essential to obtain an accurate estimate of the pitch of the voice. (The pitch is the frequency of the periodic excitation function produced by the vibration of the vocal cords.) Traditional filtering and correlation methods are subject to errors arising from the influence of the transfer function of the vocal tract. (See Chap. 8). But a speech spectrum is the

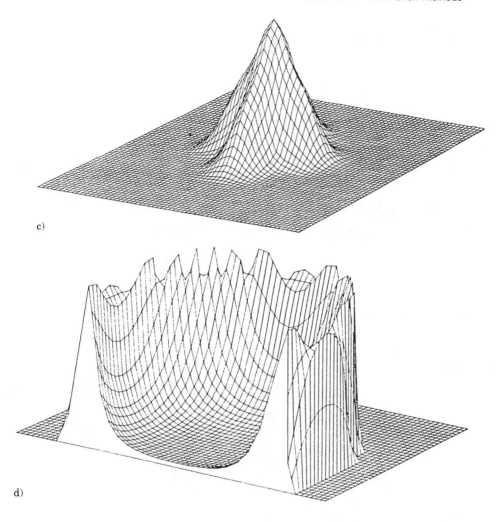

c)

d)

Figure 7-8 *Example of image restoration using Fourier transforms: (c) noisy blurred image; (d) restoring filter.*

product of the excitation spectrum, which for voiced speech consists of a train of harmonics, and the transform of the vocal-tract transfer function, which is a relatively smooth function of frequency. (See Fig. 8-4, in which the transfer function is shown superimposed on a speech spectrum.)

If the logarithm of the spectrum is taken, in order to make the excitation and transfer-function components additive, and the logged spectrum transformed again, we obtain

$$\mathfrak{F}[\ln S(\omega)] = \mathfrak{F}[\ln H(\omega)] + \mathfrak{F}[\ln F(\omega)] \qquad (7\text{-}108)$$

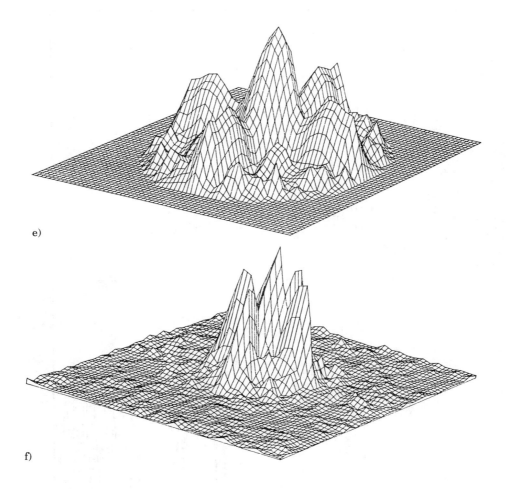

e)

f)

Figure 7-8   *Example of image restoration using Fourier transforms: (e) filtered transform; (f) restored image.*

Such a transform has been named the cepstrum (Bogert et al., 1963). A speech spectrum and its cepstrum are shown in Fig. 7-9. The spectrum was obtained using cosine (Hann) weighting; the plots show the magnitudes of the spectrum and cepstrum. (In the spectrum, the horizontal axis shows the frequency; in the cepstrum, the horizontal axis shows the *quefrency*, measured in units of time.) In Fig. 7-9(a), the harmonic peaks can be seen clearly; their frequencies are multiples of 117 Hz, the fundamental, and the envelope of their amplitudes mainly reflects the effect of the vocal-tract transfer function. In Fig. 7-9(b), the effect of logging on the multiplicative components of the spectrum can be seen clearly; the pitch harmonics are seen riding on the slowly varying envelope, and

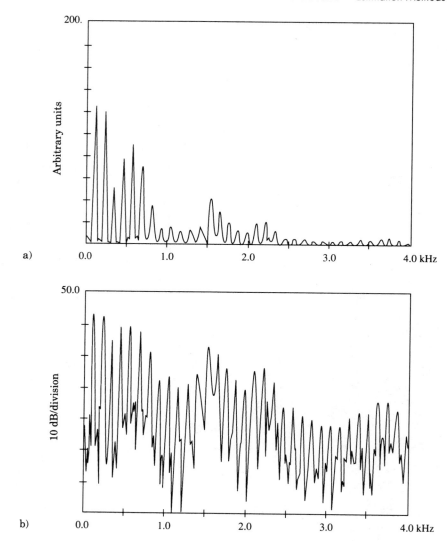

Figure 7-9 *(a) Power spectrum of voiced speech (linear); (b) logged power spectrum.*

the peak-to-valley distance of the harmonics varies only slightly over the frequency range shown.

In the cepstrum in Fig. 7-9(c), the portion near the origin corresponds to the envelope of the speech spectrum, which varies slowly with frequency. The train of harmonics, which is periodic in frequency, gives rise to the sharp peak at a quefrency of approximately 8.5 mS. This corresponds to a fundamental frequency of approximately 117 Hz, which is in agreement with the spacing of the harmonics in Fig. 7-9(a) and (b). In principle, the spectrum envelope could also be recovered from the low-

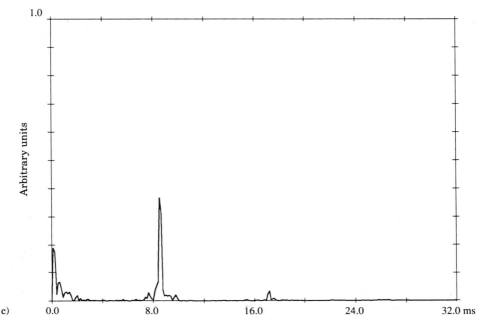

c)

Figure 7-9   *(c) cepstrum.*

quefrency portion of the cepstrum; in practice, linear prediction has been found to provide a better estimate.

Experience has shown that where noise is negligible, the cepstrum provides extremely accurate estimation of the pitch. Notice that the only feature required to separate the two functions is their different locations in the cepstrum; hence we are able to remove the effect of $H(\omega)$ without knowing its form in detail. This can be considered an example of blind deconvolution; see Sec. 7.3.4.

7.3.3 MINIMIZATION OF MEAN-SQUARED ERROR

A second approach to deconvolution also uses a restoring filter, but the filter is designed to minimize the error between the restored signal and the original in the mean-squared sense. Hence the techniques developed in Chap. 6 are applicable here. It is easiest to derive the form of this filter for a one-dimensional case; the generalization to two dimensions is not difficult.

Let the original signal be $x(n)$. Then the degraded signal is

$$b(n) = x(n) * h(n)$$

and the observation, which includes noise, is

$$y(n) = b(n) + w(n)$$

We assume that $x(n)$ and $w(n)$ are zero-mean, Gaussian random processes and that $w(n)$ is uncorrelated with $x(n)$. These assumptions are made for computational tractability; they permit us to do the deconvolution by means of linear processing. Our estimate, obtained by passing $y(n)$ through a restoring filter $q(n)$, is

$$\hat{x}(n) = q(n) * y(n)$$

so that the error is given by

$$e(n) = \hat{x}(n) - x(n)$$

$$= q(n) * y(n) - x(n)$$

The orthogonality principle tells us that for minimum mean-squared error, the error must be orthogonal to the observations:

$$\langle e(n)\bar{y}(n-j) \rangle = 0 \qquad \text{for all } n, j \qquad (7\text{-}109)$$

where the overbar indicates the complex conjugate. Expanding Eq. (7-109) gives us

$$\langle \{q(n) * [b(n) + w(n)] - x(n)\}[\bar{b}(n-j) + \bar{w}(n-j)] \rangle$$

$$= \langle q(n) * b(n)\bar{b}(n-j) \rangle + \langle q(n) * b(n)\bar{w}(n-j) \rangle$$

$$+ \langle q(n) * w(n)\bar{b}(n-j) \rangle + \langle q(n) * w(n)\bar{w}(n-j) \rangle$$

$$- \langle x(n)\bar{b}(n-j) \rangle - \langle x(n)\bar{w}(n-j) \rangle$$

$$= 0 \qquad (7\text{-}110)$$

If we expand the first term of Eq. (7-110) and represent the expected value as a sum, we get

$$\langle q(n) * b(n)\bar{b}(n-j) \rangle = \sum_n \bar{b}(n-j) \sum_k q(k)b(n-k)$$

$$= \sum_k q(k) \sum_n \bar{b}(n-j)b(n-k)$$

The inner sum is recognizable as the autocorrelation, $r_{bb}(j-k)$; hence

$$\langle q(n) * b(n)\bar{b}(n-j) \rangle = \sum_k q(k)r_{bb}(j-k)$$

$$= q(j) * r_{bb}(j)$$

and similarly for the other terms in Eq. (7-110). Thus we have

$$q(j) * r_{bb}(j) + q(j) * r_{bw}(j) + q(j) * r_{wb}(j) + q(j) * r_{ww}(j)$$

$$- r_{xb}(j) - r_{xw}(j) = 0 \qquad \text{for all } j \qquad (7\text{-}111)$$

Since w is assumed uncorrelated with x, and therefore with b as well, the terms containing r_{bw}, r_{wb}, and r_{xw} are zero. Furthermore, from the properties of linear systems, we have

$$r_{bb} = h(n) * \bar{h}(-n) * r_{xx}$$

and

$$r_{xb} = \bar{h}(n) * r_{xx}$$

Hence Eq. (7-111) reduces to

$$q(j) * h(j) * \bar{h}(-j) * R_{xx}(j) + q(j) * R_{ww}(j) - \bar{h}(j) * R_{xx}(j) = 0 \quad (7\text{-}112)$$

The solution of Eq. (7-112) becomes manageable when we enter the Fourier transform domain. Let the power spectra of the signal and noise be $S_{xx}(\omega)$ and $S_{ww}(\omega)$, respectively, and let $H(\omega)$ and $Q(\omega)$ be the transforms of $h(n)$ and $q(n)$, respectively. Then by the convolution property, we have

$$Q(\omega)H(\omega)\bar{H}(\omega)S_{xx}(\omega) + Q(\omega)S_{ww}(\omega) = H(\omega)S_{xx}(\omega) \quad (7\text{-}113)$$

which leads immediately to the solution,

$$Q(\omega) = \frac{\bar{H}(\omega)S_{xx}(\omega)}{H(\omega)\bar{H}(\omega)S_{xx}(\omega) + S_{ww}(\omega)}$$

$$= \frac{\bar{H}(\omega)}{|H(\omega)|^2 + S_{ww}(\omega)/S_{xx}(\omega)} \quad (7\text{-}114)$$

Notice that the embarrassments that attended the inverse-Fourier-transform method have disappeared. Where $S_{xx}(\omega)$ is zero, $Q(\omega)$ is likewise zero; in those regions where the denominator is dominated by $|H(\omega)|^2$, $Q(\omega)$ tends toward $1/H(\omega)$, the inverse filter. [In applications where $S_{xx}(\omega)$ and $S_{ww}(\omega)$ are not known in detail, it is often assumed that their ratio is a constant.] The only difficulty with this approach is that the phase of H must be known; in applications where only the magnitude of H is known, then some assumption must be made about its phase. There may be physical reasons for assuming H pure real, or it may be possible to assume that H is minimum phase.

In two dimensions, Eq. (7-114) can still be used by letting H, S_{xx}, and S_{ww} be functions of two spatial frequencies. An example of linear mean-squared deconvolution in two dimensions is shown in Fig. 7-10. The image, point-spread function, and signal-to-noise ratio are the same as in Fig. 7-8. Notice that the restoring filter has a shape similar to that of the restoring filter in the previous example. Where we windowed $1/H(u,v)$ by hand, taking what we hoped was an intelligent guess at the radius, it is as if the LMS algorithm had windowed the filter for us, selecting the optimum radius on the basis of signal-to-noise ratio and even including a smooth taper at its edges. Notice also that in this

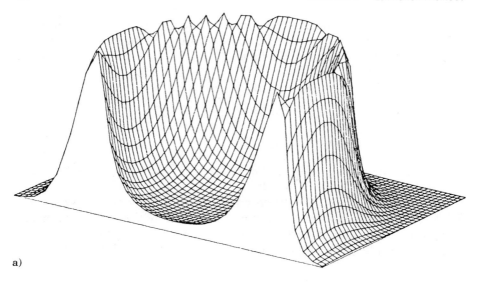

a)

Figure 7-10   *Example of image restoration using MSE techniques. (a) restoring filter.*

particular example, there is not much to choose between the results of the two methods. This was the result of chance; the Fourier transform method does not always do this well.

There are clearly limits to the power of deconvolution. Most transfer functions encountered in practical cases tend to be low-pass. If we think of deconvolution as flattening, or whitening, the overall spectrum, then clearly wherever $H(\omega)$ drives the spectrum toward zero or significantly below the noise level (which amounts to the same thing), this portion of the transform is lost forever and no amount of processing will recover it. It is easy to invent "hopeless" transfer functions of this sort;

$$H(\omega) = \begin{cases} 1 & |\omega| \le \omega_0 \\ 0 & |\omega| > \omega_0 \end{cases}$$

is a trivial example. If there is not a fair amount of high-frequency content remaining in the degraded signal, then $Q(\omega)$ has nothing on which to work and the restoration will be imperfect.

7.3.4 BLIND DECONVOLUTION

In some cases, we may not know what the transfer function $H(\omega)$ is. For example, Stockham (1975) investigated the use of deconvolution to improve the quality of acoustic recordings. The equipment used to make these recordings was passive and functioned as an acoustical transformer; it matched the radiation impedance of the horn to the mechanical

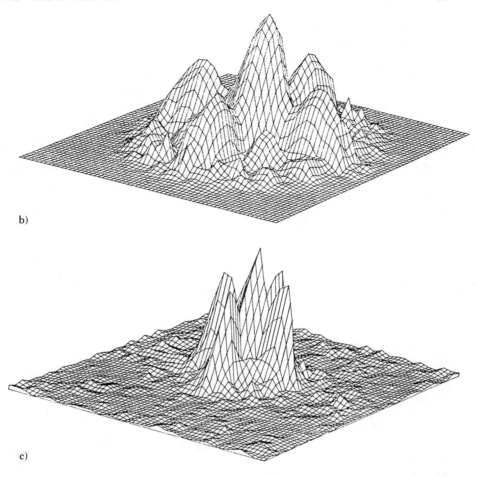

b)

c)

Figure 7-10   *Example of image restoration using MSE techniques. (b) filtered transform; (c) restored image.*

impedance of the cutting stylus. The frequency response of these recordings was affected by internal resonances in this equipment, particularly in the horn, and as an added difficulty, the recording engineers of the time varied the system parameters on each recording in order to get the best performance. Hence it was out of the question to reconstruct a recording setup with surviving equipment, if any could be found. Stockham's problem was therefore to model the equipment as a linear system with unknown parameters and to use this model to flatten the frequency response by deconvolution. This problem, which has been attacked in a number of fields, is appropriately known as *blind deconvolution.*

As stated so far, the blind deconvolution problem clearly has no solution; hence something must be missing. The missing step is to find

some way of making an intelligent guess at either the unknown system
function or the unknown signal. Hence where most signal processing is
analytical in nature, blind deconvolution incorporates a good deal of detective work that cannot be expressed in closed form. We have seen one
solution to the problem in the case of the cepstrum, where the difference
in shape between $H(\omega)$ and $F(\omega)$ was exploited in order to separate them.

Stockham's detective work consisted of finding a modern recording
whose long-term spectral characteristics could be taken as an approximation of those of the input to the acoustical recording process. In other
applications, it may similarly be feasible to use a prototype whose statistical or spectral characteristics are similar to those of the signal to be
recovered. If blind deconvolution is applied to image processing, it may
be helpful if a point source can be found in the scene being photographed.
For example, when satellites are being photographed in the night sky, a
nearby star might be used as an approximate point source. The image of
the star provides an approximation of the magnitude of the point-spread
function provided it falls close enough to the image of the object of interest that the optical characteristics of the air can be assumed essentially
constant for both objects.

7.4 SUMMARY

The field of estimation is an enormous one, and we have only scratched
its surface here. Except for Sec. 7.1 and 7.2, our emphasis has been
on one-dimensional spectral estimation. Of these techniques, Fourier
analysis and the periodogram are widely taught at the elementary level,
but the other methods (maximum-likelihood, maximum-entropy, and Pisarenko's method) are less widely known. There are, of course, many
more kinds of estimation than spectral estimation; for an in-depth study
of detection and estimation theory, the reader is referred to Van Trees
(1968). The techniques of linear prediction, discussed in App. E, provide
a means of estimating the parameters of the system which gave rise to a
particular signal; we will meet an application of this kind of estimation
in Chap. 8. (That same chapter will also provide an application for the
Markov estimation methods discussed in Sec. 7.2.)

PROBLEMS

7.1.   For a homogeneous Markov chain,

$$a_{ij} = P\{x_{n+1} = a_j | x_n = a_i\}$$

Rename this as $a_{ij}[1]$, where in general

$$a_{ij}[k] = P\{x_{n+k} = a_j | x_n = a_i\}$$

Prove the discrete form of the Chapman-Kolmogorov equation for homogeneous processes:

$$a_{ij}[m + k] = \sum_r a_{ir}[k]a_{rj}[m]$$

7.2. Show that the autocorrelation estimator of Eq. (7-50), is unbiased and that the estimator in Eq. (7-51) is biased.

7.3. Prove that $P_B(\omega)$ is everywhere nonnegative.

7.4. Refer to Eqs. (7-52) and (7-53). Let $\hat{X}(e^{j\omega})$ be the Fourier transform of the windowed sequence, $\tilde{x}(n) = x(n)$ for $n = 0, 1, ..., N - 1$. Show that $(1/N)(\hat{x}(n) * \tilde{x}(-n)) = \hat{r}(k)$. From this, show that

$$P_B(\omega) = 1/N|\hat{X}(e^{j\omega})|^2.$$

7.5. Show that

$$\phi(m) = \frac{1}{M} \sum_{j=0}^{M-1} \sum_{v=0}^{M-1} f(m + j - v) = \sum_{r=-(M-1)}^{M-1} \left(1 - \frac{|r|}{M}\right) f(m - r)$$

7.6. For $\{x(k)\}$ a zero-mean, Gaussian process, it is known (Cramer, 1957) that

$$\text{cov}\{x(j)x(j + k)x(v)x(v + \ell)\}$$
$$= r(v - j)r(v - j + \ell - k) + r(v - j + \ell)r(v - j - k)$$

Use this property and the result of Prob. 7.5 to show that

$$\text{var}\{\hat{r}(k)\} = \frac{1}{N} \sum_{v=-(N-k-1)}^{N-k-1} \left[1 - \frac{|v|}{k}\right] [r^2(v) + r(v + k)r(v - k)]$$

7.7. Show that Eq. (7-53) follows from Eq. (7-52).

7.8. Show that

$$\langle P_B(\omega)\rangle = (1/2\pi)W(\omega) * P(\omega)$$

where $W(\omega)$ is the transform of the Bartlett window

$$W(\omega) = \frac{1}{N} \left(\frac{\sin N\omega/2}{\sin \omega/2}\right)^2$$

and $P(\omega) = \mathfrak{F}\{r(k)\}$ and $P_B(\omega) = \mathfrak{F}\{\hat{r}(k)\}$.

7.9. An approximate formula for the bias in $P_B(\omega)$ of Prob. 7.8 is

$$\text{bias}\{P_B(\omega)\} = \langle P_B(\omega)\rangle - P(\omega) \approx \frac{1}{N}\frac{d^2P}{d\omega^2}$$

Derive this formula by expanding $P(\omega-\lambda)$ in a Taylor series inside the convolution $W(\omega) * P(\omega)$ and retaining second-order terms. In doing this, you must show that

$$\frac{1}{2\pi}\int_{-\pi}^{\pi} \omega^2 W(\omega)d\omega \approx \frac{2}{N} \qquad \text{for large } N$$

7.10. Blackman and Tukey (1958) showed that

$$\lim_{N\to\infty} \text{var}\{P_B(\omega)\} = \begin{cases} |P(\omega)|^2 & \omega \neq 0 \\ 2|P(0)|^2 & \omega = 0 \end{cases}$$

and $\qquad \lim_{N\to\infty} \text{cov}\{P_B(\omega)P_B(\omega+\epsilon)\} = 0 \quad \epsilon \neq 0$

These suggest that the periodogram itself can be smoothed to reduce its variance. Let the smoothed periodogram be

$$\bar{P}_B(\omega) = \tfrac{1}{2\pi}H(\omega) * P_B(\omega)$$

(a) Show that the resulting bias and variance are

$$\text{bias}\{\bar{P}_B(\omega)\} \approx M_2\frac{d^2P}{d\omega^2}$$

$$\text{var}\{\bar{P}_B(\omega)\} \approx \begin{cases} E_2|P(\omega)|^2 & \omega \neq 0 \\ 2|P(0)|^2 & \omega = 0 \end{cases}$$

where M_2 and E_2 are respectively the second moment and the energy in the smoothing filter:

$$M_2 = \frac{1}{2\pi}\int_{-\pi}^{\pi} \omega^2 H(\omega)d\omega$$

$$E_2 = \frac{1}{2\pi}\int_{-\pi}^{\pi} |H(\omega)|^2 d\omega$$

(b) Let $H(\omega)$ in (a) be the transform of a Bartlett window $h(k)$ of length L. Show that

$$\text{bias}\{\bar{P}_B(\omega)\} \approx \frac{1}{L}\frac{d^2P}{d\omega^2}$$

$$\text{var}\{\bar{P}_B(\omega)\} \approx \frac{2L}{3N}|P(\omega)|^2$$

7.11. Using the result in Eq. (7-61), show in detail how $P_L(\omega)$ is computed using the DFT.

7.12. Show that the probability density function having the greatest entropy for a fixed standard deviation σ is Gaussian. Hint: Use

$$H = -\int p(x)\ln p(x)\,dx \qquad \int x^2 p(x)\,dx = \sigma^2 \qquad \int p(x)\,dx = 1$$

7.13. Show that the entropy of a Gaussian random variable is given by

$$H = \ln(\sigma\sqrt{2\pi e})$$

7.14. Show that the entropy of n random variables having a zero-mean multivariate Gaussian density function with covariance matrix $\mathbf{C}$ is given by

$$H = \log\sqrt{(2\pi e)^n |\mathbf{C}|}$$

7.15. (Refer to App. E.) Let $\mathbf{L}$ be the lower-triangular matrix,

$$\mathbf{L} = \begin{bmatrix} 1 & 0 & 0 & 0 & \cdots \\ a_n(1) & 1 & 0 & 0 & \cdots \\ a_n(2) & a_{n-1}(1) & 1 & 0 & \cdots \\ a_n(3) & a_{n-1}(2) & a_{n-2}(1) & 1 & \cdots \\ \cdots & \cdots & \cdots & \cdots & \cdots \\ a_n(n) & a_{n-1}(n-1) & a_{n-2}(n-2) & a_{n-3}(n-3) & \cdots \end{bmatrix}$$

(a) Using Eq. (E-9), show that the product $\mathbf{RL}$ is an upper-triangular matrix whose diagonal elements are the prediction errors E_i.
(b) Show that $\mathbf{L}^T\mathbf{RL}$ is a diagonal matrix whose diagonal elements are the same as those of $\mathbf{RL}$.

7.16. Show that, if E_i is the mean-squared prediction error for an order-i linear predictor, the determinant of the autocorrelation matrix is given by

$$|\mathbf{R}| = E_0 E_1 E_2 \cdots E_p$$

7.17. Prove eq. (E-18).

7.18. Prove that when the Cholesky decomposition is applied to a Toeplitz matrix, the intermediate vector $\mathbf{x}$ contains the reflection coefficients. (Hint: Show that $x_n = a_n(n) = k_n$.)

7.19. (a) Show how Eq. (7-101) follows from Eq. (7-80). (b) Show why

$$\mathbf{e}^*\mathbf{L}_i\mathbf{L}_i^*\mathbf{e}/E_i = 1/P_{E_i}(\omega)$$

REFERENCES

H. Akaike. "Fitting autoregressive models for prediction." *Ann. Inst. Stat. Math.*, vol. 21, pp. 243–247, 1969.

L. E. Baum. "An inequality and associated maximization technique in statistical estimation for probabilistic functions of a Markov process." *Inequalities*, vol. 3, pp. 1–8, 1972.

R. B. Blackman and J. W. Tukey, *The Measurement of Power Spectra*. New York: Dover Books, 1958.

D. P. Bogert et al. "The quefrency alanysis of time series for echoes: cepstrum, pseudo-autocovariance, cross-cepstrum, and saphe cracking." M. Rosenblatt, ed. *Proc. Symp. Time Series Analysis*. New York: John Wiley, pp. 209–243, 1963.

J. P. Burg. "The relationship between maximum entropy spectra and maximum likelihood spectra." *Geophysics*, vol. 37, pp. 375–376, Apr. 1972.

———, "Maximum entropy spectral analysis." Ph.D. dissertation, Stanford University, 1975.

J. Capon. "High-resolution frequency-wavenumber spectrum analysis." *Proc. IEEE*, vol. 57, pp. 1408–1418, Aug. 1969.

D. G. Childers, ed. *Modern Spectrum Analysis*. New York: IEEE Press, 1981.

H. Cramer, *Mathematical Methods of Statistics*, Princeton: Princeton Univ. Press, 1957.

R. M. Gray. "On the asymptotic eigenvalue distribution of Toeplitz matrices." *IEEE Trans. Inf. Theory*, vol. IT-18, no. 6, pp. 725–730, Nov. 1972.

U. Grenander and G. Szegö, *Toeplitz Forms and their Applications*. Berkeley: Univ. of California Press, 1958.

F. J. Harris. "On the use of windows for harmonic analysis with the discrete Fourier transform." *Proc. IEEE*, vol. 66, no. 1, pp. 51–83, Jan. 1978.

J. L. Harris, Sr. "Image evaluation and restoration." *J. Opt. Soc. Am.*, vol. 56, pp. 569–574, May 1966.

E. T. Jaynes. "New engineering applications of information theory." J. L. Bogdanoff and F. Kozin, eds. *Proc. First Symposium on Engineering Applications of Random Function Theory and Probability*. New York: John Wiley, 1963.

S. M. Kay and S. L. Marple, Jr. "Spectrum analysis–a modern perspective." *Proc. IEEE*, vol. 69, no. 11, pp. 1380–1419, Nov. 1981.

R. T. Lacoss. "Data adaptive spectral analysis methods." *Geophysics*, vol. 36, no. 4, pp. 661–675, Aug. 1971.

H. Landau and H. Pollack. "Prolate-spheroidal wave functions, Fourier analysis, and uncertainty–II.", *BSTJ*, vol. 40, pp. 65–84, Jan. 1961.

S. E. Levinson et al. "An introduction to the application of the theory of probabilistic functions of a Markov process to automatic speech recognition." *BSTJ*, vol. 62, no. 4, pp. 1035–1074, Apr. 1983.

J. Makhoul. "Linear prediction: a tutorial review." *Proc. IEEE*, vol. 63, no. 4, pp. 561–580, Apr. 1975.

J. D. Markel and A. H. Gray, Jr, *Linear Prediction of Speech*. Berlin: Springer-Verlag, 1976.

A. V. Oppenheim et al. "Nonlinear filtering of multiplied and convolved signals." *Proc. IEEE*, vol. 56, pp. 1264–1291, Aug. 1968.

A. Papoulis. "Maximum entropy and spectral estimation: a review." *IEEE Trans. ASSP*, vol. ASSP-29, no. 6, pp. 1176–1186, Dec. 1981.

———, *The Fourier Integral and Its Applications*. New York: McGraw-Hill, 1962.

———, *Probability, Random Variables, and Stochastic Processes*. New York: McGraw-Hill, 1984.

T. W. Parsons, *Voice and Speech Processing*. New York: McGraw-Hill, 1986.

V. F. Pisarenko. "The Retrieval of Harmonics from a Covariance Function." *Geophys. J. Roy. Astr. Soc.*, vol. 33, pp. 347–366, 1973.

C. E. Shannon. "A Mathematical Theory of Communication." *BSTJ*, vol. 27, pp. 379–423, 623–656, 1948.

D. Slepian and H. Pollack. "Prolate-spheroidal wave functions, Fourier analysis, and uncertainty–I." *BSTJ*, vol. 40, pp. 43–64, Jan. 1961.

D. E. Smylie et al. "Analysis of irregularities in the earth's rotation." *Methods of Computational Physics*. New York: Academic Press, 1973.

M. M. Sondhi. "Image restoration: the removal of spatially invariant degradations." *Proc. IEEE*, vol. 60, no. 7, pp. 842–853, July 1972.

T. G. Stockham, Jr, et al. "Blind deconvolution through digital signal processing." *Proc. IEEE*, vol. 63, no. 4, pp. 678–692, Apr. 1975.

P. A. Thompson. "An adaptive spectral analysis technique for unbiased frequency estimation in the presence of white noise." *Proc. 13th Asilomar Conf. Circ., Sys., Comput.*, pp. 529–533, Nov. 1979.

T. J. Ulrych and T. N. Bishop. "Maximum entropy spectral analysis and autoregressive decomposition." *Rev. Geophys. and Space Physics*, vol. 13, no. 1, pp. 183–200, Feb. 1975.

A. van den Bos. "Alternative interpretation of maximum entropy spectral analysis." *IEEE Trans. Inf. Theory*, vol. IT-17, pp. 493–494, July 1971.

H. L. Van Trees, *Detection, Estimation, and Modulation Theory*. New York: John Wiley and Sons, 1968.

PART II
■■
APPLICATIONS

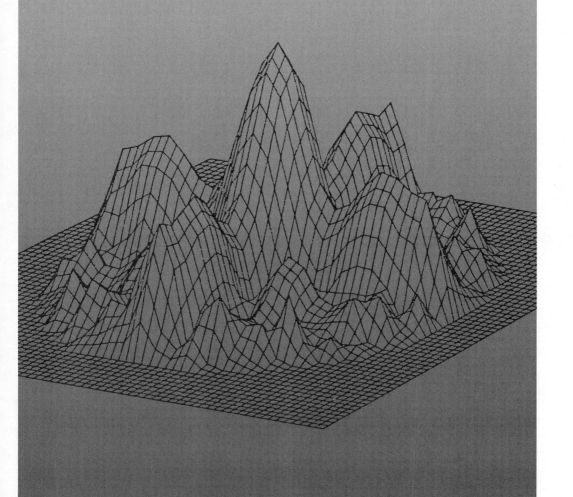

CHAPTER 8

■ ■

SPEECH ANALYSIS, SYNTHESIS, AND RECOGNITION

Voice processing has grown into such a large field that it is impossible to survey all of it adequately here. For detailed coverage, the reader is referred to Flanagan (1972), Rabiner and Schafer (1978), Parsons (1986), and O'Shaughnessy (1987).

The main areas of voice processing include encoding, synthesis, and recognition. There is a considerable amount of redundancy in the speech signal, and encoding tries to reduce the data rate of digitized speech by removing the redundancy. Encoding embraces two major efforts: compact representation of the waveform itself without regard to meaning—also called speech compression—and parameterization in a form that characterizes the speech in some linguistically or acoustically meaningful form.

Synthesis is of interest, first, as the receiving end of a digital speech-communication system whose transmitter performs compression or coding, second, for such purposes as prostheses for the blind, and third, for use as a playback device in applications like telephone-call interception. The second application includes text-to-speech synthesis, in which the input data is conventionally written, spelled, and punctuated text. In the third application area, a small market has developed for appliances that talk; at this writing, the future of this market is still uncertain.

Recognition includes recognition of both speech and the speaker. Speech recognition permits voice communication with the computer, in cases where either the hands are not free or the user wishes to dictate text to a word processor. Recognition of the speaker is of interest for authentication or, in some cases, for forensic purposes.

There is a good deal of overlap among these applications. We have already mentioned the use of synthesis in conjunction with coding. In addition, any parameterization developed for coding will almost inevitably be considered as a source of features for recognition.

8.1 NATURE OF THE SPEECH SIGNAL

Speech is generated by a complex process in which various organs of speech generate and modulate sounds. The speech signal consists mostly of periodic sounds, interspersed with bursts of wide-band noise and occasional short silences. Since the vocal organs are continually in motion, the signal is not stationary, although short segments on the order of 50 ms in length can often be treated as approximately stationary.

The primary generation process is the vibration of the vocal cords. These cords are stretched by a complicated system of muscles in the larynx; when air is forced through them, they vibrate, and the resultant sound takes the form of a pulse train, rich in harmonics. The repetition rate of these pulses determines the *pitch* of the voice. This excitation is used in generating periodic sounds. (Since the pitch and harmonic content are continually changing, these sounds are not strictly periodic, but the term is useful as a means of distinguishing between these sounds and generally noiselike sounds, and it is commonly used.) The periodic sounds comprise vowels, glides (sounds like the [w] in "we" or the [y] in "yes"), and nasal consonants like [m] and [n].[†]

For certain other types of speech sounds, the excitation comes from broadband noise generated by a turbulent flow of air past some constriction in the vocal tract. For example, the sound [s] (as in "say") is produced by moving the tip of the tongue up against the alveolar ridge (the portion of the gums just above the upper incisors) and forcing air through the narrow slit thus formed. The vibration of the vocal cords is termed *phonation*, and speech sounds in which phonation occurs are said to be *voiced*, while sounds in which it is absent are *unvoiced*. Vowels are voiced (although some phoneticians consider the [h] an unvoiced vowel); among consonants, [s] is unvoiced and [z] is voiced.

Consonants take of various forms. Those marked by a more-or-less continuous turbulence are called *fricatives* or (particularly in the case of [s] and [z]) *sibilants*. Some consonants are formed by closing off the vocal tract entirely for a brief time. During this time, pressure builds up and the sudden release causes a mild explosion. Such consonants are called *plosives*; examples are [t], [p], and [k]. Plosive sounds thus consist of a short silence followed by a short noise burst, usually with an abrupt onset. In plosives occurring between two vowels (for example, the [t] in "bottom"), the noise bursts may be weak or absent. If the vocal cords are

---

[†]Speech sounds considered as sounds only are written as phonetic symbols inside square brackets. Speech sounds considered in the context of a specific language are written as phonetic symbols inside solidi, as /i/.

allowed to vibrate during the closure, voiced plosives like [d], [b], and [g] result. Acoustically, these sounds consist of a subdued (low-amplitude) periodic sound followed by a noise burst or a louder periodic sound upon release of the closure.

In the production of vowel sounds, the excitation from the vocal cords is filtered by the action of the rest of the vocal tract. The inside of the mouth and the nasal passages act as an acoustical filter. The jaw, lips, and tongue are all movable, and by varying their positions, most notably the position of the tongue, the speaker can vary the frequency response of this filter. The vocal tract is a distributed-parameter structure and thus has many resonances. In an adult male, the vocal tract is approximately 17 cm in length, and the resonances are spaced about 1 kHz apart on the average. The principal effect of moving the tongue and lips is to change the location (in frequency) of these resonances, and there is a fairly well-defined correspondence between the locations of these resonances and the various vowel sounds. These resonances are called formants. They are numbered in order of increasing frequency; thus we speak of formant 1, formant 2, and so on. The first three formants, and especially the first two, are our main source of information in the speech sound. The correspondence between vowels and formant frequencies varies from speaker to speaker, however, and even in the same speaker over time; this fact is one of the many sources of difficulty in the problem of speech recognition.

The noise excitation of consonants is also filtered by the vocal tract to some extent, but the amount of filtering action depends greatly on the location of the turbulence. Closures at or near the lips, like [p], [f], or [t], are relatively uninfluenced by the rest of the vocal tract; the noise from closures toward the back, like [k] or German [x] (as in *Bach*), may have a formant-like structure imposed on it.

Nasalization imposes its own characteristics on the speech signal. The roof of the mouth terminates at the back in a flexible piece called the *velar flap*. This flap is under voluntary control and can be used to seal off the nasal passages. When the flap is raised and the nasal passages sealed off, they are effectively uncoupled from the transmission path. In these circumstances, the transfer function of the vocal tract is an all-pole function (Parsons, 1986). Opening the passages by relaxing the velar flap provides a parallel transmission path and introduces one or more zeroes in the frequency range of the voice; the parallel path also perturbs the positions of the poles. In some languages, most familiarly French and Portuguese, there is a systematic and meaningful difference between vowels that are nasalized and those that are not.

Nasal consonants are produced by closing off the mouth at some point. In this case, the main transmission path is through the nasal passages (although there is also some radiation from vibration of the throat and cheeks). The shape of the nasal passages is fixed; in addition, the insides of these passages are so shaped that the resonances are rather heavily damped. The closed-off portion of the mouth is coupled to the path through the nasal passages and again introduces a zero into the transfer function.

The acoustical characteristics of vowels are quite definite and are determined almost entirely by the *formant frequencies*. The acoustical characteristics of consonants is much less well-defined; the spectral content of the noise from [s], [f], and [x], for example, varies considerably, but the variation is not consistent enough to be more than marginally useful for recognition purposes. Most of the cues for consonant identification appear to come from *formant transitions* (Delattre et al., 1955) that result from the gradual movement of the speech organs toward and away from the position required by the consonant. The ear is very particular about the acoustics of vowel sounds and about the formant transitions adjacent to consonants, but very forgiving about the nature of the noise associated with consonants.

8.2 ANALYSIS AND PARAMETERIZATION

We can model the speech signal by duplicating the vocal system in hardware. We require a source of periodic excitation, a variable filter to represent the vocal tract with a side branch that can be switched in or out to model the nasal passages; fricatives can be modelled by inserting a wide-band noise source in the vocal-tract filter. The result (Fant, 1961) is shown in Fig. 8-1(a). In practice, this model is more complicated than necessary; we normally consolidate the nasal and oral filters and refer the noise source back to the glottis, adjusting the filter parameters as needed to compensate. Hence the practical model takes the form shown in Fig. 8-1(b). This model is used in analysis and as a practical system for synthesis as well.

8.2.1 FREQUENCY-DOMAIN ANALYSIS

The speech signal can be analyzed in a number of ways, but there are two approaches of particular interest. Since formant frequencies are of such great importance, frequency-domain analysis is an attractive place to start. Indeed, modern speech research could be said to date from the development and widespread use of the FFT.

Analysis using Fourier transforms requires a segment of data long enough that the individual harmonics can be resolved, but short enough that the signal is approximately stationary. We would generally like a time window that is at least two or three pitch periods long. The pitch of the human voice ranges from a minimum of approximately 80 Hz in males to a maximum of approximately 400 Hz in females. Three periods of 80-Hz voice add up to 37.5 ms; the actual data length required is usually chosen to embrace a number of samples equal to a power of 2, for convenience in computing the FFT. Figure 8-2 shows a spectrum of a vowel from a male speaker with an excitation frequency of 185 Hz;

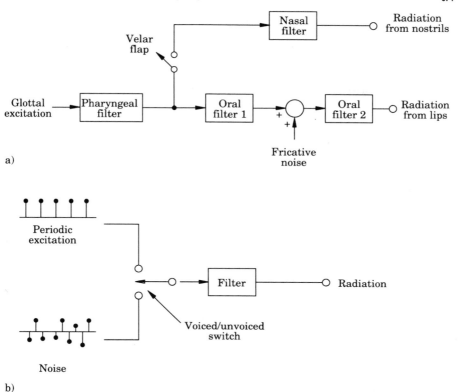

Figure 8-1 *Models of the vocal organs: (a) complete; (b) practical.*

this signal was sampled at 8 kHz and transformed with a 512-point time window. The time function was weighted with a Hanning window and padded with zeroes to a length of 1,024 points. The narrow peaks are the harmonics of the excitation; the envelope of the peaks principally reflects the resonances of the vocal tract.

Some processing of speech is occasionally done in the frequency domain; for example, filtering is easily done by multiplying the transform by the desired transfer function and then computing the inverse transform. A long segment of speech can be processed by transforming and processing overlapping windows. An example is shown in Fig. 8-3, where triangular (Bartlett) time weighting has been employed; this arrangement ensures that all samples make an equal contribution, and the overlapping weighting functions minimize any discrepancies or discontinuities between consecutive windows in the processed waveform.

8.2.2 LINEAR PREDICTION

We have seen (App. E) that linear prediction can be used to provide a model of the system from which the signal has been obtained. This

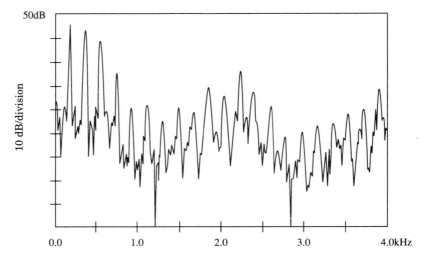

Figure 8-2 *Frequency spectrum of a vowel sound.*

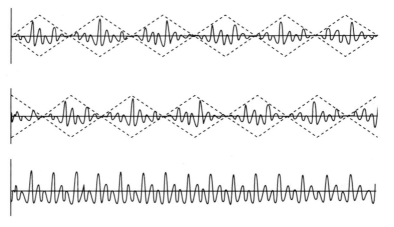

Figure 8-3 *Fourier analysis of speech by means of overlapping, tapered time windows.*

fact constitutes one of the principal attractions of linear-predictive modelling for speech applications. The all-pole predictor provides a reasonable model for unnasalized vowels and a usable model for other periodic speech sounds. If the order of the predictor is set properly, it provides an effective separation of the excitation and modulation implied by the model in Fig. 8-1(b). The order of the predictor is set to accommodate the expected number of formants. We said above that there is, on the average, one formant per kilohertz of bandwidth. Since each formant represents a conjugate complex pole pair in the transfer function of the vocal tract, and since each pole requires one predictor coefficient, the predictor must have an order $p = f_s/1,000$. In practice, the predictor is provided with some extra flexibility by providing a couple of extra poles; hence the rule of thumb is $p = f_s/1,000 + q$, where q is typically 2 or 3.

The predictor coefficients can be used to estimate the spectrum envelope of the speech signal. The model provided by the predictor is of the form,

$$\hat{H}(z) = \frac{E}{A(z)} \tag{8-1}$$

where E is the prediction error power. The corresponding spectrum can be found by evaluating Eq. (8-1) on the unit circle. This is most readily done by dividing E by the DFT of $\{a_0, a_1, ..., a_p, 0, 0, ...\}$ where the zeroes are added to make the length of the sequence a convenient power of 2. Fig. 8-4 shows a DFT of the vowel [æ] (as in "hat") with a superimposed spectrum envelope obtained from Eq. (8-1). The speech was digitized at 8 kHz; the spectrum shown is of a segment 256 samples in length, Hamming-weighted and padded with zeroes to a length of 512 points. The envelope is obtained from an order-10 predictor. The envelope tends to match the peaks of the spectrum more closely than it does the valleys. This is characteristic of spectra obtained by linear prediction; since the envelope of the peaks is of greatest interest, this is a desirable feature. The main peaks in the envelope are at 680, 1,530, and 2,170 Hz, with a smaller peak at about 3,500 Hz; these are close to the average values for the vowel [æ].

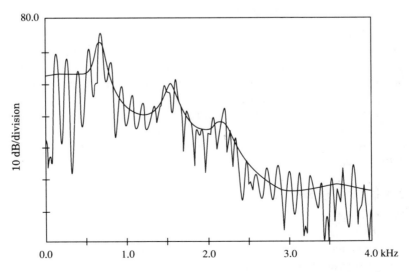

Figure 8-4 *Frequency spectrum of a vowel sound with spectrum envelope determined by linear prediction.*

8.3 COMPRESSION AND CODING

The minimum channel for a voice signal of acceptable quality requires a frequency range of approximately 300 to 3,000 Hz and a dynamic range of approximately 72 dB. Direct digitization of voice can thus be done at a sampling rate of 8 kHz and a word length of 12 bits. The resulting 96k bits/s data rate is usually too high to be acceptable.

Many compression methods seek to reduce the variance of the signal to be transmitted by taking advantage of the considerable redundancy in the speech signal, on the grounds that a small-amplitude signal can usually be transmitted with fewer bits than a large-amplitude signal. The beginner must bear in mind that the reduction in bit rate is equal to the log of the reduction in amplitude; this means that significant reductions in the bit rate tend to require drastic reductions in the variance. For example, if the variance is reduced by a factor of 16, this may seem to promise a great reduction in bit rate; but a factor of 16 is only 4 bits. If the voice was initially digitized with 12 bits, then we have reduced the bit rate by only one-third.

The wealth of methods that have been explored for voice compression and coding can be grouped into four general categories:

Waveform coding includes nonuniform, differential, and adaptive quantization.

Transform coding is transformation of the voice signal, preferably into an orthogonal system, followed by coding of the transform.

Frequency-band encoding divides the frequency range of the voice into discrete channels, which are then coded separately.

Parametric methods are based mostly on linear prediction.

In addition, coding takes two general forms. Compression attempts to preserve the actual waveform from end to end. Using vocoders attempts only to transmit information from which a simulacrum of the voice can be reconstructed at the receiving end without any attempt at detailed duplication of the waveform.

There is an enormous literature on speech compression (see, for instance, the classic work by Jayant and Noll, 1984); we can only offer a summary here; what follows should be considered in the nature of selected topics.

8.3.1 WAVEFORM CODING

The techniques discussed here include nonuniform quantization and adaptive and predictive methods.

The probability density of a sample of speech is shown in Fig. 8-5(a). (This estimate was obtained from a 40-s utterance sampled at 8 kHz.) It is clearly not Gaussian; the approximation commonly used is the modified gamma density

$$p(x) = \frac{\sqrt[4]{3}}{2\sqrt{2\pi}\sigma} \frac{\exp(-\sqrt{3}|x|/2\sigma)}{\sqrt{|x|}} \tag{8-2}$$

This is drawn as the dotted curve in Fig. 8-5(a).

This probability density is the result of two phenomena: (1) the short-term probability density of individual cycles of periodic portions of speech and of fricative noise and (2) longer-term, syllable-rate amplitude fluctuations. The latter component can be removed by scaling short blocks of the speech signal to a uniform variance. If this is done, the probability density of component (1) is approximately Gaussian, as shown in Fig. 8-5(b).

These figures suggest two possible ways of reducing the bit rate. We can use a non-uniform quantizer scaled to match the density of Fig. 8-5(a), or we can use an adaptive quantizer whose step size is varied so as to remove the long-term amplitude fluctuations as we did in producing Fig. 8-5(b). Both of these approaches have been used.

8.3.1.1 Nonuniform Quantization It is clear from Fig. 8-5(a) that an optimum nonuniform quantizer must represent the low-amplitude portions in great detail. A Lloyd-Max quantizer for this density function cannot generally be expressed in closed form, although numerical solutions have been computed by Paez and Glisson (1972).

Two widely used approximations, known as μ-law and A-law compression, have been developed. We can model a nonuniform quantization system by a cascaded nonlinear compression function, a uniform quantizer, and a nonlinear expansion function which is the inverse of the compression function, as shown in Fig. 8-6. Such a combination is known as a *compander*, and these approximations are known as μ-law and A-law *companding*.

With this model, we can conveniently represent the μ-law and A-law functions as the input-output functions of the compressor. If we normalize the input and output so that the maximum value is unity, then these functions are given by

$$y'_\mu(x') = \frac{\ln(1 + \mu|x'|)}{\ln(1 + \mu)} \operatorname{sign} x' \tag{8-3a}$$

$$y'_A(x') = \begin{cases} \frac{A|x'|}{1+\ln A} \operatorname{sign} x' & |x'| \leq 1/A \\ \frac{1+\ln(A|x'|)}{1+\ln A} \operatorname{sign} x' & \text{otherwise} \end{cases} \tag{8-3b}$$

where $x' = x/x_{\max}$ and $y' = y/y_{\max}$. Both of these functions are essentially logarithmic, adjusted to avoid the singularity at $x' = 0$. In the A-law compander, the function for small x' is a straight line, adjusted to be tangent to the log function where they join. In the μ-law compander, the log functions are shifted toward the origin so that the singularity is moved out of the domain of interest.

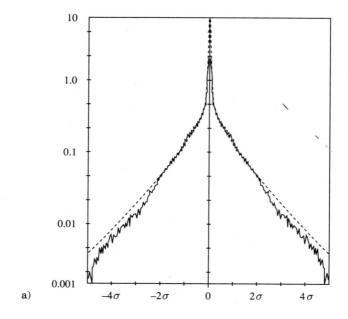

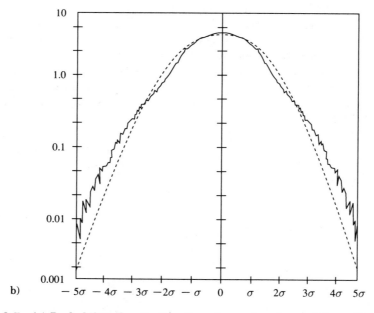

Figure 8-5 *(a) Probability density function of speech estimated from 40-s sample: solid curve, measured density; dashed curve, modified gamma density; (b) probability density function of speech corrected for syllable-length fluctuations in amplitude: solid curve, measured density; dashed curve, Gaussian density.*

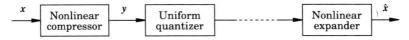

Figure 8-6 *Nonuniform speech quantizer modeled as cascaded compression, quantization, and expansion.*

The μ-law compander with $\mu = 255$ is a North American standard; A-law companding with $A = 87.56$ is a European (CCITT) standard. These functions are plotted in Fig. 8-7(a) and (b); the difference between them, plotted in Fig. 8-7(c), is microscopic, except in the neighborhood of the origin. A-law companding gives slightly better performance at high signal levels; μ-law is better at low levels. The quality of 8-bit μ-law-encoded speech is roughly comparable to that of 12-bit uniformly quantized speech. These devices can be built as 12-bit uniform quantizers followed by a digital converter which computes a piecewise-linear approximation to the companding law used. Details of this digital converter are given in Bellamy (1982) and in Jayant and Noll (1984).

8.3.1.2 Adaptive Quantization We saw in Fig. 8-5 that the great concentration of the probability density about the origin made it important to provide good resolution for small amplitudes in speech signals. Nonuniform quantization provides this resolution by companding, as we have described; another way, suggested by Fig. 8-5(b), would be to provide an automatic gain control before the quantizer so that the low-level portions could be amplified to use the full resolution of the quantizer instead of the bottom few bits. In effect, the gain control varies the step size of the quantizer to match the input. The quantizer itself could be nonuniform, to match the Gaussian shape of Fig. 8-5(b), or a uniform quantizer could be used on the grounds that the amplitude scaling will do most of the job for us; the latter approach is commonly used.

The most straightforward way to estimate the amplitude would be to examine a block of the incoming speech and note either its variance or its peak amplitude. Such a system is shown in Fig. 8-8(a); this approach is known as forward adaptive quantization. It requires storage for the block to be examined and entails a delay in transmission, since the block cannot be released for quantization until its entire contents have been examined. Furthermore, the step-size information has to be passed as side information to the receiver, which must of course make a corresponding step-size adjustment. The need to transmit side information in addition to the quantized signal increases the bit rate above what would be required for the quantized data alone. If the block is long or if the number of allowed step sizes is limited, the increase will be small.

A long block may entail an unacceptable delay in two-way speech communication, and it may also fail to catch significant amplitude fluctuations if their durations are shorter than the block length. There is another way open to us, however. Suppose we used an infinitely long

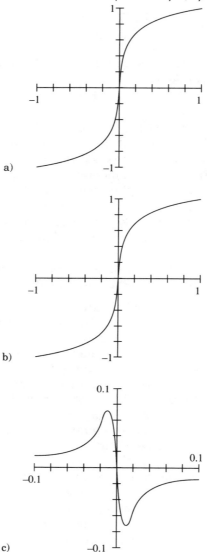

Figure 8-7   *Compression characteristics: (a) μ-law (μ = 255); (b) A-law; (c) difference.*

block with a fading memory, so that the most recent samples would make the greatest contribution. Then we may write our current variance estimate after receiving the sample, $x(n)$, as

$$\hat{\sigma}^2(n) = (1 - a) \sum_{i=0}^{\infty} a^i x^2(n - i)$$

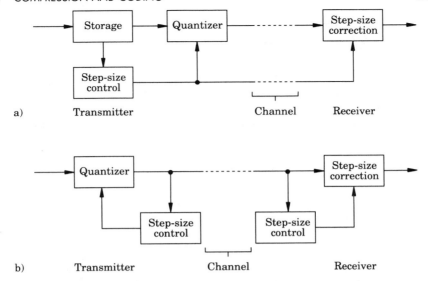

Figure 8-8   *System block diagrams for adaptive quantization. (a) forward adaptive quantization; (b) backward adaptive quantization.*

where a is a weighting factor, $0 < a < 1$, and the factor $(1 - a)$ compensates for the fact that the sum of the weights is $1/(1 - a)$. If we break out the first term in the sum, we have

$$\hat{\sigma}^2(n) = (1 - a)x^2(n) + (1 - a)\sum_{i=1}^{\infty} a^i x^2(n - i)$$

$$= (1 - a)x^2(n) + a(1 - a)\sum_{i=0}^{\infty} a^i x^2(n - 1 - i)$$

$$= (1 - a)x^2(n) + a\hat{\sigma}^2(n - 1) \tag{8-4}$$

Hence if we use an exponential fading memory, we require only storage for the previous estimate of σ^2. If the step size is made proportional to $\hat{\sigma}^2(n)$, then we can implement the step-size changes as a multiplier based on $\hat{\sigma}^2(n)/\hat{\sigma}^2(n - 1)$.

We can eliminate the side information altogether if we base the step-size decision on the output of the quantizer. Any decision logic which uses the quantizer output at the transmitter can use the incoming signal at the receiver, as shown in Fig. 8-8(b); this is known as backward adaptive quantization. The main difficulty with backward adaptation is that in the event of a mismatch between the step size and the signal level, for example, if an overload has occurred, the adaptation logic does not find out until after the damage has been done. In general, overload resulting from too small a step size is a more serious problem than quantization noise resulting from too large a step size; hence the adaptation logic is designed to respond quickly to overloads and slowly to small outputs. Jayant and Noll (1984) show that stable sets of step-size multipliers based on Eq. (8-4) have this property.

8.3.1.3 *Differential Quantization* At any adequate sampling rate, consecutive samples of speech tend to be highly correlated, except for those sounds that contain a significant amount of wide-band noise. This suggests that the bit rate could be reduced by quantizing the difference between samples. That is, we quantize

$$y_n = x_n - x_{n-1}$$

If this is done, it is not difficult to show that the variance of y is given by

$$\langle y_n^2 \rangle = 2\langle x_n^2 \rangle (1 - r_1)$$

where r_i is the autocorrelation of x at a lag of i samples. We can do better than that, however. Let

$$y_n = x_n - ax_{n-1}$$

then $$\langle y_n^2 \rangle = \langle x_n^2 \rangle - 2a\langle x_n x_{n-1} \rangle + a^2 \langle x_{n-1}^2 \rangle$$

Differentiating with respect to a, we have

$$a = r_1/r_0$$

for the minimum variance of y; then

$$\langle y_n^2 \rangle = \langle x_n^2 \rangle (1 - a^2) \tag{8-5}$$

This will be recognized as an order-1 linear predictor, and indeed Eq. (8-5) falls out automatically as a result of the derivation given in App. E. For male speakers sampled at 8 kHz, long-term values of r_1/r_0 range from 0.7 to 0.9 (Noll, 1972, Parsons, 1986). Short-term values fluctuate greatly, however, and the reduction in variance is not always as great as the long-term values, applied to Eq. (8-5), would seem to indicate. This suggests using a predictor that adapts to the moment-to-moment variation in the autocorrelation function of the incoming signal. Pursuing this strategy leads to adaptive linear-predictive coders, which will be discussed below.

8.3.2 TRANSFORM CODING

In transform coding, the signal is represented by one or another of the orthogonal transforms described in Secs. 2.4 and 2.5 and the coefficients of the transform are encoded. The ideal transform will result in uncorrelated components; in this way redundancy is removed from the signal. The Karhunen-Loève transform (KLT) is optimum in this respect; the components are orthogonal and minimally redundant. But to use the KLT, one must transmit not only the coefficients (that is, the eigenvalues) but also the eigenvectors themselves or (more realistically) the autocorrelations from which they were found. Transmitting the autocorrelations greatly increases the bit rate, and the cost of computing the KLT for a vector of length N is $O(N^4)$.

A number of other transforms in which the basis vectors are not data-dependent have been evaluated; the method of choice uses the discrete cosine transform (DCT, Sec. 2.5.2); it has been shown (Ahmed and Rao, 1975) that this expansion approaches the performance of the KLT.

In such coding methods, the main thrust is to minimize the number of bits required to transmit the coefficients. In extreme cases, this can be done using zonal sampling (Sec. 2.4); generally, the number of quantization levels is tailored to the variance of the coefficients. In the DCT, these variances depend on the spectral shape of the speech signal. Zelinski and Noll (1977) based their bit-assignment rule on the spectral shape as estimated from a smoothed form of the DCT; Tribolet and Crochiere (1979) found the spectrum envelope by means of linear prediction. As the shape of the envelope changes, so does the bit-assignment rule; hence the side information must include information from which the current bit-assignment rule can be determined.

8.3.3 FREQUENCY-BAND ENCODING

This is one of the oldest methods used; the channel vocoder (Dudley, 1939) analyzes the voice signal into a small number of contiguous channels. The transmitted data consist of the moment-to-moment variations in the output amplitudes of the channels, plus pitch and voicing information. The channel amplitudes vary slowly, and so can be sampled at a low rate. The pitch varies slowly, and voicing, which in natural speech may be present in varying degrees, is generally reduced to a yes/no decision requiring only 1 bit of information.

Frequency-band encoding uses a filter bank to separate the bands. In Dudley's channel vocoder, shown in Fig. 8-9, there are typically 8 to 10 filters; the amplitudes of the outputs of these filters are encoded and sent to the receiver, along with pitch and voicing information. At the receiving end, a wide-band excitation signal is generated using the pitch and voicing information. If the current sound is voiced, the excitation consists of a periodic signal of the appropriate frequency. In the simplest implementations this signal may be an impulse train; in more elaborate systems the impulses may be shaped so as to make the resultant speech sound more lifelike. If the current sound is unvoiced, the excitation is white noise. The receiver has a matching filter bank, and multipliers control the amplitude of the excitation each filter receives, so that the

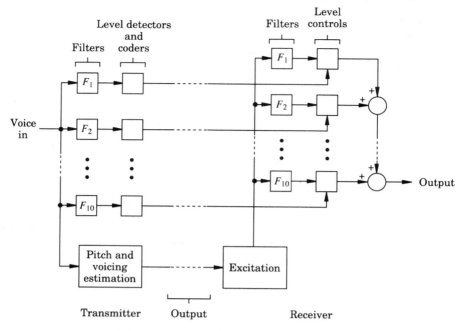

Figure 8-9 *The channel vocoder.*

output level matches the encoded value. The filter outputs are added to produce the decompressed speech.

Filter-bank coding suffered in the early days from the practical difficulty of fabricating analog filter-banks of the required uniformity. Digital filtering obviates this problem, and a number of approaches are possible, for example, the polyphase filter-banks described in Chap. 5.

The channel vocoder is crude, but very robust. The quality of the received speech is generally poor, but the intelligibility is good and the system as a whole is remarkably immune to degradation from noise or other interference. It was for many years the preferred vocoder for military communications, although it has been superseded in this area by the LPC vocoder, which will be described below. The channel vocoder has produced a number of offshoots, however, that are of considerable interest. Since they rely on principles we have discussed earlier, most notably in Chap. 5, we need describe them only briefly here.

If the filters in the transmitter and receiver were ideal, with linear phase and equal delays, then in principle we should be able to get perfect reconstruction if we did away with the encoding and simply connected the outputs of the transmitter's filters to the inputs of the corresponding receiver filters. We cannot do away with the encoding, of course, but since the center frequency of each filter is fixed and known to both the transmitter and receiver, we can provide the equivalent of the filter output by transmitting only its amplitude and phase. This reasoning gave rise to the *phase vocoder* (Flanagan and Golden, 1966). In this vocoder, the output of each filter is heterodyned down to d-c by means of a Weaver modulator, as described in Sec. 5.2.1.1; see Figs. 5-18 and 5-19. The amplitude and phase of the output can be obtained from the real and imaginary parts of the heterodyned signal. Since phase may be unbounded, it is customary to compute and encode the phase derivative; for details, see Parsons (1986).

Just as we found it awkward to deal with a complex baseband signal in multirate processing, so also the complex intermediate output of the phase vocoder is unattractive. The subband vocoder (Crochiere et al., 1976) uses the modulation scheme of Fig. 5-20 to obtain a real baseband signal, and integer-band sampling (Sec. 5.2.1.1) can be used to make the process particularly economical. Esteban and Galand (1977) used quadrature mirror filters (Sec. 5.2.6) to split the frequency range into two subbands, which were then split further as desired to obtain any 2^n subbands.

8.3.4 LINEAR-PREDICTIVE CODING

This is an explicit implementation of the excitation-modulation model of Fig. 8-1(b). At the transmitter, the filter parameters are obtained by linear prediction of the waveform, as described above in Sec. 8-2. These parameters, along with some kind of excitation information, are sent to the receiver, where they control a synthesizer of the form of Fig. 8-1(b).

Linear prediction itself has been discussed in Sec. 8.2.2 and in App. E; the issues to be discussed here are (1) how the predictor parameters are to be transmitted, (2) what form of excitation information is to be transmitted, and (3) how to code noisy speech.

If bit rate were not a problem, it would be sufficient to transmit the predictor coefficients. Since bit rate is a problem, it is important to encode the parameters economically. Errors introduced in encoding the predictor coefficients can result in an unstable filter, however. At the transmitter, the filter is an all-zero filter and stability is no problem. At the receiver, however, the filter is the inverse of the transmitter filter; it is thus an all-pole filter, and if coding (or transmission) errors move the poles outside the unit circle, the filter will be unstable. Hence it is customary to encode and transmit the PARCOR coefficients instead. It is not difficult to devise coding schemes that guarantee that the decoded values will have magnitudes < 1, and as we have seen, this is sufficient to guarantee stability. Bit assignments and coding schemes for efficient coding of the PARCOR coefficients have been drawn up by Gray and Markel (1976) and by Markel and Gray (1980).

The structure of such a system is shown in Fig. 8-10. At the transmitter, the PARCOR coefficients are computed and encoded. At the receiver, the decoded PARCOR coefficients are converted to predictor coefficients, using the Levinson recursion of Eq. (E-12). These coefficients control an inverse filter to which the excitation is applied.

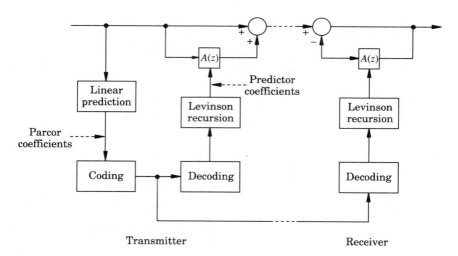

Figure 8-10   *Linear-predictive coding of speech.*

The way in which the excitation information is transmitted has a crucial effect on the quality of the reconstructed speech. The surest, and least economical, way to proceed is to transmit the prediction residual. In that case, the transmitter-receiver combination reduces to a cascade of the prediction filter and its inverse and the output speech should essentially match the input.

If the residual or any reasonable approximation to it is to be transmitted, it is important that the transmitter and receiver filters match. Hence in such a system, the transmitter also decodes the PARCOR coefficients, exactly as the receiver does, and these are used to control the predictor filter from which the residual is obtained. This is shown in the transmitter of Fig. 8-10.

At the other extreme, we can transmit only pitch and voicing information. In this case, there is no residual to be transmitted, and the predictor filter in the transmitter can be omitted, unless it is used to obtain pitch and voicing information. [For pitch estimation methods based on the prediction residual, see Atal and Hanauer (1971), Markel (1972), and Maksym (1973).] Voicing can be indicated by a 1-bit yes/no decision. Pitch varies slowly and over a limited range, and hence it requires only a small number of bits; in the government's LPC-10 algorithm, pitch and voicing are encoded together with 7 bits and transmitted once per 22.5-ms frame. At the receiver, the simplest excitation is an impulse train at the pitch period for voiced speech and white noise for unvoiced speech.

Speech that is reconstructed from an impulse train is generally of poor quality. There are several alternative solutions. The simplest recourse is to convolve the impulses with a simple approximation to a glottal pulse shape. Alternatively, the residual itself can be compressed in various ways. One way is to transmit a low-pass filtered version of the residual;

at the receiver, this residual is converted into a broadband excitation and applied to the filter. This is known as residual-excited linear prediction (RELP, Un and Magill, 1975).

There is a certain amount of redundancy between the residuals for consecutive pitch periods, and some researchers (Atal and Schroeder, 1980) have incorporated pitch-period predictors in order to take advantage of this redundancy. Other compression methods can be found in the literature; these are cited in Rabiner and Schafer (1978), Parsons (1986), and O'Shaughnessy (1987).

Linear prediction of noisy speech presents special problems. The predictor attempts to model the incoming signal; when the incoming signal includes noise, the predictor is not smart enough to ignore the noise. The resultant model is therefore not a faithful representation of the speech signal, and at the receiver the degradation of speech quality is severe. Most methods for coping with noise require preprocessing the speech before linear prediction. If the noise is approximately white, one simple recourse is to note that the noise will show up in the autocorrelation as an increase in the amplitude of r_0. Some research has attempted to find suitable ways to estimate the correct r_0 before doing linear prediction. More information on coding noisy speech may be found in Lim (1983) and in Parsons (1986).

8.3.5 VECTOR QUANTIZATION

In his classic work on information theory, Shannon (1948) showed that the most economical coding of information required a bit rate no greater than the entropy of the source and that this rate was to be achieved by coding large groups, or vectors, of samples rather than coding the individual samples.

This is done by using a *codebook*. A codebook is, in effect, a numbered list of S possible vectors. To transmit a vector, one transmits only the number of its entry in the codebook, that is, its address; this requires $k = \lceil \log_2 S \rceil$ bits. The receiver has its own copy of the codebook, and to recover the transmitted vector, it looks up the entry at the address that was transmitted.

Suppose each entry in the codebook were a vector of n B-bit numbers. If the codebook were complete, that is, if it contained every conceivable vector, there would be $S = 2^{nB}$ entries in the book, and the addresses associated with the entries would themselves be nB-bit numbers. Under these circumstances, transmitting the address would require as many bits as transmitting the entry itself, and it is not clear what advantage the use of a codebook could offer. But in vector quantization the codebook is not complete. It contains a small, but representative, sample of the vectors actually encountered in the data to be encoded, and to transmit a sequence of n samples, we select the most closely matching codebook entry and transmit its address. In such a case, $S \ll 2^{nB}$, and we obtain a correspondingly reduced bit rate at the cost of distortion in the signal arising from the mismatch between the actual vector and the selected entry.

The issues in vector quantization center around (1) the model of the signal that is embodied in the vector, (2) the construction of the codebook, and (3) the organization of the codebook. In speech compression, the signal models considered have been either the actual speech waveform (Abut et al., 1982) or a parameterized model such as a vector of LPC parameters. LPC-based vector quantizers are very popular; a detailed discussion of the process for vector-quantizing predictor parameters may be found in Buzo et al. (1980).

The construction of the codebook uses Lloyd's (1957) algorithm. This is a technique for refining an existing codebook, using new data. The new data are encoded with the existing codebook, and the average distortion is found, using some suitable distortion measure. If this average distortion is less than some predetermined limit, no improvement is possible and the algorithm terminates. Otherwise, the codebook entry for every address i is replaced by the centroid of all the input vectors that were mapped to that address. The data are then encoded with the revised codebook, and this process repeats until no further improvement is possible.

Lloyd's algorithm will always result in an optimal codebook. The set of all possible codebooks for a given ensemble of signals will typically contain many local optima that may be inferior to the global optimum, however; the optimum that Lloyd's algorithm yields will be the local optimum nearest to the initial codebook. Hence it may be desirable to try several initial codebooks in order to obtain an optimum that is close to the global value.

Organization of the codebook is an issue because it affects the time required by the transmitter to locate the best match to the input vector. Codebooks tend to be large, and there is no inherent linear ordering in the codes; furthermore, distortion measures are nonnegative numbers. Hence the classic searches (binary search, tree search, and the like), which depend on a signed error, are not applicable. Three techniques that have been used to reduce either the search time or the size of the codebook are tree-structured codebooks, multiple-stage encoders, and product-code quantizers.

Tree-structured codebooks (Buzo et al., 1980) use auxiliary codebook entries that are representative of subsets of the main codebook. The first two auxiliary entries $A_{1,0}$ and $A_{1,1}$ are based on the centroids of the first and second halves of the main codebook, and the search is directed to one half or the other depending on which centroid best matches the input vector. The next four auxiliary entries $A_{2,0}, ..., A_{2,3}$ are based on the centroids of the four quarters of the codebook; hence two preliminary probes in these auxiliary entries reduce the search time by a factor of 4. The number of these auxiliary codebook layers can be increased as desired; each layer reduces the search time by another factor of 2 at the expense of additional storage for the new layer.

Multiple-stage encoders (Juang and Gray, 1982) use two cascaded code-books. A small codebook C_1 provides a rough approximation to the input vector and a second small codebook C_2 encodes the error resulting from C_1's approximation. Product-code quantizers (Buzo et al., 1980, and Sabin and Gray, 1984) factor the input vector into an overall gain and a shape, which are then encoded separately.

8.4   SYNTHESIS

We may divide synthesis into two processes, control and generation. The generation process, as the term implies, generates the waveform of the synthetic speech; in most applications, it is by far the simpler of the two systems. The control system determines what sort of waveforms are to be generated.

8.4.1   CONTROL

We can only describe the control process in very general terms here; the problems involved are linguistic and not related to digital signal processing. For more complete treatment, consult Parsons (1986) or O'Shaughnessy (1987) and from there go to the literature. Klatt (1987) provides an excellent survey of text-to-speech synthesis. It includes a discussion of the problems and the various approaches used, a recording of more than 30 synthesizers, and an enormous bibliography.

The complexity of the control system depends on how ambitious the synthesizer is to be. At the lowest level, the synthesizer may be a simple playback device, where the control system need only read the stored waveform or its parameters from some source and pass them to the generation hardware. Text-to-speech synthesis represents the highest level; here the input to the control system is conventionally written and spelled speech. The control system must then determine the pronunciation of the word and pass the corresponding control signals to the generation hardware.

There is a good deal more to text-to-speech synthesis than this. It is not enough for synthetic speech to be pronounced correctly; it must also be pronounced naturally. Naturalness is the biggest problem in text-to-speech synthesis. It arises at several levels, but the principal ones are naturalness in pronunciation of individual words and naturalness in the manner of speaking entire sentences.

The pronunciation of a word, as given by a dictionary, represents the word in the conventional phonetic units of the language. These units are called *phonemes*. (What is commonly called a phonetic spelling of a word is actually a phonemic spelling.) A phoneme is a set of acoustically different speech sounds that are perceived by native speakers of the language as being the same sound. The members of this set are called *allophones*. (To distinguish phonemes from allophones, we will write phonemes between solidi, as in /i/, and will put allophones inside square brackets, as in [i].)

For example, we said above that French and Portuguese distinguish between nasalized and nonnasalized versions of vowels. In English, where no such distinction is made, nasalized and nonnasalized vowels are allophones. For example, a native speaker of English will normally pronounce "cat" with a nonnasalized [æ]. (We represent the a of "cat," "hat," "bad" by the symbol æ.) The velar flap must be raised in order to seal off the nasal passages for the /c/ and the /t/, and it is simply too much trouble to lower the flap for the /æ/. On the other hand, "man" is usually pronounced with a nasalized [æ̃]; here it is too much trouble to close the nasal passage for the /æ/. Since nasalization is not phonemic in English, we are quite unaware of the difference between these two sounds, and it may in fact come as a surprise to learn that they are different.

The relevance of this behavior to synthesis is this: for the synthetic speech to sound natural, it is essential that the correct allophone be produced. Despite the fact that different allophones may be perceived as the same sound, if the wrong allophone is selected, the speech will not sound natural. Translation from phonemes to allophones is generally controlled by a set of rules; these rules are well enough understood that at least one manufacturer has marketed an integrated circuit containing rules that produce the desired result more than 95 percent of the time.

It is also important to control the timing of a word in a natural-sounding manner. People do not utter the chosen allophones at a constant rate, and although this is another largely unconscious process, errors in timing can make synthetic speech sound absurd.

Naturalness in the pronunciation of sentences is a much more difficult problem and one that has not yet been completely solved. The issue here is how the speaker's pitch varies over the sentence and which words receive emphasis. (Emphasis is usually produced by a momentary increase in amplitude, frequently accompanied by a higher pitch.) These decisions are largely determined by the meaning of the sentence. Analyzing the meaning of a sentence is in itself a complex process and lies more in the domain of artificial intelligence than in that of voice processing. The analysis requires a sizeable body of knowledge, not only about the language but about the subject being talked about, and although considerable progress has been made in this area, the systems that work well are huge and impose an unacceptable cost in complexity and speed of response. Hence even the best text-to-speech synthesizers tend to speak in a near monotone relieved only by quirky and inexplicable emphases.

8.4.2 GENERATION

The generation system typically makes use of an excitation-modulation model of the sort shown in Fig. 8-1(b). The excitation generates a periodic function for vowels and similar continuous sounds or wide-band noise for fricatives and plosives; it may also provide a mixture of both for such sounds as voiced fricatives. The modulation part is a digital filter that simulates the effect of the vocal tract. Analysis and duplication of the acoustics of the vocal tract has been one of the triumphs of modern speech analysis. The analysis is carried out with the aid of linear prediction, as described above, and the transfer function of the filter is as given in Eq. (8-1).

Some investigators (Klatt, 1987) have preferred to use a parallel filter structure with one branch for each of the first three formants. The most common form employs an inverse linear-prediction filter of the form used in linear-predictive coding, however.

This filter can be synthesized directly from the predictor coefficients, as suggested by Eq. (8-1); in some applications, however, it may be preferable to use the PARCOR coefficients, in which case the lattice structures of Figs. 5-6 and 5-7 are used. For example, some synthesis techniques interpolate the predictor parameters between stored reference values. There is no guarantee that interpolated predictor coefficients will yield a stable filter; hence interpolation is done on the PARCOR coefficients instead, and the resultant values are applied to a lattice filter.

Excitation of the synthesizer has some of the problems associated with linear-predictive coding; the simplest method uses an impulse train, and the quality of the speech leaves a good deal to be desired. An alternative approach is to store one cycle of a representative residual and use this as the excitation.

8.5 RECOGNITION

Speech recognition is a particular application of the general field of pattern recognition. Pattern-recognition systems have two modes of operation: training and recognition. In training, a speech recognizer is given spoken words and told what they are. The recognizer stores the significant features of the words in a library, along with their identities. In recognition, the system compares the features of an unknown word with all the entries in the library; it identifies the unknown word with that library entry whose features are the best match to those of the unknown word.

There are two general problems in designing pattern-recognition systems; in addition, there are a number of particularly difficult problems peculiar to speech recognition. We will discuss the general problems first.

8.5.1 FEATURE SELECTION

An incoming utterance contains a wealth of information, some of which is relevant and much of which is not. For example, the acoustical data tells you what the word is, who is speaking it, how the speaker is feeling that day, whether he or she has a cold, and so on, almost endlessly. Hence in designing a recognition system, one must select the relevant information and discard the rest. The relevant information consists of *features*, observables which can be used to distinguish among the patterns. The search for features is generally the most demanding part of designing a pattern-recognition system.

There are no rules for identifying useful features; one must rely on one's knowledge of the problem domain, backed up by careful study of the data. There are statistical methods for evaluating candidates, however. The general rule is to compare the separation among the various words (or *classes*, as they are called in general) against the scatter within each class. The fact that the mean values of a feature, measured for two classes, differ by (say) 10 units is meaningless in itself. If the estimated standard deviation within each class is 3 units, the feature is a very good discriminator; if it is 30 units, it is worthless. All methods of statistical feature evaluation are based on such comparisons.

Different features are usually correlated to some degree; hence we cannot consider them in isolation but must evaluate their effectiveness in combination. The dimensionality of the feature space is the same as the number of features under consideration; since this dimensionality is virtually always greater than 1, we must resort to the covariance matrices of the features. The covariance matrix is sometimes diagonalized to obtain uncorrelated features that are linear combinations of the observed ones.

The degree of separation among classes can be estimated from the covariance matrix found from the mean values of the features. This is known as the *interclass* covariance matrix, because it considers solely how the classes are separated in the feature space. We can estimate the degree of scatter within any class by the covariance matrix of the features as measured for that word alone; this is known as the *intraclass* covariance matrix.

To be specific, suppose the recognizer must distinguish among n words. We measure the candidate features in a data base consisting of k utterances of each word. The measured values of the features of any utterance of any word can be represented as a column vector. Let the values for the ith utterance of the jth word be $\mathbf{f}_{ij}$. To simplify notation, let μ_j be the mean of all $\mathbf{f}_{ij}$ for word j:

$$\mu_j = \frac{1}{k} \sum_{i=1}^{k} \mathbf{f}_{ij} \tag{8-6}$$

and let μ (with no subscript) be the mean over all words:

$$\mu = \frac{1}{n} \sum_{j=1}^{n} \mu_j \tag{8-7}$$

Then the intraclass covariance matrix for the jth word is given by

$$\mathbf{W}_j = \frac{1}{k} \sum_{i=1}^{k} (\mathbf{f}_{ij} - \boldsymbol{\mu}_j)(\mathbf{f}_{ij} - \boldsymbol{\mu}_j)^T \tag{8-8}$$

If there is no strong evidence that the elements of $\mathbf{W}_j$ differ significantly from class to class, it is customary to average them; hence we may characterize the within-class scatter by $\mathbf{W} = \langle \mathbf{W}_j \rangle$. ($\mathbf{W}$ is generally understood as standing for "within," since it is the covariance within the class.) The interclass covariance matrix is obtained from the means of the words:

$$\mathbf{B} = \frac{1}{n} \sum_{i=1}^{n} (\boldsymbol{\mu}_j - \boldsymbol{\mu})(\boldsymbol{\mu}_j - \boldsymbol{\mu})^T \tag{8-9}$$

($\mathbf{B}$ stands for "between.")

There are three common ways of comparing the scatter described by these matrices. They are based on considering the covariance matrices in the light of the Gaussian probability density. (Features are not generally Gaussian, but the Gaussian assumption is the most convenient and is often made; in addition, with limited data it is the safest assumption, in the maximum-entropy sense.) The three ways are the ratio of the determinants of the matrices, the ratio of their traces, and tr $(\mathbf{W}^{-1}\mathbf{B})$.

We can justify the first two readily: It can be shown (Parsons, 1986) that the determinant of a covariance matrix is equal to the product of the variances of the uncorrelated variables that would result from diagonalizing the matrix and that the trace of the covariance matrix is equal to their sum. Hence $|\mathbf{B}|$ and tr $\mathbf{B}$ both provide a measure of the overall scatter between classes, and similarly $|\mathbf{W}|$ and tr $\mathbf{W}$ provide a measure of the scatter within a class. If $|\mathbf{B}|/|\mathbf{W}| > 1$, or if tr $\mathbf{B}/$ tr $\mathbf{W} > 1$, then the average separation between classes is larger than the average scatter within classes and the feature is a good one. The third measure of separability tr $(\mathbf{W}^{-1}\mathbf{B})$ cannot be justified without going rather far afield; see Fukunaga (1972), Duda and Hart (1973), Tou and Gonzalez (1974), or Parsons (1986) for further details.

8.5.2 DISTANCE MEASURES

In recognition, the feature vector for the unknown word must be compared with all the feature vectors in the pattern library. This comparison requires a measure of similarity; these measures are commonly known as distance measures. The distance from a feature vector $\mathbf{f}$ to the library entry $\boldsymbol{\mu}_i$ for Word i is most readily measured by the Euclidean metric,

$$d(\mathbf{f}, i) = (\mathbf{f} - \boldsymbol{\mu}_i)^T(\mathbf{f} - \boldsymbol{\mu}_i) \tag{8-10}$$

where n is the number of features. This is a simplification of a more general metric based on the maximum-likelihood principle, and as given it is applicable only to sets of uncorrelated features having equal variance.

Suppose the features for each class are Gaussian. Then the probability that word i will give rise to features $\mathbf{f}$ is given by the usual multivariate form,

$$p(\mathbf{f}|i) = (2\pi)^{-n/2}|\mathbf{W}_i|^{-1/2} \exp[-\frac{1}{2}(\mathbf{f} - \boldsymbol{\mu}_i)^T \mathbf{W}_i^{-1}(\mathbf{f} - \boldsymbol{\mu}_i)]$$

We wish to run this backward: given an observed vector of features $\mathbf{f}$, we wish to know the likelihood that they were produced by word i; this is $p(i|\mathbf{f})$. These two probabilities are related by Bayes' theorem,

$$p(i|\mathbf{f}) = p(\mathbf{f}|i)p(i)/p(\mathbf{f})$$

We normally assume all words equally likely, so in that case $p(i)$ is a constant. If we are comparing two words with a view to finding which word is more likely to have produced $\mathbf{f}$, then $p(\mathbf{f})$ is likewise constant; hence in this case we need only find the word which maximizes $p(\mathbf{f}|i)$. It is computationally cheaper (and conceptually clearer) to minimize $-\ln p(\mathbf{f}|i)$. This is the basis of all principal measures of similarity; it is given by

$$d(\mathbf{f}, i) = (\mathbf{f} - \boldsymbol{\mu}_i)^T \mathbf{W}_i^{-1}(\mathbf{f} - \boldsymbol{\mu}_i) + \ln |\mathbf{W}_i| \qquad (8\text{-}11)$$

where constant terms and factors have been dropped.

This is the theoretical *maximum-likelihood* distance measure. We cannot use this measure in practice unless we have such a wealth of data that reliable estimates of $\mathbf{W}_i$ for each word can be found. This rarely happens even in practical applications and almost never in research; hence we commonly pool all the $\{\mathbf{W}_i\}$ into a single mean intraclass covariance matrix $\mathbf{W}$. In this case, $\ln |\mathbf{W}|$ is constant for all words and can be omitted from the comparison. The result is the *Mahalanobis distance*

$$d(\mathbf{f}, i) = (\mathbf{f} - \boldsymbol{\mu}_i)^T \mathbf{W}^{-1}(\mathbf{f} - \boldsymbol{\mu}_i) \qquad (8\text{-}12)$$

Other simplifications are possible (including the Euclidean distance with which we started) if the features are known to be approximately uncorrelated; see the references cited above for details.

8.5.3 SEGMENTATION

The principal practical problem in recognition of spoken words is finding out where one word ends and the next word begins. If the speech is connected, as it is in ordinary discourse, it is known as the *segmentation problem*; if the words are isolated (that is, spoken—with—a—pause—between—words), the problem is known as *endpoint detection*.

The reasons for the segmentation problem should be clear; in normal speech we run our words together. In the acoustical signal it is difficult to determine just when one word ends and the next begins; but recognition systems are dependent on this information. After a lifetime of experience, our brains process speech so readily, that it is hard for us to see why this should be a problem, but even trained speech researchers have trouble locating word boundaries if they have nothing but the time waveform to go on. The clues to word boundaries are changes in the speech sound. If the changes were abrupt, the decision would be easy to make, but in a sentence like "We were away a year ago," the sounds are nearly all vowels or vowel-like sounds. In this example, the transition from the /i/ of "we" to the /w/ of "were" is smooth. (We represent the long e of "we" with the symbol i.) With actual data the point where the /i/ ends and the /w/ begins can be hard to pinpoint. The problem is aggravated by the fact that the exact pronunciation of phonemes is influenced by context; thus the /i/ in "we were" will be slightly different from the /i/ in "we are" and from the /i/ in "we asked."

If the errors were small ones, the accuracy of the recognition process would not be greatly impaired; but in fact they can be gross; this means that the recognition process starts out with flawed data.

The endpoint problem in isolated words arises from two interacting phenomena. Many words start from silence and build up gradually, and there is always background noise present, unless the data have been obtained under laboratory conditions. The endpoint detector must therefore decide when it is listening to speech rather than noise. Existing solutions to this problem make guesses at the endpoints, usually including too much signal at the start, and then refine the guess. The refinement procedures are largely ad hoc, and no endpoint detector currently performs with perfect reliability.

8.5.4 TIME NORMALIZATION

A second problem is that of timing within the word. There is a surprisingly wide variation in both the duration of a word and the detailed timing within the word. This variation occurs not only among speakers, but also among different utterances of the word by a single speaker at different times. Since the pattern of features characterizing a word is a function of time (that is, from time to feature space), temporal alignment between the library pattern and the incoming word is crucial. In word recognition, these time-function patterns are commonly called *templates*.

There are two main techniques for time alignment. One uses a procedure that distorts the time axes so as to align the features; the other makes use of a Markov model of speech production. We will discuss time-axis distortion here and defer the Markov model to a later section.

Temporal alignment of the corresponding parts of two different templates can be done by setting up a mapping from one to the other. Suppose we draw one template horizontally and the other vertically, as in

Fig. 8-11(a). For simplicity, we are assuming that the template consists of a single feature here. Then the mapping can be represented by a curve like the one labelled $f(x)$ in the figure. We could straighten out the curve by suitably distorting the time axes of the two templates; hence this technique is known as time warping. The curve is constrained so as to align the endpoints of the two templates, as shown. There is usually an added constraint which prevents the curve from wandering too far away from a straight line; this constraint is frequently a limit on $|x - y|$; in the figure, this limit is shown by the dotted lines A and B.

We start in the lower left-hand corner and compare all permissible points (x, y). For every such point, we compute a cost function that measures the degree of discrepancy between the values of the two templates at these two points. If the cost is $c(x, y)$ and the two templates are p_1 and p_2, then

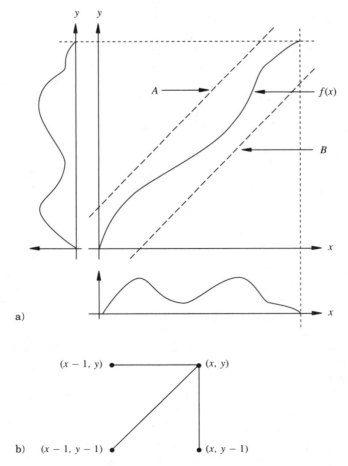

a)

b) $(x - 1, y)$ (x, y)

      $(x - 1, y - 1)$ $(x, y - 1)$

Figure 8-11   *Time normalization by dynamic time warping (DTW).*

$$c(x, y) = \text{dist}(p_1(x), p_2(y)) \tag{8-13}$$

The cost function dist (·) is normally one of the distance measures discussed above. The total cost of the mapping is the sum of the costs for the individual point pairs along the path. Since the individual costs are additive, we can find the desired mapping economically by means of dynamic programming (Bellman, 1957).

The cost of the cheapest path to any point (x, y) is the cost of (x, y) itself plus the cost of the cheapest admissible predecessor of (x, y). We require that $f(x)$ be a monotone nondecreasing function of x and that $f^{-1}(y)$ be similarly a monotone nondecreasing function of y; hence the admissible predecessors of (x, y) are usually $(x - 1, y)$, $(x - 1, y - 1)$, and $(x, y - 1)$, as shown in Fig. 8-11(b). (Other admissible predecessors have been considered; for details, see Parsons, 1986.)

The time-warping program thus starts in the lower left-hand corner and works up each column of the diagram; for each new point (x, y), the program finds the cheapest admissible predecessor point and makes that point the chosen predecessor of (x, y). The output of the program is an array of chosen predecessors. We can represent this graphically by the tree structure shown in Fig. 8-12. The path can be found by working backward along the tree, starting from the upper right-hand corner; this is the heavy line in Fig. 8-12. The total cost is the cost associated with this path. The total cost is also normally taken as a distance measure for the two templates being compared.

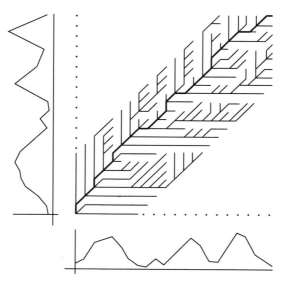

Figure 8-12   *Tree diagram showing history of dynamic time warping. The heavy line shows the final warping function.*

Execution time for this program is minimized by three constraints. First, no more than three predecessors need be considered for any point under consideration; second, the width constraint imposed by the lines A and B limits the number of points to be considered; and finally, the entire process may be aborted if the accrued costs exceed some predetermined limit. In particular, in word recognition we may already have found a better match to some previous library pattern, in which case the warping program may be aborted as soon as it is clear that the current template is going to be a poorer match.

8.5.5 THE HIDDEN MARKOV MODEL

This model is best approached by means of an oversimplification. As we speak, we move the vocal organs into various positions in order to produce various speech sounds. Let us assign a state to each position of the organs. With each state we can associate an output, that is, the sound that is produced with the organs in that position. In natural speech, there is a certain imprecision in the position of the organs, so our first refinement is to associate a state with a group of closely-related configurations. There is also a corresponding degree of variation in the resultant sound, so we will also associate a set of possible outputs with each state.

As a word or sentence is spoken, the vocal organs move from one position to another; this is represented in our model by transitions from one state to another. With each such transition, we associate a probability. A process consisting of a number of different states with various transition probabilities between states, where the probabilities depend only on the current state, is a Markov process. In using this process to model speech, we have access only to the resultant sounds, and from these we must infer the transition probabilities during the training phase, and, during the recognition phase, which states were entered. Since the underlying model is thus never explicitly visible to us, it is said to be hidden; this explains the term "hidden Markov model" (HMM).

In natural speech, the organs may take on a vast number of different configurations; in speech recognition, we assume a relatively small number of states. This is where our oversimplification fails; for recognition of isolated words, a model with five to seven states has been found adequate. Furthermore, the model is normally made to be very simple. A general state diagram might look like Fig. 8-13(a); practical recognizers tend to use a left-to-right model like that of Fig. 8-13(b). In this model, the only options are to remain in the same state, to advance to the next state, or to skip a state. The parameters of such a simple model are cheaper to estimate during training; in addition, Rabiner et al. (1983) found that the left-to-right model yielded better recognition accuracy than more ambitious models.

We represent the model essentially as we did in Chap. 7. Let the model have N states, $\{q_1, q_2, ..., q_N\}$. Then the probability of a transition

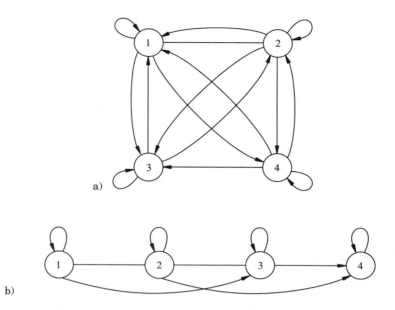

a)

b)

Figure 8-13   *State diagrams for Markov models: (a) general, (b) left to right.*

from q_i to q_j is given by $P\{q_j|q_i\}$. We hold these probabilities in an $n \times n$ transition matrix, **A**, where $a_{ij} = P\{$ next state $= q_j|$ current state $= q_i\}$. The model starts out in an initial state and makes transitions as the utterance proceeds. At any time t, the model may be in any one of several states; hence we show the current state as a vector $\mathbf{p}(t)$, where $p_i(t) = P\{$state at time t is $q_i\}$. The starting-state probabilities are given by $\mathbf{p}(1)$.

We also have a set of M possible outputs $\{z_1, z_2, ..., z_M\}$. As in Chap. 7, the $N \times M$ matrix **B** gives the probabilities of each output for each state: that is, $b_{ij} = P\{z_j|q_i\}$. The ith row vector of **B**, $\mathbf{b}_i T$, gives the distribution of probabilities for the various outputs for state q_j.

We can now trace the progress of the model through an utterance. This trace takes the form of a sequence of **p** vectors. The probability of being in state q_j at time t is the product of the probability of having been in state q_i at time $(t-1)$ times the probability of a transition from q_i to q_j, summed over all states q_i. This is

$$\mathbf{p}(t) = \mathbf{A}\mathbf{p}(t-1)$$

Since the starting probabilities were $\mathbf{p}(1)$, this gives us

$$\mathbf{p}(t) = \mathbf{A}^{t-1}\mathbf{p}(1)$$

At time t, the probability of an output z_k is given by the product of the model being in state q_i times the probability of z_k given q_i; this is

$$P\{z_k \text{ at time } t\} = \mathbf{b}_k^T \mathbf{p}(t)$$

$$= \mathbf{b}_k^T \mathbf{A}^{t-1} \mathbf{p}(1) \qquad (8\text{-}14)$$

The quadruple $\mathbf{M} = [N, \mathbf{p}(1), \mathbf{A}, \mathbf{B}]$ thus constitutes a model of the word. In training the recognizer, we must find such a model for each word in the vocabulary; these models go in the pattern library. In recognition, the system receives a sequence of observations $\mathbf{O}$ whose elements are drawn from the outputs $\{z_i\}$. The recognizer identifies the word by selecting the model that is most likely to have produced the sequence $\mathbf{O}$. It does this for each model by tracing the state transitions and outputs that best explain $\mathbf{O}$—that is, that are most likely to have occurred and to have resulted in $\mathbf{O}$.

Going through the states of a left-to-right model is a little like going through the columns of a dynamic time warping (DTW) tableau, but this similarity must not be exaggerated. The most important difference is that the HMM is more compact and runs faster. In a comparison between HMM and DTW recognizers, Rabiner et al. (1983) found that the HMM required only one tenth as much storage and ran 17 times as fast. The other difference of interest is that training the HMM is very expensive and requires a vast quantity of data.

The theory of training the model has already been discussed in Sect. 7.1.3; see also the classic papers by Levinson et al. (1983) and Rabiner et al. (1983) or the books by Parsons (1986) and O'Shaughnessy (1987). Here we will give only a brief discussion of the recognition phase.

The HMM is, in effect, a model of speech production, or perhaps of speech sequencing. We assume that the observed utterance can be explained by such a model; hence the recognition process consists of considering for each model in the library how it could have produced the sequence of observations and how likely it is to have done so. We have the same Bayesian reasoning we encountered in the consideration of distance measures: we want

$$P\{\mathbf{M}|\mathbf{O}\} = P\{\mathbf{O}|\mathbf{M}\} P\{\mathbf{M}\} / P\{\mathbf{O}\}$$

Again, we have only one observation $\mathbf{O}$ to consider and we consider all models equally likely; hence we can look for the maximum $P\{\mathbf{O}|\mathbf{M}\}$. We must now find an expression for this probability.

Transitions among the states of such a model are commonly shown by a lattice diagram of the sort shown in Fig. 8-14. This diagram applies to the left-to-right model of Fig. 8-13(b). For simplicity, the diagram assumes only a single starting state and a single final state; the lattice can easily be generalized to accommodate more than one of each. The states are represented by the numbered circles and the transitions by the arrows; the arrows reflect the choice of transitions permitted by Fig. 8-13(b). Time runs from left to right, and moving from the top down takes us through the states in sequence. This diagram includes all possible

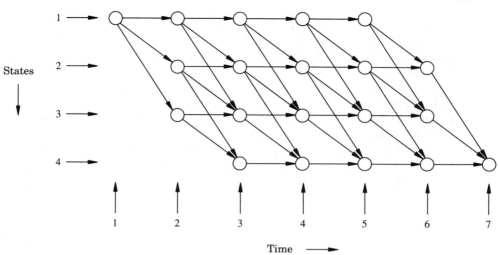

States

Time ⟶

Figure 8-14 *Lattice diagram showing possible state transitions in the Markov model.*

paths of length 7 from state 1 to state 5. Any path taken by following the arrows from the starting state at the left to the final state at the right is a possible explanation of how the model gave rise to the sequence of observations **O**.

Suppose the sequence of observations is of length T. We need to find the probability that the model under consideration can go from a starting state to a final state in T steps and emit the observed outputs. This probability can be readily obtained with the aid of the forward recursive relation among partial sequences of states and observations, that we derived in Sect. 7.1.3. The forward recursion was

$$\alpha_{t+1}(j) = \sum_{i=1}^{N} \alpha_t(i) a_{ij} b_{jk} \tag{8-15}$$

with initial conditions $\alpha_1(j) = p_j(1), j = 1, ..., M$. Running this recursion out to $\alpha_T(j)$ gives us the probability that the model gave rise to the observation sequence **O** and ended in state j. Then the probability that that **M** produced **O** is simply the sum of $\alpha_T(j)$ over all states:

$$P\{\mathbf{O}|\mathbf{M}\} = \sum_{j=1}^{N} \alpha_T(j) \tag{8-16}$$

This provides our answer: we perform recognition by computing $P\{\mathbf{O}|\mathbf{M}_j\}$ for every model $\mathbf{M}_j$ and selecting the word j for which this probability is greatest.

We should emphasize that although the theory behind the HMM is a good deal less accessible than that for DTW, the computational burden, involving typically no more than five or six states, is significantly lighter than it is for DTW. Hence, while training is slow and expensive and usually requires more data than are at hand, recognition proceeds quickly. Performance figures for the two techniques are comparable. In one of the earliest trials of the HMM, Rabiner et al. (1983) compared two talker-independent digit recognition systems using the HMM and DTW. Although the training data used for the HMM were "woefully inadequate," the recognition accuracies of the two methods differed by only a fraction of a percentage point.

8.6 SUMMARY

This chapter has provided a superficial survey of the main problems and techniques used in the field that has come to be known as voice processing. From the early sections of this chapter, it should be clear that most of the mechanics and characteristics of speech are very well understood; but from the rest of the chapter, it should be equally clear that our successes in this field have not been commensurate with our understanding.

The invention of the sound spectrograph in 1941 provided the first easily readable and intuitively appealing representation of the acoustical content of speech sounds. In the first flush of enthusiasm at this development, people predicted that the "phonetic typewriter" would be only a few years off. Nearly 50 years have elapsed since then, and the phonetic typewriter still lies off in the future, a tantalizing goal, realized so far only under very restricted conditions. A similar story could be told about speech synthesis and to a lesser extent about speech compression and coding.

In this chapter, we have tried to show what the difficulties are, but perhaps two general observations may yet be appropriate. First, speech and language are human activities, and humans—and living creatures generally—are much less predictable and much less susceptible of analysis than are non-living systems. Successful analysis and a successful solution of a hard problem are often the result of an ingenious simplification, and living systems are notoriously resistant to simplification.

Second, it is generally very difficult in technology to predict what things will be done easily and what things will be difficult. We may instance two examples from other fields here. We have been similarly disappointed in our expectations for the generation of electric power by fusion: this is another area that has been just around the corner for about as long as the phonetic typewriter. It has simply turned out to be much harder to do than most people initially expected, and in this field we do not have the excuse of human cussedness. The second example is computer tomography, a problem that yielded as soon as it occurred to someone to try to solve it. It would appear that nobody thought it would be that easy.

This unpredictability is uncomfortable to live with, but it is well to realize it and face it early on: it will spare us the embarrassing consequences of undue optimism on the one hand, and on the other, it may occasionally embolden us to attempt the seemingly impossible.

REFERENCES

H. R. Abut et al. "Vector quantization of speech and speech-like wave-forms." *IEEE Trans. ASSP*, vol. ASSP-30, no. 3, pp. 423–435, June 1982.

N. T. Ahmed and K. R. Rao. *Orthogonal Transforms for Digital Signal Processing.* New York: Springer-Verlag, 1975.

B. S. Atal and S. L. Hanauer. "Speech analysis and synthesis by linear prediction of the speech wave." *JASA*, vol. 50, no. 2, pp. 637–655, Aug. 1971.

J. C. Bellamy, *Digital Telephony.* New York: John Wiley, 1982.

R. E. Bellman, *Dynamic programming.* Princeton: Princeton University Press, 1957.

A. A. Buzo et al. "Speech coding based on vector quantization." *IEEE Trans. ASSP*, vol. ASSP-28, no. 5, pp. 562–574, Oct. 1980.

R. E. Crochiere. "On the design of sub-band coders for low bit-rate speech communication." *BSTJ*, vol. 56, pp. 747–770, May-June 1977.

R. E. Crochiere et al. "Digital coding of speech in sub-bands." *BSTJ*, vol. 55, pp. 1069–1085, Oct. 1976.

P. C. Delattre et al. "Acoustic loci and transitional cues for consonants." *J. Acous. Soc. Am.*, vol. 27, no. 4, pp. 769–775, July, 1955.

R. O. Duda and P. E. Hart. *Pattern Classification and Scene Analysis.* New York: Wiley-Interscience, 1973.

H. Dudley. "The vocoder." *Bell Labs. Record*, vol. 18, pp. 122–126, Dec. 1939.

D. Esteban and C. Galand. "Applications of quadrature mirror filters to split band voice coding schemes." *ICASSP-77*, pp. 191–195, Apr. 1977.

C. G. M. Fant. "The acoustics of speech." L. Cremer ed. *Proc. 3rd Int. Con. Acous.*, Stuttgart, 1961; reprinted in C. G. M. Fant, *Speech Sounds* and Features, Cambridge, Mass.: MIT Press, 1973.

J. L. Flanagan. *Speech Analysis, Synthesis, and Perception.* Berlin: Springer-Verlag, 1972.

J. L. Flanagan and R. M. Golden. "Phase vocoder." *BSTJ*, vol. 45, pp. 1493–1509, Nov. 1966.

K. Fukunaga, *Introduction to Statistical Pattern Recognition.* New York: Academic Press, 1972.

A. H. Gray, Jr, and J. D. Markel. "Quantization and bit allocation in speech processing, *IEEE Trans. ASSP*, vol. ASSP-24, no. 6, pp. 459–473, Dec. 1976.

N. S. Jayant and P. Noll. *Digital Coding of Waveforms*. Englewood Cliffs, N. J.: Prentice-Hall, 1984.

B.-H. Juang and A. H. Gray, Jr. "Multiple stage vector quantisation for speech coding, *Proc. ICASSP,* pp. 597–600, Apr. 1982.

D. H. Klatt. "Review of text-to-speech conversion for English." *JASA*, vol. 82, no. 3, pp. 737–793, Sept. 1987.

S. E. Levinson et al. "An introduction to the application of the theory of probabilistic functions of a Markov process to automatic speech recognition." *BSTJ*, vol. 62, no. 4, pp. 1035–1074, Apr. 1983.

J. S. Lim ed. *Speech Enhancement*, Englewood Cliffs, N. J.: Prentice-Hall, 1983.

S. P. Lloyd. "Least-squares quantization in PCM." Bell Laboratories Technical Note, 1957; reprinted in *IEEE Trans. Inf. Theory*, vol. IT-28, no. 2, pp. 129–137, Mar. 1982.

J. N. Maksym. "Real-time pitch extraction by adaptive prediction of the speech waveform." *IEEE Trans. Audio and Electroacous.*, vol. AU-21, no. 3, pp. 149–154, June 1973.

J. D. Markel. "The SIFT algorithm for fundamental frequency estimation." *IEEE Trans. Audio and Electroacous.*, vol. AU-20, no. 5, pp. 367–378, Dec. 1972.

J. D. Markel and A. H. Gray, Jr. "Implementation and comparison of two transformed reflection coefficient scalar quantization methods." *IEEE Trans. ASSP*, vol. ASSP-28, no. 5, pp. 575–583, Oct. 1980.

P. Noll. "Non-adaptive and adaptive DPCM of speech signals." *Overdruk uit Polytech. Tijdschr. Ed. Electrotech. / Elektron.*, no. 19, 1972.

D. O'Shaughnessy. *Speech Communication: Human and Machine*. Reading, Mass.: Addison-Wesley, 1987.

M. D. Paez and T. H. Glisson. "Minimum mean-squared-error quantization in speech PCM and DPCM systems." *IEEE Trans. Comm.*, vol. COM-20, pp. 225–230, Apr. 1972.

T. W. Parsons, Voice and Speech Processing. New York: McGraw-Hill, 1986.

L. R. Rabiner and R. W. Schafer. *Digital Processing of Speech Signals*. Englewood Cliffs, N. J.: Prentice-Hall, 1978.

L. R. Rabiner et al. "On the application of vector quantization and hidden Markov models to speaker-independent, isolated word recognition." *BSTJ*, vol. 62, no. 4, pp. 1075–1105, Apr. 1983.

M. J. Sabin and R. M. Gray. "Product code vector quantizers for waveform and voice coding." *IEEE Trans. ASSP*, vol. ASSP-32, no. 3, pp. 474–488, June 1984.

C. E. Shannon. "A mathematical theory of communication." *BSTJ*, vol. 27, pp. 379–423, 623–656, 1948.

J. T. Tou and R. C. Gonzalez, *Pattern Recognition Principles*. Reading, Mass.: Addison-Wesley, 1974.

J. M Tribolet and R. E. Crochiere. "Frequency domain coding of speech, *IEEE Trans. ASSP*, vol. ASSP-27, no. 5, pp. 512–530, Oct. 1979.

C. K. Un and D. T. Magill. "The residual-excited linear-prediction vocoder with transmission rate below 9.6 kbits/s." *IEEE Trans. Comm.*, vol. COM-23, no. 12, pp. 1466–1474, Dec. 1975.

R. Zelinski and P. Noll. "Adaptive transform coding of speech signals." *IEEE Trans. ASSP*, vol ASSP-25, no. 4, pp. 299–309, Aug. 1977.

CHAPTER 9

■ ■

IMAGE PROCESSING
APPLICATIONS

In Chap. 3, we developed the analytic tools for representing two-dimensional (2D) signals and systems in spatial and frequency domains. We described the design of 2D FIR and IIR filters from a frequency-domain standpoint. Random signals were characterized by spectral densities and correlation functions on the one hand and by difference equations on the other.

Here we apply these analytic techniques to various problems in image processing. Our intent is to provide an introduction to some 2D image-processing problems and their proposed solutions and to give the reader guideposts for further study and reference in these areas.

The three areas of application of 2D signal processing described in this chapter are image enhancement, image restoration, and image coding and compression.

Image enhancement tries to pick out or focus on features of an image. These include such things as sharpening the image to bring out the details on an x-ray film; edge enhancement to pick out man-made objects in an aerial photograph; filtering, so as to accentuate desired spectral components of an image; and contrast enhancement.

Two approaches to enhancement are shown in Fig. 9-1. In the top branch, linear filtering is used to emphasize certain spectral regions of the signal. In the bottom branch, nonlinear histogram modification on a pixel-by-pixel basis is used for contrast equalization or enhancement.

Image restoration deals with techniques for reconstructing an image that may have been blurred by some sensor operation or camera motion, and in which additive noise may be present, as shown in Fig. 9-2. In the first instance, with no noise, the restoration algorithm attempts to deconvolve the observed image by passing the degraded image through

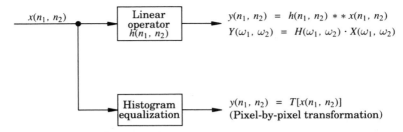

$$y(n_1, n_2) = h(n_1, n_2) ** x(n_1, n_2)$$
$$Y(\omega_1, \omega_2) = H(\omega_1, \omega_2) \cdot X(\omega_1, \omega_2)$$

$$y(n_1, n_2) = T[x(n_1, n_2)]$$
(Pixel-by-pixel transformation)

Figure 9-1 *Image enhancement. Top branch: linear filtering (for example, low-pass, high-pass). Bottom branch: nonlinear gray-level enhancement for contrast or equalization.*

an inverse filter. With additive noise present, the least-squares (or Wiener) algorithm effectively filters the noise while deblurring the degraded image. The filter design is based on appropriate spatial and spectral-domain models for the original image, the blurring filter, and noise.

Image compression, or more precisely image data compression, deals with the reduction in the number of bits needed to store or transmit images with negligible or acceptable levels of distortion. In this chapter, we discuss transform coding, predictive signal coding, and subband coding as three broad approaches having compression capabilities.

The approaches to signal coding are shown in block-diagram form in Fig. 9-3. In Fig. 9-3(a), the simplest coding technique, pulse-code modulation (PCM), shows the sampler followed by a quantizer that can be optimized by the Lloyd-Max method described in Sec. 2.2. The transmitter-receiver is relatively simple. But there is scant possibility for compression in PCM, since the signal is not conditioned prior to quantization. The differential pulse-code modulation (DPCM) system shown in Fig. 9-3(b) achieves compression by quantizing the prediction error to fewer bits than the number needed in the PCM signal. The DPCM structure is clearly more complex than simple PCM. Transform coding is illustrated in Fig. 9-3(c); here the input array is transformed into a coefficient array with unequal variances. An optimum bit-allocation scheme is then used to encode the coefficients in a bit-efficient manner. This

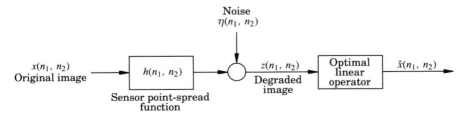

Figure 9-2 *Model for restoration of degraded image.*

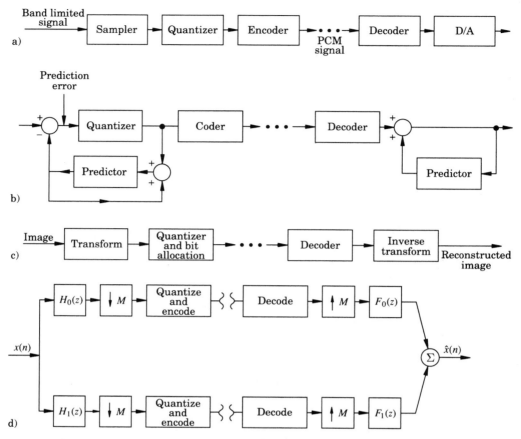

Figure 9-3 *PCM, DPCM, transform, and subband coding block diagrams. (a) PCM transmitter and receiver; (b) DPCM transmitter and receiver; (c) transform coding transmitter and receiver; (d) subband coding in one dimension, M = 2.*

latter scheme is the most computationally complex of the three, but it has the best compression capability.

More recently, subband coding of images has emerged as a viable contender for ascendancy in the data-compression competition. Figure 9-3(d) shows a two-channel, one-dimensional coder. The signal is split into frequency bands of unequal energy and encoded using an efficient bit-allocation algorithm. Subband coding is comparable to transform coding in complexity and performance.

9.1 IMAGE COMPRESSION

The amount of data in a visual image is very large. A simple black-and-white still picture represented by a 512×512 array of picture elements (pixels, or pels) using 8 bits per pixel requires 2.1×10^6 bits of information. A television camera scanning such a scene at 30 frames per second generates information at a rate of 6.3×10^7 bits/s. Three color channels require about three times this rate, and HDTV (high-definition television) requirements are even greater. Storage or transmission of these images requires large memory capacity and bandwidths. Data compression by image coding provides a viable mechanism for reducing the storage and throughput needs.

Four image coding techniques suitable for data compression are:

1. Transform coding and optimal bit allocation for storage and transmission of data [Fig. 9-3(c)]

2. Predictive coding for transmission [Fig. 9-3(b)]

3. Hybrid coding techniques using transform coding in one dimension and predictive coding in the other

4. Subband coding

We now describe these four methods in what follows.

Predictive coding, as noted in Chap. 8, relies on the redundancy of the data for compression, as in one-dimensional DPCM. Transform coding, on the other hand, transforms correlated image data into an almost uncorrelated array and repacks the energy in the transformed array into a smaller band. Coding bits can then be allocated on the basis of the energy, or variance, of the transformed coefficients.

Subband coding, on the other hand, achieves compaction by dividing the signal into uncorrelated subbands by filtering. Coding bits are allocated according to the signal energy in the various subbands. The subband and transform coders are therefore conceptually similar.

The reader is referred to the literature for more detailed descriptions and evaluations of data-compression methods. Particularly noteworthy are the review papers by Jain (1981a), by Netravali and Limb (1980), and by Forchheimer and Kronander (1989).

9.1.1 TRANSFORM CODING OF IMAGES

The transform coding of images is a conceptually straightforward extension of the 1D orthogonal expansions introduced in Chap. 2 (see Secs. 2.3-2.5). Recall that for a signal $\mathbf{f}$ with components $\{f_1, ..., f_n\}$, the spectral expansion is

$$\boldsymbol{\theta} = \boldsymbol{\Phi}\mathbf{f} \qquad \text{or} \qquad \theta_r = \sum_{n=1}^{N} \phi(r,n)f(n) \qquad (9\text{-}1)$$

and inversely

$$\mathbf{f} = (\mathbf{\Phi}^*)^T \boldsymbol{\theta} \qquad \text{or} \qquad f(n) = \sum_{r=1}^{N} \phi^*(r,n)\theta_r \qquad (9\text{-}2)$$

where $\mathbf{\Phi}$ is a unitary matrix with the property

$$\mathbf{\Phi}^{-1} = (\mathbf{\Phi}^*)^T \qquad (9\text{-}3)$$

and the transformation is energy-preserving (i.e., Parseval's theorem applies):

$$\mathbf{f}^T \mathbf{f}^* = \boldsymbol{\theta}^T \boldsymbol{\theta}^* \qquad (9\text{-}4)$$

The components of the signal vector $\mathbf{f}$ are typically correlated, and each component has the same variance. That is, the correlation matrix for $\mathbf{f}$ is nondiagonal Toeplitz. The purpose of the transformation is to decorrelate the signal samples and to repack the energy as implied by Eq. (9-4) into a relatively small number of spectral coefficients $\{\theta_r\}$. Each component of $\{\theta_r\}$ is then separately and optimally quantized using the Lloyd-Max quantization procedure outlined in Sec. 2.2. Additionally, the number of bits assigned to each spectral component is optimally allocated on the basis of the logarithm of its variance. A 1D version of this transform coding system is shown in Fig. 9-4.

As pointed out in Sec. 2.5.7, the Karhunen-Loève transform (KLT) is the optimal transform for decorrelating the signal and repacking the energy. The KLT, however, depends on the input signal correlation matrix and is therefore signal-dependent. The fixed unitary transforms described in Chap. 2–the DFT, DCT, MHT, and DWHT–are suboptimal, but they have the advantage that they are signal-independent and can be employed using fast computational algorithms.

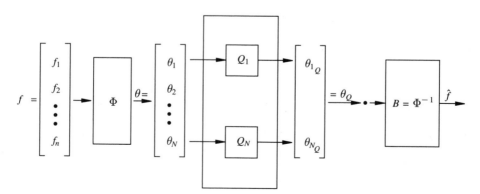

Figure 9-4   *Transform coding in one dimension.*

The transform coding problem of Fig. 9-4 may be stated as follows: For a given unitary transformation Φ and Gaussian input signal $\mathbf{f}$, find the optimal quantizer Q and inverse matrix $\mathbf{B}$ to minimize D, the mean-squared reconstruction error (or distortion):

$$D = E\{\tilde{\mathbf{f}}^T \tilde{\mathbf{f}}\} \qquad \tilde{\mathbf{f}} = \mathbf{f} - \hat{\mathbf{f}} \tag{9-5}$$

The solution is as follows:

1. Use the Lloyd-Max quantizer on each coefficient, as indicated in Fig. 9-4.

2. $\mathbf{B} = \Phi^{-1} = (\Phi^*)^T.$ $\qquad\qquad$ (9-6)

Under the conditions of Eq. (9-6), it can be shown (see Prob. 9-1) that the mean-squared reconstruction error is equal to the mean-squared quantization error, that is,

$$D = E\{\tilde{\mathbf{f}}^T \tilde{\mathbf{f}}\} = E\{\tilde{\boldsymbol{\theta}}^T \tilde{\boldsymbol{\theta}}\}, \tag{9-7}$$

where $\qquad\qquad \tilde{\boldsymbol{\theta}} = \boldsymbol{\theta} - \boldsymbol{\theta}_Q$ $\qquad\qquad\qquad\qquad$ (9-8)

is the quantization-error vector. Thus optimizing the quantizer also optimizes the overall reconstruction error (in the absence of channel noise or transmission errors).

The resulting mean-squared distortion depends on the spectral coefficient variances and probability density functions, the quantizer used, and the number of bits B_k allocated to the kth coefficient θ_k. This MSE can be expressed by

$$D = \sum_{k=1}^{N} \sigma_k^2 f(B_k) \tag{9-9}$$

$$\sigma_k^2 = E\{\theta_k^2\}$$

where $f(B_k)$ is the quantizer distortion function for a unity-variance input. Typically,

$$f(B_k) = r_k 2^{-2B_k} \tag{9-10}$$

where $r_k > 0$ depends on the pdf for θ_k and on the specific quantizer. For the Lloyd-Max quantizer, Jayant (1984) reports values of $r = 1.0$, 2.7, 4.5, and 5.7 for uniform, Gaussian, Laplacian, and gamma pdfs, respectively.

Finally, there is the question of bit allocation. With B as the average number of bits per coefficient,

$$B = \frac{1}{N} \sum_{k=1}^{N} B_k \tag{9-11}$$

the mean-squared distortion D is minimized by selecting B_k as

$$B_k = B + \frac{1}{2} \log_2 \frac{\sigma_k^2}{\left(\prod_{j=1}^{N} \sigma_j^2 \right)^{1/N}} \tag{9-12}$$

This result is due to Huang and Schultheiss (1963) and Segall (1976); see also Prob. 9.2. The number of bits is seen to be proportional to the logarithm of the coefficient variance. It can also be shown (Prob. 9.3) that the allocation of Eq. (9-12) results in the same average quantization error for each coefficient, and thus the distortion is spread out evenly among all the coefficients.

The 2D version of transform coding is easily extrapolated from the foregoing. As shown in Fig. 9-5, the image array is divided into sub-blocks, and each subblock is separately transform coded. Typically, these blocks are square, with 4×4, 8×8, and 16×16 as representative sizes. The $N \times N$ sub-block image array is denoted by

$$\mathbf{F} = [f(m,n)] \qquad 0 \le m, n \le N - 1 \tag{9-13}$$

The direct transform is

$$\theta(i,j) = \sum_{m=0}^{N-1} \sum_{n=0}^{N-1} \alpha(i,j,m,n) f(m,n) \qquad 0 \le i,j \le N - 1 \tag{9-14}$$

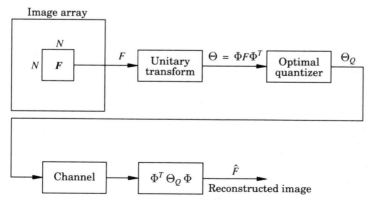

Figure 9-5 *Two-dimensional transform coding.*

and the inverse is

$$f(m,n) = \sum_{i=0}^{N-1}\sum_{j=0}^{N-1} \beta(i,j,m,n)\theta(i,j) \qquad (9\text{-}15)$$

where $\alpha(\cdot)$ and $\beta(\cdot)$ are the forward and inverse transform kernels. (Note that in this 2D representation, we are using the index set $[0, N-1]$ with the notation employed in Chap. 3.)

For our applications, the kernels are separable and symmetric: that is, the 2D kernels are represented as the product of two 1D orthogonal basis functions,

$$\alpha(i,j,m,n) = \phi(i,m)\phi(j,n) = \phi_i(m)\phi_j(n) \qquad 0 \le i,j,m,n \le N-1. \quad (9\text{-}16)$$

As in the 1D formulation, the basis functions $\{\phi_r(n)\}$ constitute the rows of the unitary matrix $\mathbf{\Phi} = [\phi(i,j)]$, and the direct and inverse transformations take the form,

$$\mathbf{\Theta} = \mathbf{\Phi F}(\mathbf{\Phi}^*)^T \qquad (9\text{-}17)$$

$$\mathbf{F} = (\mathbf{\Phi}^*)^T \mathbf{\Theta \Phi} \qquad (9\text{-}18)$$

The Parseval relation takes the form

$$\sum_m \sum_n |f(m,n)|^2 = \sum_i \sum_j |\theta(i,j)|^2 \qquad (9\text{-}19)$$

We encountered a transformation of the form of Eq. (9-17) in Chap. 3 (Sec. 3.5.1), where the 2D DFT was developed. There we showed that the image transformation could be performed in two stages: first one took the unitary transform for each row of the array to obtain an intermediate array; then we applied the unitary transform to each column of the intermediate array to obtain the final transformed image array. The same procedure is applicable here.

For our separable transformation, the 1D results carry over with only obvious modifications. Thus we apply the Lloyd-Max quantizer to each pixel within the transform mask, and the optimal bit allocation formula becomes

$$B_{i,j} = B + \frac{1}{2}\log_2 \frac{\sigma^2(i,j)}{\left(\displaystyle\prod_{k,l=0}^{N-1}\sigma^2(k,l)\right)^{1/N^2}} \qquad (9\text{-}20)$$

where $\sigma^2(i,j)$ is the variance in the coefficient $\theta(i,j)$. The resulting minimum reconstruction error in each pixel is

$$r(2^{-2B_k}) \left[\prod_{i,j=0}^{N-1} \sigma^2(i,j) \right]^{1/N^2} \tag{9-21}$$

The variance needed in the optimal bit allocation can be determined from an a priori image model, or it can be found empirically by computing $\sigma^2(k,l)$ from a prototype image.

Netravali and Limb (1981) have evaluated the mean-squared error performance of different transforms for a 2D Markov-1 signal source (see also Sec. 3.9) with equal horizontal and vertical correlation coefficients of 0.95. These are displayed in Fig. 9-6.

It is noted that the DCT is very close to the optimal KLT for ρ on the order of 0.95. The decision must be weighed on the basis not only of performance but also of speed of computation. In this regard, only fast transforms are of any practical value, and the trade-off between the DCT and KLT, for example, gives the nod to the DCT.

Finally, for further compression, only those coefficients falling within a desired spectral zone, or those whose variances exceed some threshold, need be encoded. The zonal mask is determined a priori from an ensemble of images of a certain genre and is constant from picture to picture. Coefficients outside the mask are simply not calculated.

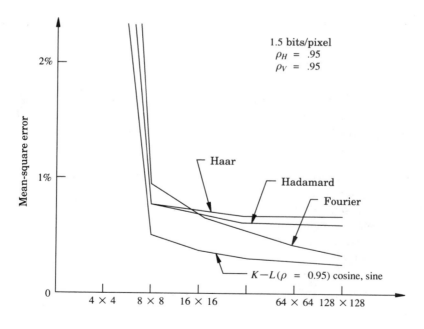

Figure 9-6 *MSE performance of various kinds of transform coding. [After Netravali and Limb (1980), copyright IEEE, Inc.]*

Threshold coding, on the other hand, has an adaptive aspect. Coefficients with the largest calculated values are retained, along with their addresses. Transform coding is then applied to the reduced set. For a typical 256×256 image array with 8 bits per pixel, compression ratios of 4:1 to 8:1 are common. Adaptive 2D image transform coding can be expected to yield even higher compression ratios, up to 16:1.

For additional material, the reader may consult the references listed earlier.

9.1.2 PREDICTIVE CODING

An alternative method for data compression is the predictive coding technique that was introduced in Chap. 3 in Example 3.12 and Fig. 3-31, redrawn here as Fig. 9-7.

The key idea underlying predictive coding is that successive samples of a 1D signal (for example speech) or of a 2D signal (for example an image) tend to be correlated. Rather than quantize and encode each successive sample $x(n_1, n_2)$ as in PCM, we instead quantize the *difference* between an input sample and some prediction $\hat{x}_p(n_1, n_2)$ of it. This difference, or residual $\epsilon(n_1, n_2)$ will have a smaller amplitude than the individual samples of the input and hence can be represented by fewer bits than the original sample. Furthermore if the prediction is optimal, it can be shown that the residuals resemble a white-noise sequence; that is, successive residuals are spatially uncorrelated and have equal variances. The quantizer can then be optimized to match the probability density function of the residual.

Note the difference between these two approaches. DPCM tries to generate a sequence of uncorrelated residuals with reduced but equal variances. These residuals can then be quantized with fewer bits than required by a comparable PCM system. Transform coding, on the other hand, tries to create a set of uncorrelated spectral coefficients where the total energy is constant but repacked among the coefficients so that the coefficient variances are unequal. Bit allocation can now be distributed among the spectral coefficients according to their individual variances.

To fix ideas, we will analyze the one-dimensional DPCM codec (coder-decoder) shown in Fig. 9-8. First we assume that the signal $x(n)$ is modeled as a Markov sequence (Sec. 7.1)

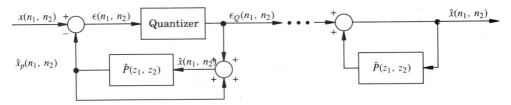

Figure 9-7 *Block diagram of two-dimensional predictive coding.*

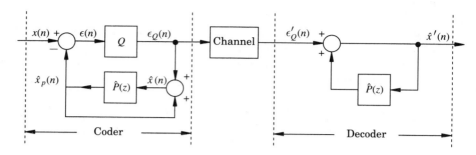

Figure 9-8 *DPCM coder-decoder in one dimension.*

$$x(n) = a_1 x(n-1) + \cdots + a_N x(n-N) + \xi(n) \qquad (9\text{-}22)$$

where $\{\xi(n)\}$ is a sequence of zero-mean uncorrelated random variables with variance σ^2, that is, discrete stationary white noise. This signal is shown as the front end in the block diagram of Fig. 9-9, where

$$P(z) = \sum_{i=1}^{N} a_i z^{-i} \qquad (9\text{-}23)$$

If a Kalman-like filter were used to estimate or predict the value of $x(n)$, based on past values $\{x(n-1), x(n-2), ...\}$, it would have the form of the predictor-corrector structure of Fig. 9-9. In that case, the Kalman gain K(n) would be optimized on the basis of signal and noise statistics and of the dynamic signal model. The DPCM structure of Fig. 9-8 can be thought of as a lineal descendant of the Kalman filter where the Kalman gain is replaced by the quantizer Q and where the predictor

$$\hat{P}(z) = \sum_{i=1}^{N} \hat{a}_i z^{-i} \qquad (9\text{-}24)$$

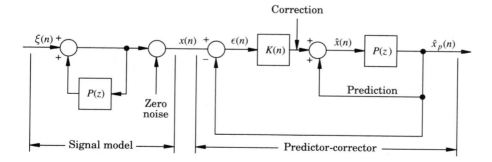

Figure 9-9 *Predictor-corrector structure for estimating $\hat{x}_p(n)$.*

is used in place of $P(z)$. The predictor $\hat{P}(z)$ uses our best guesstimate $\{\hat{a}_i\}$ for the actual signal parameters $\{a_i\}$. With these substitutions, the predictor-corrector structure of Fig. 9-9 can be redrawn to take the form of the DPCM coder of Fig. 9-8. The Kalman "innovations" $\varepsilon(n)$ become the DPCM residuals $\epsilon(n)$. Just as the Kalman filter generates a white sequence for its innovations, so does the DPCM coder try to "whiten" its residuals.

For analysis purposes, the DPCM coder of Fig. 9-8 can be approximately linearized by taking the quantizer Q outside the loop, as shown in Fig. 9-10. This approximate representation, valid for Q having 2 or more bits, is useful in designing the quantizer and analyzing the DPCM performance. In effect, the nonlinear DPCM residuals are approximated by the linear "innovations."

From the coder side of the DPCM structure of Fig. 9-8, the transfer function from e_Q to $\hat{x}_p$ is

$$\hat{X}_p(z) = \frac{\hat{P}(z)}{1 - \hat{P}(z)} E_Q(z) \tag{9-25}$$

It is also clear that

$$\hat{X}(z) = \frac{1}{\hat{P}(z)} \hat{X}_p(z) = \frac{1}{1 - \hat{P}(z)} E_Q(z) \tag{9-26}$$

and
$$E(z) = X(z) - \hat{X}_p(z) \tag{9-27}$$

At the receiver or decoder, if there are no channel errors,

$$E'_Q(z) = E_Q(z) \tag{9-28}$$

Consequently,
$$\hat{X}'(z) = \frac{1}{1 - \hat{P}(z)} E_Q(z) = \hat{X}(z) \tag{9-29}$$

Hence the reconstructed signal $\hat{x}'(n)$ at the decoder is equal to $\hat{x}(n)$ at the coder. The reconstruction error $\tilde{x}(n)$ therefore can be evaluated by analysis of the coder loop:

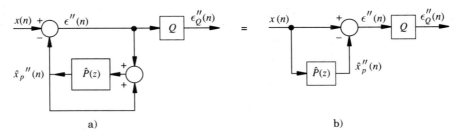

a)   b)

Figure 9-10 (a) Approximate DPCM coder for analysis purposes; (b) structure equivalent to (a).

$$\tilde{x}(n) = x(n) - \hat{x}'(n) = x(n) - \hat{x}(n) \qquad (9\text{-}30)$$

Combining Eqs. (9-27) and (9-26) with Eq. (9-30) gives

$$\tilde{X}(z) = X(z) - \hat{X}(z)$$

$$= E(z) + \frac{\hat{P}(z)}{1 - \hat{P}(z)} E_Q(z) - \frac{1}{1 - \hat{P}(z)} E_Q(z)$$

Hence $\qquad \tilde{X}(z) = E(z) - E_Q(z) \qquad\qquad\qquad\qquad (9\text{-}31)$

or $\qquad \tilde{x}(n) = \epsilon(n) - \epsilon_Q(n) = q(n)$

This last equation shows that the reconstruction error $\tilde{x}(n)$ equals the quantization error $q(n)$. The same result holds in transform coding, under the same assumption of zero transmission errors. Hence, to minimize the mean-squared reconstruction error or distortion $E\{|\tilde{x}(n)|^2\}$, we can focus on reducing

$$E\{|q(n)|^2\} = \sigma_q^2 \qquad (9\text{-}32)$$

For any quantizer, the distortion was expressed in Eq. (9-9), which we rewrite here in the form

$$\sigma_q^2 = \sigma_e^2 \cdot f(B) \qquad (9\text{-}33)$$

where σ_e^2 is the variance $\epsilon(n)$ of the residual input to the quantizer and $f(B)$ is the quantizer distortion function Eq. (9-10), which depends on the quantizer, the input probability density function, and the number of bits.

The two terms in Eq. (9-33) are, of course, interdependent, since the quantizer is actually within the coder feedback loop. The simple expedient of moving the quantizer outside the loop, as in Fig. 9-10, permits us to evaluate the two components of σ_q^2 in Eq. (9-33) separately. Thus $\epsilon''(n)$ in Fig. 9-10 is the linear approximation to $\epsilon(n)$ in Fig. 9-8. The two parts of a DPCM, the predictor and the quantizer, can now be separately designed and analyzed.

Using the Markov representation of $x(n)$ given in Eq. (9-22), repeated here, and the coder model of Fig. 9-10(b), we have

$$x(n) = \sum_{i=1}^{N} a_i x(n - i) + \xi(n) \qquad (9\text{-}22)$$

$$\hat{x}_p''(n) = \sum_{i=1}^{N} \hat{a}_i x(n - i) \qquad (9\text{-}34)$$

Subtracting Eq. (9-34) from Eq. (9-22),

$$\epsilon''(n) = x(n) - \hat{x}_p''(n) = \sum_{i=1}^{N}(a_i - \hat{a}_i)x(n-i) + \xi(n)$$

$$\approx \epsilon(n) \tag{9-35}$$

If we further pretend that $\hat{a}_i = a_i$, then Eq. (9-35) reduces to

$$\epsilon(n) \approx \xi(n) \tag{9-36}$$

This demonstrates that under coefficient-matched conditions, the residuals $\epsilon(n)$ equal the white-noise process $\xi(n)$ and that the DPCM loop decorrelates the signal $x(n)$. Thus

$$\sigma_e^2 = E\{\epsilon^2(n)\} = E\{\xi^2(n)\} = \sigma^2$$

Another point of interest is to compare σ_e^2 with σ_x^2, the variance or power of the original signal. From Eq. (9-22), we see that

$$S_{xx}(z) = H(z)H(z^{-1})\sigma^2 \leftrightarrow R_{xx}(k) \tag{9-37}$$

$$H(z) = \frac{1}{1 - P(z)} \tag{9-38}$$

If the $\{a_i\}$ are known, the power spectral density $S_{xx}(z)$ and autocorrelation $R_{xx}(k)$ can be found, from which we can find $\sigma_x^2 = R_{xx}(0)$.
For a Markov-1 model, $a_1 = \rho$ and $a_j = 0$ for $j \geq 2$, and we find that

$$R_{xx}(k) = \frac{\sigma^2}{1 - \rho^2}\rho^{|k|} \tag{9-39}$$

and
$$\sigma_x^2 = R_{xx}(0) = \sigma^2(1 - \rho^2)^{-1} \tag{9-40}$$

The measure of the variance reduction in the signal applied to the quantizer is thus

$$\frac{\sigma_x^2}{\sigma_e^2} = \frac{1}{1 - \rho^2} \tag{9-41}$$

We are now in a position to compare the distortion measures for PCM and DPCM as applied to the Markov-1 model:

$$(\sigma_q^2)_{PCM} = \sigma_x^2 f(B) = \frac{\sigma^2 f(B)}{1 - \rho^2} \tag{9-42}$$

$$(\sigma_q^2)_{DPCM} = \sigma_e^2 f(B) \approx \sigma^2 f(B)$$

The quantity $f(B)$ in this last equation is the quantizer distortion function in Eq. (9-10), which depends on the quantizer used and on the pdf of the signal. We can also compare PCM with DPCM on a SNR basis,

$$(SNR) = 10 \log_{10} \sigma_x^2 / \sigma_q^2 \qquad (9\text{-}43)$$

Using the distortion formulas of Eq. (9-42), for the same quantizer and an equal number of bits, we find

$$(SNR)_{DPCM} - (SNR)_{PCM} = 10 \log_{10} \frac{(\sigma_q^2)_{PCM}}{(\sigma_q^2)_{DPCM}}$$

$$= 10 \log_{10} \frac{1}{1 - \rho^2} \qquad (9\text{-}44)$$

Typical values of ρ are 0.85 for speech and 0.95 (in each dimension) for images. Hence Eq. (9-44) suggests SNR improvements on the order of 6 to 10 dB for speech and images, respectively, assuming line-by-line scanning for the latter.

Rate distortion theory (Berger, 1971) states that the distortion in block quantization of independent Gaussian samples of variance σ^2 and B bits is given by

$$D = \sigma^2 2^{-2B}$$

or

$$B = \frac{1}{2} \log_2(\sigma^2 / D) \qquad (9\text{-}45)$$

In our case, $D = \sigma_q^2$. For the same distortion, the number of bits needed in PCM and DPCM for a first-order Markov model is

$$B_{PCM} - B_{DPCM} = \frac{1}{2} \log_2(\sigma_x^2 / \sigma_e^2)$$

$$= \frac{1}{2} \log_2 \frac{1}{1 - \rho^2} = \begin{cases} 0.93 & \rho = 0.85 \text{ (speech)} \\ 1.7 & \rho = 0.95 \text{ (images)} \end{cases} \qquad (9\text{-}46)$$

Hence, for the same distortion, this example shows that speech can be coded to 1 bit less per sample, and images to almost 2 fewer bits per pixel, as compared with PCM.

With this one-dimensional model as background, we can now easily explain the two-dimensional DPCM. First, the image is represented as a 2D Markov random field as in Sec. 3.9 by the causal difference equation Eq. (3-135), repeated here as

$$x(n_1, n_2) = \rho_1 x(n_1 - 1, n_2) + \rho_2 x(n_1, n_2 - 1)$$

$$- \rho_1 \rho_2 x(n_1 - 1, n_2 - 1) + \xi(n_1, n_2) \qquad (9\text{-}47)$$

The "driving" signal $\xi(n_1, n_2)$ is zero-mean, stationary white noise with variance σ^2. (Note that the normalization used here differs from that of Sec. 3.9, but otherwise the models are the same.) The image source signal $x(n_1, n_2)$ is therefore modelled as the output of a 2D filter with separable transfer function $H(z_1, z_2)$ driven by white noise, where

$$H(z_1, z_2) = \frac{1}{(1 - \rho_1 z_1^{-1})(1 - \rho_2 z_2^{-1})} \tag{9-48}$$

Following the discussion in Sec. 3.9 and Example 3.12 (which the reader is advised to review at this point), we see that the predictor structure is just

$$\hat{P}(z_1, z_2) = \hat{\rho}_1 z_1^{-1} + \hat{\rho}_2 z_2^{-1} - \hat{\rho}_1 \hat{\rho}_2 z_1^{-1} z_2^{-1} \tag{9-49}$$

where $\hat{\rho}_1$ and $\hat{\rho}_2$ are guesstimates of the correlation coefficients between adjacent horizontal and vertical pixels, respectively. Explicitly, the predictor equation is

$$\hat{x}_p(n_1, n_2) = \hat{\rho}_1 \hat{x}(n_1 - 1, n_2) + \hat{\rho}_2 \hat{x}(n_1, n_2 - 1) - \hat{\rho}_1 \hat{\rho}_2 \hat{x}(n_1 - 1, n_2 - 1) \tag{9-50}$$

Note that the predictor transfer function is not separable, even though $H(z_1, z_2)$ is.

The 2D extensions of Eqs. (9-39) to (9-42) on a per-pixel basis are simply

$$\sigma_x^2 = \frac{1}{(1 - \rho_1^2)(1 - \rho_2^2)}$$

$$\sigma_e^2 = 1 \tag{9-51}$$

$$\sigma_q^2 = \begin{cases} \sigma_e^2 f(B) & \text{DPCM} \\ \sigma_x^2 f(B) & \text{PCM} \end{cases}$$

and $$(SNR)_{DPCM} - (SNR)_{PCM} = 10 \log_{10} \frac{1}{(1 - \rho_1^2)(1 - \rho_2^2)} \tag{9-52}$$

and finally $$B_{PCM} - B_{DPCM} = \tfrac{1}{2} \log_2 \frac{1}{(1 - \rho_1^2)(1 - \rho_2^2)} \tag{9-53}$$

For $\rho_1 = \rho_2 = 0.95$, these last two equations show that the 2D-DPCM system is 20 dB better than PCM for equal numbers of bits; alternatively, for the same performance, the DPCM coder requires 3.4 fewer bits per pixel than PCM.

Figure 9-11 compares the performances of several DPCM methods with PCM. It may be seen that the 2D DPCM curve runs 20 dB above that for PCM, as predicted by our analysis. Extrapolation of these curves

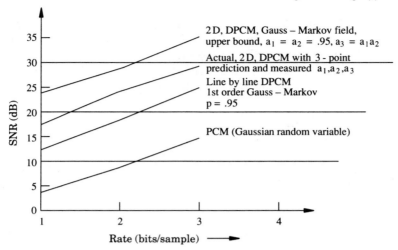

Figure 9-11 *Comparison of PCM with several DPCM methods. [After Jain (1981a), copyright IEEE, Inc.]*

also bears out the argument that the compression achieved by DPCM for typical 8-bit/pixel data is about 3.4 bits.

We close this section with a comparison between DPCM and transform coding.

2D DPCM: This technique is simple, on-line, and easily implemented. It is quite sensitive to channel errors and to variations in image statistics from the model assumed.

2D transform coding: This method achieves more compression than DPCM. It is more robust than predictive coding and tends to distribute quantization and channel errors over the whole image. The major drawbacks are the complexity of the algorithms and hardware. Much more memory is needed and many more operations must be performed.

9.1.3 HYBRID IMAGE CODING

The term, "hybrid image coding," as used here refers to a compression method that uses 1D transform coding in one direction and 1D predictive coding in the other. The goal is an optimum combination of the best features of each 1D coding method. The technique is particularly suited to the semi-causal source model introduced in Sec. 3.9, Eqs. (3-151) to (3-155).

In that representation, a separable 2D Markov correlation function

$$R_x(k_1, k_2) = \rho_1^{|k_1|} \rho_2^{|k_2|} \tag{9-54}$$

was represented as the output of a semicausal system with transfer function

$$H(z_1, z_2) = \frac{1}{[1 - \alpha(z_1^{-1} + z_1)](1 - \rho_2 z_2^{-1})} \tag{9-55}$$

$$\alpha = \frac{\rho_1}{1 + \rho_1^2}$$

This system was driven by a random input $\xi(n_1, n_2)$, whose power spectral density was white noise in the causal direction but non-white in the other, noncausal, variable, as indicated in Fig. 9-12. This input psd is therefore

$$S_\xi(z_1, z_2) = A[1 - \alpha(z_1 + z_1^{-1})] \tag{9-56}$$

where A is a scale factor.

One-dimensional predictive coding is used in the recursive causal variable, while 1D transform coding is used on the non-causal variable. The process is illustrated in Fig. 9-13 for an $N \times N$ image array. First, each column of the image $[x(n_1, n_2)]$ is transform-coded by a unitary matrix (typically the DCT) to give

$$\theta_j = \Phi x_j$$

where x_j is the jth column of $[x(n_1, n_2)]$. This operation has the effect of decorrelating the components of each vector θ_j. Next, each row of the resulting intermediate array $[\theta(i, j)]$ is predictively coded by the 1D DPCM algorithm. This technique is seen to be an attempt to blend the robustness of transform coding with the relative hardware simplicity of (at most) N DPCM channels. In fact, the number of channels employed may be less than N, since those rows whose coefficients have small variances could be allocated zero bits and thus not transmitted at all.

It should be noted, however, that the DPCM channels are not identical. Each row of the intermediate transformed array can be represented

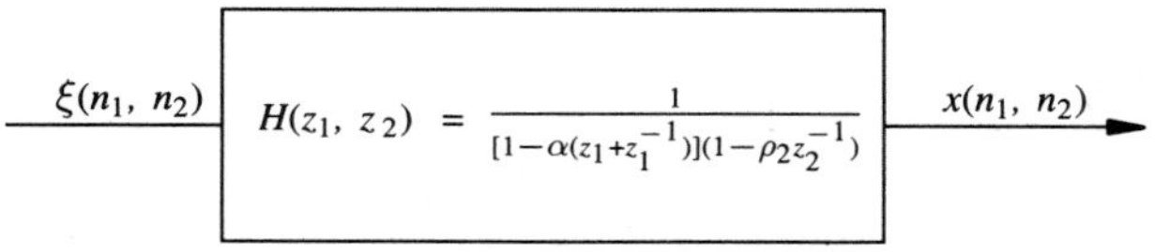

$$\xi(n_1, n_2) \longrightarrow \boxed{\; H(z_1, z_2) \;=\; \frac{1}{[1-\alpha(z_1+z_1^{-1})](1-\rho_2 z_2^{-1})} \;} \longrightarrow x(n_1, n_2)$$

$$R_\xi(n_1, n_2) \;=\; A\{\delta(n_1, n_2) - \alpha[\delta(n_1 - 1, n_2) + \delta(n_1 + 1, n_2)]\}$$

Figure 9-12   *Semicausal source model for hybrid coding.*

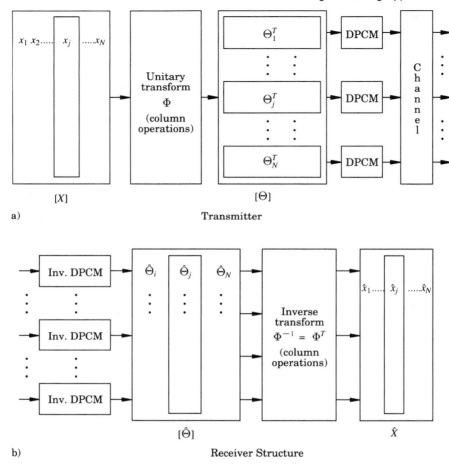

a)   Transmitter

b)   Receiver Structure

Figure 9-13   *Block diagram of hybrid coder: (a) transmitter section; (b) receiver structure.*

as a stationary 1D Markov sequence with constant variance along that row, but the variance will vary from row to row.

To fix ideas, we will use the example discussed in Jain (1981a). The random field is the semicausal system driven by zero-mean white noise represented by

$$x(n_1, n_2) = \alpha[x(n_1 + 1, n_2) + x(n_1 - 1, n_2)]$$
$$+ \beta x(n_1, n_2 - 1) + \xi(n_1, n_2) \tag{9-57}$$

where $\alpha < 1/2$ $|2\alpha + \beta| < 1$ and $S_\xi(z_1, z_2) = \sigma^2$

This system is non-causal in n_1, but causal and recursive in n_2. [Note that this example differs from the separable model of Eqs. (9-54) to (9-56).] The details of the hybrid processing are as follows:

1. Holding j fixed[†], we apply the unitary transformation Φ to the jth column of $[x(i, j)]$:

$$\boldsymbol{\theta}_j = \Phi \mathbf{x}_j \qquad j = 1, 2, ..., N \tag{9-58}$$

Successive column transformations give the intermediate array $[\theta(i, j)]$.

2. With i fixed, the row-recursive model for the ith row is first-order Markov:

$$\theta(i, j) = \rho_i \theta(i, j - 1) + \eta(i, j) \tag{9-59}$$

In this formulation, the row index i is fixed and j is the the running index. The driving noise $\eta(i, j)$ is stationary and white in the j index. The coefficient ρ_i and the variances of the driving noise will vary from row to row but will be constant along any one row. Thus if we determine ρ_i and the variance of $\eta(i, j)$, Eq. (9-59) is completely determined and fits the mold of the 1D DPCM system analyzed earlier. For a DCT transform and the semicausal representation of Eq. (9-57), Jain shows that

$$\rho_i = \beta \lambda_i$$
$$\lambda_i = 1 - 2\alpha \cos[(i - 1)\pi/N] \tag{9-60}$$

and $\qquad E\{\eta^2(i, j)\} = \sigma^2/\lambda_i^2 \triangleq \sigma_i^2$

Therefore, it follows that

$$\text{var}\{\theta(i, j)\} = \sigma^2(i, j) = \frac{\sigma_i^2}{1 - \rho_i^2} \tag{9-61}$$

We can apply the 1D DPCM results of the previous section to design and assess the performance of the predictive coder. Following that development, we can now summarize the DPCM transmitter equations for the ith row:

Predictor: $\qquad\qquad \hat{\theta}_P(i, j) = \rho_i \hat{\theta}(i, j - 1)$

Quantizer input
(or residual): $\quad \epsilon(i, j) = \theta(i, j) - \hat{\theta}_P(i, j)$

Quantizer output: $\qquad \epsilon_Q(i, j) = Q[\epsilon(i, j)]$

Reconstruction: $\qquad \hat{\theta}(i, j) = \epsilon_Q(i, j) + \hat{\theta}_P(i, j)$

[†]The variables i, j are used instead of the subscripted variables n_1, n_2 for ease in notation.

The DPCM distortion in each channel is just

$$\sigma_q^2 = \sigma_e^2(i)f(B_i) \tag{9-62}$$

where $f(B_i)$ is the quantizer distortion function and B_i is the number of bits optimally assigned to the ith channel on the basis of the coefficient variance for that row:

$$\sigma_e^2 = \sigma_i^2 = \sigma^2/\lambda_i^2$$

Hence B_i can be selected in accordance with the optimal allocation specified by Eq. (9-12).

As mentioned earlier, hybrid coders try to exploit the advantages of each of the coding modalities–the robustness and high performance of transform coding and the hardware simplicity of DPCM. It is not surprising, then, to find that the performance of a properly tuned hybrid coder lies somewhere between that of transform coding and that of DPCM, and that the sensitivity and robustness lie between these extremes as well. These hybrid coders have been effectively employed in the compression and transmission of images from remotely piloted vehicles.

9.1.4 SUBBAND CODING

The concept and purpose of subband coding of signals were developed in Sec. 5.2 for one-dimensional processes. Here we extend those notions to images and compare the efficiency of this kind of coding with that of the others–transform coding and DPCM–developed in this section.

Recall that the basic objective of subband coding (SBC) is to divide the image signal into uncorrelated frequency bands by filtering and then to code each subband by using a bit allocation matched to the signal energy in that subband. The actual signal coding of the subband signals could be done using some of the techniques already discussed, such as PCM or DPCM. In this sense, the SBC is a hybrid coder.

The SBC tries to achieve energy compaction via filtering, whereas transform coding tries to obtain compaction by block transformations. We can demonstrate this SBC compaction capability by analysis of the ideal digital filter bank (Jayant and Noll, 1984) shown in Fig. 9-14.

The spectrum of the full-band sampled signal $X(e^{j\omega})$ is split up into $M = 4$ nonoverlapping regions by the bank of ideal band-pass filters. Each signal is down-sampled or decimated by a factor of M, quantized, encoded, and transmitted. At the receiver, the process is reversed. Each subband signal is decoded, up-sampled, and passed through another ideal filter whose purpose is to remove the images created by the up-sampling. Figure 9-15 shows the frequency characteristics of the filters, the filtered signals $X_i(e^{j\omega})$, and the up-sampled signal $V_i(e^{j\omega})$. The reconstructed signal is obtained by summing all the reconstituted subband signals.

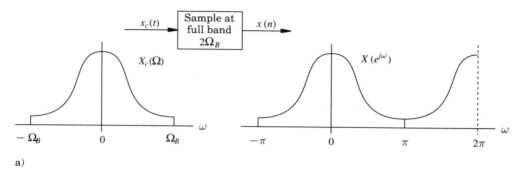

a)

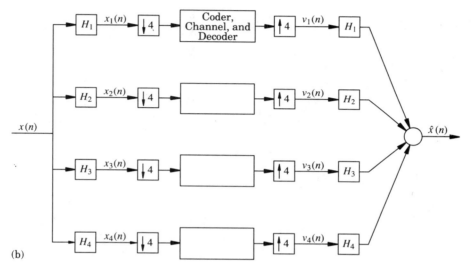

(b)

Figure 9-14   *Idealized subband coder.*

With the assumption implicit in Figs. 9-14 and 9-15, that is, ideal filters, it can be seen that in the absence of encoding (quantization) and channel errors, the reconstructed signal $\hat{x}(n)$ is exactly equal to the original signal. This was also the case for transform coding. Hence in the idealized case considered, the total reconstruction error is due to quantization errors alone; the channel is assumed to be error-free.

We can now evaluate the compaction performance of this ideal SBC. Let the input signal be a random sequence; hence the spectra shown in Figs. 9-14 and 9-15 are interpreted as power spectral densities. The power (or variance) in the kth subband is just the area under the psd $S_k(e^{j\omega})$ in that band, or

$$E_k = \sigma_k^2 = 2\left(\frac{1}{2\pi}\right)\int_{w_k}^{w_{k+1}} S_k(e^{j\omega})\,d\omega \qquad (9\text{-}63)$$

where $\omega_{k+1} - \omega_k = \pi/M$

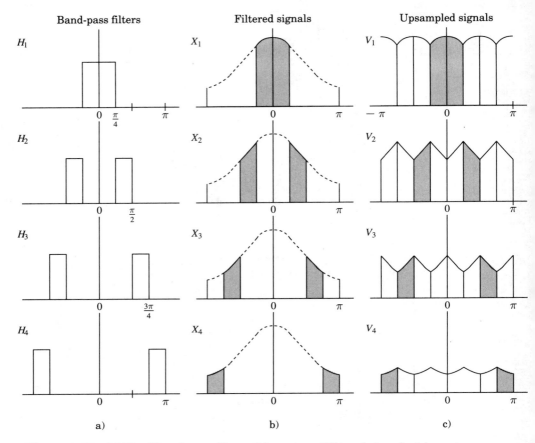

Figure 9-15 *(a) Ideal band-pass filters; (b) spectra of filtered signals; (c) spectra of up-sampled signals.*

Following our analysis of quantization errors in transform coding, we can say that the mean-squared error (or distortion) in the kth subband is

$$D_k = r_k \sigma_k^2 2^{-2B_k} \qquad (9\text{-}64)$$

where B_k is the number of bits allocated to signals in the kth subband and r_k is a parameter depending on the pdf of the signal. The total distortion is therefore

$$D = \sum_{k=1}^{M} r_k 2^{-2B_k} \sigma_k^2 \qquad (9\text{-}65)$$

Equations (9-64) and (9-65) have exactly the same form as Eq. (9-9) and (9-10). Hence we can draw upon previous results and assert that the

optimal bit allocation should follow the formula given by Eq. (9-12). Substituting Eq. (9-12) into Eq. (9-65) and letting $r_k \equiv r$ a constant, we find
that the distortion in each band is equal and is given by

$$D_k = r2^{-2B}\sigma_y^2 \tag{9-66}$$

and
$$D = MD_k$$

where σ_y^2 is the geometric mean of the variances:

$$\sigma_y^2 = \left(\prod_{j=1}^{M}\sigma_j^2\right)^{1/M} \tag{9-67}$$

We obtained similar results for optimally encoded transform coding; thus
the performance measures of these techniques reduce essentially to the
energy compaction induced by the filters or by the transformations.
      The subband image coder of Fig. 9-16 is a 2D version of the 1D
QMF bank introduced in Sec. 5.2.6. The basic operation is similar in
principle to the one-dimensional SBC. The analysis filter bank partitions
the signal into M frequency bands. These reduced-bandwidth signals
are subsampled at the reduced Nyquist frequency to produce the subband signals. At the receiver, the inverse operations are performed as
indicated.
      The 2D SBC is not a simple extension of the 1D case, except in the
separable case to be considered shortly. The first complication concerns

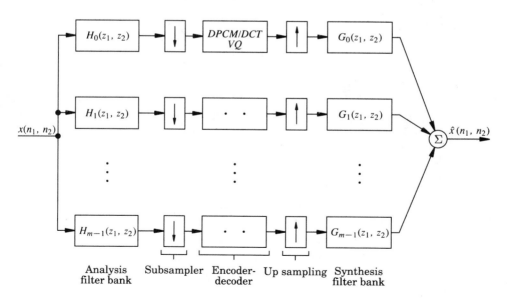

Figure 9-16   *Structure of subband image coder.*

the subsampling lattice to be used in the down-sampler. Three patterns are shown in Fig. 9-17. The checkerboard lattice down-samples by a factor of 2, while the rectangular and hexagonal grids represent a one-out-of-four decimation.

The advantages of an oriented grid such as the hexagon are discussed in Vetterli (1989). For our purposes, we consider the simpler separable case of the rectangular grid shown in two alternative representations in Fig. 9-18. This decimator is now embedded in the four-band analysis/synthesis structure shown in Fig. 9-19 (Woods and O'Neil, 1986). Ideal band-pass filters $H_0, ..., H_3$ would partition the frequency domain into four equal subbands as shown in Fig. 9-20. Then the operation and analysis essentially parallel those for the 1D case. In this idealized situation, the band-pass filters completely eliminate the aliasing induced by sampling.

For the nonideal case, 2D quadrature mirror filters are used in the analysis and synthesis stages. As explained in Sec. 5.2.6, these eliminate

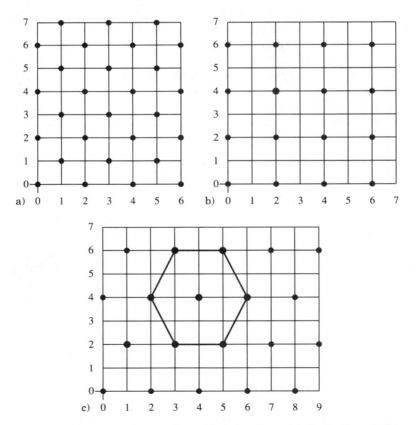

Figure 9-17   *Subsampling lattices for subband coding: (a) checkerboard, (b) rectangular pattern, (c) hexagonal pattern.*

Figure 9-18 *Alternative representation of the rectangular subsampler of Fig. 9-17(b).*

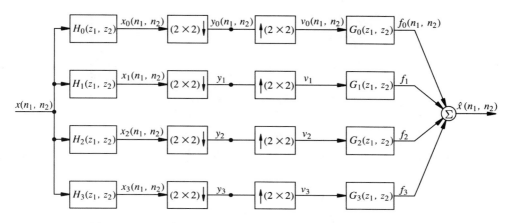

Figure 9-19 *Four-band rectangularly decimated SBC.*

aliasing by cancellation rather than by sharp cutoffs. First, we need to extend the 1D decimation/interpolation formulas of Sec. 5.2.6 to two dimensions.

The representation of these 2D rectangular lattices is shown in Fig. 9-21. Recall that in one dimension, we had

$$y(n) = x(Mn) \qquad \text{decimator output} \qquad (9\text{-}68)$$

$$v(n) = f(n)x(n) \qquad \text{interpolator output}$$

where

$$f(n) = \sum_{r=-\infty}^{\infty} \delta(n - rM) \qquad (9\text{-}69)$$

The corresponding transforms were found to be

$$Y(z) = \frac{1}{M} \sum_{r=0}^{M-1} X(W_M^r z^{1/M}) \qquad W_M \triangleq e^{j2\pi/M} \qquad (9\text{-}70)$$

$$V(z) = Y(z^M) = \frac{1}{M} \sum_{r=0}^{M-1} X(zW_M^r) \qquad (9\text{-}71)$$

In the 2D case, rectangular, separable sampling is defined by

$$y(n_1, n_2) = x(M_1 n_1, M_2 n_2) \qquad (9\text{-}72)$$

$$v(n_1, n_2) = f(n_1, n_2)x(n_1, n_2)$$

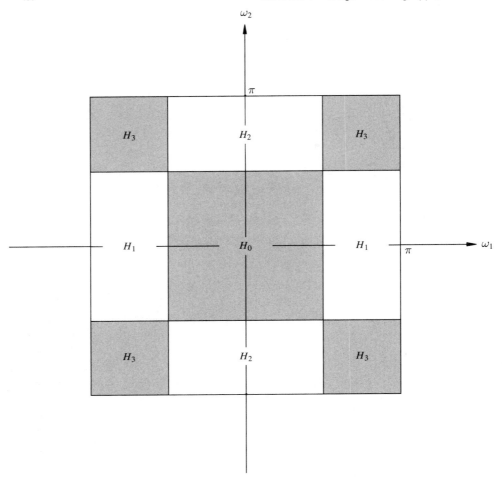

Figure 9-20 *Idealized partition of the frequency domain.*

Figure 9-21 *Decimation and interpolation: (a) one-dimensional; (b) two-dimensional.*

$$v(n_1, n_2) = f(n_1, n_2)x(n_1, n_2)$$

where
$$f(n_1, n_2) = \sum_{\ell}\sum_{r}\delta(n_1 - rM_1, n_2 - \ell M_2) \qquad (9\text{-}73)$$

The reader can verify that the 2D transforms are simple extrapolations,

$$V(z_1, z_2) = Y(z_1^{M_1}, z_2^{M_2}) \tag{9-75}$$

The nonseparable sampling lattice leads to more complex expressions. The interested reader can consult Vetterli (1989) for a detailed exposition. For the case in Fig. 9-19, we have $M_1 = M_2 = 2$ and

$$V_i(z_1, z_2) = \frac{1}{4}[X_i(z_1, z_2) + X_i(-z_1, z_2) + X_i(z_1, -z_2) + X_i(-z_1, -z_2)] \tag{9-76}$$

$$X_i(z_1, z_2) = H_i(z_1, z_2)X(z_1, z_2)$$

$$F_i(z_1, z_2) = G_i(z_1, z_2)V_i(z_1, z_2) \tag{9-77}$$

The interpolated subbands are summed at the output to yield

$$\hat{X}(z_1, z_2) = \sum_{i=0}^{3} F_i(z_1, z_2) \tag{9-78}$$

Substituting Eqs. (9-76) and (9-77) into Eq. (9-78), we obtain an expression of the form

$$\hat{X}(z_1, z_2) = T(z_1, z_2)X(z_1, z_2) + L_1(z_1, z_2)X(-z_1, z_2)$$
$$+ L_2(z_1, z_2)X(z_1, -z_2) + L_3(z_1, z_2)X(-z_1, -z_2) \tag{9-79}$$

The evaluation of the terms $L_i(z_1, z_2)$ is straightforward but messy and is left as an exercise for the reader.

Our objective is to make $L_i = 0$, $i = 1, 2,$ and 3, and thus eliminate aliasing from the reconstructed signal. If we choose the analysis/synthesis bank to be quadrature mirror filters

$$
\begin{aligned}
H_0(z_1, z_2) &\triangleq H(z_1, z_2) &&= G_0(z_1, z_2) \\
H_1(z_1, z_2) &= H(-z_1, z_2) &&= -G_1(z_1, z_2) \\
H_2(z_1, z_2) &= H(z_1, -z_2) &&= -G_2(z_1, z_2) \\
H_3(z_1, z_2) &= H(-z_1, -z_2) &&= G_3(z_1, z_2)
\end{aligned}
\tag{9-80}
$$

then the L_1 and L_2 terms are zero, leaving us with

$$T(z_1, z_2) = \frac{1}{4}[H^2(z_1, z_2) - H^2(-z_1, z_2)$$
$$- H^2(z_1, -z_2) + H^2(-z_1, -z_2)]$$

$$L_3(z_1, z_2) = \frac{1}{2}[H(z_1, z_2)H(-z_1, -z_2)$$
$$- H(z_1, -z_2)H(-z_1, z_2)] \tag{9-81}$$

It is easily verified that a *sufficient* condition for elimination of the remaining aliasing terms is that the filter be separable:

$$H(z_1, z_2) = H_1(z_1)H_2(z_2) \tag{9-82}$$

This leaves us with

$$\hat{X}(z_1, z_2) = T(z_1, z_2)X(z_1, z_2) \tag{9-83}$$

where $T(z_1, z_2)$, the "distortion" function, separates into

$$T(z_1, z_2) = \frac{1}{4}[H_1^2(z_1) - H_1^2(-z_1)][H_2^2(z_2) - H_2^2(-z_2)]$$
$$= T_1(z_1)T_2(z_2) \tag{9-84}$$

This is recognized as the simple extension of the 1D result in Eq. (5-47). As discussed in Sec. 5.2.6, we can choose $H(z)$ to be linear phase and FIR,

$$H(e^{j\omega}) = e^{-j(L-1)\omega/2}A(\omega) \tag{9-85}$$

This generates

$$T(e^{j\omega}) = \tfrac{1}{2}e^{-j(L-1)\omega}[A^2(\omega) + A^2(\omega - \pi)] \tag{9-86}$$

Similarly to the 1D case, the condition for distortionless reconstruction is that the sum of the squared magnitudes in Eq. (9-86) equal a constant—a condition which can be approximated, but not met exactly, by FIR filters. Nevertheless, such an approximation provides a practical solution, which also gives sharp cutoffs to mitigate the resulting quantization errors in the encoder.

Another advantage of the separable filter, separable sampler is that the 2D subband coder of Fig. 9-19 can be realized by 1D QMFs operating in the row-column filter tree shown in Fig. 9-22. It is readily shown that the $\{y_i(n_1, n_2)\}$ in both structures are equal. This tree structure can be continued by repeated decimation. Each of the outputs y_i in Fig. 9-22 can be iterated once to obtain the 16-subband structure shown in Fig. 9-23.

The nonseparable perfect-reconstruction image filter bank has been developed by Karlsson, Vetterli, and Kovacevic (1988). Both sampler and filter are non-separable. These are 2D generalizations of the perfect reconstruction filters described in Sec. 5.2.7.

As indicated in Fig. 9-16, the 2D subband signal itself must be encoded prior to transmission. This coder could be PCM, DPCM, DCT, or even a vector quantizer. We will pretend that the DPCM codec shown in Fig. 9-7 and analyzed in Sec. 9.1.2 is used. The allocation of bits to each subband via Eq. (9-12) depends on the signal energy in the subband; the

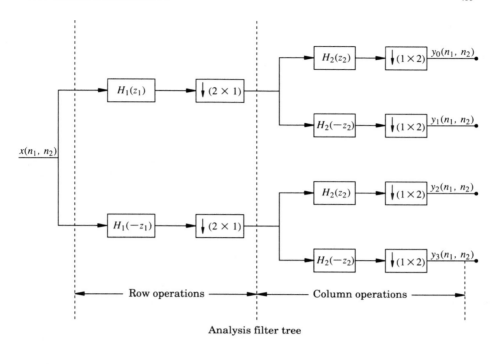

Analysis filter tree

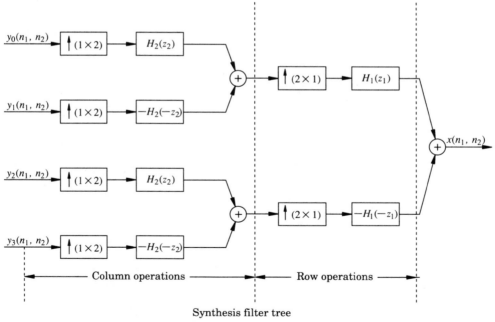

Synthesis filter tree

Figure 9-22   *Separable 4-subband structure.*

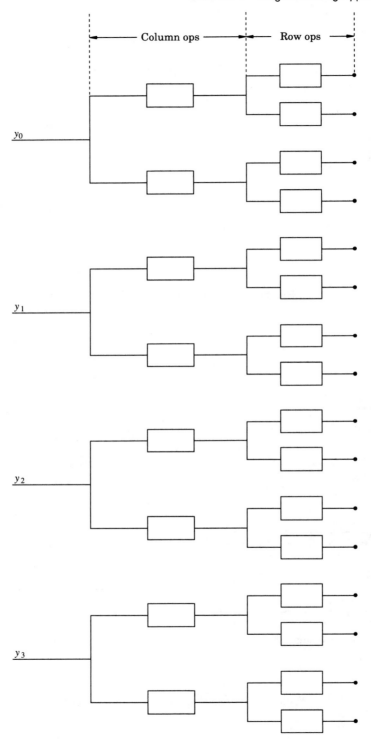

Figure 9-23   *Separable 16-subband tree structure.*

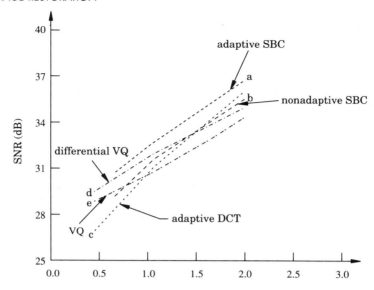

Figure 9-24   *Signal-to-noise ratio vs. bit rate for "lady" image using: (a) adaptive SBC, (b) nonadaptive SBC, (c) adaptive DCT, (d) differential VQ, and (e) VQ. [From Woods and o'Neil, 1986.Copyright © IEEE, Inc.]*

latter can be determined from the input signal spectrum and the filter bank used.

Woods and O'Neil (1986) reported coding simulations for SBC and other techniques for standard test images. These results are shown in Fig. 9-24; they indicate that adaptive SBC gives the best SNR performance at the bit rates tested. Also, fixed (that is, non-adaptive) SBC is comparable to the other coders tested, including the adaptive DCT.

In closing this section, we comment on coding sequences of images, as might be needed in teleconferencing or packet video. These involve three dimensions—two spatial and one temporal. Motion estimation between frames and spatial intraframe coding provide a rich source for hybrid 3D coding. Typically, predictive interframe and subband or transform intraframe coding combinations have been studied; even 3D SBC has been reported. The jury is still out, and current research continues in this field. The interested reader may consult Forchheimer and Kronander (1989) for additional discussion of these possibilities.

9.2 IMAGE RESTORATION

Image restoration deals with reconstructing an image that has been degraded by blurring or noise or both, as indicated in Fig. 9-2. The problem defined by that configuration is a classical one, studied by Gauss, Wiener, and Kalman in one dimension. The techniques of least-squares, Wiener, and Kalman filtering provide optimal linear reconstruction algorithms for linearly distorted signals, that is, those where $h(n_1, n_2)$ is a linear operator and the noise is additive (and usually Gaussian as well).

The image reconstruction algorithm depends on blur representation, the noise model, and the error measure used. Therefore, we first develop two examples of spatially invariant blur models from simple first principles. The noise models are drawn from the random field representations developed in Chap. 3. These are then followed by two error measures: least squares, primarily a deterministic problem formulation, which calls for a blur model for implementations, and the mean-squared error criterion, which requires signal and noise statistics as well as a blur model. We will show that the least-squares and Wiener filters are structurally similar. Furthermore, we will show that efficient implementation of both of these algorithms in the frequency domain is the preferred configuration.

9.2.1 BLURRING MODELS

For our purposes in this section, we pretend to know the a priori blur model $h(x, y)$ and the statistical characteristics of the noise. The reconstruction algorithm then depends on the models of these processes.

The degradation of an image called blurring can be modeled as a convolution of the original image with a spatially invariant impulse response $h(x, y)$. That is, as indicated in Fig. 9-25,

$$g(x, y) = h(x, y) ** f(x, y)$$

or
$$G(\omega_1, \omega_2) = H(\omega_1, \omega_2)F(\omega_1, \omega_2) \tag{9-87}$$

where $f(x, y)$, $h(x, y)$, and $g(x, y)$ represent the image, the blurring function, and the degraded image, respectively. Degradations caused by camera motion, atmospheric turbulence, and scanners that integrate pixel intensities over a local window are examples of blur that can be represented by Eq. (9-87). The distortion induced by a nonlinear transformation of the amplitude of each pixel does not fit this model. This topic is discussed in Sec. 9.2.6 on nonlinear techniques.

The use of an inverse filter to deblur the degraded image was addressed in Sec. 7.3, dealing with deconvolution. Here we describe some simple models that are representative of a class of linear blur operators. These models are then presumed known when used in a constrained least-squares approach to image restoration in which deconvolution and frequency-band emphasis are accounted for simultaneously.

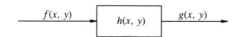

Figure 9-25   *Blurring represented as a linear operator.*

An example of a simple blurring model is that caused by relative motion between camera and scene. Suppose that a fixed image is photographed by a camera moving in the x direction at constant speed V_1. Further assume that the shutter remains open for a T-second interval and opens and closes instantaneously. Then the recorded image $g(x, y)$ appears as a time integration of the apparently moving image $f(x - V_1 t, y)$:

$$g(x, y) = \frac{1}{T} \int_{-T/2}^{T/2} f(x - V_1 t, y)\, dt \tag{9-88}$$

The Fourier transform (Sec. 3.2.3) of $g(x, y)$ is

$$G(\omega_1, \omega_2) = \frac{1}{T} \int_{-T/2}^{T/2} \mathfrak{F}\{f(x - V_1 t, y)\}\, dt \tag{9-89}$$

But from Table 3.2, with $x_0 = V_1 t$,

$$\mathfrak{F}\{f(x - x_0, y)\} = e^{-j\omega_1 x_0} F(\omega_1, \omega_2)$$

Hence

$$G(\omega_1, \omega_2) = F(\omega_1, \omega_2) \frac{1}{T} \int_{-T/2}^{T/2} e^{-j\omega_1 V_1 t}\, dt$$

$$= \frac{\sin \omega_1 V_1 T/2}{\omega_1 V_1 T/2} F(\omega_1, \omega_2) \tag{9-90}$$

from which we conclude that $H(\omega_1, \omega_2)$ is the separable transform

$$H(\omega_1, \omega_2) = \left(\operatorname{sinc} \frac{a_1 \omega_1}{2}\right) \cdot 1 = H_1(\omega_1) H_2(\omega_2) \tag{9-91}$$

where

$$H_1(\omega_1) = \operatorname{sinc} \frac{a_1 \omega_1}{2} \qquad H_2(\omega_2) = 1 \tag{9-92}$$

$$a_1 = V_1 T$$

Hence

$$h(x, y) = h_1(x) h_2(y)$$

$$= \frac{1}{a_1} \operatorname{rect}\left(\frac{x}{a_1}\right) \cdot \delta(y) \tag{9-93}$$

It is now easy to extend Eq. (9-93) to the case of relative motion at speed V_1 in the x direction and V_2 in the y direction. The result is[†]

$$H(\omega_1, \omega_2) = \operatorname{sinc} \frac{a_1 \omega_1}{2} \operatorname{sinc} \frac{a_2 \omega_2}{2} \tag{9-94}$$

$$h(x, y) = \frac{1}{a_1 a_2} \operatorname{rect}\left(\frac{x}{a_1}, \frac{y}{a_2}\right)$$

[†] $\operatorname{rect}\left(\dfrac{x}{a}, \dfrac{y}{b}\right) = \begin{cases} 1, & |x/a| < 1/2, |y/b| < 1/2 \\ 0, & \text{otherwise} \end{cases} = \operatorname{rect}\left(\dfrac{x}{a}\right) \operatorname{rect}\left(\dfrac{y}{b}\right)$

Sondhi (1972) uses this degradation model in developing deblurring algorithms (that is, inverse filters) for images degraded by uniform linear motion.

Another common example of blur is that due to a rectangular scanning aperture in the spatial domain. The output of such a device effectively integrates the image over a *spatial* window. The model is

$$g(x, y) = \frac{1}{ab} \int_{y-b/2}^{y+b/2} \int_{x-a/2}^{x+a/2} f(\xi, \eta) \, d\xi \, d\eta \tag{9-95}$$

This is easily recognized as the convolution of $f(x, y)$ with the 2D rectangular function

$$h(x, y) = \frac{1}{ab} \, \text{rect}\left(\frac{x}{a}, \frac{y}{b}\right) \tag{9-96a}$$

Therefore
$$H(\omega_1, \omega_2) = \text{sinc}\,\frac{a\omega_1}{2}\,\text{sinc}\,\frac{b\omega_2}{2} \tag{9-96b}$$

Still another blur model is that due to diffraction-limited coherent systems (Papoulis, 1968). This blur is represented by the ideal low-pass filter,

$$H(\omega_1, \omega_2) = \text{rect}(\omega_1/a, \omega_2/b)$$
$$h(x, y) = \frac{ab}{4\pi^2}\,\text{sinc}\,ax\,\text{sinc}\,by \tag{9-97}$$

and is seen to be the dual of the blur model for the rectangular scanning aperture, Eq. (9-96). Other blur phenomena and models are described in Jain (1989), and in Papoulis (1968), who treats the phenomenon from a Fourier optics standpoint.

These blur representations are summarized in Table 9-1.

For the least-squares and Wiener filter applications of this chapter, we will assume that $h(x, y)$, or its sampled version $\{h(n_1, n_2)\}$, is known.

9.2.2 NOISE MODELS

Random field representations of the image itself were described in Sec. 3.9. These Markov representations were found to be useful in DPCM and transform coding. In signal restoration, we simply assume that an appropriate model with spectral density is known a priori. As indicated in Fig. 9-26, the noise component of the degraded image consists of two types, additive and multiplicative. The specific type of noise depends on the physical process by which the image is formed. For example, wideband thermal noise in electrooptic systems is additive, Gaussian, and white; speckle noise in coherent imaging is multiplicative, white, and exponentially distributed (Jain, 1989). We will assume the noise is additive, stationary, and representable in the discrete variable form by the

Table 9-1. Representative Point-Spread Functions for Blur

Phenomenon	Point-Spread Function $h(x, y)$	Fourier Transform $H(\omega_1, \omega_2)$
Moving Image $\dot{x} = V_1, \dot{y} = V_2,$ $a_1 = V_1 T, a_2 = V_2 T$	$\frac{1}{a_1 a_2} \text{rect}\left(\frac{x}{a_1}, \frac{y}{a_2}\right)$	$\text{sinc}\,\frac{a_1 \omega_1}{2} \text{sinc}\,\frac{a_2 \omega_2}{2}$
Scanning Aperture (a, b)	$\frac{1}{ab} \text{rect}\left(\frac{x}{a}, \frac{y}{b}\right)$	$\text{sinc}\,\frac{a \omega_1}{2} \text{sinc}\,\frac{b \omega_2}{2}$
Diffraction-Limited Coherent	$\frac{ab}{4\pi^2} \text{sinc}\, ax\, \text{sinc}\, by$	$\text{rect}\left(\frac{\omega_1}{a}, \frac{\omega_2}{b}\right)$
Turbulence	$e^{-(x^2 + y^2)/r^2}$	$\pi r^2 e^{-(\omega_1^2 + \omega_2^2) r^2 / 4}$

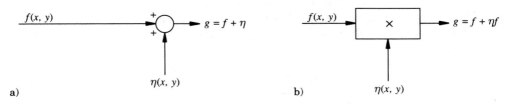

a)   b)

Figure 9-26 *Models of additive (a) and multiplicative noise (b).*

correlation sequence $R_n(k_1, k_2)$ or the power spectral density $\mathcal{S}_n(\omega_1, \omega_2)$. Typically, these are white and Gaussian, so that

$$R_n(k_1, k_2) = \sigma^2 \delta(k_1, k_2) \leftrightarrow \mathcal{S}_n(\omega_1, \omega_2) = \sigma^2 \tag{9-98}$$

9.2.3 LINEAR RESTORATION METHODS

The standard formulation of the image restoration problem is shown in Fig. 9-27. The original image $x(n_1, n_2)$ is blurred by the known, spatially invariant filter $h(n_1, n_2)$ and then contaminated by additive noise $\eta(n_1, n_2)$. Our objective is to obtain the spatially invariant reconstruction

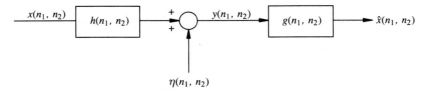

Figure 9-27 *Formulation of image-reconstruction problem.*

filter $g(n_1, n_2)$ that generates the optimal reconstruction $\hat{x}(n_1, n_2)$ from the degraded image $y(n_1, n_2)$. We employ two approaches in determining the optimal filter. The first is a constrained least-squares approach based on a deterministic error measure involving the sum of squared residuals and a high-frequency penalty term. In this formulation, no statistical model for signal or noise is used. The second approach minimizes a mean-squared error measure that leads to the Wiener filter. This formulation is predicated on a representation of the power spectral densities of signal and noise.

In both cases, the processing and matrix inversions are prohibitive if executed in the spatial domain. Instead, direct filtering in the frequency domain, using Fourier transforms, is the preferred implementation.

9.2.4 CONSTRAINED LEAST SQUARES

Our approach to 2D constrained least-squares filtering is depicted in Fig. 9-28. An unconstrained least-square approach seeks to determine the restoration filter $g(n_1, n_2)$ that minimizes the sum of the squared residuals

$$J = \sum_{n_1} \sum_{n_2} |y(n_1, n_2) - \hat{y}(n_1, n_2)|^2 \qquad (9\text{-}99)$$

This formulation often results in a wide-band filter that accentuates the high-frequency terms in the image. This could be detrimental if we know a priori that the image is smooth. In order to circumvent this, we can modify the criterion to penalize the high-frequency components, and thereby reduce the roughness. This can be achieved by imagining a high-frequency emphasis filter $f(n_1, n_2)$ operating on $\hat{x}(n_1, n_2)$, as shown in Fig. 9-28. Now it is mathematically tractable to keep the

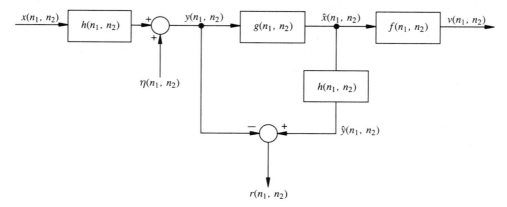

Figure 9-28   *Constrained least-squares problem formulation.*

residuals bounded while minimizing the high-frequency components in the restored image. Thus we minimize

$$\sum_{n_1} \sum_{n_2} |v(n_1, n_2)|^2 \qquad (9\text{-}100)$$

subject to the residual constraint

$$\sum_{n_1} \sum_{n_2} |y(n_1, n_2) - \hat{y}(n_1, n_2)|^2 = \epsilon^2 \qquad (9\text{-}101)$$

The bound ϵ^2 depends on a guesstimate of the noise measure $\|\eta\|^2 = \sum \sum |\eta(n_1, n_2)|^2$. Typically, $\epsilon^2 \approx \|\eta^2\|$.

The one-dimensional version of this problem can be solved following the procedures developed in Chap. 6. This version is shown in Fig. 9-29. The input sequence $\{x(n), 0 \le n \le N - 1\}$ is represented by the vector $\mathbf{x}$. The impulse response $h(n)$ is represented by its transmission matrix $\mathbf{H}$ [see Eqs. (6-75) to (6-77)]:

$$\mathbf{H} = \begin{bmatrix} h_0 & 0 & 0 & \cdots & 0 \\ h_1 & h_0 & 0 & \cdots & 0 \\ \cdots & \cdots & \cdots & \cdots & \\ h_N & h_{N-1} & h_{N-2} & \cdots & 0 \\ \cdots & \cdots & \cdots & \cdots & \\ h_L & h_{L-1} & h_{L-2} & \cdots & h_{L-N} \end{bmatrix} \qquad (9\text{-}102)$$

Similarly, $\mathcal{G}$ and $\mathbf{F}$ are transmission matrices for the impulse responses $g(n)$, $f(n)$, respectively. Thus with no noise, we have the collection

$$\mathbf{y} = \mathbf{Hx} \qquad \mathbf{v} = \mathbf{F\hat{x}}$$
$$\mathbf{\hat{x}} = \mathcal{G}\mathbf{y} \qquad \mathbf{r} = \mathbf{y} - \mathbf{H\hat{x}} \qquad (9\text{-}103)$$

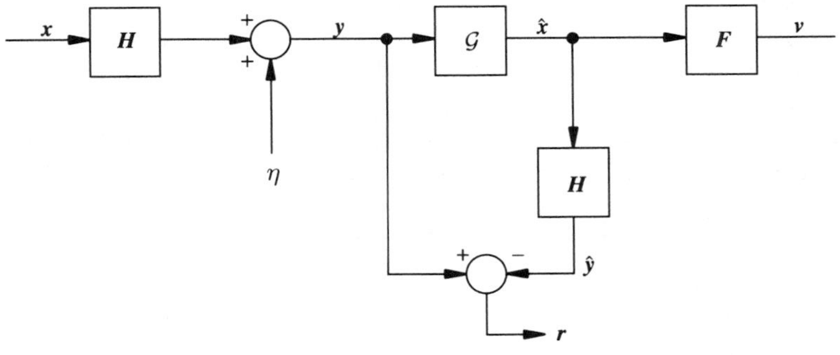

Figure 9-29 *Constrained one-dimensional least squares.*

Therefore we want to determine the $\mathcal{G}$ that minimizes

$$J = \mathbf{v}^T \mathbf{v} \tag{9-104}$$

subject to
$$\mathbf{r}^T \mathbf{r} = \epsilon^2 \tag{9-105}$$

This is a classic problem in constrained minimization and can be solved by the method of Lagrange multipliers. Let

$$J = \mathbf{v}^T \mathbf{v} + \lambda \{ \mathbf{r}^T \mathbf{r} - \epsilon^2 \} \tag{9-106}$$

where λ is the Lagrangian multiplier. J is minimized by setting $\frac{\partial J}{\partial \lambda}$ and the gradient $\frac{\partial J}{\partial \hat{\mathbf{x}}}$ to zero. Following the techniques described in Sec. 6.1.2, we obtain

$$\left[\frac{1}{\lambda} \mathbf{F}^T \mathbf{F} + \mathbf{H}^T \mathbf{H} \right] \hat{\mathbf{x}} = \mathbf{H}^T \mathbf{y} \tag{9-107}$$

and
$$\mathbf{r}^T \mathbf{r} = \epsilon^2$$

Once the value of λ is known, Eq. (9-107) can be solved for $\hat{\mathbf{x}}$:

$$\boxed{\hat{\mathbf{x}} = \left[\frac{1}{\lambda} \mathbf{F}^T \mathbf{F} + \mathbf{H}^T \mathbf{H} \right]^{-1} \mathbf{H}^T \mathbf{y}} \tag{9-108}$$

The value of λ is obtained iteratively. First one tries a value of λ, computes $\hat{\mathbf{x}}$ from Eq. (9-108), and then calculates $\mathbf{r}^T \mathbf{r}$ for that value of λ. If $\mathbf{r}^T \mathbf{r}$ exceeds the desired value ϵ^2, we decrease λ (and conversely) for the next iteration. The value of λ reflects a trade-off between signal smoothing and high-frequency rolloff. Also observe that if $\mathbf{F} = \mathbf{0}$, then Eq. (9-108) reduces to (6-24), the unconstrained least squares of Chap. 6:

$$\hat{\mathbf{x}} = (\mathbf{H}^T \mathbf{H})^{-1} \mathbf{H}^T \mathbf{y} \tag{9-109}$$

Computation of the two-dimensional counterpart of Eq. (9-108) would be prohibitive, since matrices of large dimensions must be multiplied together, added, and inverted. We may recognize that the problem formulation implicit in Fig. 9-29 and the proposed solution, Eq. (9-108) are spatial-domain formulations. We know from the treatment of transform coding that signals can be processed directly in the frequency domain and that fast transforms can be used to do the processing efficiently. In particular, the FFT can be employed to do the filtering provided the data are properly formatted.

Consider the linear convolution of the two finite-extent sequences $x(n)$ and $h(n)$, where

$$x(n) = 0 \qquad \text{for } n < 0, n > P$$

$$h(n) = 0 \qquad \text{for } n < 0, n > Q \tag{9-110}$$

The convolution gives the finite-extent sequence

$$y(n) = x(n) * h(n) = 0 \qquad \text{for } n < 0, n > P + Q \qquad (9\text{-}111)$$

Now let us select N as the least power of 2 such that $N > P + Q$. Define the zero-padded sequences

$$\tilde{x}(n) = \begin{cases} x(n) & 0 \le n \le P \\ 0 & P < n \le N - 1 \end{cases}$$

$$\tilde{h}(n) = \begin{cases} h(n) & 0 \le n \le Q \\ 0 & Q < n \le N - 1 \end{cases} \qquad (9\text{-}112)$$

We thus ensure that the circular convolution $\tilde{y}(n)$,

$$\tilde{y}(n) = \tilde{x}(n) \otimes \tilde{h}(n) = \sum_{k=0}^{N-1} \tilde{h}(n-k)\tilde{x}(k) \qquad (9\text{-}113)$$

and the linear convolution $y(n)$ are equal on $[0, N-1]$. Hence we can use DFTs to compute $y(n)$ via

$$\begin{aligned} X(k) &= DFT\{\tilde{x}(n)\} \\ H(k) &= DFT\{\tilde{h}(n)\} \\ Y(k) &= H(k)X(k) \\ \tilde{y}(n) &= IDFT\{Y(k)\} \end{aligned} \qquad (9\text{-}114)$$

The circular convolution of the zero-padded sequences Eq. (9-112) can be written in matrix form as

$$\begin{bmatrix} y(0) \\ y(1) \\ \cdots \\ y(N-1) \end{bmatrix} = \begin{bmatrix} h_0 & h_{N-1} & \cdots & h_1 \\ h_1 & h_0 & \cdots & h_2 \\ \vdots & \vdots & \ddots & \vdots \\ h_{N-1} & h_{N-2} & \cdots & h_0 \end{bmatrix} \begin{bmatrix} x(0) \\ x(1) \\ \cdots \\ x(N-1) \end{bmatrix} \qquad (9\text{-}115)$$

or $\qquad\qquad \tilde{\mathbf{y}} = \tilde{\mathbf{H}}\tilde{\mathbf{x}} \qquad\qquad (9\text{-}116)$

where $\tilde{\mathbf{H}}$ is the transmission matrix corresponding to the zero-padded, but *repetitive*, $\tilde{h}(n)$. This matrix, known as a *circulant* matrix, has the property that each column is simply a circularly shifted version of the column before it. The key property of a circulant matrix is that it can be diagonalized by a DFT transformation, that is,

$$\tilde{\mathbf{H}} = \mathbf{\Phi}\mathbf{D}_H\mathbf{\Phi}^{-1} \qquad (9\text{-}117)$$

where $\qquad\qquad \mathbf{\Phi} = [W^{nk}] \qquad W = e^{j2\pi/N} \qquad (9\text{-}118)$

is the matrix of a DFT transformation. Consequently, the inverse is simply[†]

$$\Phi^{-1} = (1/N)\Phi^* \tag{9-119}$$

and

$$\Phi\Phi^* = N\mathbf{I}.$$

The matrix $\mathbf{D}_H$ is a diagonal matrix whose entries are the N-point DFT of the sequence $\{\tilde{h}(n)\}$, that is,

$$\mathbf{D}_H = \text{diag}\{H(0), H(1), \ldots, H(N-1)\} \tag{9-120}$$

and

$$H(k) = DFT\{h(n)\} = \sum_{n=0}^{N-1} h(n)W^{-nk} \tag{9-121}$$

The proof of Eq. (9-117) is easily established with the aid of Fig. 9-30, which shows a linear shift-invariant $\{h(n)\}$ excited by a complex sinusoidal input $\tilde{x}(n) = W^{nk}$. From App. B, the output is also a complex sinusoid of the form

$$\tilde{y}(n) = \tilde{h}(n) * W^{nk} = H(k)W^{nk} \tag{9-122}$$

where

$$H(k) = DFT\{h(n)\} \tag{9-123}$$

The right-hand side of Eq. (9-122) can be written in vector form as

$$\tilde{\mathbf{y}} = \begin{bmatrix} y(0) \\ \vdots \\ y(N-1) \end{bmatrix} = \begin{bmatrix} W^0 \\ \vdots \\ W^{(N-1)k} \end{bmatrix} H(k) = \Phi_k H_k \tag{9-124}$$

where Φ_k is the kth column of Φ.

But by direct circular convolution, we can express the left-hand side of Eq. (9-122) as

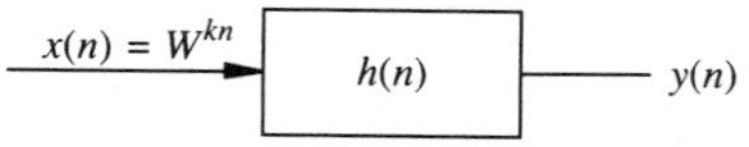

Figure 9-30 *Complex sinusoid applied to linear, shift-invariant system.*

[†]We remind the reader that an asterisk * appearing as a superscript denotes a complex conjugate, while the asterisk * appearing on the same level between two functions indicates a convolution.

$$\tilde{y}(n) = \sum_{i=0}^{N-1} \tilde{h}(n-i)W^{ki}$$

$$= [\tilde{h}(n), \tilde{h}(n-1), ..., \tilde{h}(0), ..., \tilde{h}(n-N+1)] \begin{bmatrix} W^0 \\ W^k \\ \vdots \\ W^{(N-1)k} \end{bmatrix}$$

Stacking the last equation for $n = 0, 1, ..., N-1$ gives

$$\tilde{\mathbf{y}} = \tilde{\mathbf{H}}\mathbf{\Phi}_k \tag{9-125}$$

Equating Eq. (9-124) and (9-125) gives

$$\tilde{\mathbf{H}}[\mathbf{\Phi}_0, ..., \mathbf{\Phi}_{N-1}] = [\mathbf{\Phi}_0, ..., \mathbf{\Phi}_{N-1}] \begin{bmatrix} H(0) & & & \\ & H(1) & & \\ & & \ddots & \\ & & & H(N-1) \end{bmatrix}$$

or $\qquad\qquad \tilde{\mathbf{H}}\mathbf{\Phi} = \mathbf{\Phi}\mathbf{D}_H \tag{9-126}$

This verifies Eq. (9-117).

We can now use the computationally powerful FFT methods to solve the least-squares equation, Eq. (9-108). In that equation, we may replace $\hat{\mathbf{x}}$ and $\mathbf{y}$ by their zero-padded counterparts, provided that $\mathbf{H}$ and $\mathbf{F}$ are replaced by the circulants $\tilde{\mathbf{H}}$, $\tilde{\mathbf{F}}$. Hence

$$\tilde{\mathbf{x}} = [\lambda^{-1}\tilde{\mathbf{F}}^T\tilde{\mathbf{F}} + \tilde{\mathbf{H}}^T\tilde{\mathbf{H}}]^{-1}\tilde{\mathbf{H}}^T\tilde{\mathbf{y}} \tag{9-127}$$

Next, we can show that

$$\tilde{\mathbf{H}}^T = \mathbf{\Phi}^{-1}\mathbf{D}_H\mathbf{\Phi} = \mathbf{\Phi}\mathbf{D}_H^*\mathbf{\Phi}^{-1}$$

so that $\qquad\qquad \tilde{\mathbf{H}}^T\tilde{\mathbf{H}} = (\mathbf{\Phi}\mathbf{D}_H^*\mathbf{\Phi}^{-1})(\mathbf{\Phi}\mathbf{D}_H\mathbf{\Phi}^{-1})$

$$= \mathbf{\Phi}\mathbf{D}_{|H|^2}\mathbf{\Phi}^{-1} \tag{9-128}$$

where $\qquad\qquad \mathbf{D}_{|H|^2} = \mathbf{D}_H^*\mathbf{D}_H \tag{9-129}$

The details are left as an exercise for the reader: see Probs. 9.5 and 9.6. Similarly,

$$\tilde{\mathbf{F}}^T\tilde{\mathbf{F}} = \mathbf{\Phi}\mathbf{D}_{|F|^2}\mathbf{\Phi}^{-1} \tag{9-130}$$

Then Eq. (9-127) becomes

$$\mathbf{\Phi}(\mathbf{D}_{|H|^2} + \lambda^{-1}\mathbf{D}_{|F|^2})(\mathbf{\Phi}^{-1}\hat{\mathbf{x}}) = \mathbf{\Phi}\mathbf{D}_H^*(\mathbf{\Phi}^{-1}\tilde{\mathbf{y}})$$

or $\qquad\qquad (\mathbf{D}_{|H|^2} + \lambda^{-1}\mathbf{D}_{|F|^2})\hat{\mathbf{X}} = \mathbf{D}_H^*\mathbf{Y} \tag{9-131}$

Each matrix in Eq. (9-131) is diagonal; $\hat{\mathbf{X}}$ and $\mathbf{Y}$ are the vector DFTs of $\tilde{\hat{\mathbf{x}}}$ and $\tilde{\mathbf{y}}$, respectively. The inversion is now quite trivial, and

$$\hat{\mathbf{X}} = \mathbf{GY}$$

$$\mathbf{G} = [\mathbf{D}_{|H|^2} + \lambda^{-1}\mathbf{D}_{|F|^2}]^{-1}\mathbf{D}_H^* \qquad (9\text{-}132)$$

The filtering matrix $\mathbf{G}$ is clearly diagonal:

$$\mathbf{G} = \text{diag}[..., G(k), ...] \qquad (9\text{-}133)$$

where
$$G(k) = \frac{H^*(k)}{|H(k)|^2 + (1/\lambda)|F(k)|^2} \qquad (9\text{-}134)$$

The proposed frequency-domain filter is shown in Fig. 9-31. The signal processing is simple, direct, and fast. The data signal $\mathbf{y}$ is zero-padded and transformed by a FFT to produce $\mathbf{Y}$. Each element of $\mathbf{Y}$ is *scalar* multiplied by the corresponding diagonal element to produce

$$\hat{X}(k) = G(k)Y(k) \qquad (9\text{-}135)$$

The element $G(k)$ is given explicitly by Eq. (9-134). In that formula, $H(k)$ is just the DFT of the blurring sequence $h(n)$, which is presumed known. $F(k)$ is the DFT of the high-frequency preemphasis filter, and λ is a parameter whose value is usually empirically set, although it may be calculated from the error measure. Finally, the restored signal $\hat{\mathbf{x}}$ is obtained by an inverse FFT routine.

A typical high-frequency, or roughness accentuator operator is the second-difference operator ∇^2 (see App. A):

$$v(n) = \nabla^2\hat{x}(n) = \hat{x}(n) - 2\hat{x}(n-1) + \hat{x}(n-2)$$

$$= f(n) * \hat{x}(n) \qquad (9\text{-}136)$$

$$f(n) = \delta(n) - 2\delta(n-1) + \delta(n-2) \qquad (9\text{-}137)$$

Following the procedure just described, we represent $v(n) = f(n) * \hat{x}(n)$ as the circular convolution of the zero-padded sequence $\tilde{f}(n)$ and $\tilde{\hat{x}}(n)$. Hence

$$\tilde{\mathbf{v}} = \tilde{\mathbf{F}}\tilde{\hat{\mathbf{x}}} \qquad (9\text{-}138a)$$

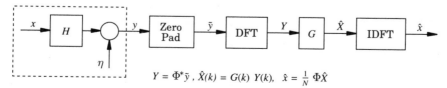

$$Y = \Phi^*\tilde{y}, \quad \hat{X}(k) = G(k)\,Y(k), \quad \hat{x} = \frac{1}{N}\,\Phi\hat{X}$$

Figure 9-31 *One-dimensional, least-squares signal restoration in transform domain.*

where $\tilde{\mathbf{F}}$ is the circulant

$$\tilde{\mathbf{F}} = \begin{bmatrix} 1 & 0 & \cdots & 1 & -2 \\ -2 & 1 & \cdots & 0 & 1 \\ 1 & -2 & \cdots & & \\ 0 & 1 & \cdots & & \\ & \cdot & \cdot & \cdots & \cdot & \cdot \\ \cdots & \cdots & \cdots & 0 & \cdots \\ 0 & 0 & \cdots & 1 & 0 \\ 0 & 0 & \cdots & -2 & 1 \end{bmatrix} \qquad (9\text{-}138\text{b})$$

As before, $\qquad\qquad \tilde{\mathbf{F}} = \mathbf{\Phi}\mathbf{D}_F\mathbf{\Phi}^{-1}$

so that $\qquad\qquad \tilde{\mathbf{F}}^T\tilde{\mathbf{F}} = \mathbf{\Phi}\mathbf{D}_{|F|^2}\mathbf{\Phi}^{-1}$ $\qquad\qquad (9\text{-}139)$

$$\mathbf{D}_F = \text{diag}[..., F(k), ...] \qquad\qquad (9\text{-}140)$$

The specific value of $F(k)$ is easily obtained from Eq. (9-137). Another exercise for the reader (Problem 9.7) will demonstrate that

$$|F(k)|^2 = (2\sin k\pi/N)^4 \qquad\qquad (9\text{-}141)$$

This value of $|F(k)|^2$ can be used in Eq. (9-134). This completes the constrained least-squares restoration filter in one dimension.

As shown in Fig. 9-32, the 2D version of this constrained least-squares filter is a straightforward extrapolation of the 1D structure. The degraded image $y(n_1, n_2)$ and the blur function $h(n_1, n_2)$ are each zero-padded to form

$$\tilde{y}(n_1, n_2) = \begin{cases} y(n_1, n_2), & 0 \le n_1 \le P_1, 0 \le n_2 \le P_2 \\ 0, & P_1 < n_1 \le M - 1, P_2 < n_2 \le N - 1 \end{cases}$$

$$\tilde{h}(n_1, n_2) = \begin{cases} h(n_1, n_2), & 0 \le n_1 \le Q_1, 0 \le n_2 \le Q_2 \\ 0 & Q_1 < n_1 \le M - 1, Q_2 < n_2 \le N - 1 \end{cases} \qquad (9\text{-}142)$$

where M and N are each the least power of $2 > (P_1 + Q_1)$ and $(P_2 + Q_2)$ respectively.

The 2D high-frequency emphasis filter is provided by the discrete Laplacian operator. Here we will use the noncausal version

$$v(n_1, n_2) = x(n_1 + 1, n_2) - 2x(n_1, n_2) + x(n_1 - 1, n_2)$$
$$+ x(n_1, n_2 + 1) - 2x(n_1, n_2) + x(n_1, n_2 - 1) \qquad (9\text{-}143)$$

Using transform notation,

$$V(z_1, z_2) = F(z_1, z_2)\hat{X}(z_1, z_2)$$
$$F(z_1, z_2) = (z_1 - 2 + z_1^{-1}) + (z_2 - 2 + z_2^{-1}) \qquad (9\text{-}144)$$

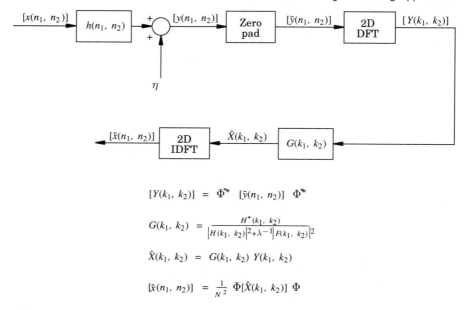

$$[Y(k_1, k_2)] = \Phi^*[\tilde{y}(n_1, n_2)]\,\Phi^*$$

$$G(k_1, k_2) = \frac{H^*(k_1, k_2)}{|H(k_1, k_2)|^2 + \lambda^{-1}|F(k_1, k_2)|^2}$$

$$\hat{X}(k_1, k_2) = G(k_1, k_2)\,Y(k_1, k_2)$$

$$[\hat{x}(n_1, n_2)] = \frac{1}{N^2}\,\Phi[\hat{X}(k_1, k_2)]\,\Phi$$

Figure 9-32 *Two-dimensional constrained least-squares image restoration.*

Similarly, the 2D DFT $H(k_1, k_2)$ can be determined from the blur model. All terms needed for Fig. 9-32 are now known. These are grouped together as Eq. (9-145) below. As with the 1D case, the Lagrangian multiplier can be set empirically or after an iterative solution of the constraint equation.

$$[Y(k_1, k_2)] = \Phi^*[\tilde{y}(n_1, n_2)]\Phi^*$$

$$G(k_1, k_2) = \frac{H^*(k_1, k_2)}{|H(k_1, k_2)|^2 + \lambda^{-1}|F(k_1, k_2)|^2}$$

$$\hat{X}(k_1, k_2) = G(k_1, k_2)Y(k_1, k_2) \tag{9-145}$$

$$[\hat{x}(n_1, n_2)] = (1/N^2)\Phi[\hat{X}(k_1, k_2)]\Phi$$

This frequency-domain formulation provides a simple and fast procedure for least-squares image restoration. No matrix inversions are needed, and the only matrix multiplications can be done by fast transforms. The actual filtering is even simpler: only a scalar multiplication of $Y(k_1, k_2)$ by $G(k_1, k_2)$ is needed.

The least-squares formulation does not account explicitly for noise, nor does it develop a statistical image model. The Wiener filter, which as we shall see, has a transform-domain structure similar to the least-squares filter, incorporates signal and noise random-field representations in its algorithm.

9.2.5 THE WIENER FILTER

The Wiener filter provides a mechanism for image restoration that takes into account statistical descriptions of the original image and the additive noise, as well as the blur. The theoretical basis for Wiener, or mean-squared error (MSE) filtering was laid down in Sec. 7.3.3. Here we rederive the Wiener filter, compare it with the least-squares filter, and demonstrate how this algorithm can also be implemented efficiently in the transform domain. Again, we do the details for the 1D case and extrapolate to two dimensions.

The problem statement is depicted in Fig. 9-33. The key difference between the least-squares and MSE (Wiener) filters is the error measure employed. In the least-squares formulation of Fig. 9-29, the error measure is the sum-squared *measurement* residuals $\|\mathbf{y} - \hat{\mathbf{y}}\|^2$, a deterministic quantity. The mean-squared error, however, minimizes the expected value of what may be called the *state* residuals $E\{\|\mathbf{x} - \hat{\mathbf{x}}\|^2\}$. Introducing the state $\mathbf{x}$ explicitly into the error measure requires a concomitant explication of the noise η and hence a statistical problem formulation.

For the 1D case, the Wiener filter determines the best estimate $\hat{\mathbf{x}} = \mathcal{G}\mathbf{y}$ that minimizes

$$J = E\{(\mathbf{x} - \hat{\mathbf{x}})^T(\mathbf{x} - \hat{\mathbf{x}})\} \tag{9-146}$$

The orthogonality principle asserts that $\hat{\mathbf{x}}$ must be such that the estimation error $(\mathbf{x} - \hat{\mathbf{x}})$ is orthogonal to the data, that is,

$$E\{(\mathbf{x} - \hat{\mathbf{x}})\mathbf{y}^T\} = 0 \tag{9-147}$$

Expanding Eq. (9-147) and defining the covariance matrices as

$$\mathbf{R}_{xy} = E\{\mathbf{x}\mathbf{y}^T\} \qquad \mathbf{R}_{yy} = E\{\mathbf{y}\mathbf{y}^T\} \tag{9-148}$$

lead to
$$\mathcal{G}\mathbf{R}_{yy} = \mathbf{R}_{xy} \tag{9-149}$$

But
$$\mathbf{R}_{yy} = E\{(\mathbf{H}\mathbf{x} + \eta)(\mathbf{H}\mathbf{x} + \eta)^T\}$$

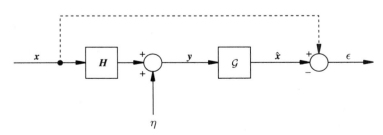

Figure 9-33 *Mean-square error problem formulation.*

If the signal $\mathbf{x}$ and the noise η are uncorrelated, the cross product-terms in this last expression will be zero, leaving us with

$$\mathbf{R}_{yy} = \mathbf{HR}_{xx}\mathbf{H}^T + \mathbf{R}_{nn} \tag{9-150}$$

where
$$\mathbf{R}_{xx} = E\{\mathbf{xx}^T\} \qquad \mathbf{R}_{nn} = E\{\eta\eta^T\} \tag{9-151}$$

are the signal and noise covariance matrices. Also,

$$\mathbf{R}_{xy} = E\{\mathbf{x}(\mathbf{Hx} + \eta)^T\} = \mathbf{R}_{xx}\mathbf{H}^T \tag{9-152}$$

Substituting Eq. (9-152) and (9-150) into Eq. (9-149) and solving for $\mathcal{G}$ gives

$$\boxed{\mathcal{G} = \mathbf{R}_{xx}\mathbf{H}^T(\mathbf{R}_{nn} + \mathbf{HR}_{xx}\mathbf{H}^T)^{-1}} \tag{9-153}$$

This last equation for $\mathcal{G}$ is a spatial-domain formula. In order to compare this MSE result with the least-squares formula, we use the matrix-inversion lemma, Eq. (6-56). After considerable manipulation, this allows us to write $\mathcal{G}$ in the alternative form

$$\mathcal{G} = (\mathbf{H}^T\mathbf{R}_{nn}^{-1}\mathbf{H} + \mathbf{R}_{xx}^{-1})\mathbf{H}^T\mathbf{R}_{nn}^{-1} \tag{9-154}$$

Now suppose the noise is white. Then $\mathbf{R}_{nn} = \sigma^2\mathbf{I}$, and Eq. (9-154) becomes

$$\mathcal{G} = (\mathbf{H}^T\mathbf{H} + \sigma^2\mathbf{R}_{xx}^{-1})\mathbf{H}^T \qquad \text{(MSE)} \tag{9-155}$$

whereas the least-squares gain was

$$\mathcal{G} = [\mathbf{H}^T\mathbf{H} + 1/\lambda(\mathbf{F}^T\mathbf{F})]^{-1}\mathbf{H}^T \qquad \text{(LS)} \tag{9-156}$$

Comparison of these last two equations shows that σ^2, the noise-power parameter, interchanges roles with $1/\lambda$, the inverse Lagrange multiplier. If λ is large (or $1/\lambda$ small), the least-squares error measure Eq. (9-106) emphasizes the residual and pays little attention to the high-frequency penalty function $\mathbf{v}^T\mathbf{v}$. Now a large λ corresponds to a small σ^2 in MSE, and a small σ^2 means little weight is given to the noise term, which is the high-frequency component in the MSE. On the other hand, one would choose a large value for $1/\lambda$ in the least-squares gain if a high level of noise were expected.

One can also compare $\sigma^2\mathbf{R}_{xx}^{-1}$ with $\lambda^{-1}\mathbf{F}^T\mathbf{F}$. The former is recognized as a form of noise-to-signal ratio (inverse SNR); the latter is a direct measure of the high-frequency component in the least-squares reconstructed signal and therefore corresponds to the wide-band noise. The MSE filter tries to use explicit signal and noise statistical characterizations; the least-squares filter attempts to fudge these by an empirical choice of $\mathbf{F}$ and λ.

The implementation of the Wiener filter in the spatial domain via Eq. (9-153) is fraught with difficulty, since the matrix products and inversions are horrendous in the 2D case. The transform-domain filter provides a basis for a practical algorithm, just as in the least-squares case.

The sequences $\mathbf{x}$ and $\mathbf{y}$ are zero-padded to form $\tilde{\mathbf{x}}$ and $\tilde{\mathbf{y}}$; the circulant matrix $\tilde{\mathbf{H}}$ replaces the transmission matrix $\mathbf{H}$. Then $\tilde{\mathbf{H}}$ is diagonalized by the DFT matrix Φ as in Eq. (9-117), repeated here:

$$\tilde{\mathbf{H}} = \Phi \mathbf{D}_H \Phi^{-1} \tag{9-117}$$

and $\mathbf{D}_H$ is the diagonal matrix of Eq. (9-120), (9-121). Let us also define the transformations:

$$\mathcal{G} = \Phi \mathbf{G} \Phi^{-1}$$
$$\mathbf{R}_{xx} = \Phi \mathbf{S}_{xx} \Phi^{-1} \tag{9-157}$$
$$\mathbf{R}_{nn} = \Phi \mathbf{S}_{nn} \Phi^{-1}$$

$\mathcal{G}$, $\mathbf{R}_{xx}$, and $\mathbf{R}_{nn}$ are spatial-domain gain, signal, and noise covariance matrices, respectively, while $\mathbf{G}$, $\mathbf{S}_{xx}$, and $\mathbf{S}_{nn}$ are their corresponding transforms. Thus $\mathbf{G}$ is the transfer-function matrix, and $\mathbf{S}_{xx}$, $\mathbf{S}_{nn}$ are spectral density matrices. Note that $\mathbf{G}$, $\mathbf{S}_{xx}$, $\mathbf{S}_{nn}$ are not necessarily diagonal. The DFT diagonalizes the circulant matrix $\tilde{\mathbf{H}}$, but not an arbitrary covariance matrix.

Substituting the transforms of Eq. (9-157) into Eq. (9-153) and making use of the properties of Φ in Eq. (9-119) leads to the transform-domain gain formula

$$\boxed{\mathbf{G} = \mathbf{S}_{xx} \mathbf{D}_H^* [\mathbf{S}_{nn} + \mathbf{D}_H \mathbf{S}_{xx} \mathbf{D}_H^*]^{-1}} \tag{9-158}$$

This equation can be compared with the least-squares transform gain formula, Eq. (9-132), where all matrices were diagonal. In the present instance, only $\mathbf{D}_H$ is exactly diagonal.

If the noise is white, $\mathbf{R}_{nn} = \sigma^2 \mathbf{I}$; therefore $\mathbf{S}_{nn} = \sigma^2 \mathbf{I}$ and Eq. (9-158) reduces to

$$\mathbf{G} = \mathbf{S}_{xx} \mathbf{D}_H^* [\sigma^2 \mathbf{I} + \mathbf{D}_H \mathbf{S}_{xx} \mathbf{D}_H^*]^{-1} \tag{9-159}$$

The gain matrix $\mathbf{G}$ is still not diagonal, because of the presence of $\mathbf{S}_{xx}$. Now, the DFT matrix Φ was selected to diagonalize $\tilde{\mathbf{H}}$; hence it cannot also be required to diagonalize $\mathbf{R}_{xx}$. We also know that the KLT could be used to diagonalize $\mathbf{R}_{xx}$, but then it would not diagonalize $\tilde{\mathbf{H}}$. However, our study of transform coding demonstrated that a unitary transformation tends to decorrelate a vector, and, within a scale factor, the DFT matrix is unitary. Therefore, we can argue that the DFT matrix $\mathbf{\Phi}$ *approximately* diagonalizes $\mathbf{R}_{xx}$. With this understanding, we have

$$\mathbf{S}_{xx} \approx \mathrm{diag}[S_{xx}(0), ..., S_{xx}(k), ..., S_{xx}(N - 1)] \qquad (9\text{-}160)$$

where $S_{xx}(k)$ is the kth diagonal element of $\mathbf{S}_{xx}$. Now all the matrices in Eq. (9-159) are diagonal, so that $\mathbf{G}$ reduces to a diagonal matrix, with

$$G(k) = \frac{S_{xx}(k)H^*(k)}{\sigma_n^2 + |H(k)|^2 S_{xx}(k)}$$

$$= \frac{H^*(k)}{|H(k)|^2 + \sigma_n^2/S_{xx}(k)} \qquad (9\text{-}161)$$

This equation may now be compared with the least-squares gain in Eq. (9-134). The comparable terms are the spectral-domain inverse SNRs

$$\sigma_n^2/S_{xx}(k) \qquad \text{Wiener filter}$$

$$|F(k)|^2/\lambda \qquad \text{least-squares filter}$$

which have already been discussed in their earlier spatial-domain forms of Eq. (9-154) and (9-156). Hence the only difference between the least-squares filter and the approximate Wiener filter is the SNR term in the gain formula. Therefore Fig. 9-31 represents the Wiener filter as well as the least-squares filter, with the least-squares $G(k)$ of Eq. (9-134) replaced by the Wiener $G(k)$ of Eq. (9-161).

The extrapolation to 2D follows easily. In fact, the 2D least-squares filter of Fig. 9-32 applies to the Wiener filter as well. The explicit 2D algorithms are

$$[Y(k_1, k_2)] = \mathbf{\Phi}^*[\tilde{y}(n_1, n_2)]\mathbf{\Phi}^* = DFT\{[\tilde{y}(n_1, n_2)]\}$$

$$G(k_1, k_2) = \frac{H^*(k_1, k_2)}{|H(k_1, k_2)|^2 + \sigma_n^2/S_{xx}(k_1, k_2)}$$

$$\hat{X}(k_1, k_2) = G(k_1, k_2)Y(k_1, k_2) \qquad (9\text{-}162)$$

$$[\hat{x}(n_1, n_2)] = IDFT\{[\hat{X}(k_1, k_2)]\}$$

$$= \frac{1}{N^2}\mathbf{\Phi}[\hat{X}(k_1, k_2)]\mathbf{\Phi}$$

9.2.6 NONLINEAR NOISE REMOVAL–THE MEDIAN FILTER

The median filter is a nonlinear restoration technique designed to remove impulse-type noise from an image while preserving edges. The one-dimensional median filter was introduced in Sec. 5.4.1. The 2D version is not greatly different, except that here we can define the shape of the neighborhood or mask around the pixel. These masks are typically square, with 3×3 and 5×5 being representative sizes. Mathematically,

$$y(n_1, n_2) = \text{med}\{x(n_1 - k_1, n_2 - k_2), k_1, k_2 \in R\} \tag{9-163}$$

where R specifies the mask. The median filter is known to have good impulse rejection and good edge preservation, but it does not filter Gaussian noise well and does not have frequency-band selectivity. It may also lose fine details, such as lines.

The need to do more than just impulse rejection has led to the development of composite structures embodying the features of a linear filter for signal shaping and a median filter for spike elimination. The ranked-order filters introduced in Sec. 5.4.2 are examples of such a class of non-linear structures. Two promising approaches to the composite structure are the modified trimmed mean (MTM) filter (Lee and Kassam, 1985) and the FIR median hybrid filter (Niemenen et al., 1987).

The former (MTM) was briefly described earlier in Sec. 5.3.2. Basically, the filter first determines the median m_k inside its window and then chooses an interval $(m_k - g, m_k + g)$, using a preselected constant g. Within the window, data samples outside the range $(m_k \pm g)$ are discarded, and the average of the remaining data is taken as the output. Note that the number of samples in the average is not fixed a priori. The value of g dramatizes the dual role of this filter. For small g the MTM approaches the median filter, while for large g it behaves very much like a running-mean filter.

By proper choice of g, the MTM filter can suppress white Gaussian noise and simultaneously preserve edges. The proper value of g equals the maximum value for which edges are to be preserved. Thus we may choose $g = H - 2\sigma$, where H is the minimum height of the edges to be preserved and σ is the standard deviation of the Gaussian noise. The MTM preserves an ideal edge, as does a median filter, provided $g < H$. It is better than a median filter when the edge is shrouded in Gaussian white noise.

The double-window modified trimmed mean (DW MTM) filter is a variation on the MTM, employing two windows. First, the sample median m_k is computed from a small window of size $(2N + 1)$. Then an interval $(m_k \pm g)$ is selected, as before. Finally, the mean of data lying within the $(m_k \pm g)$ interval among the samples in a large window of size $(2L + 1)$ is taken as the output. Test results for a signal (dashed curve) corrupted by white Gaussian noise $(\sigma = 2)$ and three impulses are shown in Fig. 9-34. In this test, the DW MTM performs the best; it preserves edges, smoothes the Gaussian noise, and completely rejects the impulses.

Two-dimensional median filters can smooth noisy images while retaining edge structures. They do suffer from an inability to retain fine

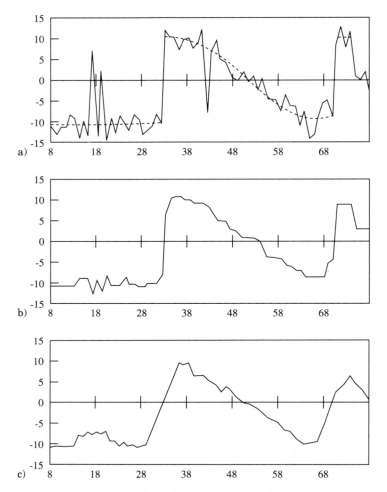

Figure 9-34 *Comparison of median, MTM, and DW MTM filters: (a) test input, additive Gaussian noise, and three impulses; (b) median filter output; (c) running mean. [After Lee and Kassam (1985) copyright IEEE, Inc.]*

details, such as lines. The k-nearest-neighbor averaging operation, or KAVE filter (Davis and Rosenfeld, 1978) is a nonlinear filter designed to suppress both Gaussian noise and impulse noise and to retain line details better than the median filter. The center point P of a $(2k + 1) \times (2k + 1)$ mask is replaced by the average gray level of the k nearest neighbors of P whose gray levels are closest to that of P. For k small, the KAVE filter preserves details and provides some noise attenuation. A large k would yield more noise reduction but at the cost of a loss of details.

The FIR-median hybrid (FMH) filter was devised by Heinonen and Neuvo (1987, 1988) in one dimension and extended to two dimensions by Niemenen, Heinonen, and Neuvo (1987). The 1D filter shown in Fig. 9-35

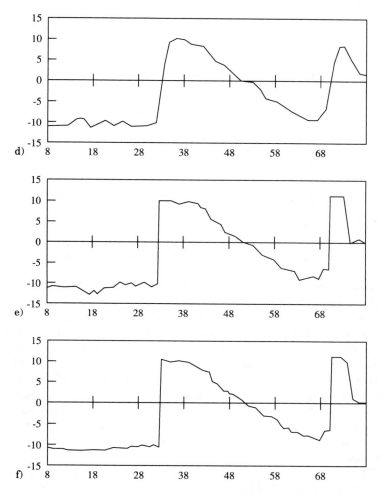

Figure 9-34   *Comparison of median, MTM, and DW MTM filters: (d) combination of (b) and (c); (e) MTM filter output; (f) DW MTM filter output. [After Lee and Kassam (1985) copyright IEEE, Inc.]*

consists of a linear FIR filter substructure functioning as a pre-processor feeding into a median filter.

The backward (causal) and forward (noncausal) FIR substructures can be tailored to the class of expected signal. In the first case considered, $y_F(n)$ and $y_B(n)$ are simple forward and backward averaging filters:

$$y_B(n) = \sum_{j=n-k}^{n-1} x(j), \qquad y_F(n) = \sum_{j=n+1}^{n+k} x(j) \tag{9-164}$$

and

$$y(n) = \mathrm{med}\{y_B(n), y_F(n), x(n)\} \tag{9-165}$$

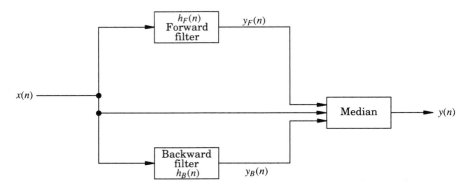

Figure 9-35   *Structure of the FIR-median hybrid (FMH) filter.*

This structure was found to be comparable to and in fact slightly better than the standard median filter when tested on step discontinuous functions contaminated with Gaussian white noise. The great advantage of the FMH is computational efficiency. The authors report a speedup in computer time by a factor of 450 as compared with that of a median filter with the same performance. The FIR averager, however, tends to cut the peaks of a signal. To remedy this, the FIR filter can be designed to follow polynomial signals in Gaussian noise—a standard subject that has received much attention in the past (Morrison, 1969). This class of polynomial extrapolators can track polynomial signals as root signals. (See Sec. 5.4.1 for a discussion of root signals.) Simulations have demonstrated that the polynomial (ramp) FMH filter is superior to the standard median in noise suppression, ramp signal tracking, and computational efficiency.

The basic structure of the two-dimensional version of this FMH filter is shown in Fig. 9-36. The east and west filters are FIR averaging filters of length k in the respective directions. Their averages are combined with the center pixel in the median filter to form the horizontal signal $y_H(n)$. Similarly, north and south filters can be combined to create the vertical mask. These masks are then rotated 45° and combined to give the multilevel structure of Fig. 9-37. These filters are designed to suppress additive noise while preserving subtle details such as lines and edges.

The three-level 5×5 FMH filter was compared in simulations with the 5×5 square median filter and the KAVE algorithm for $k = 14$. The FMH is, as expected, much faster than the other two structures, and it preserves details better. The FMH structure preserves two-pixel-wide lines regardless of their orientation and can preserve one-pixel-wide lines in certain directions. The median filter, on the other hand, loses details of lines that are one or two pixels wide. Finally, the FMH preserves edges in noisy images better than either the median or the KAVE filter.

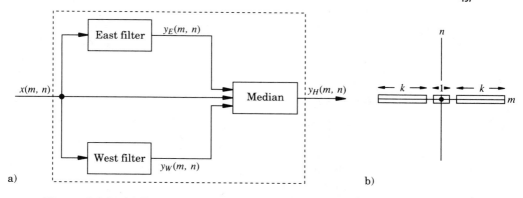

a)   b)

Figure 9-36   *(a) Basic structure of the 2D FMH filter; (b) east-west mask.*

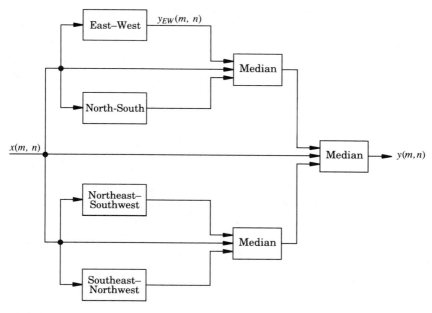

Figure 9-37   *Multilevel structure of FMH filter with directional masks.   The bottom tier uses horizontal masks rotated 45°.*

9.3 IMAGE ENHANCEMENT

We close this chapter on image-processing applications with two examples of image enhancement, one linear and one nonlinear. While image restoration tries to remove blur and noise, enhancement tries to improve the image itself—to accentuate features, provide contrast, or highlight edges and lines.

The nonlinear technique considered here is histogram equalization, in which the shading of a picture is changed to provide more contrast.

The linear technique consists of filtering images by 2D FIR filters to highlight certain spectral domains of the signal. These linear operations are called smoothing and sharpening.

9.3.1 HISTOGRAM EQUALIZATION

A picture or image sometimes lacks definition or contrast. It may be too light or too dark. An image with a preponderance of dark tones would have a probability density function (pdf) similar to the one shown in Fig. 9-38(a), and a generally light image would have a pdf skewed in the opposite direction, as shown in Fig. 9-38(b). Histogram equalization is a nonlinear, pixel-by-pixel transformation that attempts to redistribute the gray levels in an image so that the histogram (or pdf) of the equalized image is constant over the area covered by the image.

Assume that u, the gray level of a pixel, is a random variable normalized to the range $[0, 1]$ with pdf $f_u(u)$. As indicated in Fig. 9-39, the random variable u is transformed into v by

$$v = g(u) \tag{9-166}$$

The resulting pdf for v is (Papoulis, 1984)

$$f_v(v) = \frac{f_u(u)}{|g'(u)|}\bigg|_{u=g^{-1}(v)} \tag{9-167}$$

where $g^{-1}(\cdot)$ is the functional inverse of $g(\cdot)$. Another result from probability theory is that if we choose the transformation $g(\cdot)$ to be the cumulative distribution function of the input, that is, if

$$g(u) = F_u(u) = \int_{-\infty}^{u} f_u(\tau)\, d\tau \tag{9-168}$$

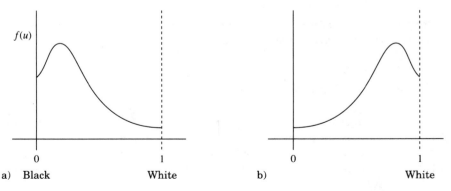

Figure 9-38 *Probability density functions of gray levels in an image: (a) dark image; (b) light image.*

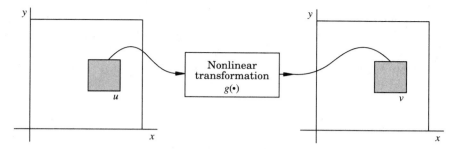

Figure 9-39 *Gray-level transformation: u = gray legel of pixel; v = gray level of transformed pixel.*

then the random variable $v = g(u)$ is uniformly distributed on $[0, 1]$:

$$f_v(v) = \begin{cases} 1 & 0 \le v \le 1 \\ 0 & \text{otherwise} \end{cases} \tag{9-169}$$

For a quantized image, the random variable is discrete-valued, and the associated pdf or histogram is a bar graph determined from the data. Suppose the pixels are quantized to L levels on $[0, 1)$. Then the probability that the gray level equals u_j is taken to be the relative frequency

$$f_u(u_j) = \frac{n_j}{N} \qquad j = 0, 1, ..., L-1 \tag{9-170}$$

where n_j is the number of times level u_j appears in the image and N is the total number of pixels. The transformation or mapping is equal to the cumulative discrete density function, or

$$v_k = g(u_k) = F_u(u_k) = \sum_{j=0}^{k} \frac{n_j}{N} \qquad k = 0, 1, ..., L-1 \tag{9-171}$$

The $\{v_k\}$ determined from Eq. (9-171) do not necessarily correspond to the quantization levels for a pixel. Each v_k must then be requantized to its nearest level $(v_k)_Q$. The actual transformation used must reflect this, so that

$$(v_k)_Q = g_Q(u_k) \tag{9-172}$$

is the quantized version of Eq. (9-171).

This histogram equalization increases the contrast in images with a preponderance of light or dark pixels by stretching out the distribution of gray levels. The resulting image is usually far more pleasing to the eye, and details can be more readily perceived.

As a final point, probability theory tells us how to choose a mapping $g(\cdot)$ which transforms a random variable u with a given pdf to another random variable w with any desired pdf $f_w(w)$. The technique is shown in Fig. 9-40. First, the random variable u is mapped into the uniformly distributed variable v. Then the mapping $h(\cdot)$ can be selected so that w has the desired distribution. It can be shown that

$$w = h(v) = F_w^{-1}(v) \tag{9-173}$$

is the mapping that transforms a uniform density into a specified density $f_w(w)$. Combining the two transformations gives

$$w = g(u) = F_w^{-1}(F_u(u)) \tag{9-174}$$

Implementation of the transformation in Eq. (9-174) for quantized random variables $\{u_k\}$ and $\{w_k\}$ requires histogram evaluation and re-quantization of the results of the first pass, that is,

$$(w_k)_Q = Q[g(u_k)] \tag{9-175}$$

where Q is a quantization operator.

9.3.2 ENHANCEMENT BY SIMPLE LINEAR OPERATORS

We have previously studied optimal linear least-squares and Wiener filters for image restoration. Linear FIR filters can also be used to enhance or emphasize certain features of an image. We will discuss two broad classes of filters: smoothing or low-pass, and sharpening or high-pass, filters.

The low-pass filter is used for noise smoothing, for interpolation, and for bandwidth reduction prior to down-sampling in a subband coder. While smoothing or averaging suppresses Gaussian noise, it also tends to blur the image, so restraint must be exercised in applying these filters.

The simplest low-pass filter is the neighborhood-averaging filter

$$y(n_1, n_2) = \frac{1}{N} \sum_{(k_1,k_2) \in R} \sum x(k_1, k_2) \tag{9-176}$$

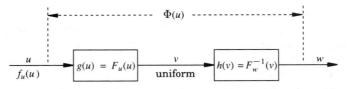

Figure 9-40 *Generation of desired histogram.* $f_u(u)$ *is given input pdf;* $f_w(w)$ *is desired output pdf.*

where R specifies the neighborhood or mask and N is the number of points within this mask. Typically the mask is square, although directional masks can be devised to prevent blurring at edges.

A weighted averaging filter has the FIR form

$$y(n_1, n_2) = \sum_{(k_1, k_2) \in R} \sum h(k_1, k_2) x(n_1 - k_1, n_2 - k_2) \qquad (9\text{-}177)$$

The most common mask R is an $M \times M$ square. The weights are normalized to make the d-c gain unity, or

$$\sum_{k_1} \sum_{k_2} h(k_1, k_2) = 1 \qquad (9\text{-}178)$$

Two typical spatial averaging masks are shown in Fig. 9-41(a) and (b). The transforms of these filters are easily shown to be

$$H'_a(z_1, z_2) = \frac{1}{8}[(z_1^{\frac{1}{2}} + z_1^{-\frac{1}{2}})^2 + (z_2^{\frac{1}{2}} + z_2^{-\frac{1}{2}})^2]$$

$$H_a(\omega_1, \omega_2) = \frac{1}{2}[\cos^2 \omega_1 / 2 + \cos^2 \omega_2 / 2] \qquad (9\text{-}179)$$

and

$$H'_b(z_1, z_2) = \frac{1}{16}[(z_1^{\frac{1}{2}} + z_1^{-\frac{1}{2}})^2 (z_2^{\frac{1}{2}} + z_2^{-\frac{1}{2}})^2]$$

$$H_b(\omega_1, \omega_2) = [(\cos \omega_1 / 2)^2 (\cos \omega_2 / 2)^2] \qquad (9\text{-}180)$$

The latter filter may be recognized as the binomial filter (with a scale factor) introduced in Secs. 2.5.4, 3.8, and 5.3

The second kind of linear enhancement operation is image sharpening, or high-pass filtering. Most of these operators are based on digital approximations to the analog differentiation operation.

The difference operator is the simplest digital approximation to differentiation. In the temporal domain,

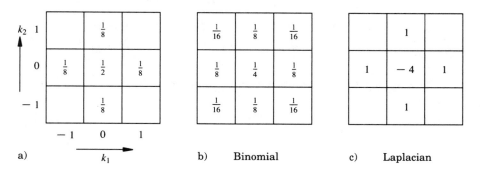

a)   b)   Binomial   c)   Laplacian

Figure 9-41 *Masks of linear operators: (a) lowpass, (b) binomial lowpass, (c) Laplacian.*

$$\frac{df}{dt} \approx \frac{f(n) - f(n-1)}{T} \tag{9-181}$$

From Fourier theory, we know that the transfer function of a differentiator is

$$H_a(j\omega) = j\omega \tag{9-182}$$

The transform of the difference operator of Eq. (9-181) is

$$H_z(z) = \frac{1 - z^{-1}}{T}$$

or
$$H_D(\omega) = \frac{1 - e^{-j\omega T}}{T} \tag{9-183}$$

For low-frequency signals, $\omega T \ll 1$, $H_D(\omega)$ can be approximated by

$$H_D(\omega) \approx \frac{1 - (1 - j\omega T)}{T} = j\omega = H_a(\omega) \tag{9-184}$$

This last result demonstrates that the difference operator is a low-frequency differentiator.

A symmetric second-derivative approximation is

$$\frac{d^2 f}{dt^2} \approx \frac{f(n+1) - 2f(n) + f(n-1)}{T^2} \leftrightarrow \frac{(1 - z^{-1})(z - 1)}{T^2} \tag{9-185}$$

Extrapolating to two dimensions, replacing time by spatial variables, and normalizing the sample spacings to unity gives forward and backward differences as

$$\nabla_x f = f(n_1, n_2) - f(n_1 - 1, n_2) \leftrightarrow (1 - z_1^{-1})$$
$$\nabla_y f = f(n_1, n_2) - f(n_1, n_2 - 1) \leftrightarrow (1 - z_2^{-1})$$
$$\text{and} \qquad \Delta_x f = f(n_1 + 1, n_2) - f(n_1, n_2) \leftrightarrow (z_1 - 1)$$
$$\Delta_y f = f(n_1, n_2 + 1) - f(n_1, n_2) \leftrightarrow (z_2 - 1) \tag{9-186}$$

The analog Laplacian operator, denoted[†] by $\nabla^2 f$, is

$$\frac{\partial^2 f}{\partial x^2} + \frac{\partial^2 f}{\partial y^2} \tag{9-187}$$

[†]We use the symbol ∇ to represent a difference operator. This same symbol is also used to denote a gradient, as in Chap. 6.

Equations (9-185) and (9-186) suggest that the digital Laplacian operator be taken as

$$y(n_1, n_2) = x(n_1 + 1, n_2) - 2x(n_1, n_2) + x(n_1 - 1, n_2)$$
$$+ x(n_1, n_2 + 1) - 2x(n_1, n_2) + x(n_1, n_2 - 1) \quad (9\text{-}188)$$

or $\quad H_z(z_1, z_2) = (z_1 - 2 + z_1^{-1}) + (z_2 - 2 + z_2^{-1})$

$$= (z_1^{\frac{1}{2}} - z_1^{-\frac{1}{2}})^2 + (z_2^{\frac{1}{2}} - z_2^{-\frac{1}{2}})^2 \quad (9\text{-}189)$$

The frequency response of this Laplacian is

$$H(\omega_1, \omega_2) = -4[(\sin \omega_1/2)^2 + (\sin \omega_2/2)^2] \quad (9\text{-}190)$$

At low frequencies $\omega_1 \ll \pi$ and $\omega_2 \ll \pi$, the frequency response is approximately

$$H(\omega_1, \omega_2) \approx -[\omega_1^2 + \omega_2^2] = (j\omega_1)^2 + (j\omega_2)^2, \quad (9\text{-}191)$$

which demonstrates the second-derivative operation. The mask of this Laplacian is shown in Fig. 9-41(c).

This Laplacian can be recognized as the high-frequency emphasis filter used in constrained least squares in Sec. 9.2.4. The Laplacian can provide several image-processing enhancements, two of which are described here. First, it is a high-pass filter that accentuates edges and lines. Specifically, the zero crossings of the output of the Laplacian filter locate the edges of the input image. In Fig. 9-42, the sketch of the second derivative of a function with an edgelike slope demonstrates the relationship between zero crossing and edge location.

The Laplacian, like all derivative operators, is sensitive to noise. Therefore the image can be smoothed first and then doubly differentiated by the Laplacian as shown in Fig. 9-43. A suitable combination is the binomial mask of Fig. 9-41(b) followed by the Laplacian of Fig. 9-41(c).

A second use for the Laplacian is in deblurring an image where the blur was caused by a diffusion-type operation, that is by neighborhood averaging. Suppose the image $x(n_1, n_2)$ is blurred by leakage from neighboring pixels:

$$\bar{x}(n_1, n_2) = x(n_1, n_2) + \lambda[x(n_1 + 1, n_2) + x(n_1 - 1, n_2)$$
$$+ x(n_1, n_2 + 1) + x(n_1, n_2 - 1)] \quad (9\text{-}192)$$

or $\quad \bar{X}(z_1, z_2) = [1 + \lambda(z_1 + z_1^{-1} + z_2 + z_2^{-1})]X(z_1, z_2)$

But from Eq. (9-189), the Laplacian operating on $x(n_1, n_2)$ gives

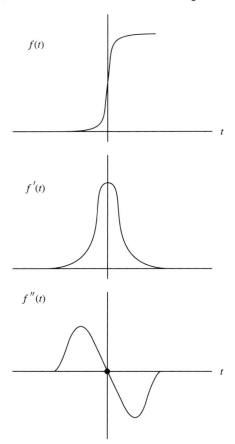

Figure 9-42 *Relationship between an edge and the zero crossing of the second derivative.*

Figure 9-43 *Laplacian operator preceded by low-pass filter.*

$$y(n_1, n_2) = \nabla^2 x(n_1, n_2)$$

or

$$Y(z_1, z_2) = -4[1 - \frac{1}{4}(z_1 + z_1^{-1} + z_2 + z_2^{-1})]X(z_1, z_2)$$

or

$$\frac{Y(z_1, z_2)}{-4} = \left[1 - \frac{1}{4}(z_1 + z_1^{-1} + z_2 + z_2^{-1})\right] X(z_1, z_2) \qquad \text{(9-193)}$$

Adding $\bar{x}(n_1, n_2)$ to a scaled $y(n_1, n_2)$ gives

$$\bar{x}(n_1, n_2) + k \frac{\nabla^2 x(n_1, n_2)}{-4}$$

$$\leftrightarrow \left[(1 + k) + (\lambda - k/4)(z_1 + z_1^{-1} + z_2 + z_2^{-1}) \right] X(z_1, z_2)$$

If one could choose $k = 4\lambda$, one would have

$$\bar{x} - \lambda \nabla^2 x = (1 + 4\lambda)x$$

or
$$x(n_1, n_2) = \frac{1}{1 + 4\lambda} [\bar{x}(n_1, n_2) - \lambda \nabla^2 x(n_1, n_2)] \qquad (9\text{-}194)$$

However, only the blurred image is available for processing. The cancellation explicit in Eq. (9-194) can be approximated by

$$x(n_1, n_2) \approx \hat{x}(n_1, n_2) = \frac{1}{1 + 4\lambda} [\bar{x}(n_1, n_2) - \lambda \nabla^2 \bar{x}(n_1, n_2)]$$

$$= \frac{1 - \lambda \nabla^2}{1 + 4\lambda} \bar{x}(n_1, n_2) \qquad (9\text{-}195)$$

This last equation demonstrates that by subtracting a scaled Laplacian of the blurred image from the blurred image itself, we can approximately deblur, that is restore, the original image. This simple approach can be contrasted sharply with the complex least-squares deblurring filters in the preceding section.

9.4 SUMMARY

We have described three areas of application of 2D signal processing concepts of substantial current interest: image coding, restoration, and enhancement. Where possible, 2D concepts are linked to more familiar 1D signal theory. We invoked separability—of signal, filter, and decimator—to simplify the mathematics and to ease the transition to two dimensions. The reader can consult the references for the more difficult cases, skimmed or skipped over here. The much more difficult topic of applications of 2D IIR filters to image processing is beyond the scope of this chapter. Thus we described the Wiener FIR filter but not the Kalman IIR structure. A survey of these topics can be found in Jain (1989). Exposition of the theory of 2D Kalman filtering can be found in the seminal papers by Woods and Radevan (1977) and Woods and Ingle (1981).

The topic of image compression is currently of great concern, particularly with regard to potential HDTV applications. We have tried to present the basic ideas and measures of performance. It remains for you, the jury, to test the evidence and return the verdict as to which of these approaches is to be the dominant image-compression technique of the 1990s and beyond.

PROBLEMS

9.1.   Show that for orthogonal transformations, the mean-squared re-
        construction error in transform coding equals the mean-squared
        quantization error; that is, derive Eq. (9-7).

9.2.   Derive Eq. (9-12) and show that it can also be expressed as

$$\sigma_k^2 2^{-2B_k} = C \qquad \text{a constant}$$

where $\qquad C = 2^{-2A} \qquad A = B - \dfrac{1}{2N}\sum_{j=1}^{N} \log \sigma_j^2$

and σ_k^2 is the variance of the coefficient θ_k being quantized.

9.3.   From Eqs. (9-9) and (9-10), the formula for the quantization error
        in θ_k for any quantizer with B_k bits is given by

$$\sigma_{Q_k}^2 = \sigma_k^2 f(B_k) 2^{-2B_k}$$

where $f(B_k)$ depends on the pdf for x and on the specific quantizer,
and σ_k^2 is the variance of θ_k. Show that if $f(B_k)$ is approximately
the same for all k, and if the bits B_k are allocated according to
Eq. (9-12), then $\sigma_{Q_k}^2$ is the same for all k; that is, the quantization
error is the same for all coefficients. (Use the results of Prob. 9.2.)

9.4.   Demonstrate the equivalence of the block diagrams shown in Fig. 9-
        10 (a) and (b).

9.5.   Consider a linear time-invariant system $\{h(n)\}$. Show that the
        input W^{kn} is an eigenfunction of this operator in the sense that

$$W^{kn} * h(n) = H(k)W^{kn}$$

where $H(k) = \text{DFT}\{h(n)\}$ is the associated eigenvalue. (See also
Fig. 9-30.)

9.6.   Prove that the DFT matrix diagonalizes H; that is, show that

$$\tilde{\mathbf{H}}^T = \mathbf{\Phi}^{-1}\mathbf{D}_H\mathbf{\Phi} = \mathbf{\Phi}\mathbf{D}_H^*\mathbf{\Phi}^{-1}$$

From this derive the expression for $\tilde{\mathbf{H}}^T\mathbf{H}$ given by Eq. (9-128) and
(9-129).

9.7.   Evaluate and plot $|F(k)|^2$, where $F(k) = DFT\{f(n)\}$ and $f(n) = \delta(n) - 2\delta(n-1) + \delta(n-2)$.

9.8.   Provide all the intermediate steps required to arrive at Eq. (9-154) for the optimum gain matrix $\mathcal{G}$.

9.9.   Verify the transform gain formula given by Eq. (9-158).

9.10. The down-sampling performed by the checkerboard subsampling lattice is shown in Fig. 9-17(a). Show that the output of the up-sampler can be expressed as the modulation

$$v(n_1, n_2) = f(n_1, n_2)x(n_1, n_2)$$

Show that $f(n_1, n_2) = \frac{1}{2}[1 + (-1)^{n_1 + n_2}]$ and use this result to obtain $V(z_1, z_2)$ in terms of $X(z_1, z_2)$.

9.11. Expand the modulation function $f(n_1, n_2)$ of Eq. (9-73) in a discrete Fourier series. Use this result to derive Eq. (9-74).

9.12. Evaluate explicitly all terms in T, L_1, L_2, and L_3 in Eq. (9-79). Show that the choice of the QMF in Eq. (9-80) makes L_2 and L_3 zero.

9.13. Consider the four-subband analysis section of the SBC shown in Fig. 9-19. Demonstrate that that structure can be represented by the row/column 1D structure of Fig. 9-22 if the 2D filters are separable.

9.14. Hybrid coding problem: Given an $N \times N$ image $x(n_1, n_2)$ defined by Eq. (9-47) and (9-48) with separable autocorrelation functions and equal vertical and horizontal correlation coefficients

$$R(k_1, k_2) = \rho^{|k_1| + |k_2|}$$

Find the Karhunen-Loève transformation Φ used in transforming each column of $x(n_1, n_2)$ by

$$\boldsymbol{\theta}_i = \Phi \boldsymbol{x}_i$$

Show that the components of each vector $\boldsymbol{\theta}_i$ are uncorrelated:

$$E\{\theta_j(m)\theta_k(n)\} = \begin{cases} 0 & m \neq n \\ \lambda_m \rho^{|j-k|} & m = n \end{cases}$$

and evaluate λ_m. Next, show that the row-recursive model for the ith row has the first-order autoregressive form

$$\theta_i(j) = \rho \theta_i(j-1) + \eta_i(j)$$

and $$E\{\eta_i^2(j)\} = (1 - \rho^2)\lambda_i$$

9.15. Consider an image $x(n_1, n_2)$ with separable source model

$$R(k_1, k_2) = \rho^{|k_1| + |k_2|}$$

Let
$$\mathbf{A} = 1/\sqrt{2} \begin{bmatrix} 1 & 1 \\ 1 & -1 \end{bmatrix}$$

be the Hadamard matrix of order 2. With

$$\Theta = \mathbf{A}\mathbf{X}\mathbf{A}^T$$

show that the variances of the transformed coefficients are

$$\begin{bmatrix} E\{\theta_{11}^2\} & E\{\theta_{12}^2\} \\ E\{\theta_{21}^2\} & E\{\theta_{22}^2\} \end{bmatrix} = \begin{bmatrix} (1 + \rho)^2 & 1 - \rho^2 \\ 1 - \rho^2 & (1 - \rho)^2 \end{bmatrix}$$

9.16. Determine a blurring model for an image induced by uniform acceleration in the x direction according to $x(t) = x_0 + \frac{1}{2}at^2$.

REFERENCES

R. L. Baker and R. M. Gray. "Image compression using non-adaptive spatial vector quantization," in *Proc. 16th Asilomar Conf.*, Nov. 1982.

T. Berger. *Rate Distortion Theory*. Englewood Cliffs, N. J.: Prentice-Hall, 1971.

R. J. Clarke. *Transform Coding of Images*. New York: Academic Press, 1985.

L. S. Davis and A. Rosenfeld. "Noise cleaning by iterated local averaging." *IEEE Trans. Man, Cybern.*, vol. SMC-8, pp. 705–710, Sept. 1978.

R. Forchheimer and T. Kronander. "Image coding–from waveforms to animation." *IEEE Trans. ASSP*, vol. ASSP-37, no. 12, pp. 2008–2023, Dec. 1989.

H. Gharavie and A. Tratabai. "Sub-band coding of digital images using two-dimensional quadrature mirror filtering." *Proc. SPIE Conf. Visual Comm. Sig. Proc.*, vol. 707, pp. 51–61, Sept. 1986.

P. Heinonen and Y. Neuvo. "FIR-median hybrid filters." *IEEE Trans. ASSP*, vol. ASSP-35, no. 6, pp. 832–838, June, 1987.

———. "FIR-median hybrid filters with predictive FIR substructures." *IEEE Trans. ASSP*, vol. ASSP-36, no. 6, June, 1988.

Y. Huang and P. M. Schultheiss. "Block quantization of correlated Gaussian random variables." *IEEE Trans. Comm. Sys.*, vol CS-11, no. 3, pp. 289–296, Sept. 1963.

A. K. Jain. "Image data compression: a review." *Proc. IEEE*, vol. 69, no. 3, pp. 349–389, Mar. 1981(a).

———. "Advances in mathematical models for image processing." *Proc. IEEE*, vol. 69, no. 5, pp. 502–508, May 1981(b).

————. *Fundamentals of Digital Image Processing*. Englewood Cliffs, N. J.: Prentice-Hall, 1989.

N. S. Jayant and P. Noll. *Digital Coding of Waveforms*. Englewood Cliffs, N. J.: Prentice-Hall, 1984.

G. Karlsson et al. "Non-separable two dimensional perfect reconstruction filter banks." *Proc. SPIE, Visual Comm.*, vol. 1001, pt. 1, pp. 189–199, Nov. 1988.

Y. H. Lee and S. A. Kassam. "Generalized median filtering and related nonlinear filtering techniques." *IEEE Trans. ASSP*, vol. ASSP-33, no. 3, pp. 672–683, June, 1985.

J. S. Lim. *Two Dimensional Signal and Image Processing*. Englewood Cliffs, N. J.: Prentice-Hall, 1990.

N. Morrison. *Introduction to Sequential Smoothing and Prediction*. New York: McGraw-Hill, 1969.

A. N. Netravali and J. O. Limb. "Picture coding: a review." *Proc. IEEE*, vol. 68, no. 3, pp. 366–406, March, 1980.

A. Niemenen et al. "A new class of detail-preserving filters for image processing." *IEEE Trans. Pattern Rec. and Machine Intel.*, vol. PAMI-9, no. 1, pp. 74–90, Jan. 1987.

A. Papoulis. *Probability, Random Variables, and Stochastic Processes*. New York: McGraw-Hill, 1984.

————. *Systems and Transforms with Applications in Optics*. New York: McGraw-Hill, 1968.

A. Segall. "Bit allocation and encoding for vector sources." *IEEE Trans. Inf. Theory,* vol. IT-22, no. 2, pp. 162–169, Mar. 1976.

D. Slepian. "Least-squares filtering of distorted images." *J. Opt. Soc. Am.*, vol. 57, mo. 7, pp., 918–922, July 1967.

M. M. Sondhi. "Image restoration: the removal of spatially-invariant degradation." *Proc. IEEE*, vol. 60, no. 7, pp. 842–853, Aug. 1972.

M. Vetterli. "Multi-dimensional sub-band coding: some theory and algorithms." *Sig. Proc.*, vol. 6, pp. 97–112, Apr. 1984.

G. Wackersreuther. "On two-dimensional polyphase filter banks." *IEEE Trans. ASSP*, vol. ASSP-34, no. 1, pp. 192–199, Feb. 1986.

J. W. Woods and V. K. Ingle. "Kalman filtering in two dimensions: further results." *IEEE Trans. ASSP*, vol. ASSP-29, pp. 188–197, Apr. 1981.

J. W. Woods and S. D. O'Neil. "Subband coding of images." *IEEE Trans. ASSP*, vol. ASSP-34, no. 5, pp. 1278–1288, Oct. 1986.

J. W. Woods and C. H. Radewan. "Kalman filtering in two dimensions." *IEEE Trans. Inf. Theory*, vol. IT-23, pp. 473–482, July 1977.

PART III

HARDWARE
REALIZATIONS

CHAPTER 10

■ ■

HARDWARE REALIZATIONS

Stanley H. Smith
Stevens Institute of Technology

The development of the microprocessor and recent innovations in very large scale integrated (VLSI) circuits and very high speed integrated Circuits (VHSIC) have led to the inevitable fabrication of specialized processors, dedicated to signal processing. These processors are called digital signal processors (DSPs), and they have some design features that are convenient for signal processing. Microprocessors have been available since 1969, but only relatively recently have processors been made with sufficiently high clock speeds and short instruction execution times to provide for the processing of signals with reasonably high bandwidths. A list of some of the potential applications of DSPs is shown in Table 10.1.

10.1 DATA ACQUISITION SYSTEMS FOR DIGITAL SIGNAL PROCESSING

The hardware realization of a digital signal-processing system depends upon the nature of the input signal. Generally, the input signal is in analog form and must be converted to digital for processing. In order to accomplish this, a system of the type shown in Fig. 10-1 is required. Multiple analog inputs may be serviced by using an analog multiplexer (MUX) at the input, but this will reduce the conversion rate for each analog channel. In order to accomplish the conversion, it is first necessary to limit the bandwidth so that aliasing does not introduce significant errors. (See Chap. 2.) An anti-aliasing filter is used to accomplish this.

Table 10.1. Typical Applications of the DSPs

General-Purpose DSP	Graphics Imaging	Instrumentation
Digital filtering	3-D rotation	Spectrum analysis
Convolution	Robot vision	Function generation
Correlation	Image transmission/compression	Pattern matching
Hilbert transforms	Compression	Seismic processing
Fast fourier transform	Pattern recognition	Transient analysis
Adaptive filtering	Image enhancement	Digital filtering
Windowing	Homomorphic processing	Phase-locked loops
Waveform generation	Work stations	

Voice/Speech	Control	Military
Voice mail	Disk control	Secure communications
Speech vocoding	Servo control	Radar processing
Speech recognition	Robot control	Sonar processing
Speaker verification	Laser printer control	Image processing
Speech enhancement	Engine control	Navigation
Speech synthesis	Motor control	Missile guidance
Text-to-speech	Kalman filtering	Radio frequency modems
Neural networks		Sensor fusion

Telecommunications		Automotive
Echo cancellation	FAX	Engine control
ADPCM telecoders	Cellular telephones	Vibration analysis
Digital PBXs	Speaker phones	Antiskid Brakes
Line repeaters	Digital speech	Adaptive ride control
Channel multiplexing	Interpolation (DSI)	Global positioning
1200 to 9200-bps modems	X.25 packet switching	Navigation
Adaptive equalizers	video conferencing	Voice commands
DTMF encoding/decoding	Spread spectrum	Digital radio
Data encryption	Communications	Cellular telephones

Consumer	Industrial	Medical
Radar detectors	Robotics	Hearing aids
Power tools	Numeric control	Patient monitoring
Digital audio/TV	Security access	Ultrasound equipment
Music synthesizer	Power line monitors	Diagnostic tools
Toys and games	Visual inspection	Prosthetics
Solid-state answering	Lathe control	Fetal monitors
Machines	Cam	NMR imaging

Source: *Texas Instruments TMS320 User's Guide* (SPRU031) (1988), Table 1-1, p. 1-5.

This type of filter is usually implemented in analog form and is generally a low-pass filter to remove the high frequency components from the signal. Note from Chap. 2 that the cutoff frequency for this filter must be less than one-half the sampling frequency.

Analog-to-digital converters (ADCs) generally operate under the premise that the input signal does not change during the conversion process. If the signals change very rapidly, the output from the ADC may

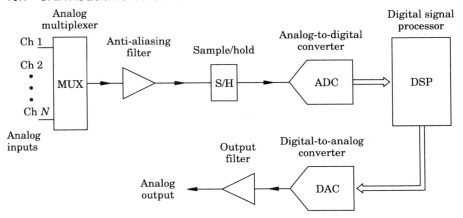

Figure 10-1   *Data acquisition system for digital processing.*

not represent the analog input. To prevent this, a sample/hold (S/H) amplifier is required. This S/H is generally a unity-gain amplifier that samples the signal and holds it during the conversion. The signal is held on an analog memory, a capacitor. Some ADCs have S/H circuits on the chip.

The output of the S/H is sent to an analog-to-digital converter (ADC) for quantization (Chap. 2). The conversion rate (number of conversions per second) must be greater than twice the required bandwidth. Several types of ADCs exist, and each has its application based upon the required conversion rate and the resolution (number of bits). Generally, the higher the conversion rate, the lower the resolution, so that the fastest ADCs are generally limited to 8-bit resolution. The types of ADC that are available for moderate and high conversion rates are the following:

Successive approximation converters [Fig. 10-2(a)] are often called bit-at-a-time converters and use an approximation algorithm to convert the analog signal. This algorithm is basically one that modifies a bit (sets it equal to a logic 1) and uses feedback; that is, the generated digital word is provided to a digital-to-analog converter (DAC) and the output of the DAC is one input of a comparator (the other input contains the input analog signal). The output of the comparator provides the test result. If the DAC output is greater than the input signal, it sets the bit to 0. It starts with the most significant bit of the digital output. This testing process continues until all the bits have been obtained. Note that with this process, the conversion is serial and the results may be outputted serially as well. Generally, n-bit successive approximation converters require as few as $(n + 1)$ clock cycles for conversion.

Tracking converters [Fig. 10-2(b)] utilize counters to generate the digital word. As with the successive approximation type, they utilize feedback to compare the generated digital word (converted to analog with an internal DAC) with the input signal. They are distinguished from other counter types that count up only, down only, or up/down depending

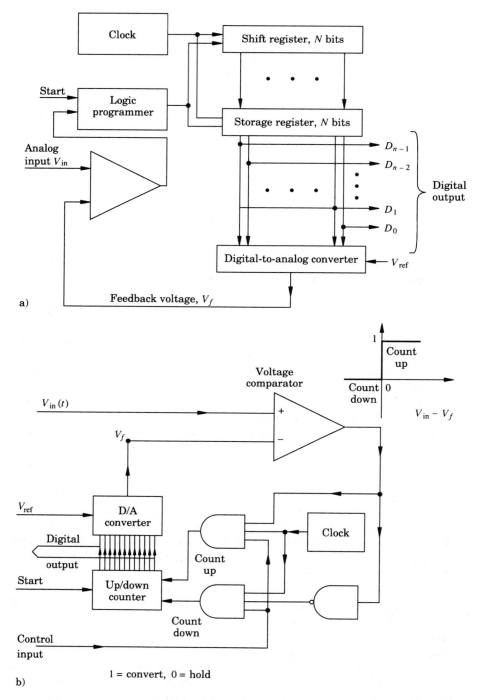

Figure 10-2 *Analog-to-digital conversion methods: (a) successive approximation; (b) tracking.*

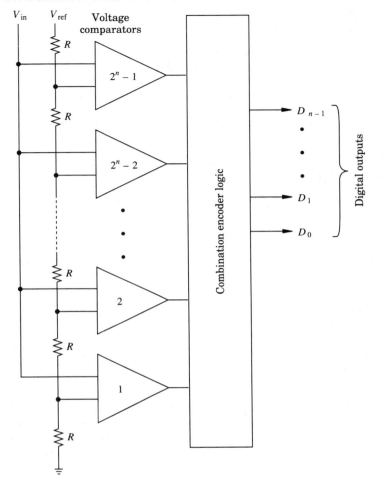

Figure 10-2   *Analog-to-digital conversion methods: (c) flash.*

on the sign of the error in that the tracking converter saves the result of the previous conversion and uses this as the starting point for the next conversion. For continuously varying analog inputs, this can be much faster than the ordinary counter types.

Flash converters [Fig. 10-2(c)] are the fastest type of ADC. They do not require clocks for conversion as do the previous types. Instead they rely on an array of comparators (essentially configured as window detectors) to perform the conversion in parallel. An n-bit flash converter requires $(2^n - 1)$ comparators. For an 8-bit converter, 255 comparators with their associated resistive divider networks are required. The large number of comparators required for longer word lengths makes them prohibitive in size and cost so that most are made as 8-bit converters.

Half-flash converters [Fig. 10-2(d)] provide a compromise between component density (lower) and speed (slower) to the flash converter.

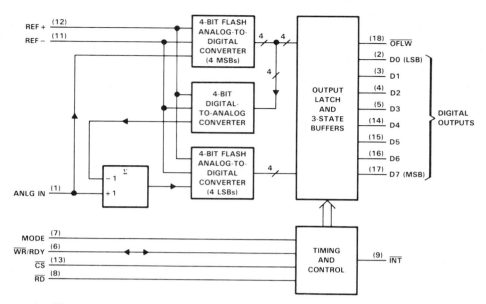

Figure 10-2   *Analog-to-digital conversion methods: (d) half-flash.*

They generally convert half the number of bits (the most significant first) using a flash converter, convert this to analog with a resistor network, subtract this from the input signal, and convert the difference using another flash converter for the least significant bits. They generally contain a S/H circuit internally for the subtraction process required before the second conversion. A typical half-flash converter requires $2^{(1+n/2)}$ comparators so that for example, an 8-bit converter requires only 32 comparators.

While other types of ADCs are available with higher resolution, such as the dual-slope converter, they have very low conversion rates and are not generally used in DSP applications. With most available converters, the higher the resolution the lower the maximum conversion rate. Special converters are available to provide outputs such as the μ-law and A-law quantizations for speech conversion described in Sec. 8.3.1.

The outputs of the converters are generally ordinary binary representations

$$[D] = b_{n-1}2^{-1} + b_{n-2}2^{-2} + \cdots + b_1 2^{-(n-1)} + b_0 2^{-n} \qquad (10\text{-}1)$$

where b_{n-1} (0 or 1) is the most significant bit (MSb) and b_0 is the least significant bit (LSb). This code is used to represent unipolar analog inputs (inputs that are positive). Many converters contain provisions for bipolar analog inputs (which may be positive or negative) and for encoding the digital output. For unipolar codes, either the binary or BCD (binary-coded decimal) codes are used. For bipolar inputs, the codes may be either offset binary, sign magnitude, 2's-complement binary, or signed

BCD. In offset binary, 0 represents the maximum negative value; in sign magnitude, the MSb is the sign and the remainder of the bits represent the magnitude of the input sample in binary. In 2's-complement binary if the input is positive, the MSb is 0 and the remaining bits represent the magnitude; if the input is negative, the bits contain the 2's complement of the magnitude. In signed BCD, the MSb represents the sign and the remainder of the bits represent the magnitude in BCD.

In most digital signal processing applications either ordinary binary (for unipolar inputs) or 2's complement binary (for bipolar inputs) is used. This is because of the fact that digital signal processors and microprocessors are designed to perform unsigned and/or signed arithmetic.

ADCs are generally ratiometric. They require an external reference voltage V_{ref}, although some have built-in references. Regardless of whether the reference is internal or external, the digital output $[D]$ represents the ratio of the analog voltages V_{in} and V_{ref}:

$$[D] = V_{in}/V_{ref} \qquad (10\text{-}2)$$

This reference voltage should be accurate and stable to within the resolution of the converter. In some applications, the use of an external reference voltage has the advantage of not requiring an accurate reference. For example, in a strain-gauge bridge application, the output from the bridge is proportional to its source voltage so that if the source voltage is used as the ADC reference voltage, the output of the ADC is independent of the source voltage.

For accurate waveform reproduction in the absence of a S/H, the conversion rate required by the sampling theorem (Chap. 2) may not be adequate. To illustrate this, assume that the ADC has the following characteristics:

Resolution = n bits.
Conversion time = τ.
Reference Voltage = V_{ref}.

We shall assume that the ADC is unipolar (input voltage is between 0 and $+V_{ref}$) and the input signal utilizes the full dynamic range of the ADC (for greatest accuracy). Therefore, a sinusoidal analog input signal with a frequency f may be expressed as

$$V_{in} = (V_{ref}/2)\sin(2\pi f t) + V_{ref}/2 \qquad 0 \leq V_{in} \leq V_{ref}$$

The maximum rate of change in the input signal is $\pi f V_{ref}$ (the derivative of V_{in} with respect to t, evaluated at $t = 0$). For maximum resolution with no errors (within the resolution of the ADC), the change in this signal during the conversion must not be greater than the resolution of the ADC ($V_{ref}/2^n$); that is, the signal should not change by more than the equivalent of one bit during the conversion process. Therefore,

$$\pi f V_{ref} \tau \leq V_{ref}/2^n$$

This implies that

$$f \leq \frac{1}{\pi \tau 2^n} \qquad\qquad (10\text{-}3)$$

so that using an 8-bit ADC with a conversion time of 1 μs, for example, the input signal should have a frequency less than 1,243 Hz for maximum waveform conversion accuracy.

The control of the system may be performed by the digital signal processor, microprocessor, or digital logic. For the system shown in Fig. 10-1, the sequence of digital control signals, including delays to allow analog signals to settle, will generally be the following:

Input channel select—to the multiplexer (MUX)
Sample—to the S/H amplifier to sample the signal
Delay—to allow for MUX and S/H to settle
Hold—to the S/H amplifier to store the analog signal
Start Conversion—to the Analog-to-Digital Converter
Delay—to allow time for conversion or wait for ADC end-of-conversion
    signal
Read ADC data

Digital signal processing (DSP) hardware is discussed in the following sections. If analog output is required, a digital-to-analog converter (DAC) is used (Fig. 10-1). DACs consist of a resistive ladder network and switches that are controlled by the digital input (Fig. 10-3). As with ADCs, DACs require a voltage or current reference that may be internal or, more often, external. DACs are also characterized by the digital input and analog output requirements; for example, a four-quadrant multiplying DAC implies that the analog reference and analog output may be both positive and negative, whereas a two-quadrant DAC generally requires a positive reference but the output may be positive and negative (bipolar). In general, with an input reference I_{ref} and digital input $[D]$ as in Eq. (10-1), the output current may be written as (Fig. 10-3)

$$I_{out1} - I_{out2} = I_{ref}[D]$$

It may be necessary to convert the output to a voltage. This is generally performed with an operational amplifier. The reference current may be generated by using a resistance in series with a reference voltage.

The DAC in Fig. 10-3 is a multiplying DAC with internal digital input data latches for processor applications. It may be used as unipolar (two-quadrant multiplication) or bipolar (four-quadrant multiplication).

The output of the DAC generally must be smoothed and this may be done with a low-pass interpolating filter (Chap. 2).

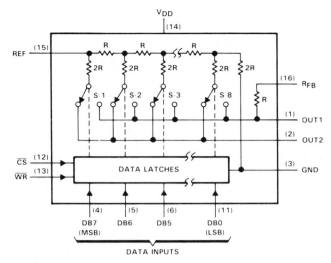

Figure 10-3 *8-bit multiplying digital-to-analog converter.*

10.2 INTRODUCTION TO DIGITAL SIGNAL PROCESSING HARDWARE

The types of digital VLSI hardware the digital signal processing engineer has to consider fall primarily into the following categories:

Microprocessors

Single-chip microcomputers or microcontrollers

Digital signal processors (DSPs)

Reduced-instruction-set computers (RISC)

Specialized, custom logic—for example, application-specific integrated circuits (ASIC)

The first four items in this list are not mutually exclusive, since they all involve the use of microprocessors. In general, microprocessors may be considered to be central processing units which contain ALUs (arithmetic and logical units), control, and a limited number of internal registers. They require external or peripheral devices: RAM (random-access memory—read/write) for storage of data (operands), ROM (read-only memory) for program storage, and I/O (input/output) devices. Single-chip microcomputers contain all of these elements of a computer on one chip. In addition to parallel I/O ports, some contain serial I/O and internal timers (for example, Motorola's MC6801, MC68HC11, some MC68HC05 versions, and Intel's i8051). Some also have internal analog-to-digital converters (for example, the MC68HC11, some MC68H05 versions, the i8098, and OKI's MSM66301). Digital signal processors are single-chip microcomputers with special features as described in the next section.

RISC processors operate with a reduced, relatively elementary set of instructions to perform operations very fast—in at most one or two clock cycles. Some of these are also available with floating-point coprocessors to extend the instruction set. While a detailed discussion of these is beyond the scope of this chapter, new RISC processors have been announced, such as Motorola's 88000, Intel's 80860, and SUN's SPARC series, that may outperform some of the current DSPs. Generally, the development of RISC computers involves many components and consequently, they are not as easy to design and fabricate as is a single-chip DSP. Fortunately the need for high-speed processors and DSP integrated circuits is still great, and the development of new devices continues.

The last category includes VLSI devices that are developed for specific digital signal-processing applications and specific functions for other processors. These include, for example, multiplier-accumulators, FIR filters, histogramming and Hough-transform chips for image processing, and address generators and coefficient (twiddle-factor) generators for FFTs.

10.3 WHAT DSPS ARE AND HOW THEY DIFFER FROM CONVENTIONAL MICROPROCESSORS AND SINGLE-CHIP MICROCOMPUTERS

Digital signal processors are basically high-speed, single-chip microcomputers. Their internal architectures are different, and in most cases, instruction execution times are considerably shorter than in conventional microprocessors. DSP chips such as the TMS 320 series from Texas Instruments have an internal architecture which is called a Harvard-type architecture. They have two separate address spaces for program and data memory, which permit an overlap of instruction fetch and execution. That is, they have separate internal buses for instructions and data, which allow them to fetch both simultaneously instead of sequentially and thereby reduce instruction execution times. Perhaps the most distinctive characteristic of DSPs is that they have a single-instruction multiply-and-accumulate (MAC) operation. This is particularly useful for digital signal processing, since most algorithms involve the sum-of-products arithmetic operation. Other features are described with the specific DSPs considered in the following sections.

The material presented in the chapter is current as of this date and should be useful for future developments. Only some of the available DSP chips are described in this chapter. The selection has been limited to DSP chips with on-chip capabilities for floating-point arithmetic, with the exception of the TMS32025, which performs integer arithmetic primarily. The ability to perform floating-point arithmetic enables the insertion of coefficients and computations over a wide dynamic range without some of the effects of truncation or rounding errors that are more prevalent in integer processors. It also alleviates problems of data scaling while performing computations. This chapter is provided as an overview, and it is not intended to be comprehensive. For further information, the reader may consult the manufacturers and the references at the end of this chapter.

The DSPs considered in this chapter are from Texas Instruments, AT&T, and Motorola. A comparison of some of their features is shown in Table 10.2. An overview and the comparative performance of a large number of DSPs using benchmark programs may be found in Shear (1988) and in Lee (1988, 1989). A more detailed description is given

Table 10.2. Summary of some characteristics of Texas Instruments, AT&T, and Motorola DSPs

Company	Part	Intro. date	MAC time (ns)	No. Bits fixed pt.	No. Bits float. pt.	Inst. cache	Internal memory	External space
	TMS32010	1982	390	16/32	N/A	N/A	144W RAM (D) 1.5KW ROM (P)	4KW (P)
Texas Instruments	TMS32020	1985	195	16/32	N/A	1W	288W RAM (D) 256W ROM (D/P)	64KW (D) 64KW (P)
Texas	TMS32025	1987	97	16/32	N/A	1W	288W RAM (D) 256W RAM (D/P) 4KW ROM (P)	64W (D) 64KW (P)
	TMS320C30	1988	60	24/32	32/40	64W	(2)1KW RAM (D) 4KW ROM (D/P)	16MW (D/P)
	DSP16	1987	55	16/36	N/A	15W	512 RAM (D) 2KW ROM (D/P)	64KW (P)
AT&T	DSP16A	1988	33	16/36	N/A	15W	2KW RAM (D) 4KW ROM (D/P)	64KW (P)
	DSP32	1984	160	16	32/40	N/A	(2)512W RAM (D) 512W ROM (P)	14KW (D/P)
	DSP32C	1988	80	16 or 24	32/40	N/A	(2)512W RAM (D) 512W RAM (P) or 1KW ROM (P)	4MW (D/P)
Motorola	DSP56001	1987	74.1	24/56	N/A	N/A	(2)256W RAM (D) (2)256W ROM (D) 512W RAM (P)	(3)64KW (D/P)
	DSP96002	1989	75 (est.)	32/64	44/96	N/A	512W RAM (P) (2)512W RAM (D) (2)512W ROM (D)	4MW (D/P)

NOTES:
For the number of bits, where denoted by a/b, a is the number of bits available for multiplier operands and b is number of bits for the adder or ALU operations
N/A means not available.
(D) denotes DATA memory, (P) denotes PROGRAM memory, and (D/P) denotes either usage.
W denotes words. KW is kilowords where 1K = 1,024.
Source: Modified and corrected by S. H. Smith from E. A. Lee, "Programmable DSP Architecture," *IEEE AASP MAgazine*, vol. 5, no. 4, pp. 4–19, Oct. 1988.

for the TMS320C25 and the TMS320C30 in TMS320C25 User's Guide (1986), Digital Signal Processing Applications with the TMS320 Family (1986), Third-Generation TMS320 User's Guide (1988), and Lin (1987).

10.4 THE TEXAS INSTRUMENTS TMS320 FAMILY OF DSPS

The TMS family of digital signal processors from Texas Instruments, of which the TMS32010 was one of the first of the dedicated digital signal processors configured as single-chip microcomputers with on-chip RAM, mask-programmable ROM, and I/O ports. A summary of this family is shown in Table 10.3. Applications involving the TMS32010, TMS32020, and TMS320C25 may be found in Digital Signal Processing (1986) and in Lin (1987). In the following section, only the TMS320C25 and the TMS320C30 are considered as examples of fixed-point (integer) and floating-point processors, respectively.

10.4.1 TMS320C25 FEATURES AND ARCHITECTURE

The functional block diagram of the TMS320C25 is shown in Fig. 10-4. One of the main features is that it contains two separate buses. This configuration is called a Harvard-type architecture, since it has some features that differ from the basic Harvard architecture. For example, variations include fetching operands from the program memory and supplementing the program/data memory with an instruction cache. The program bus carries the instruction codes and immediate operands. The data bus connects elements such as the central arithmetic and logic unit (CALU) and the auxiliary register file (AR0-AR7) to the data RAM. For additional speed, it can execute instructions in parallel in both the CALU (the grouping of the ALU, multiplier, accumulator, and scaling shifter) and the auxiliary register arithmetic unit (ARAU).

The TMS320C25 also has a pipelined architecture so that the current instruction is executed while the next instruction is decoded and the one following that is prefetched. The number of cycles may vary, depending upon whether the next data fetch is from internal or external memory, with the highest throughput being obtained when the data are in the on-chip memory.

Some of the other key features of the TMS320C25 family are:

100-ns instruction cycle time (An 80-ns version, the TMS320C25-50, has been announced.)

544 bytes of on-chip data RAM

4K words of on-chip masked ROM [An EPROM (erasable programmable read-only memory) version, the TMS320E25, has been announced.]

128K words of data/program space

Single-cycle multiply/accumulate instructions

Table 10.3. TMS320 Family of Digital Signal Processors (Courtesy, Texas Instruments, Inc.

Generation	Device	Technology	Cycle time (ns)	Typical power (W)	Data type	RAM	ROM	EPROM	Cache	Total memory space	Parallel	Serial	DMA	Timers	High-level language
1st	TMS32010	NMOS	200	.9	16-bit integer	144	1.5K	-	-	4K	8	-	-	-	3rd party
1st	TMS3102010-14	NMOS	320	.9	16-bit integer	144	1.5K	-	-	4K	8	-	-	-	3rd party
1st	TMS3102010-25	NMOS	160	.9	16-bit integer	144	1.5K	-	-	4K	8	-	-	1	3rd party
1st	TMS3102011	NMOS	200	.9	16-bit integer	144	1.5K	-	-	1.5K	6	-	-	-	3rd party
1st	TMS320C10	CMOS	200	.165	16-bit integer	144	1.5K	-	-	4K	8	-	-	-	3rd party
1st	TMS320C10-25	CMOS	160	.2	16-bit integer	144	1.5K	-	-	4K	8	2	-	-	3rd party
1st	TMS320E15	CMOS	200	.3	16-bit integer	256	-	4K	-	4K	8	-	-	-	3rd party
1st	TMS320C15	CMOS	200	.225	16-bit integer	256	4K	-	-	4K	8	-	-	-	3rd party
1st	TMS320C15-25	CMOS	160	.25	16-bit integer	256	4K	-	-	4K	8	-	-	-	3rd party
1st	TMS320dE17	CMOS	200	.325	16-bit integer	256	-	4K	-	4K	6	2	-	1	3rd party
1st	TMS320dC17	CMOS	200	.25	16-bit integer	256	4K	-	-	4K	6	2	-	1	3rd party
1st	TMS320dC17-25	CMOS	160	.275	16-bit integer	256	4K	-	-	4K	6	2	-	1	3rd party
2nd	TMS32020	NMOS	200	1.5	16-bit integer	544	-	-	-	128K	16	1	§	1	C compiler
2nd	TMS320C25	CMOS	100	1.0	16-bit integer	544	4K	-	-	128K	16	1	§	1	C compiler
3rd	TMS320C30	CMOS	60	1.0	32-bit F/I†	2K	4K	-	64	16M	‡	2	||	2	C compiler

†Floating point integer   §External DMA
‡Unlimited ||Internal/external DMA
Source: Texas Instruments TMS320C30 DSP Preview Bulletin (SPRT036) (1971), p. 7.

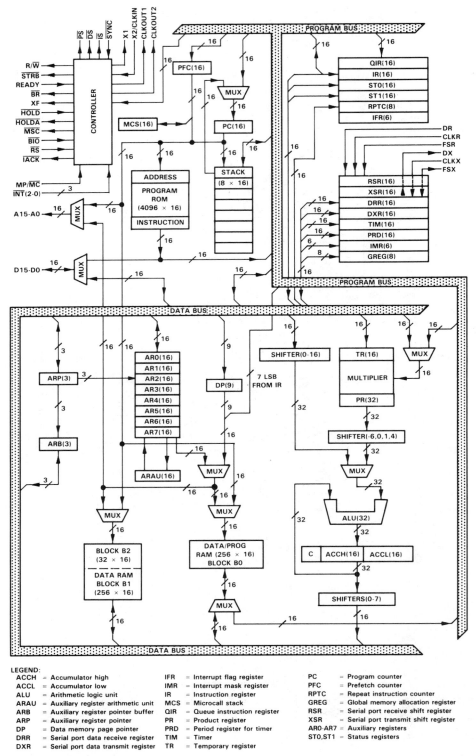

Figure 10-4   *TMS320C25 block diagram. (Courtesy Texas Instruments, Inc.)*

16-bit instruction and data words

32-bit ALU/accumulator

16-bit parallel shifter

Block moves for data/program management

Instructions to support adaptive filtering, FFTs, and extended-precision arithmetic

Bit-reversed indexed addressing mode for radix-2 FFTs

Serial port for direct CODEC interface

Wait states for communication to slow off-chip memories and peripherals

Synchronization capability between multiple processors

On-chip clock generator

On-chip timer for control operations

10.4.2 TMS320C25 ADDRESSING MODES

The TMS320C25 is object-code upward compatible with the TMS32020 so that programs developed for the latter can run unmodified on the TMS320C25. Both assemblers (including cross-assemblers) and C compilers, as well as simulators and emulators, are available for software development.

In addition to the instruction set, the number and types of addressing modes determine the way the processor accesses data and determines the flexibility and efficiency of the instruction set. In the TMS320C25, three memory addressing modes are available: direct, indirect, and immediate; these are shown in Table 10.4. The direct addressing mode is principally to access data. In this mode, 7 bits of the instruction are concatenated with the 9 bits of the data-memory page pointer (DP), as shown in Fig. 10-4, to form a 16-bit address. This permits 512 data memory pages of length 128 to obtain a 64k total data memory space.

Indirect addressing is provided by eight auxiliary registers (AR0 through AR7). These registers contain the addresses of data, and the particular register is selected by inserting its corresponding number in the auxiliary register pointer (ARP). The last two indexed modes permit convenient evaluation of FFTs. These bit-reversed modes allow efficient I/O to be performed for the resequencing of data points in a radix-2 FFT program.

In the immediate addressing mode for short immediate instructions (8- and 13-bit constants), the instruction word contains the value of the immediate operand. For long immediate instructions (16-bit constants), the word following the instruction opcode is used as the immediate operand. Immediate operands therefore reside in the program memory.

Table 10.4.   TMS320C25 Addressing Modes

Addressing Mode	Operation
OP A	Direct addressing.
OP *(,NARP)	Indirect; no change to AR.
OP *+(,NARP)	Indirect; current AR is incremented.
OP *−(,NARP)	Indirect; current AR is decremented.
OP *0+(,NARP)	Indirect; AR0 is added to current AR.
OP *0−(,NARP)	Indirect; AR0 is subtracted from current AR.
OP *BRO+(,NARP)	Indirect; AR0 is added to current AR (with reverse carry propagation).
OP *BRO−(,NARP)	Indirect; AR0 is subtracted from current AR (with reverse carry propagation).

Notes: The optional NARP field specifies a new value of the auxiliary register pointer (ARP). OP is the appropriate operation mnemonic code.

Source: Texas Instruments TMS 320C25 User's Guide (SPRU012) (1986), Table 2-2, p. 2-16

10.4.3 TMS320C25 INSTRUCTION SET

The instruction set of the TMS320C25 is rather comprehensive, and in addition to the instructions available in a conventional microprocessor, it provides more instructions for multiplies and accumulates, barrel shifting, and I/O control. It also provides new instructions for extended-precision arithmetic, adaptive filtering, control, and I/O, and it offers new accumulator and register instructions that were not available in earlier versions of the TMS320 family. For example, two new instructions (MPYA and ZALR) allow a least-mean-squares (LMS) adaptive filter tap and update to be performed in 4 machine cycles. As in the TMS32020, it also has the capability to perform FIR filter computations in a more compact form. The instructions

    RPTK   ⟨constant⟩
    MACD   m1,m2

permit the MACD (multiply-and-accumulate with data move) instruction to be repeated. The repeat operation is specified by the RPTK (repeat instruction as specified by immediate value) instruction, and the number of times it is repeated is determined by the ⟨constant⟩, which can range from 0 to 255.

A description of the symbols used in the table of instructions appears in Table 10.5. A listing of the TMS320C25 instruction set, with a brief description of the operations performed, is given in Table 10.6, which is organized according to function. These tables are provided for reference and to give the reader an indication of the type of instructions that are

Table 10.5.   Instruction Symbols

Symbol	Meaning
ACC	Accumulator
ARB	Auxiliary register pointer buffer
ARn	Auxiliary register n (AR0 through AR7 are predefined assembler symbols equal to 0 through 7, respectively.)
ARP	Auxiliary register pointer
BIO	Branch control input
C	Carry bit
CM	2-bit field specifying compare mode
CNF	On-chip RAM configuration control bit
dma	Data memory address
DP	Data page pointer
FO	Format status bit
FSM	Frame synchronization mode bit
HM	Hold mode bit
INTM	Interrupt mode flag bit
>nn	Indicated nn is a hexadecimal number. (All others are assumed to be decimal values.)
OV	Overflow flag bit
OVM	Overflow mode bit
P	Product register
PA	Port address (PA0 through PA15 are predefined assembler symbols equal to 0 through 15, respectively.)
PC	Program counter
PM	2-bit field specifying P register output shift code
pma	Program memory address
Preg	Product register
RPTC	Repeat counter
STn	Status Register n (ST0 or ST1)
SXM	Sign-extension mode bit
T	Temporary register
TC	Test control bit
TOS	Top of stack
Treg	Temporary register
TXM	Transmit mode bit
Usgn	Unsigned value
XF	XF pin status bit
$\rightarrow$	Is assigned to
||	An absolute value
[]	Optional items
()	Contents of

Source: Texas Instruments TMS320C2S User's Guide (SPRU012) (1986), Table 2-3, p. 2-17.

Table 10.6.   TMS320C25 Instructions

Accumulator memory reference instructions

Mnemonic	Description	No. words	Operation
ABS	Absolute value of accumulator	1	$|(ACC)| \rightarrow ACC$
ADD	Add to accumulator with shift	1	$(ACC) + [(dma) \times 2^{shift}] \rightarrow ACC$
ADDC‡	Add to accumulator with carry	1	$(ACC) + (dma) + (C) \rightarrow ACC$
ADDH	Add to high accumulator	1	$(ACC) + [(dma) \times 2^{16}] \rightarrow ACC$
ADDK‡	Add to accumulator short immediate	1	$(ACC) + 8\text{-bit constant} \rightarrow ACC$
ADDS	Add to low accumulator with sign extension suppressed	1	$(ACC) + (dma) \rightarrow ACC$
ADDT†	Add to accumulator with shift specified by T register	1	$(ACC) + [(dma) \times 2^{(Treg)}] \rightarrow ACC$
ADLK†	Add to accumulator long immediate with shift	2	$(ACC) + [16\text{-bit constant} \times 2^{shift}] \rightarrow ACC$
AND	AND with accumulator	1	$(ACC(15 - 0)), \text{ AND, } (dma) \rightarrow ACC(15 - 0), \quad 0 \rightarrow ACC(31 - 16)$
ANDK†	AND immediate with accumulator with shift	2	$(ACC(30 - 0)). \text{ AND } . [16\text{-bit constant} \times 2^{shift}] \rightarrow ACC(30 - 0), \quad 0 \rightarrow ACC(30 - 0)$
CMPL†	Complement accumulator	1	$\overline{(ACC)} \rightarrow ACC$
LAC	Load accumulator with shift	1	$(dma) \times 2^{shift} \rightarrow ACC$
LACK	Load accumulator immediate short	1	$8\text{-bit constant} \rightarrow ACC$
LACT†	Load accumulator with shift specified by T register	1	$(dma) \times 2^{(Treg)} \rightarrow ACC$
LALK†	Load accumulator long immediate with shift	2	$(16\text{-bit constant}) \times 2^{16} \rightarrow ACC$
NEG†	Negate accumulator	1	$-ACC \rightarrow ACC$
NORM†	Normalize contents of accumulator	1	
OR	OR with accumulator	1	$(ACC(15 - 0)).OR.(dma) \rightarrow ACC(15 - 0)$
ORK†	OR immediate with accumulator with shift	2	$(ACC(30 - 0)).OR.[16\text{-bit constant} \times 2^{shift}] \rightarrow ACC(30 - 0)$
ROL‡	Rotate accumulator left	1	$(ACC(30 - 0)) \rightarrow ACC(31 - 1), (C) \rightarrow ACC(0), \quad (ACC(31)) \rightarrow C$
ROR‡	Rotate accumulator right	1	$(ACC(31 - 1)) \rightarrow ACC(30 - 0), (C) \rightarrow ACC(31), \quad (ACC(0)) \rightarrow C$
SACH	Store high accumulator with shift	1	$[(ACC) \times 2^{shift}] \rightarrow dma$
SACL	Store low accumulator with shift	1	$[(ACCL) \times 2^{shift}] \rightarrow dma$
SBLK†	Subtract from accumulator long immediate with shift	2	$(ACC) - [16\text{-bit constant} \times 2^{shift}] \rightarrow ACC$

†These instructions are not included in the TMS32010 instruction set.
‡These instructions are not included in the TMS32020 instruction set.
Source: *Texas Instruments TMS320C25 User's Guide* (SPRU012), Table 2-4, pp. 2-18–2-21

Table 10.6.   TMS320C25 Instructions (continued)

Accumulator memory reference instructions

Mnemonic	Description	No. words	Operation
SFL[†]	Shift accumulator left	1	$(ACC(30-0)) \to ACC(31-1)$, $0 \to ACC(0)$
SFR[†]	Shift accumulator right	1	$(ACC(31-1)) \to ACC(30-0)$, $(ACC(31)) \to ACC(31)$
SUB	Subtract from accumulator with shift	1	$(ACC) - [(dma) \times 2^{shift}] \to ACC$
SUBB[†]	Subtract from accumulator with borrow	1	$(ACC) - (dma) - (\overline{C}) \to ACC$
SUBC	Conditional subtract	1	
SUBH	Subtract from high accumulator	1	$(ACC) - [(dma) \times 2^{16}] \to ACC$
SUBK[‡]	Subtract from accumulator short immediate	1	$(ACC) - $ 8-bit constant $\to ACC$
SUBS	Subtract from low accumulator with sign extension suppressed	1	$(ACC) - (dma) \to ACC$
SUBT[†]	Subtract from accumulator with shift specified by T register	1	$(ACC) - [(dma) \times 2^{(Treg)}] \to ACC$
XOR	Exclusive-OR with accumulator	1	$(ACC(15-0)).XOR.(dma) \to ACC(15-0)$
XORK[†]	Exclusive-OR immediate with accumulator with shift	2	$(ACC(30-0)).XOR.$ [16-bit constant $\times 2^{shift}] \to ACC(30-0)$
ZAC	Zero accumulator	1	$0 \to ACC$
ZALH	Zero low accumulator and load high accumulator	1	$(dma) \times 2^{16} \to ACC$
ZALR[†]	Zero low accumulator and load high accumulator with rounding	1	$(dma) \times 2^{16} + > 8000 \leftarrow ACC$
ZALS	Zero low accumulator and load low accumulator with sign extension suppressed	1	$(dma) \to ACCL, 0 \to ACCH$

Auxiliary Registers and Data Page Pointer Instructions

Mnemonic	Description	No. words	Operation
ADRK[‡]	Add to auxiliary register short immediate	1	$(ARn) + $ 8-bit constant $\to ARn$
CMPR[†]	Compare auxiliary register with auxiliary register AR0	1	If ARn $|CM|AR0$, then $1 \to TC$; else $0 \to TC$
LAR	Load auxiliary register	1	$(dma) \to (ARn)$
LARK	Load auxiliary register short immediate	1	8-bit constant $\to ARn$

[†] These instructions are not included in the TMS32010 instruction set.
[‡] These instructions are not included in the TMS32020 instruction set.

Table 10.6.   TMS320C25 Instructions (continued)

Auxiliary Registers and Data Page Pointer Instructions (continued)

Mnemonic	Description	No. words	Operation
LARP	Load auxiliary register pointer	1	3-bit constant $\rightarrow$ ARP, (ARP) $\rightarrow$ ARB
LDP	Load data memory page pointer	1	(dma) $\rightarrow DP$
LDPK	Load data memory page pointer immediate	1	9-bit constant $\rightarrow DP$
LRLK[†]	Load auxiliary register long immediate	2	16-bit constant $\rightarrow$ ARn
MAR	Modify auxiliary register	1	
SAR	Store auxiliary register	1	(ARn) $\rightarrow$ dma
SBRK[‡]	Subtract from auxiliary register short immediate	1	(ARn) $-$ 8-bit constant $\rightarrow$ ARn

T Register, P Register, And Multiply Instructions

Mnemonic	Description	No. words	Operation
APAC	Add P register to accumulator	1	(ACC) + (shift Preg) $\rightarrow$ ACC
LPH[†]	Load high P register	1	(dma) $\rightarrow$ Preg(31 $-$ 16)
LT	Load T register	1	(dma) $\rightarrow$ Treg
LTA	Load T register and accumulate previous product	1	(dma) $\rightarrow$ Treg, (ACC) + (shifted Preg) $\rightarrow$ ACC
LTD	Load T register, accumulate previous product, and move data	1	(dma) $\rightarrow$ Treg, (dma) $\rightarrow$ dma + 1, (ACC) + (shifted Preg) $\rightarrow$ ACC
LTP[†]	Load T register and store P register in accumulator	1	(dma) $\rightarrow$ Treg, (shifted Preg.) $\rightarrow$ ACC,
LTS[†]	Load T register and subtract previous product	1	(dma) $\rightarrow$ Treg, (ACC) $-$ (shifted Preg) $\rightarrow$ ACC
MAC[†]	Multipy and accumulate	2	(ACC) + (shifted Preg) $\rightarrow$ ACC (pma) $\times$ (dma) $\rightarrow$ Preg
MACD[†]	Multiply and accumulate with data move	2	(ACC) + (shifted Preg) $\rightarrow$ ACC (pma) $\times$ (dma) $\rightarrow$ Preg, (dma) $\rightarrow$ dma h + 1
MPY	Multiply (with T register, store product in P register)	1	(Treg) $\times$ (dma) $\rightarrow$ Preg
MPYA[‡]	Multiply and accumulate previous product	1	(ACC) + (shifted Preg) $\rightarrow$ ACC (Treg) $\times$ (dma) $\rightarrow$ Preg
MPYK	Multiply immediate	1	(Treg) $\times$ 13-bit constant $\rightarrow$ Preg
MPYS[‡]	Multiply and subtract previous product	1	(ACC) $-$ (shifted Preg) $\rightarrow$ ACC, (Treg) $\times$ (dma) $\rightarrow$ Preg
MPYU[‡]	Multiply unsigned	1	Usgn(Treg) $\times$ Usgn(dma) $\rightarrow$ Preg
PAC	Load accumulator with P register	1	(shifted Preg) $\rightarrow$ ACC

[†] These instructions are not included in the TMS32010 instruction set.
[‡] These instructions are not included in the TMS32010 instruction set.

Table 10.6.   TMS320C25 Instructions (continued)

T Register, P Register, And Multiply Instructions (continued)

Mnemonic	Description	No. words	Operation
SPAC	Subtract P register from accumulator	1	$(ACC) - $ (shifted Preg) $\rightarrow$ ACC
SPH‡	Store high P register	1	(shifted Preg$(31 - 16)$) $\rightarrow$ dma
SPL†	Store low P register	1	(shifted Preg$(15 - 0)$) $\rightarrow$ dma
SPM†	Set P register output shift mode	1	2-bit constant $\rightarrow$ PM
SQRA†	Square and accumulate	1	$(ACC) + $ (shifted Preg) $\rightarrow$ ACC, (dma) $\times$ (dma) $\rightarrow$ Preg
SQRS†	Square and subtract previous product	1	$(ACC) - $ (shifted Preg) $\rightarrow$ ACC (dma) $\times$ (dma) $\rightarrow$ Preg

Branch/Call Instructions

Mnemonic	Description	No. words	Operation
B	Branch unconditionally	2	pma $\rightarrow$ PC
BACC†	Branch to address specified by accumulator	1	$(ACC(15 - 0)) \rightarrow$ PC
BANZ	Branch on auxiliary register not zero	2	If (AR(ARP)) $\neq 0$, then pma $\rightarrow$ PC ; else (PC) $+ 2 \rightarrow$ PC
BBNZ†	Branch if TC bit $\neq 0$	2	If (TC) $= 1$, then pma $\rightarrow$ PC ; else (PC) $+ 2 \rightarrow$ PC
BBZ†	Branch if TC bit $= 0$	2	If (TC) $= 0$, then pma $\rightarrow$ PC; else (PC) $+ 2 \rightarrow$ PC
BC‡	Branch on carry	2	If (C) $= 1$, then pma $\rightarrow$ PC; else (PC) $+ 2 \rightarrow$ PC
BGEZ	Branch if accumulator ≥ 0	2	If (ACC) ≥ 0, then pma $\rightarrow$ PC; else (PC) $+ 2 \rightarrow$ PC
BGZ	Branch if accumulator > 0	2	If (ACC) > 0, then pma $\rightarrow$ PC; else (PC) $+ 2 \rightarrow$ PC
BIOZ	Branch on I/O status $= 0$	2	If ($\overline{BIO}$) $= 0$, then pma $\rightarrow$ PC; else (PC) $+ 2 \rightarrow$ PC
BLEZ	Branch if accumulator ≤ 0	2	If (ACC) ≤ 0, then pma $\rightarrow$ PC; else (PC) $+ 2 \rightarrow$ PC
BLZ	Branch if accumulator < 0	2	If (ACC) < 0, then pma $\rightarrow$ PC; else (PC) $+ 2 \rightarrow$ PC
BNC‡	Branch on no carry	2	If (C) $= 0$, then pma $\rightarrow$ PC; else (PC) $+ 2 \rightarrow$ PC
BNV†	Branch if no overflow	2	If (OV) $\neq 0$, then pma $\rightarrow$ PC; else (PC) $+ 2 \rightarrow$ PC
BNZ	Branch if accumulator $\neq 0$	2	If (ACC) $\neq 0$, then pma $\rightarrow$ PC; else (PC) $+ 2 \rightarrow$ PC
BV	Branch on overflow	2	If (OV) $= 0$, then pma $\rightarrow$ PC; else (PC) $+ 2 \rightarrow$ PC
BZ	Branch if accumulator $= 0$	2	If (ACC) $= 0$, then pma $\rightarrow$ PC; else (PC) $+ 2 \rightarrow$ PC
CALA	Call subroutine indirect	1	$(ACC(15 - 0)) \rightarrow$ PC , (PC) $+ 1 \rightarrow$ TOS
CALL	Call subroutine	2	(PC) $+ 2 \rightarrow$ TOS, pma $\rightarrow$ PC
RET	Return from subroutine	1	(TOS) $\rightarrow$ PC

Table 10.6.   TMS320C25 Instructions (continued)

I/O And Data Memory Operations

Mnemonic	Description	No. Words	Operation
BLKD[†]	Block move from data memory to data memory	2	(dma1, addressed by PC) → dma2
BLKP[†]	Block move from program to data memory	2	(pma, addressed by PC) → dma
DMOV	Data move in data memory	1	(dma) → dma + 1
FORT[†]	Format serial port registers	1	1-bit constant → FO
IN	Input data from port	1	(data bus, addressed by PA) → dma
OUT	Output data to port	1	(dma) → data bus, addressed by PA
RFSM[†]	Reset serial port frame synchronization mode	1	0 → FSM
RTXM[†]	Reset serial port transmit mode	1	0 → TXM
RXF[‡]	Reset external flag	1	0 → XF
SFSM[‡]	Set serial port frame synchronization mode	1	1 → FSM
STXM[†]	Set serial port transmit mode	1	1 → TXM
SXF[†]	Set external flag	1	1 → XF
TBLR	Table read	1	(pma, addressed by ACC (15 − 0)) → dma
TBLW	Table write	1	(dma) → pma, addressed by ACC (15-0)

Control Instructions

Mnemonic	Description	No. words	Operation
BIT[†]	Test bit	1	(dma bit at (16-bit code)) → TC
BITT[†]	Test bit specified by T register	1	(dma bit at 15-Treg)) → TC
CNFD[†]	Configure block as data memory	1	0 → CNF
CNFP[†]	Configure block as program memory	1	1 → CNF
DINT	Disable interrupt	1	1 → INTM
EINT	Enable interrupt	1	0 → INTM
IDLE[†]	Idle until interrupt	1	(PC) + 1 → PC, powerdown
LST	Load status register ST0	1	(dma) → ST0
LST1[†]	Load status register ST1	1	(dma) → ST1
NOP	No operation	1	(PC) + 1 → PC
POP	Pop top of stack to low accumulator	1	(TOS) → ACC
POPD[†]	Pop top of stack to data memory	1	(TOS) → dma
PSHD[†]	Push data memory value onto stack	1	(dma) → TOS
PUSH	Push low accumulator onto stack	1	(ACCL) → TOS

Table 10.6.   TMS320C25 Instructions (continued)

Control Instructions (continued)

Mnemonic	Description	No. words	Operation
RC‡	Reset carry bit	1	$0 \rightarrow C$
RHM‡	Reset hold mode	1	$0 \rightarrow HM$
ROVM	Reset overflow mode	1	$0 \rightarrow OVM$
RPT†	Repeat instruction as specified by data memory value	1	$(dma) \rightarrow RPTC$
RPTK†	Repeat instruction as specified by data immediate value	1	8-bit constant $\rightarrow RPTC$
RSXM†	Reset sign-extension mode	1	$0 \rightarrow SXM$
RTC‡	Reset test/control flag	1	$0 \rightarrow TC$
SC‡	Set carry bit	1	$1 \rightarrow C$
SHM‡	Set hold mode	1	$1 \rightarrow HM$
SOVM	Set overflow mode	1	$1 \rightarrow OVM$
SST	Store status register ST0	1	$ST0 \rightarrow dma$
SST1†	Store status register ST1	1	$ST1 \rightarrow dma$
SSXM†	Set sign-extension mode	1	$1 \rightarrow SXM$
STC‡	Set test/control flag	1	$1 \rightarrow TC$
TRAP†	Software interrupt	1	$(PC) + 1 \rightarrow TOS, 30 \rightarrow PC$

† These instructions are not included in the TMS32010 instruction set.
‡ These instructions are not included in the TMS32020 instruction set.

available. For a considerably more detailed description of these instructions, the reader should refer to TI (1986a), from which these tables were obtained.

10.4.4 EXAMPLE OF TMS320C25 SIGNAL-PROCESSOR USE

Many examples of the use of the TMS320 family exist. For example, applications of the TMS32010 and TMS32020 are given in Digital Signal Processing (1986) and Lin (1987). As stated in Sec. 10.4.2, the TMS320C25 is object-code compatible with the TMS32020 and source-code compatible with the TMS32010, so that the examples in these references can be applied to the TMS320C25, although not as efficiently as when the program is written using the enhanced instructions. For many DSP applications, the input data are provided from the analog world to the digital system through an A/D converter. For the examples included here, only the digital portion is considered. Generally, benchmark programs (programs meant to evaluate the relative performance of DSPs) consider the FFT and FIR and IIR filters. A description of some benchmarks and their performance with many DSPs appears in Shear (1988).

In this section, only one application will be demonstrated, the software for a FIR filter. The response $y(n)$ of a FIR filter to input samples $\{x(n)\}$ may be given by

$$y(n) = \sum_{i=0}^{N-1} x(n-i)h(i) \qquad (10\text{-}1)$$

where N is the length of the filter and $h(i)$ are the filter coefficients. As an example, a demonstration program is presented. This program was written for a 40-term ($N = 40$) FIR band-pass filter with cutoff frequencies at 0.3 and 0.6 times one-half the sampling frequency, using a Hamming window. The frequency response of this filter is shown in Fig. 10-5.

The program for this FIR filter was assembled using the Texas Instruments cross-assembler for the IBM PC; the assembled program listing is shown in Fig. 10-6. In this program, control for an A/D converter is provided. The basic system is illustrated in Fig. 10-1 where the clock for the ADC and the start conversion may be determined by the DSP software and hardware. (See Sec. 10.1.) This clock determines the conversion rate. The end of conversion (from the ADC) is sensed when the BIO pin on the DSP is brought low (to a logic 0). The input sample $x(n)$ from the ADC is input through port 2 (PA2). The output $y(n)$ to a DAC is provided through the same port. Since the ADC is accessed with a *read* instruction and the DAC data are output with a *write* instruction, the same port may be used provided the read/write (R/W) output pin from the DSP is used to select the ADC or the DAC, respectively.

A brief description of this program follows. Note that text on lines beginning with an asterisk (*) are comments, as is all text to the right of a semicolon.

The first column contains the line numbers referring to the source program; the second column contains the address (hexadecimal) in memory in which the instruction will be placed; and the third column contains the machine code instruction in hexadecimal. The remaining columns are the assembly language (source) program, consisting of label, operation, operand(s), and comments.

Lines 39-40 define the memory (RAM) addresses for the filter output $y(n)$ and the data input x_i, respectively.

Line 42 defines the absolute origin (AORG) as 0 for the cross-assembler.

Line 43 is an unconditional branch to the start of the program (START).

Line 53 defines the beginning of the coefficient table (CTABLE) at memory location 32 (20 in hexadecimal).

Lines 56-95 define the filter coefficients. (> denotes a hexadecimal number.)

Line 96 indicates the start of the program.

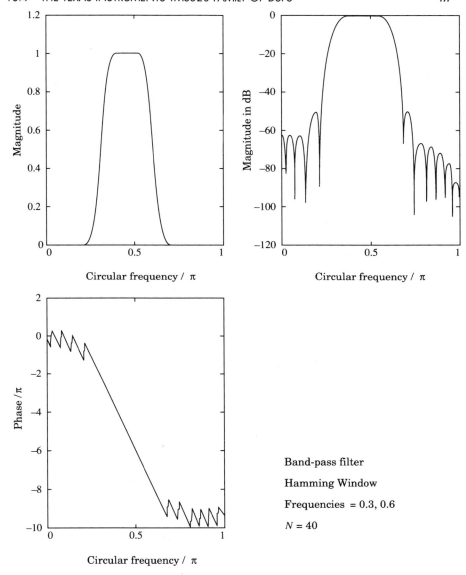

Figure 10-5   *FIR filter response.*

Line 104 defines the auxiliary register AR0 for indirect addressing.

Line 106 loads the address 200 (in hexadecimal) into AR0. This address marks the start of the area where the filter coefficients will be moved.

Line 107 sets the processor to repeat the next instruction 40 times.

Line 108 moves a word of data from the coefficient table in the program memory and places it in the data memory. The address in AR0 is incremented by 1.

```
FIRDEMO     32020 FAMILY MACRO ASSEMBLER   PC 1.0 85.157

0001          ************************************************************
0002          *
0003          *        Finite-Impulse-Response (FIR) Bandpass Filter
0004          *                   Using Hamming Window
0005          *
0006          *  Program courtesy of Texas Instruments, Inc. (Modified)
0007          *
0008          *        Sampling frequency = 10 KHz - Determined externally.
0009          *
0010          *           FILTER CHARACTERISTICS (See text)
0011          *
0012          *        Lower design cut-off frequency  =  1.5 KHz.
0013          *        Upper design cut-off frequency  =  3.0 KHz.
0014          *
0015          ************************************************************
0016          ************************************************************
0017          *
0018          *                     FILTER STRUCTURE
0019          *
0020          *                         N-1
0021          *                 y(n) = Σ  h(i)•x(i)
0022          *                         i=0
0023          *
0024          *              -1        -1       -1              -1
0025          *            z         z        z              z
0026          * o——>——o——>——o——>——o——> - - —o——>—┐
0027          * x(n)   |       |        |           |       |
0028          *        |       |        |           |       |
0029          *      | h(0)  | h(1)  | h(2)       | h(N-2) | h(N-1)
0030          *        V       V        V           V       V
0031          *        |       |        |           |       |
0032          *        └——>——o——>——o——> - - —o——>——o——>——o
0033          *                                                    y(n)
0034          *
0035          ************************************************************
0036          ************************************************************
0037          *
0038                        IDT 'FIRDEMO'
0039    002D  YN    EQU     45       ;Output memory address - y(n).
0040    002E  XI    EQU     46       ;Input sample memory address.
0041          *
0042 0000              AORG 0
0043 0000 FF80         B    START
     0001 0048
0044          *                    FILTER COEFFICIENT TABLE
0045          * Coefficients are signed, 2's complement numbers, scaled
```

Figure 10-6 *Assembled FIR program for the TMS320C25.*

```
0046                 * by multiplying decimal values by 100,000.
0047                 *
0048                 * Note: scale depends on the signal input Analog-to-Digital
0049                 * Converter (ADC) reference voltage, output Digital-to-Analog
0050                 * Converter (DAC) voltage reference and desired output level
0051                 * requirements.
0052                 *
0053 0020           CTABLE  AORG    32
0054                 *
0055                 *               HEX         DECIMAL
0056 0020 FFD2      CH0     DATA    >FFD2   * -0.00046 *
0057 0021 0064      CH1     DATA    >0064   *  0.00100 *
0058 0022 0142      CH2     DATA    >0142   *  0.00322 *
0059 0023 FF89      CH3     DATA    >FF89   * -0.00119 *
0060 0024 FDA7      CH4     DATA    >FDA7   * -0.00601 *
0061 0025 FFD8      CH5     DATA    >FFD8   * -0.00040 *
0062 0026 0064      CH6     DATA    >0064   *  0.00100 *
0063 0027 FF00      CH7     DATA    >FF00   * -0.00256 *
0064 0028 05C7      CH8     DATA    >05C7   *  0.01479 *
0065 0029 073F      CH9     DATA    >073F   *  0.01855 *
0066 002A F6C4      CH10    DATA    >F6C4   * -0.02364 *
0067 002B F403      CH11    DATA    >F403   * -0.03069 *
0068 002C 036A      CH12    DATA    >036A   *  0.00874 *
0069 002D FDC1      CH13    DATA    >FDC1   * -0.00575 *
0070 002E 018A      CH14    DATA    >018A   *  0.00394 *
0071 002F 298A      CH15    DATA    >298A   *  0.10634 *
0072 0030 0F5C      CH16    DATA    >0F5C   *  0.03932 *
0073 0031 AE39      CH17    DATA    >AE39   * -0.20935 *
0074 0032 C87F      CH18    DATA    >C87F   * -0.14209 *
0075 0033 5828      CH19    DATA    >5828   *  0.22568 *
0076 0034 5828      CH20    DATA    >5828   *  0.22568 *
0077 0035 C87F      CH21    DATA    >C87F   * -0.14209 *
0078 0036 AE39      CH22    DATA    >AE39   * -0.20935 *
0079 0037 0F5C      CH23    DATA    >0F5C   *  0.03932 *
0080 0038 298A      CH24    DATA    >298A   *  0.10634 *
0081 0039 018A      CH25    DATA    >018A   *  0.00394 *
0082 003A FDC1      CH26    DATA    >FDC1   * -0.00575 *
0083 003B 036A      CH27    DATA    >036A   *  0.00874 *
0084 003C F403      CH28    DATA    >F403   * -0.03069 *
0085 003D F6C4      CH29    DATA    >F6C4   * -0.02364 *
0086 003E 073F      CH30    DATA    >073F   *  0.01855 *
0087 003F 05C7      CH31    DATA    >05C7   *  0.01479 *
0088 0040 FF00      CH32    DATA    >FF00   * -0.00256 *
0089 0041 0064      CH33    DATA    >0064   *  0.00100 *
0090 0042 FFD8      CH34    DATA    >FFD8   * -0.00040 *
0091 0043 FDA7      CH35    DATA    >FDA7   * -0.00601 *
0092 0044 F89B      CH36    DATA    >F89B   * -0.00119 *
0093 0045 0142      CH37    DATA    >0142   *  0.00322 *
0094 0046 0064      CH38    DATA    >0064   *  0.00100 *
0095 0047 FFD2      CH39    DATA    >FFD2   * -0.00046 *
```

Figure 10-6   *Assembled FIR program for the TMS320C25.*

```
0096      0048    START    EQU      $
0097                       *
0098                       * NOTE: It is assumed that the Analog-to-Digital converter is run
0099                       *       by an external clock which is set for the required
0100                       *       conversion rate.
0101                       *
0102                       * LOAD FILTER COEFFICIENTS
0103                       *
0104 0048 5588             LARP     AR0              ; Use AR0 for indirect
0105                       *                         ;   addressing.
0106 0049 D000             LRLK     AR0,>200         ; Point to block B0.
     004A 0200
0107 004B CB27             RPTK     >27              ; 40 coefficients (0 to 39).
0108 004C FCA0             BLKP     CTABLE,*+
     004D 0020
0109                       *
0110 004E CE05             CNFP                      ; Use block B0 as program area.
0111                       *
0112 004F FA80    WAIT     BIOZ     NXTPT            ; BIO pin goes low when a new
     0050 0053
0113 0051 FF80             B        WAIT             ;    sample is available (ADC
     0052 004F
0114                       *                         ;    end-of-conversion output).
0115                       *
0116 0053 822E    NXTPT    IN       XI,PA2           ; Get new input sample x(i).
0117                       *
0118 0054 D100             LRLK     AR1,>3FF         ; Point to the bottom of block
     0055 03FF
0119 0056 5589             LARP     AR1              ;    B1.
0120                       *
0121 0057 A000             MPYK     0
0122 0058 CA00             ZAC
0123                       *
0124 0059 CB27             RPTK     >27              ; Repeat 40 times.
0125 005A 5C90             MACD     >FF00,*-
     005B FF00
0126                       *
0127 005C CE15             APAC
0128 005D 692D             SACH     YN,1
0129                       *
0130 005E E22D             OUT      YN,PA2           ; Output the filter response
0131                       *                                     y(n).
0132 005F FF80             B        WAIT             ; Go get the next point.
     0060 004F
0133                       *
0134                       *                         End of demonstration program.
0135                       END
NO ERRORS, NO WARNINGS
```

LINKER OUTPUT

```
K0000FIR       90000BFF80B00489002BBFFD2B0064B0142BFF89BFDA7BFFD87F1DFF    FIR 0001
B0064BFF00B05C7B073FBF6C4BF403B036ABFDC1B018AB298AB0F5CBAE39BC87F7F0BCF    FIR 0002
B5828B5828BC87FBAE39B0F5CB298AB018ABFDC1B036ABF403BF6C4B073FB05C77F0C4F    FIR 0003
BFF00B0064BFFD8BFDA7BF89BB0142B0064BFFD2B5588BD000B0200BCB27BFCA07F0CBF    FIR 0004
B0020BCE05BFA80B0053BFF80B004FB822EBD100B03FFB5589BA000BCA00BCB277F118F    FIR 0005
B5C90BFF00BCE15B692DBE22DBFF80B004F7F7A0F                                 FIR 0006
:     FIR                        XLNKPC   v2.4    85.158                   FIR 0007
```

Figure 10-6   *Assembled FIR program for the TMS320C25.*

Line 110 configures the on-chip RAM block 0 as program memory.

Line 112 tests to see whether the BIO pin on the DSP has gone low, signalling an end-of-conversion from the ADC. If the pin is low, the program continues on line 116; otherwise the next instruction is executed.

Line 113 is an unconditional branch to line 112.

Line 116 reads a new value of the input sample from the ADC (port PA2) and places it in the memory location defined by the symbol XI (line 40).

Line 118 loads AR1 with the number 3FF.

Line 119 defines AR1 for indirect addressing in Block 1.

Line 121 multiplies the contents of the T register by 0 and places the result in the P register (that is, clears the P register).

Line 122 clears the accumulator.

Line 124 sets the processor to repeat the next instruction 40 times.

Line 125 multiplies a data-memory value in the address specified in AR1 by a program-memory value and adds the result to the previous value in the accumulator. Both addresses are decremented by 1.

Line 127 shifts the contents of the P register and adds it to the contents of the accumulator.

Line 128 copies the contents of the accumulator to a shifter, then shifts the entire 32-bit number by 1, and stores the result in the data memory whose address is defined by YN (the output).

Line 130 outputs the contents of YN to port PA2.

Line 132 returns program control to line 112 to get the next input sample. The program will continue to acquire and process new data until the processor is reset.

The last set of data is the linker output. This contains the complete machine-code set of instructions and data that may be down-loaded to the processor.

10.4.5 TMS320C30 FEATURES AND ARCHITECTURE

The TMSC320C30 provides an example of a DSP with floating-point capability that may be compared with the fixed-point TMS320C25. The TMS320C30 satisfies the need for more accuracy than was available on the TMS320C25 without the need to scale numerical values or to be concerned about overflows. As may be seen from Table 10.2, this DSP also provides for a much larger external memory space (16MW) and a larger internal data memory. In addition, the TMS320C30 provides for direct memory access (DMA) internally as well as externally. The DMA feature allows faster data transfers than is possible with the instruction set.

Some of the key features of the TMS320C30 are

60-ns single-cycle instruction execution time; 16.7 MIPS (million instructions per second); 33.3 MFLOPS (million floating-point operations per second) (These figures are due to the fact that some instructions, such as multiply-accumulate (MAC), are performed in one instruction cycle.)

One 4k block of 32-bit single-cycle, dual-access, on-chip ROM

Two 1k blocks of 32-bit single-cycle, dual-access, on-chip RAM

64 × 32-bit instruction cache

32-bit instruction and data words, 24-bit addresses

40/32-bit floating-point/integer multiplier and ALU

32-bit barrel shifter

Eight extended-precision accumulators

Two address generators with eight auxiliary registers and two auxiliary-register arithmetic units

On-chip DMA (direct memory access) controller for concurrent I/O and CPU operations

Integer, floating-point, and logical operations

Two- and three-operand instructions

Parallel ALU and multiplier instructions in a single cycle

Block repeat capability

Zero-overhead loops with single-cycle branches

Conditional calls and returns

Interlocked instructions for multiprocessing support

Two serial ports to support 8/16/32-bit transfers

Two 32-bit timers

Two general-purpose external flags and four external interrupts

A block diagram of the architecture is shown in Fig. 10-7. As with the TMS320C25, a Harvard-type architecture is used. The multiplier operates in a single cycle and performs multiplications on 24-bit integer and 32-bit floating-point operands. To provide even higher throughput, multiplication and arithmetic and logic unit (ALU) operations may be performed in a single cycle. For floating-point multiplication the product of two 32-bit operands is represented as a 40-bit floating-point number, whereas for 24-bit integer multiplication, the product is represented as a 32-bit number.

The ALU performs single-cycle arithmetic and logical operations on 32-bit integers and 40-bit floating-point data. The barrel shifter is used to shift up to 32 bits up (multiplication by powers of 2) or down (division by powers of 2) in a single cycle. Two auxiliary-register arithmetic units (ARAU0 and ARAU1) can generate two addresses in a single cycle and operate in parallel with the multiplier and ALU. A CPU register file provides 28 multiport registers, which are listed and briefly described in Table 10.7. The multiplier and ALU can operate on all these registers.

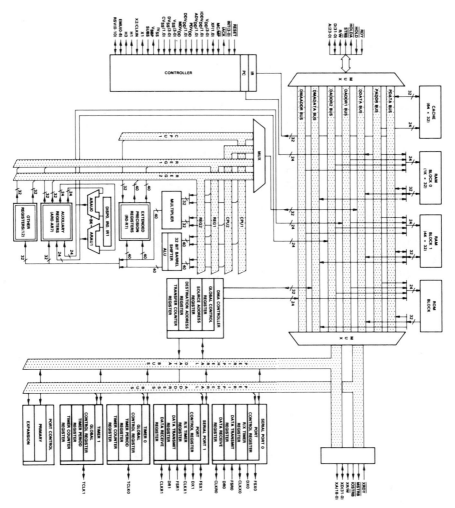

Figure 10-7   *TMS320C30 block diagram. (Courtesy Texas Instruments, Inc.)*

The internal bus structure contributes to the high performance of the TMS320C30. Two program buses (PADDR and PDATA), three data buses (DADDR1, DADDR2, and DDATA), and two DMA buses (DMAADDR and DMADATA) allow for parallel program fetches, data accesses, and direct memory access (DMA) for fast data or program fetches. These buses connect on- and off-chip memory and on-chip peripherals, such as the two timers and the two serial ports.

10.4.6 TMS320C30 ADDRESSING MODES AND INSTRUCTION SET

For comparison with the TMS320C25 (Sec. 10.4.2) and for reference purposes, the addressing modes are shown in Table 10.8. It may be seen

Table 10.7. CPU Registers

CPU register address	Assembler syntax	Assigned function
00h	R0	Extended-precision register
01h	R1	Extended-precision register
02h	R2	Extended-precision register
03h	R3	Extended-precision register
04h	R4	Extended-precision register
05h	R5	Extended-precision register
06h	R6	Extended-precision register
07h	R7	Extended-precision register
08h	AR0	Auxiliary register
09h	AR1	Auxiliary register
0Ah	AR2	Auxiliary register
0Bh	AR3	Auxiliary register
0Ch	AR4	Auxiliary register
0Dh	AR5	Auxiliary register
0Eh	AR6	Auxiliary register
0Fh	AR7	Auxiliary register
10h	DP	Data page pointer
11h	IR0	Index register 0
12h	IR1	Index register 1
13h	BK	Block size
14h	SP	Active stack pointer
15h	ST	Status register
16h	IE	CPU/DMA interrupt enable
17h	IF	CPU interrupt flags
18h	IOF	I/O flags
19h	RS	Repeat start address
1Ah	RE	Repeat end address
1Bh	RC	Repeat counter

Source: *Texas Instruments TMS320 User's Guide* (SPRU031) (1988), Table 3-1, p.3-6.

that there are considerably more addressing modes on the TMS320C30; this improves the efficiency of the instruction set. The instruction set is also more comprehensive than that of the TMS320C25 (Sec. 10.4.3) and is shown in alphabetical order in Table 10.9. A legend of symbols used for addressing and registers also appears at the bottom of each table. From Table 10.9, it may be seen that some operations may be performed

Table 10.8.   TMS320C30 Addressing Modes

General addressing modes
        Register. The operand is a CPU register.
        Short immediate. The operand is a 16-bit immediate value.
        Direct. The operand is the contents of a 24-bit address.
        Indirect. An auxiliary register indicates the address of
                the operand.
Three-operand addressing modes
        Register. Same as for general addressing mode.
        Indirect. Same as for general addressing mode.
Parallel addressing modes
        Register. The operand is an extended precision register.
        Indirect. Same as for the general addressing mode.
Long-immediate addressing mode
        Long-immediate. The operand is a 24-bit immediate value
Conditional branch addressing modes
        Register. Same as for general addressing mode.
        PC-relative. A signed 16-bit displacement is added to the PC
                (program counter)

Source: *Texas Instruments TMS320 User's Guide* (SPRU031) (1988), p. 3-12.

in parallel, for example arithmetic operations with store and multiply with add or subtract.

10.4.7 EXAMPLE OF TMS320C30 SIGNAL PROCESSOR USE

To provide a brief illustration of the software for the TMS320C30, a short program is provided in Figure 10-8. This program is different from that shown for the TMS320C25 (Sec. 10.4.4) in that it is written as a subroutine and the input data are assumed to be already in memory. It should be noted how short the program is because of the addressing modes and instruction set.

It should also be noted that because of the parallel operations available in this architecture, it is possible to have a multiply in parallel with an addition or subtraction and stores in parallel with any multiply or ALU operation. The latter permits processing data while acquiring new data. This has particular significance when performing a fast Fourier transform in which all the data are present in memory. New data may be acquired while processing the previous data set thereby providing almost continuous spectral analysis (depending upon the sampling rate and other processing requirements).

Table 10.9.   TMS320C30 Instruction Set Summary

Mnemonic	Description	Operation
ABSF	Absoute value of a floating-point number	$|src| \rightarrow Rn$
ABSI	Absolute value of an integer	$|src| \rightarrow Dreg$
ADDC	Add integers with carry	$src + Dreg + C \rightarrow Dreg$
ADDC3	Add integers with carry (3-operand)	$src1 + src2 + C \rightarrow Dreg$
ADDF	Add floating-point values	$src + Rn \rightarrow Rn$
ADDF3	Add floating-point values (3-operand)	$src1 + src2 \rightarrow Rn$
ADDI	Add integers	$src + Dreg \rightarrow Dreg$
ADDI3	Add integers (3-operand)	$src1 + src2 + \rightarrow Dreg$
AND	Bitwise logical-AND	$Dreg \text{ AND } src \rightarrow Dreg$
AND3	Bitwise logical-AND (3-operand)	$src1 \text{ AND } src2 \rightarrow Dreg$
ANDN	Bitwise logical-AND with complement	$Dreg \text{ AND } \overline{src} \rightarrow Dreg$
ANDN3	Bitwise logical-ANDN (3-operand)	$src1 \text{ AND } \overline{src2} \rightarrow Dreg$
ASH	Arithmetic shift	If count $\geq$ 0: (Shift Dreg left by count) $\rightarrow$ Dreg Else: (Shift Dreg right by $|count|$) $\rightarrow$ Dreg

LEGEND

src	general addressing modes	Dreg	register address (any register)
src1	three-operand addressing modes	Rn	register address (RO-R7)
src2	three-operand addressing modes	Daddr	destination memory address
Csrc	conditional-branch addressing modes	ARn	auxiliary register n (AR0-AR7)
Sreg	register address (any register)	addr	24-bit immediate address (label)
count	shift value (general addressing modes)	cond	condition code (see Section 11)
SP	stack pointer	ST	status register
GIE	global interrupt enable register	RE	repeat interrupt register
RM	repeat mode bit	RS	repeat start register
TOS	top of stack	PC	program counter

Source: *Texas Instruments TMS320C30 User's Guide* (SPRU031) (1988), Table 3-2, pp. 3-13–3-19

Table 10.9. TMS320C30 Instruction Set Summary (continued)

Mnemonic	Description	Operation
ASH3	Arithmetic shift (3-operand)	If count $\geq$ 0: Shift src left by count) $\rightarrow$ Dreg Else: Shift src right by |count|) $\rightarrow$ Dreg
B*cond*	Branch conditionally (standard)	If cond = true: If Csrc is a register, Csrc $\rightarrow$ PC If Csrc is a value, Csrc + PC $\rightarrow$ PC Else, PC + 1 $\rightarrow$ PC
B*cond*D	Branch conditionally (delayed)	If cond = true: If Csrc is a register, Csrc $\rightarrow$ PC If Csrc is a value, Csrc + PC + 3 $\rightarrow$ PC Else, PC + 1 $\rightarrow$ PC
BR	Branch unconditionally (standard)	Value $\rightarrow$ PC
BRD	Branch unconditionally (delayed)	Value $\rightarrow$ PC
CALL	Call subroutine	PC+1 $\rightarrow$ TOS Value $\rightarrow$ PC
CALL*cond*	Call subroutine conditionally	If cond=true: PC+1 $\rightarrow$ TOS If Csrc is a register, Csrc $\rightarrow$ PC If Csrc is a value, Csrc+PC $\rightarrow$ PC Else, PC+1 $\rightarrow$ PC
CMPF	Compare floating-point values	Set flags on Rn-src
CMPF3	Compare floating-point values (3-operand)	Set flags on src1-src2
CMPI	Compare integers	Set flags on Dreg-src
CMPI3	Compare integers (3-operand)	Set flags on src1-src2
DB*cond*	Decrement and branch conditionally (standard)	ARn-1 $\rightarrow$ ARn If cond=true and ARn $\geq$ 0: If Csrc is a register, Csrc $\rightarrow$ PC If Csrc is a value, Csrc+PC $\rightarrow$ PC Else, PC+1 $\rightarrow$ PC
DB*cond*D	Decrement and branch conditionally (delayed)	ARn-1 $\rightarrow$ ARn If cond=true and ARn $\geq$ 0: If Csrc is a register, Csrc $\rightarrow$ PC If Csrc is a value, Csrc+PC+3 $\rightarrow$ PC Else, PC+1 $\rightarrow$ PC

Table 10.9. TMS320C30 Instruction Set Summary (continued)

Mnemonic	Description	Operation
FIX	Convert floating-point value to integer	Fix (src) → Dreg
FLOAT	Convert integer to floating-point value	Float (src) → Rn
IDLE	Idle until interrupt	PC+1 → PC Idle until next interrupt
LDE	Load floating-point exponent	src(exponent) → Rn(exponent)
LDF	Load floating-point value	src → Rn
LDFcond	Load floating-point value conditionally	If cond=true, src → Rn Else, Rn is not changed
LDFI	Load floating-point value, interlocked	Signal interlocked operation src → Rn
LDI	Load integer	src → Dreg
LDIcond	Load integer, conditionally	If cond=true, src → Dreg Else, Dreg is not changed
LDII	Load integer, interlocked	Signal interlocked operation src → Rn
LDM	Load floating-point mantissa	src(mantisa) → Rn(mantissa)
LSH	Logical shift	If count ≥ 0: (Dreg left-shifted by count) → Dreg Else: (Dreg right-shifted by |count|) → Dreg
LSH3	Logical shift (3-operand)	If count ≥ 0: (src left-shifted by count) → Dreg Else: (src right-shifted by |count|) → Dreg
MPYF	Multiply floating-point values	src × Rn → Rn
MPHF3	Multiply floating-point values (3-operand)	src1 × src2 → Rn
MPYI	Multiply integers	src × Dreg → Dreg
MPYI3	Multiply integers (3-operand)	src × src2 → Dreg
NEGB	Negate integer with borrow	0-src-C → Dreg

Table 10.9. TMS320C30 Instruction Set Summary (continued)

Mnemonic	Description	Operation
NEGF	Negate floating-point value	0-src → Rn
NEGI	Negate integer	0-src → Dreg
NOP	No operation	Modify src if specified
NORM	Normalize floating-point value	Normalize (src) → Rn
NOT	Bitwise logical-complement	$\overline{\text{src}}$ → Dreg
OR	Bitwise logical-OR	Dreg OR src → Dreg
OR3	Bitwise logical-OR (3-operand)	src1 OR src2 → Dreg
POP	Pop integer from stack	*SP-- → Dreg
POPF	Pop floating-point value from stack	*SP-- → Rn
PUSH	Push integer on stack	Sreg → * + + SP
PUSHF	Push floating-point value on stack	Rn → * + + SP
RETI*cond*	Return from interrupt conditionally	If cond=true or missing: *SP-- → PC 1 → ST(GIE) Else, continue
RETS*cond*	Return from subroutine conditionally	If cond=true or missing: *SP-- → PC Else, continue
RND	Round floating-point value	Round (src) → Rn
ROL	Rotate left	Dreg rotated left 1 bit → Dreg
ROLC	Rotate left through carry	Dreg rotated left 1 bit through carry → Dreg
ROR	Rotate right	Dreg rotated right 1 bit → Dreg
RORC	Rotate right through carry	Dreg rotated right 1 bit thru carry → Dreg
RPTB	Repeat block of instructions	src → RE 1 → ST (RM) Next PC → RS

Table 10.9.   TMS320C30 Instruction Set Summary (continued)

Mnemonic	Description	Operation
RPTS	Repeat single instruction	src $\rightarrow$ RC 1 $\rightarrow$ ST (RM) Next PC $\rightarrow$ RS Next PC $\rightarrow$ RE
SIGI	Signal, interlocked	Signal interlocked operation Wait for interlock acknowledge Clear interlock
STF	Store floating-point value	Rn $\rightarrow$ Daddr
STFI	Store floating-point value, interlocked	Rn $\rightarrow$ Daddr Signal end of interlocked operation
STI	Store integer	Sreg $\rightarrow$ Daddr
STII	Store integer, interlocked	Sreg $\rightarrow$ Daddr Signal end of interlocked operation
SUBB	Subract integers with borrow	Dreg-src-C $\rightarrow$ Dreg
SUBB3	Subract integers with borrow (3-operand)	src1-src2-c $\rightarrow$ Dreg
SUBC	Subtract integers conditionally	If Dreg-src $\geq$ 0: [(Dreg-src)<<1] OR 1 $\rightarrow$ Dreg Else, Drg << 1 $\rightarrow$ Dreg
SUBF	Subtract floating-point values	Rn-src $\rightarrow$ Rn
SUBF3	Subtract floating-point values (3-operand)	src1-src2 $\rightarrow$ Rn
SUBI3	Subtract integers	Dreg-src $\rightarrow$ Dreg
SUBI	Subtract integers (3-operand)	src1-src2 $\rightarrow$ Dreg
SUBRB	Subtract reverse integer with borrow	src-Dreg-C $\rightarrow$ Dreg
SUBRF	Subtract reverse floating-point value	src-Rn $\rightarrow$ Rn
SUBRI	Subtract reverse integer	src-Dreg $\rightarrow$ Dreg
SWI	Software interrupt	Perform emulator interrupt sequence

Table 10.9. TMS320C30 Instruction Set Summary (continued)

Mnemonic	Description	Operation
Trap*cond*	Trap conditionally	If cond=true or missing: Next PC $\rightarrow$ * ++ SP Trap vector N $\rightarrow$ PC $0 \rightarrow$ ST (GIE) Else, continue
TSTB	Test bit fields	Dreg AND src
TSTB3	Test bit fields (3-operand)	src1 AND src2
XOR	Bitwise exclusive-OR	Dreg XOR src $\rightarrow$ Dreg
XOR3	Bitwise exclusive-OR (3-operand)	src1 XOR src2 $\rightarrow$ Dreg

	Parallel Arithmetic With Store Instructions	
ABSF || STF	Absolute value of a floating-point	|src2| $\rightarrow$ dst1 || src3 $\rightarrow$ dst2
ABSI || STI	Absolute value of an integer	|src2| $\rightarrow$ dst1 || src3 $\rightarrow$ dst2
ADDF3 || STF	Add floating-point	src1+src2 $\rightarrow$ dst1 || src3 $\rightarrow$ dst2
ADDI3 || STI	Add integer	src1+src2 $\rightarrow$ dst1 || src3 $\rightarrow$ dst2
AD3 || STI	Bitwise logical-AND	src1 AND src2 $\rightarrow$ dst1 || src3 $\rightarrow$ dst2
ASH3 || STI	Arithmetic shift	If count $\geq$ 0: src2 << count $\rightarrow$ dst1 || src3 $\rightarrow$ dst2 Else: src2 >> |count| $\rightarrow$ dst1 || src3 $\rightarrow$ dst2
FIX || STI	Convert floating-point to integer	Fix (src2) $\rightarrow$ dst1 || src3 $\rightarrow$ dst2
FLOAT || STF	Convert integer to floating-point	Float (src2) $\rightarrow$ dst1 || src3 $\rightarrow$ dst2
LDF || STF	Load floating-point	src2 $\rightarrow$ dst1 || src3 $\rightarrow$ dst2

src1 register addr (R0-R7)		src2 indirect addr (disp = 0, 1, IR0, IR1)
src3 register addr (R0-R7)		src4 indirect addr (disp = 0, 1, IR0, IR1)
dst1 register addr (R0-R7)		dst2 indirect addr (disp = 0, 1, IR0, IR1)

Table 10.9. TMS320C30 Instruction Set Summary (continued)

Parallel Arithmetic With Store Instructions

Mnemonic	Description	Operation
LDI || STI	Load integer	src2 → dst1 || src3 → dst2

LSH3
|| STI
	Logical shift	If count ≥ 0:
		  src2 << count → dst1
		|| src3 → dst2
		Else:
		  src2 >> |count| → dst1
		|| src3 → dst2

MPYF3 || STF	Multiply floating-point	src1 × src2 → dst1 || src3 → dst2
MPY13 || STI	Multiply integer	src1 × src2 → dst1 || src3 → dst2
NEGF || STF	Negate floating-point	0-src2 → dst1 || src3 → dst2
NEGI || STI	Negate integer	0-src2 → dst1 || src3 → dst2
NOT3 || STI	Complement	$\overline{src1}$ → dst1 || src3 → dst2
OR3 || STI	Bitwise logical-OR	src1 OR src2 → dst1 || src3 → dst2
STF || STF	Store floating-point	src1 → dst1 || src3 → dst2
STI || STI	Store integer	src1 → dst1 || src3 → dst2
SUBF3 || STI	Subtract floating-point	src1-src2 → dst1 || src3 → dst2
SUBI3 || STI	Subtract integer	src1-src2 → dst2 || src3 → dst2
XOR3 || STI	Bitwise exclusive-OR	src1 XOR src2 → dst1 || src3 → dst2

src1 register addr (R0-R7)   src2 indirect addr (disp = 0, 1, IR0, IR1)
src3 register addr (R0-R7)   src4 indirect addr (disp = 0, 1, IR0, IR1)
dst1 register addr (R0-R7)   dst2 indirect addr (disp = 0, 1, IR0, IR1)
op3   register addr (R0 or R1)   op6   register addr (R2 or R3)
op1, op2, op4, op5—Two of these operands must be specified using register addr
and two must be specified using indirect

Table 10.9. TMS320C30 Instruction Set Summary (continued)

Parallel Load Instructions

Mnemonic	Description	Operation
LDF || LDF	Load floating-point	src2 → dst1 || src4 → dst2
LDI || LDI	Load integer	src2 → dst1 || src4 → dst2

Parallel Multiply and Add Subtract Instructions

MPYF3 || ADDF3	Multiply and add floating-point	op1 × op2 → op3 || op4+op5 → op6

Parallel Multiply and Add Subtract Instructions (continued)

MPYF3 || SUBF3	Multiply and subtract floating-point	op1 × op2 → op3 || op4-op5 → op6
MPYI3 || ADDI3	Multiply and add integer	op1 × op2 → op3 || op4+op5 → op6
MPYI3 || SUBI3	Multiply and subtract integer	op1 × op2 → op3 || op4-op5 → op6

10.5 ALTERNATIVE DSP HARDWARE APPROACHES

Alternative hardware approaches for digital signal processing involve the use of microprocessors such as the Motorola 68xx (8 bits) or 680xx (16 and 32 bits) families, the Intel 80xx (8 bits) or 80xxx (16 or 32-bits) or RISC processors (see Sec. 10.2). With these approaches, many additional components (RAM, ROM, timers, and serial and parallel I/O) are required. Also, with the possible exception of RISC processors, the instruction execution times are not short enough to ensure sufficiently high bandwidth digital signal processing. Using these processor systems with peripheral devices such as those described in Sec. 10.5.2 can improve execution times. Generally, microprocessors should be considered for cases where (1) low effective bandwidth signals are to be processed, (2) digital signal processing is required as part of a relatively large system, and (3) low cost is essential. It should be noted that while the DSPs may be operated with external memories (with reduced efficiency), they are essentially single-chip computers with limited on-board ROM and RAM.

Higher speed integer and floating-point operations with microprocessors may be performed using an arithmetic coprocessor such as the Motorola MC68882 or the Intel i80387. Other floating-point processors using the ANSI/IEEE floating-point standard are available for 16- and

TMS320C30 Assembler   Version 1.23
 (c) Copyright 1987, 1989, Texas Instruments Incorporated

```
0001                          *  TITL  FIR FILTER   (From Texas Instruments, Inc.,
0003                          *                            Reference 7)
0004                          *  SUBROUTINE     F I R
0005                          *
0006                          * EQUATION:   y(n)  = h(0) * X(n) + h(1) * x(n-1) +
0007                          *                            ... + h(N-1) * x(n-(N-1))
0008                          *
0009                          * TYPICAL CALLING SEQUENCE:
0010                          *
0011                          *      LOAD      AR0
0012                          *      LOAD      AR1
0013                          *      LOAD      RC
0014                          *      LOAD      BK
0015                          *      CALL      FIR
0016                          *
0017                          *
0018                          * ARGUMENT ASSIGNMENTS:
0019                          *
0020                          *    AR0          ADDRESS  OF h(N-1)
0021                          *    AR1          ADDRESS  OF x(N-1)
0022                          *    RC           LENGTH  OF FILTER  - 2  (N-2)
0023                          *    BK           LENGTH  OF FILTER  (N)
0024                          *
0025                          * REGISTERS USED  AS INPUT: AR0, AR1, RC, BK
0026                          * REGISTERS MODIFIED: R0, R2, AR0, AR1, RC
0027                          * REGISTER    CONTAINING   RESULT:  R0
0028                          *
0029                          *
0030                          *  CYCLES:  11 + (N-1)            WORDS:   6
0031                          *
0032                          *
0033                               .global  FIR
0034                                              ; Initialize  R0:
0035 000000 24e03120    FIR      MPYF3    *AR0++(1),*AR1++(1)%,R0
0036                          *                    ;  h(N-1) * X(n-(N-1)) -> R0
0037 000001 07628000         LDF      0.0,R2       ;  Initialize  R2.
0038                          *
0039                          * FILTER  ( 1 ≤ i < N)
0040                          *
0041 000002 139b001b         RPTS     RC           ; Setup the repeat cycle.
0042 000003 80103120         MPYF3    *AR0++(1),*AR1++(1)%,R0  ;  h(N-1-i) * x(n-(N-1-i)) -> R0
0043                     ||   ADDF3    R0,R2,R2    ;  Multiply and add operation
0044                          *
0045 000004 20800200         ADDF     R0,R2,R0    ;  Add last product
0046                          *
0047                          * RETURN   SEQUENCE
0048                          *
0049 000005 78800000         RETS                 ;  Return
0050                          *
0051                          * end
0052                          *
0053                               .end
```

No Errors, No Warnings

Figure 10-8 *Assembled FIR subroutine for the TMS320C30.*

32-bit processing. The newer 32-bit microprocessors such as the Motorola
MC68040 and Intel i80486 have floating-point coprocessors on the chip
and are challenging some of the RISC machines and DSP processors in
speed.

10.5.1 SINGLE-CHIP MICROCOMPUTERS WITH ON-BOARD ANALOG-TO-DIGITAL CONVERTERS

As a cost effective approach to data acquisition and signal processing, single-chip microcomputers or microcontrollers (MCUs) offer the possibility of having an on-chip analog input multiplexer (MUX), an analog-to-digital converter, ROM or EPROM, RAM, timers, and serial and parallel I/O. Examples of this are the Motorola 8-bit microcontrollers MC68HC11 and MC68HC05 families (see Microprocessor, Microcontroller, and Peripheral Data, 1988). These are available with all of the above features including an 8-bit analog-to-digital converter with up to eight analog inputs. Some, such as the MC68HC05B6, also have analog output in the form of a pulse-width modulation (PWM) in which the output pulse width may be a representation of an analog signal. This MC86HC05 family is not readily expandable in terms of off-chip memory (RAM or ROM) and has a limited amount of on-chip memory.

The MC68HC11 family has all of the above features as well as the capability of off-chip memory expansion with an addressing capability of up to 64k bytes and an on-chip EEPROM (electrically erasable programmable read-only memory). A functional block diagram of the MC68HC11A8 is shown in Fig. 10-9. The processing times are not nearly as short as with DSPs. The fastest clock speed is 2 MHz; however, instructions generally take from two to seven clock cycles so that it is considerably slower than the DSPs. For example, an 8-bit unsigned integer multiply with a 16-bit product takes 10 clock cycles.

Other MCUs such as the Intel i8098 offer 16-bit MCUs with 10-bit ADC and are several times faster than the MC68HC11 although not nearly as fast as the DSPs. Other sources of MCUs include, for example, Texas Instruments, OKI, and Mitsubishi.

10.5.2 EXAMPLES OF DEDICATED VLSI SIGNAL PROCESSORS

Dedicated DSP chips may be used in a digital system or in combination with a microprocessor or RISC computer system. The early processing chips were primarily multipliers to provide fast multiplication for microprocessors that did not provide for this operation in their instruction set or that performed multiplication very slowly. These chips were then expanded to provide for the most often used operations in digital signal processing, the arithmetic sum of products. These multiplier-accumulators (MACs) are very fast (on the order of 50 million operations per second for 16-bit MAC).

More complex digital signal processing function circuits have been developed that perform functions for applications such as filtering, video processing, and image processing. An example of a multibit filter (MFIR) is the L64240 from LSI Logic Corp. (Data Sheets), shown in the functional block diagram in Fig. 10-10. This is a 64-tap high-speed (20 MHz) transversal filter that consists of two 32-tap sections with 8-bit wide coefficients and data. This is a FIR filter, typical of the filters used in image

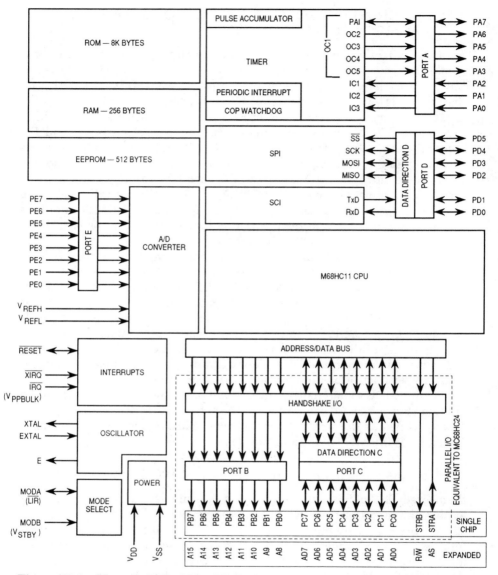

Figure 10-9   *Motorola MC68HC11AB block diagram. (Courtesy Motorola Corp.)*

processing. For image processing, it can be configured for window sizes up to 8 × 8 pixels.

An example of a digital filter for image processing is the median filter, described in detail in Secs. 5.4.1 and 9.2.6. As observed in these sections, linear smoothing methods such as averaging pixel intensities over a window tend to blur edges and other sharp details. Median filtering, which replaces the intensity of each pixel with the median of

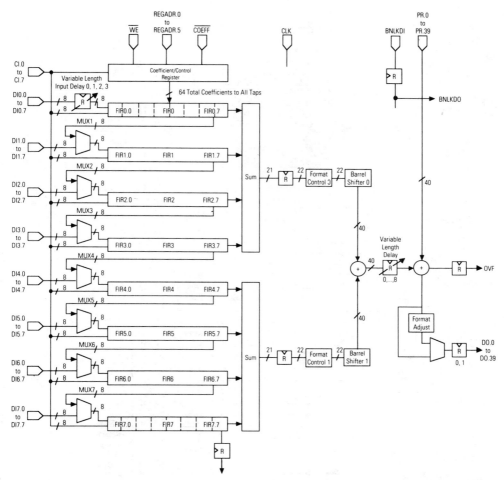

Figure 10-10   *LSI Devices L64240 multibit filter block diagram.(Courtesy LSI Logic Corp.)*

the neighboring intensities, is effective in removing impulse noise while preserving the sharpness of edges. (See also Gonzalez and Wintz, 1987.) One way of performing this in real time, with clock rates up to 20 MHz, is to use a *rank-value filter* (RVF) or rank-order filter (ROF, Sec. 5.4.2) such as the L64220, shown in Fig. 10-11. This RVF is similar to a transversal filter except that the output is chosen from a sorted list of the input values rather than a weighted sum. For example, rank values may include maximum, minimum, or median.

Referring to Fig. 10-11, the CI.x ($x = 0$ to 7) 8-bit inputs are the coefficient/control input data lines, DIY.x ($Y = 0$ to 8 are the data bus inputs, $x = 0$ to 11 are the individual bits) inputs are data inputs, DO.x

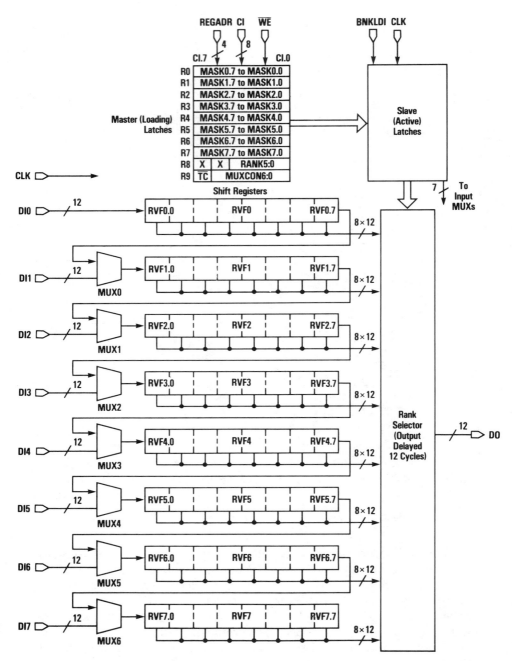

Figure 10-11 *Block diagram of the L64220 rank-value filter. (Courtesy LSI Logic Corp.)*

REGADR	CI7	CI6	CI5	CI4	CI3	CI2	CI1	CI0	
D0	0	0	0	0	0	1	1	1	Mask0
D1	0	0	0	0	0	1	1	1	Mask1
D2	0	0	0	0	0	1	1	1	Mask2
D3	0	0	0	0	0	0	0	0	Mask3
D4	0	0	0	0	0	0	0	0	Mask4
D5	0	0	0	0	0	0	0	0	Mask5
D6	0	0	0	0	0	0	0	0	Mask6
D7	0	0	0	0	0	0	0	0	Mask7
D8	X	X	1	1	1	0	1	1	RANK
D9	1	1	1	1	1	1	1	1	$\overline{TC}$, MUXCON

3×3 Median Filter for a 512×512 Image

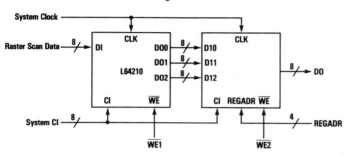

Figure 10-12   *Median filter for image processing using a rank-value filter (RVF). (Courtesy LSI Logic Corp.)*

($x = 0$ to 11) is a 12-bit output bus, REGADR.x ($x = 0$ to 3) are the 4-bit address locations into which the data are provided at CI and the remainder are control lines. This RVF may be used to process a 2×32, 4×16, or 8×8 window with 12-bit resolution.

An example of the realization of a 3×3 median filter for a 512×512 pixel image is shown in Fig. 10-12. In this figure, the L64210 is a *variable-length video shift register* that has four individual 8-bit shift registers, each with a length of up to 1,032. It is set for 512-pixel operation and to output three rows (three outputs are used) of the image. This is accomplished by configuring each shift register for 512 bytes (8 bits). The input to the L64210 (DI) is a raster-scanned, digitized (using an 8-bit ADC) image. Its outputs DO0, DO1, and DO2) contain three lines of a raster-scanned image that are input to the RVF. The output of the RVF (DO, 8 most significant bits, delayed by 13 clock cycles relative to the input data) is the median filter output. The control inputs of the L64220, shown in Fig. 10-12, are set for the coefficient/control registers and mask to configure the RVF as a 3×3 window as shown in the table. The rank was selected to find the median in this window as follows:

$$RANK = 64 - (\text{window size}) + (\text{desired RANK } [0 = \text{minimum}])$$
$$= 64 - (3 \times 3) + 4(\text{median}) = 64 - 9 + 4$$
$$= 59(\text{hexadecimal 3B})$$

Other function circuits from LSI Logic Corp. include a histogram/ Hough transform processor (HHP) for image processing (L64250). The HHP is capable of computing histograms and modified Hough transforms for data sets up to 2^{24} points or $4,096 \times 4,096$ pixels at data rates up to 20 MHz with 9-bit data.

REFERENCES

Data sheets. LSI Logic Corp., n/d.

Digital Signal Processing Applications with the TMS320 Family: Theory, Algorithms, and Implementations, TI Document No. SPRA012A. Texas Instruments, 1986.

M. J. Fuccia et al., "The DSP32C: AT&T's second-generation floating-point digital signal processor." *IEEE Micro*, vol. 8, no. 6, pp. 30–48, Dec. 1988.

R. C. Gonzalez and P. Wintz. *Digital Image Processing*, second edition. Reading. Mass.: Addison-Wesley, 1987.

E. A. Lee. "Programmable DSP Architectures: Part I." *IEEE AASP Magazine*, vol. 5, no. 4, pp. 4–19, Oct. 1988.

——. "Programmable DSP Architectures: Part II." *IEEE AASP Magazine*, vol. 6, no. 1, pp. 4–14, Jan. 1989.

K.-S. Lin, ed. *Digital Signal Processing Applications with the TMS320 Family*, vol. 1. Englewood Cliffs, N. J.: Prentice-Hall, 1987.

Motorola Microprocessor, Microcontroller, and Peripheral Data, vol. 2. Motorola, 1988 .

P. Papamichalis and R. Simar, Jr. "The TMS320C30 floating-point digital signal processor." *IEEE Micro*, vol. 8, no. 6, pp. 13–29, Dec., 1988.

D. Shear. "EDN's DSP Benchmarks." *EDN*, pp. 126–148, Sept. 29, 1988.

G. R. L. Sohie and K. L. Kloker. "A digital signal processor with IEEE floating-point arithmetic." *IEEE Micro*, vol. 8, no. 6, pp. 49–67, Dec. 1988.

Technical Summary: 96-bit General-Purpose Floating-Point Digital Signal Processor (DSP), Document BR575/D. Motorola, 1988.

Third-Generation TMS320 User's Guide, TI Document No. SPRU031. Texas Instruments, 1988.

TMS320C25 User's Guide, TI Document No. SPRU012. Texas Instruments, 1986.

WE DSP32 Digital Signal Processor Information Manual. AT&T, 1986.

WE DSP32-SL Support Software Library User Manual, AT&T, 1987

APPENDIX A

■ ■

DISCRETE-TIME CALCULUS

Experience has shown that while most readers are adept at integrating, differentiating, and expanding functions of a continuous variable, they are ill at ease with the corresponding discrete operations. This appendix tries to bridge the gap between the continuous and discrete domains by reviewing certain features of the calculus of finite differences.

A.1 OPERATORS

The unit delay E^{-1} and unit forward-shift E operators on a discrete-time sequence $\{f(k)\}$ are defined by

$$E^{-1}\{f(k)\} \triangleq \{f(k-1)\} \qquad \text{delay or backward shift}$$
$$Ef(k) \triangleq \{f(k+1)\} \qquad \text{forward shift} \tag{A-1}$$

The braces identify the sequence $\{f(k)\}$ and are often omitted when no ambiguity will result; it will usually be clear from context whether $E\{f(k)\}$ means the forward shift or the expected value.

These shifting operators can also be extended to continuous-time functions $f(t)$ as follows:

$$E^{-1}f(t) \triangleq f(t-T)$$
$$Ef(t) \triangleq f(t+T) \tag{A-2}$$

where T is a constant spacing. In Eq. (A-2) E represents a pure prediction operator, while E^{-1} is realizable in continuous time as a delay line. At sampling instants $t = kT$, Eq. (A-2) reduces to Eq. (A-1).

The forward and backward difference operators are

$$\Delta f(k) \triangleq f(k+1) - f(k) = (E-1)f(k)$$
$$\nabla f(k) \triangleq f(k) - f(k-1) = (1 - E^{-1})\{f(k)\}$$

(A-3)

where 1 is the identity operator.

Higher-order shift and difference operators are defined by iteration; for example,

$$\nabla^2 f(k) \triangleq \nabla\{\nabla f(k)\} = \nabla\{f(k+1) - f(k)\}$$
$$= f(k+2) - 2f(k+1) + f(k)$$
$$E^2 f(k) \triangleq E\{Ef(k)\} = f(k+2)$$

(A-4)

These operators are linear and commutative. For example for any two functions $f(k)$ and $g(k)$ and any two constants a and b, we have

$$\nabla\{af(k) + bg(k)\} = a\nabla f(k) + b\nabla g(k)$$

and

$$\nabla\{Ef(k)\} = E\{\nabla f(k)\}$$

Furthermore, a linear combination of operators, for example $L = aE + b\nabla$, is also linear and commutative and follows the index rule that we associate with exponentiation:

$$L^k L^j = L^{k+j}$$

(A-5)

and consequently

$$L^0 = 1$$

Finally, two operators L_1 and L_2 are equal if and only if

$$L_1 f(k) = L_2 f(k) \qquad \text{for all } f(k)$$

(A-6)

The mth forward and backward difference operators can be expressed as a superposition of forward and backward shift operators, respectively, where the weights are the binomial coefficients:

$$\Delta^m = (E-1)^m = \sum_{j=0}^{m} (-1)^{m+j} \binom{m}{j} E^j$$

$$\nabla^m = (1 - E^{-1})^m = \sum_{j=0}^{m} (-1)^j \binom{m}{j} E^{-j}$$

(A-7)

where $\binom{m}{j}$ is the binomial coefficient

$$\binom{m}{j} = \frac{m!}{j!(m-j)!} \tag{A-8}$$

Hence

$$\Delta^m f(t) = \sum_{j=0}^{m} (-1)^{m+j} \binom{m}{j} E^j f(t)$$

$$= \sum_{j=0}^{m} (-1)^{m+j} \binom{m}{j} f(t + jT) \tag{A-9}$$

$$\nabla^m f(t) = \sum_{j=0}^{m} (-1)^{j} \binom{m}{j} f(t - jT)$$

For $T = 1$ and $k = nT$, these become

$$\Delta^m f(k) = \sum_{j=0}^{m} (-1)^{m+j} \binom{m}{j} f(k + j)$$

$$\tag{A-10}$$

$$\nabla^m f(k) = \sum_{j=0}^{m} (-1)^{j} \binom{m}{j} f(k - j)$$

From a signal-processing point of view, Δ^m is a noncausal operator that generates the weighted sum of $f(t), f(t + T), ..., f(t + mT)$ from the input signal $f(t)$.

The difference operators Δ and ∇ are analogous to the derivative operator $D = d/dt$ in the sense that they reduce the degree of a polynomial. Let $f(t)$ be a polynomial of degree m

$$f(t) = \sum_{j=0}^{m} a_i t^i$$

Successive differentiation of $f(t)$ leads to

$$D^m f(t) = m! a_m$$

$$D^{m+k} f(t) = 0 \qquad k > 0$$

Thus the operator D^m *annihilates* all polynomials of degree $< m$. Similarly, the operators Δ^m and ∇^m annihilate all polynomials of degree $< m$.

Differences of products These are similar to the analog form and can be shown to be

$$\Delta f(k)g(k) = f(k + 1)\Delta g(k) + g(k)\Delta f(k)$$

$$\nabla f(k)g(k) = f(k)\nabla g(k) + g(k - 1)\nabla f(k) \tag{A-11}$$

These relations are used in deriving the formula for summation by parts.

Inverse operators In calculus, $D^{-1}f(t)$ is any function whose derivative is $f(t)$. $D^{-1}f(t)$ is not a unique function, since one can add an arbitrary constant—that is,

$$D(D^{-1}f(t) + C) = D(D^{-1}f(t)) = f(t) \tag{A-12}$$

Accordingly, Δ^{-1} is defined as

$$\Delta(\Delta^{-1}f(k)) \triangleq f(k)$$

or

$$\Delta(\Delta^{-1}) \triangleq 1$$

Similarly, the inverse of the backward difference operator is defined by

$$\nabla(\nabla^{-1}f(k)) \triangleq f(k)$$

A.2 FACTORIAL FUNCTIONS AND EXPANSIONS

In the calculus, the function t^n plays an important rôle by virtue of the fact that $D(t^n) = nt^{n-1}$. In the difference calculus, the *factorial functions* have a similar property. These functions are polynomials in t. The forward factorial function is sketched in Fig. A-1; it is defined as[†]

$$t^{(m)} \triangleq \begin{cases} t(t-1)\cdots(t-m+1) & m \geq 1 \\ 1 & m = 0 \end{cases} \tag{A-13}$$

an mth-degree polynomial in t with uniformly-spaced zeroes at $t = 0, 1, ...,$ $m - 1$; it has the following properties:

1. $\Delta t^{(m)} = mt^{(m-1)}$

2. $\Delta^r t^{(m)} = \begin{cases} m^{(r)}t^{(m-r)}, & r \leq m \\ 0 & r > m \end{cases}$ $\tag{A-14}$

3. $\dfrac{t^{(m)}}{m!} = \binom{t}{m}$

Using the factorial function, we can construct Taylor expansions in forward differences by writing

---

[†]We assume that the time spacing T is normalized to unity. Otherwise we would replace t by t/T everywhere.

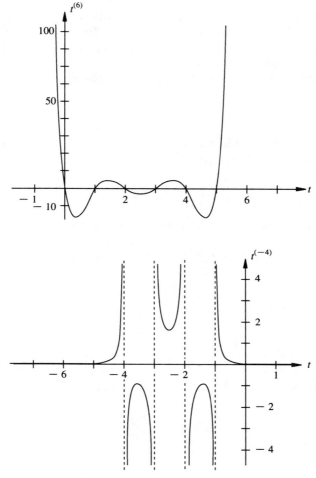

Figure A-1   *Examples of the factorial function: (a)* $t^{(6)}$; *(b)* $t^{(-4)}$.

$$f(t) = \sum_{k=0}^{\infty} \alpha_k t^{(k)}$$

where
$$\alpha_k = \Delta^k f(t)/k! \Big|_{t=0} \tag{A-15}$$

The purely discrete version is obtained by substituting n for t in the foregoing.

When $f(t)$ is a polynomial of degree N, $\Delta^k f(n) = 0$ for $k > N$; hence the Taylor series terminates after N terms and results in

$$f(t) = \sum_{k=0}^{N} \Delta^k f(0) \binom{t}{k} \tag{A-16}$$

This latter expression is Newton's interpolation formula, which asserts that $f(\cdot)$ can be determined at any time t from the $(N + 1)$ samples $f(0), f(1), ..., f(N)$ via Eq. (A-16). [When $f(t)$ is not a polynomial, Eq. (A-16) is an approximate interpolation.]

The factorial function for negative exponents is defined as

$$t^{(-m)} = 1/[(t + m)^{(m)}] \tag{A-17}$$

This function has simple poles at $t = -1, -2, ..., -m$, whereas $t^{(m)}$ has simple zeroes at $t = 0, 1, ..., m - 1$, as indicated in Fig. A-1. Also,

$$\Delta t^{(-m)} = -mt^{(-m-1)} \tag{A-18}$$

Expansions in backward differences are defined similarly. Thus, if

$$t^{[m]} = \begin{cases} t(t + 1) \cdots (t + m - 1) & m > 0 \\ 1 & m = 0 \end{cases} \tag{A-19}$$

with key property

$$\nabla^r t^{[m]} = \begin{cases} m^{(r)} t^{[m-r]} & r \le m \\ 0 & r > m \end{cases} \tag{A-20}$$

then the expansion will be

$$f(t) = \sum_{k=0}^{\infty} \frac{\nabla^k f(0)}{k!} t^{[k]} \tag{A-21}$$

A.3 THE SUMMATION CALCULUS

General techniques for the evaluation of sums are presented in this section.

In integral calculus $\int_a^b f(t)\, dt$ can be evaluated if $f(t)$ can be represented as dF/dt, for then

$$\int_a^b f(t)\, dt = \int_a^b dF(t) = F(b) - F(a)$$

Similarly, $\sum_{t=a}^{b} f(t)$ can be easily determined if $f(t) = \Delta F(t)$:

$$\sum_{t=a}^{b} f(t) = \sum_{t=a}^{b} \Delta F(t) = F(b + 1) - F(a) \tag{A-22}$$

For example,

$$\sum_{n=M}^{N} a^n = \sum_{n=M}^{N} \frac{1}{a-1}\Delta a^n = \frac{1}{a-1}(a^{N+1} - a^M) \qquad \text{(A-23)}$$

For the factorial function,

$$\sum_{n=M}^{N} n^{(m)} = \sum_{n=M}^{N} \frac{1}{m+1}\Delta n^{(m+1)} = \frac{1}{m+1}[(N+1)^{(m+1)} - M^{(m+1)}] \qquad \text{(A-24)}$$

and for negative exponents

$$\sum_{n=M}^{N} t^{(-m)} = \sum_{n=M}^{N} \frac{-1}{m-1}\Delta t^{(-m+1)}$$

$$= \frac{1}{m-1}[(N+1)^{(-m+1)} - M^{(-m+1)}] \qquad \text{(A-25)}$$

The discrete counterpart to integration by parts is summation by parts. Since $\Delta(u_n v_n) = u_n \Delta v_n + v_{n+1}\Delta u_n$,

$$\sum_{n=M}^{N} u_n \Delta v_n = \sum_{n=M}^{N} \Delta(u_n v_n) - \sum_{n=M}^{N} v_{n+1}\Delta u_n$$

$$= u_n v_n \Big|_M^{N+1} - \sum_{n=M}^{N} v_{n+1}\Delta u_n \qquad \text{(A-26)}$$

We conclude this appendix with the following table of summations. The forms listed here are the ones most likely to be encountered in digital signal processing; for more exhaustive lists, consult Gradshteyn et al. (1980), Tuma (1987), or especially Jolley (1925).

REFERENCES

I. S. Gradshteyn et al. *Table of Integrals, Series, and Products*. New York: Academic Press, 1980

L. Jolley. *Summation of Series*. London: Chapman and Hall, 1925. Reprinted New York: Dover Books, 1961.

J. J. Tuma. *Engineering Mathematics Handbook*. New York: McGraw-Hill, 1987.

Table A-1. Sums of Series

---

1. $\displaystyle\sum_{k=0}^{n} a = (n+1)a$

2. $\displaystyle\sum_{k=1}^{n} k = n(n+1)/2$

3. $\displaystyle\sum_{k=1}^{n} k^2 = \frac{n(n+1)(2n+1)}{6} = \frac{n^3}{3} + \frac{n^2}{2} + \frac{n}{6}$

4. $\displaystyle\sum_{k=1}^{n} k^3 = \left[\frac{n(n+1)}{2}\right]^2$

5. $\displaystyle\sum_{k=1}^{n} k^4 = \frac{n(n+1)(2n+1)(3n^2+3n-1)}{30} = \frac{n^5}{5} + \frac{n^4}{2} + \frac{n^3}{3} - \frac{n}{30}$

6. $\displaystyle\sum_{k=0}^{n-1} r^{kx} = \frac{1 - r^{nx}}{1 - r^x}$

7. $\displaystyle\sum_{k=0}^{\infty} r^{kx} = \frac{1}{1 - r^x} \qquad |r^x| < 1$

8. $\displaystyle\sum_{k=0}^{n-1} kr^k = \frac{r(1 - r^n)}{(1-r)^2} - \frac{nr^n}{1-r}$

9. $\displaystyle\sum_{k=0}^{\infty} kr^k = \frac{r}{(1-r)^2} \qquad |r| < 1$

10. $\displaystyle\sum_{k=1}^{\infty} \frac{r^k}{k!} = e^r$

11. $\displaystyle\sum_{k=1}^{\infty} (-1)^{k+1} \frac{x^k}{k} = \ln(x+1) \qquad -1 < x \le 1$

12. $\displaystyle\sum_{k=1}^{n} \sin k\theta = \frac{[\sin(n+1)\theta/2]\sin n\theta/2}{\sin \theta/2}$

13. $\displaystyle\sum_{k=1}^{n} \cos k\theta = \frac{[\cos(n+1)\theta/2]\sin n\theta/2}{\sin \theta/2}$

14. $\displaystyle\sum_{k=1}^{n} \sin(2k-1)\theta = \frac{\sin^2 n\theta}{\sin \theta}$

15. $\displaystyle\sum_{k=1}^{n} \cos(2k-1)\theta = \frac{\sin 2n\theta}{2\sin \theta}$

16. $\displaystyle\sum_{k=1}^{n-1} k\sin k\theta = \frac{\sin n\theta}{4\sin^2 \theta/2} - \frac{n\cos(2n-1)\theta/2}{2\sin \theta/2}$

17. $\displaystyle\sum_{k=1}^{n-1} k\cos k\theta = \frac{n\sin(2n-1)\theta/2}{2\sin \theta/2} - \frac{1-\cos n\theta}{4\sin^2 \theta/2}$

Table A-1.   Sums of Series (*continued*)

18. $\displaystyle\sum_{k=1}^{n+1}(-1)^{k-1}\sin(2k-1)\theta = (-1)^n\frac{\sin(2n+2)\theta}{2\cos\theta}$

19. $\displaystyle\sum_{k=1}^{n}(-1)^k\cos k\theta = -\frac{1}{2}+(-1)^n\frac{\cos(2n+1)\theta/2}{2\cos\theta/2}$

20. $\displaystyle\sum_{k=1}^{n}\sin^2 k\theta = \frac{n}{2}-\frac{\cos(n+1)\theta\sin n\theta}{2\sin\theta}$

21. $\displaystyle\sum_{k=1}^{n}\cos^2 k\theta = \frac{n}{2}+\frac{\cos(n+1)\theta\sin n\theta}{2\sin\theta}$

22. $\displaystyle\sum_{k=1}^{n-1}a^k\sin k\theta = \frac{a\sin\theta(1-a^n\cos n\theta)-(1-a\cos\theta)a^n\sin n\theta}{1-2a\cos\theta+a^2}$

23. $\displaystyle\sum_{k=0}^{n-1}a^k\cos k\theta = \frac{1-a\cos\theta+a^{n+1}\cos(n-1)\theta-a^n\cos n\theta}{1-2a\cos\theta+a^2}$

24. $\displaystyle\sum_{k=1}^{\infty}a^k\sin k\theta = \frac{a\sin\theta}{1-2a\cos\theta+a^2}\qquad a^2<1$

25. $\displaystyle\sum_{k=1}^{\infty}a^k\cos k\theta = \frac{1-a\cos\theta}{1-2a\cos\theta+a^2}\qquad a^2<1$

26. $\displaystyle\sum_{k=1}^{\infty}\frac{\sin k\theta}{k} = \frac{\pi-\theta}{2}\qquad 0<\theta\le 2\pi$

29. $\displaystyle\sum_{k=1}^{\infty}\frac{\cos k\theta}{k} = -\ln(2\sin\theta/2)\qquad 0<\theta\le 2\pi$

27. $\displaystyle\sum_{k=1}^{\infty}\left(\frac{\sin k\theta}{k}\right)^2 = \frac{\theta(\pi-\theta)}{2}\qquad 0<\theta\le\pi$

28. $\displaystyle\sum_{k=1}^{\infty}(-1)^k\frac{\sin k\theta}{k} = -\frac{\theta}{2}\qquad |\theta|<\pi$

APPENDIX B

■ ■

Z AND FOURIER TRANSFORMS

The Z transform was originally introduced in the engineering literature of the 1950s to represent sampled-data control systems. Today, it and discrete-Fourier methods are the foundations of transform representation of digital signals. Definitions and key properties are summarized here.

B.1 THE TWO-SIDED Z TRANSFORM: DEFINITION

For a sequence $\{f(n)\}$ defined on $-\infty < n < \infty$, the two-sided (or bilateral) Z transform is defined as

$$F(z) = \sum_{n=-\infty}^{\infty} f(n)z^{-n} = Z\{f(n)\} \qquad \text{(B-1)}$$

or
$$F(z) = \sum_{n=-\infty}^{-1} f(n)z^{-n} + \sum_{n=0}^{\infty} f(n)z^{-n} \qquad \text{(B-2)}$$

We abbreviate this by writing, $F(z) = Z\{f(n)\}$. The first sum in Eq. (B-2) is a power series in z and converges for $|z| < \rho_2$; the second sum is a series in z^{-1} and converges for $|z| > \rho_1$. Hence Eq. (B-2) converges in the annular region $\rho_1 < |z| < \rho_2$. When the limits are $-\infty$ and ∞ as in Eq. (B-1), we obtain the bilateral or two-sided Z transform. The one-sided Z transform will be discussed in Sec. B.6.

   If the unit circle is within this annular region, we can evaluate $F(z)$ on $z = e^{j\omega}$ and obtain the Fourier transform

$$F(e^{j\omega}) = \sum_{n=-\infty}^{\infty} f(n)e^{-jn\omega} \tag{B-3}$$

B.2   PROPERTIES OF THE Z TRANSFORM

Some key properties of the bilateral Z transform are tabulated below.

1. **Analyticity.** In the region of convergence, $F(z)$ is analytic and converges uniformly and absolutely.

2. **Linearity.** For any constants a and b,

$$Z\{af(n) + bg(n)\} = aZ\{f(n)\} + bZ\{g(n)\} \tag{B-4}$$

3. **Uniqueness.** For each $\{f(n)\}$ there is one and only one $F(z)$, and conversely. [Note that the specification of $F(z)$ must include the region of convergence.]

B.3   INVERSION OF THE Z TRANSFORM

The sequence $\{f(n)\}$ can be obtained from the transform by Laurent series expansion, by contour integration, or by a partial-fraction expansion.

1. **Laurent series.** Let

$$F(z) = F_1(z) + F_2(z) \tag{B-5}$$

where $F_1(z)$ converges for $|z| > \rho_1$ and $F_2(z)$ converges for $|z| < \rho_2$. Expanding $F_1(z)$ in powers of z^{-1} and F_2 in powers of z gives

$$F(z) = \sum_{n=0}^{\infty} a_n z^{-n} + \sum_{n=1}^{\infty} b_n z^n$$

as the Laurent series expansion in $\rho_1 < |z| < \rho_2$. Hence

$$f(n) = \begin{cases} a_n & n \geq 0 \\ b_n & n < 0 \end{cases} \tag{B-6}$$

2. **Partial-fraction expansion.** This method is the most direct route to obtaining a closed-form representation for $\{f(n)\}$ when $F(z)$ is rational in z. For example,

$$F(z) = \frac{z(z-7)}{(z-2)(z-3)} \qquad 2 < |z| < 3$$

is expanded into

$$F(z) = \frac{5z}{z-2} + \frac{4(-z)}{z-3}$$

The ring of convergence implies that the first term corresponds to a positive-time function, while the second is the transform of $f(n)$ for $n < 0$. From a look-up table, we get

$$f(n) = 5 \cdot 2^n \cdot u(n) + 4 \cdot 3^n \cdot u(-n-1)$$

Note that the Fourier transform of this sequence does not exist.

3. **Contour integration.** The contour integral is the most general way of extracting $\{f(n)\}$ from $F(z)$; in fact, it defines the inverse transform. For $F(z)$ convergent in $\rho_1 < |z| < \rho_2$,

$$f(n) = \frac{1}{2\pi j} \oint_P F(z) z^{n-1} \, dz \tag{B-7}$$

where P is a closed contour within the ring of convergence. Then

$$f(n) = \begin{cases} \sum_j R_j & n \geq 0 \\ -\sum_j R'_j & n < 0 \end{cases} \tag{B-8}$$

where $R_j (R'_j)$ is the residue of $F(z)z^{n-1}$ in the jth pole of $F(z)$ inside (outside) of P.

B.4 CONVOLUTION AND PARSEVAL'S THEOREM

The fundamental property of the Z transform (as well as the Laplace and Fourier transforms) is that convolution in the time domain can be replaced by a product of transforms in the frequency domain. Thus

$$y(n) = \sum_{k=-\infty}^{\infty} h(n-k)x(k) \tag{B-9}$$

transforms into $Y(z) = H(z)X(z)$

where X, Y, and H are Z transforms of x, y, and h, respectively.
    On the other hand, if

$$y(n) = x(n)h(n)$$

then $Y(z) = \dfrac{1}{2\pi j} \oint H(\sigma)X\left(\dfrac{z}{\sigma}\right) \dfrac{d\sigma}{\sigma} \qquad \gamma_1 < |z| < \gamma_2$ (B-10) where $\gamma_1 = \alpha_1\beta_1$, $\gamma_2 = \alpha_2\beta_2$, and where (α_1, α_2) and (β_1, β_2) define the rings of convergence for $X(z)$ and $H(z)$, respectively.

If $Y(z)$ in Eq. (B-10) converges on $|z| = 1$, then

$$Y(1) = \sum_{n=-\infty}^{\infty} x(n)h(n) = \frac{1}{2\pi j} \oint_\Gamma H(\sigma)X(1/\sigma)\frac{d\sigma}{\sigma} \qquad \text{(B-11)}$$

where Γ lies in the common ring of convergence. Furthermore, if $h(n) = \bar{x}(n)$, then[†]

$$\sum_{n=-\infty}^{\infty} |x(n)|^2 = \frac{1}{2\pi j} \oint_\Gamma X(\sigma)\overline{X}(1/\bar{\sigma})\frac{d\sigma}{\sigma} \qquad \text{(B-12)}$$

On the unit circle, $\sigma = e^{j\omega}$. Hence the latter expression for the energy of the signal can be written as the integral

$$\sum_{n=-\infty}^{\infty} |x(n)|^2 = \frac{1}{2\pi} \int_{-\pi}^{\pi} |X(e^{j\omega})|^2 d\omega \qquad \text{(B-13)}$$

Equations Eq. (B-12) and (B-13) are two versions of Parseval's theorem.

B.5 TRANSFORMS OF SIGNALS AND OPERATORS, AND OTHER PROPERTIES

We give the transforms of special signals and operators here.

1. The Kronecker delta and unit-step sequences.

$$Z\{\delta(n)\} = 1$$
$$Z\{\delta(n - n_0)\} = z^{-n_0} \qquad \text{(B-14)}$$
$$Z\{u(n)\} = z/(z - 1) \qquad |z| > 1$$

2. Operators. Here we emphasize that these properties are valid for the bilateral transform and not for the one-sided transform. With

$$x(n) \leftrightarrow X(z), \qquad \text{(B-15)}$$

[†]In this appendix the overbar denotes the comples conjugate.

a bilateral transform pair convergent in $\rho_1 < |z| < \rho_2$, we find

$$Z\{x(n+r)\} = z^{+r}X(z) \tag{B-16}$$

Also

$$Z\{\nabla^r x(n)\} = (1 - z^{-1})^r X(z) \tag{B-17}$$

and

$$Z\{\Delta^r x(n)\} = (z - 1)^r X(z) \tag{B-18}$$

3. **Transform differentiation and moments.** From the defining equations implicit in Eq. (B-15), it can be shown that if $x(n) \leftrightarrow X(z)$, then

$$nx(n) \leftrightarrow -z\frac{d}{dz}X(z) \tag{B-19}$$

and

$$n^{[m]}x(n) \leftrightarrow (-1)^m z^m \frac{d^m}{dz^m} X(z) \tag{B-20}$$

Furthermore, if the transforms in Eq. (B-20) converge on $|z| = 1$, we obtain the moments

$$\sum_{n=-\infty}^{\infty} x(n) = X(1) \tag{B-21a}$$

$$\sum_{n=-\infty}^{\infty} n^{[m]}x(n) = (-1)^m \frac{d^m X}{dz^m}\bigg|_{z=1} \tag{B-21b}$$

4. **Modulation.** With $x(n) \leftrightarrow X(z)$,

$$\lambda^n x(n) \leftrightarrow X\left(\frac{z}{\lambda}\right) \qquad \rho_1 < \left|\frac{z}{\lambda}\right| < \rho_2 \tag{B-22}$$

5. **Time reversal.** If $f(n)$ is a given time sequence, then $f(-n)$ is that sequence reversed in time. Let $F(z)$ be $Z\{f(n)\}$ and $F^R(z)$ be $Z\{f(-n)\}$. Then we have

$$F^R(z) = \sum_{n=-\infty}^{\infty} f(-n)z^{-n} = \sum_{k=-\infty}^{\infty} f(k)z^k = \sum_{k=-\infty}^{\infty} f(k)(1/z)^{-k}$$

$$= F(1/z) \tag{B-23}$$

As a corollary of this property, if $f(n)$ is an even function of n, then $f(n) = f(-n)$ and so $F(z) = F^R(z) = F(1/z)$. It is immediately apparent that any pole or zero of $F(z)$ will also be a pole or zero of $F(1/z)$; hence the singularities of $F(z)$ either lie on the unit circle or will occur in reciprocal pairs, one at $z_i e^{j\theta}$ and the other at $1/z_i e^{j\theta}$.

B.6 THE ONE-SIDED Z TRANSFORM

This transform is used mainly in solving transient problems and difference equations, when signals are turned on at $t = 0$. It is defined as

$$X_I(z) = \sum_{n=0}^{\infty} x(n)z^{-n} \qquad |z| > \rho \qquad \text{(B-24)}$$

denoted as $X_I(z) \leftrightarrow x(n)$

or $X_I(z) = Z_I\{x(n)\}$

The modifications to earlier results are as follows:

$$Z_I\{x(n-1)\} = z^{-1}X_I(z) + x(-1)$$

$$Z_I\{x(n-m)\} = z^{-m}X_I(z) + \sum_{k=1}^{m} x(-k)z^k \qquad \text{(B-25)}$$

Also, it follows that

$$Z_I\{x(n+m)\} = z^m[X_I(z) - \sum_{k=0}^{m-1} x(k)z^{-k}] \qquad \text{(B-26)}$$

Finally, $Z_I\{\nabla x(n)\} = (1 - z^{-1})X(z) - x(-1) \qquad \text{(B-27)}$

$Z_I\{\Delta x(n)\} = (z - 1)X(z) - zx(0) \qquad \text{(B-28)}$

B.7 THE FOURIER INTEGRAL AND FOURIER SERIES FOR ANALOG SIGNALS

The Fourier transform of the continuous-time signal $x(t)$ is the integral

$$X(\omega) = \int_{-\infty}^{\infty} x(t)e^{-j\omega t}\, dt \triangleq \mathfrak{F}\{x(t)\} \qquad \text{(B-29)}$$

and the inverse transform is

$$x(t) = \frac{1}{2\pi} \int_{-\infty}^{\infty} X(\omega)e^{-jt\omega}\, d\omega \qquad \text{(B-30)}$$

The notation $x(t) \leftrightarrow X(\omega)$ denotes the Fourier transform pair given explicitly in Eqs. (B-29) and (B-30). From the defining integrals, we can derive several theorems and transform pairs. These are summarized in Tables B.1 and B.2. Note particularly the symmetrical form of the direct and inverse transformations. This symmetry property, entry 1 in Table B.1, allows us to interchange time and frequency variables to obtain a dual set of relations, as evidenced by the other entries in this table.

Application of the symmetry theorem also allows us to extend the transform pairs in Table B.2. For example,

$$x(t) = e^{-a|t|} \leftrightarrow X(\omega) = \frac{2a}{\omega^2 + a^2}$$

Table B.1. Fourier Transform Relationships[†]

1.	Symmetry	$2\pi x(-\omega) \leftrightarrow X(t)$
2.	Scaling	$x(at) \leftrightarrow \dfrac{1}{|a|} X(\omega/a)$
3.	Time shift	$x(t - \tau) \leftrightarrow e^{-j\tau\omega} X(\omega)$
4.	Frequency shift	$e^{j\beta t} x(t) \leftrightarrow X(\omega - \beta)$
5.	Time differentiation	$\dfrac{d^m}{dt^m} x(t) \leftrightarrow (j\omega)^m X(\omega)$
6.	Frequency differentiation	$(-jt)^m x(t) \leftrightarrow \dfrac{d^m}{d\omega^m} X(\omega)$
7.	Moments	$(-j)^m \displaystyle\int t^m x(t)\, dt \leftrightarrow \left. \dfrac{d^m x}{d\omega^m} \right|_{\omega=0}$
8.	Convolution	$x_1(t) * x_2(t) \leftrightarrow X_1(\omega) X_2(\omega)$ $x_1(t) x_2(t) \leftrightarrow \dfrac{1}{2\pi} X_1(\omega) * X_2(\omega)$
9.	Parseval's theorem	$\displaystyle\int_{-\infty}^{\infty} |x(t)|^2 dt = \dfrac{1}{2\pi} \int_{-\infty}^{\infty} |X(\omega)|^2\, d\omega$
10.	System function	$e^{j\omega t} * x(t) \leftrightarrow X(\omega) e^{j\omega t}$

[†]$x(t) \leftrightarrow X(\omega)$ means $X(\omega) = \mathfrak{F}x(t)$.

implies the pair,

$$2\pi x(-\omega) = e^{-a|\omega|} \leftrightarrow X(t) = \frac{2a}{t^2 + a^2}$$

Periodic Signals. A periodic signal

$$x_p(t) = x_p(t + \nu T) \qquad \nu \text{ an integer} \tag{B-31}$$

can be represented as the weighted sum of the exponential sinusoids,

$$x_p(t) = \sum_{k=-\infty}^{\infty} a_k e^{jk\omega_0 t} \qquad \omega_0 = 2\pi/T \tag{B-32}$$

where
$$a_k = \frac{1}{T} \int_{-\infty}^{\infty} x(t) e^{jk\omega_0 t}\, dt \tag{B-33}$$

The set $\{a_k\}$ are the Fourier series (FS) coefficients. The proof of the inversion formula Eq. (B-33) rests on the orthogonality of the FS basis functions, $\{\phi_k(t) = e^{jk\omega_0 t}\}$ on $[-T/2, T/2]$:

$$\int_{-\infty}^{\infty} \phi_n(t) \bar{\phi}_m(t) dt = T\delta_{n-m} \tag{B-34}$$

Table B.2. A Short Table of Fourier Transform Pairs

Time Function $x(t)$	Transform $X(\omega)$
1. $\delta(t)$	1
2. $\delta(t-\tau)$	$e^{-j\tau\omega}$
3. $e^{-j\beta t}$	$2\pi\delta(\omega-\beta)$
4. $e^{-a\lvert t\rvert}$	$\dfrac{2a}{\omega^2+a^2}$
5. $e^{-at}u(t)$	$\dfrac{1}{j\omega+a}$
6. $\operatorname{sgn}t = \begin{cases} 0, & t=0 \\ 1, & t>0 \\ -1, & t<0 \end{cases}$	$\dfrac{2}{j\omega}$
7. $\displaystyle\sum_{n}\delta(t-nT)$	$\displaystyle\omega_0\sum_{n}\delta(\omega-n\omega_0)\qquad \omega_0 T=2\pi$
8. $\operatorname{rect}(t/T) = \begin{cases} 1, & \lvert t/T\rvert\le\frac{1}{2} \\ 0, & \text{otherwise} \end{cases}$	$T\operatorname{sinc}\dfrac{\omega T}{2}\triangleq\dfrac{T\sin(\omega T/2)}{\omega T/2}$
9. $\Lambda(2t/T) = \begin{cases} 1-2\lvert t\rvert/T & \lvert t\rvert<T/2 \\ 0, & \text{otherwise} \end{cases}$	$T/2\operatorname{sinc}^2(\omega T/4)$
10. $\dfrac{1}{\sqrt{2\pi}}e^{-t^2/2}$	$e^{-\omega^2/2}$

A consequence of Eq. (B-34) is Parseval's theorem for periodic signals

$$\frac{1}{T}\int_{-T/2}^{T/2}\lvert x_p(t)\rvert^2 dt = \sum_{k=-\infty}^{\infty}\lvert a_k\rvert^2 \tag{B-35}$$

The Poisson Sum Formula. (Papoulis 1962, 1977). This formula is the link between Fourier series for periodic signals and the transform of aperiodic signals. It can be argued that this formula, in its various forms, summarizes all Fourier theory.

Starting with the transform pair $x(t)\leftrightarrow X(\omega)$, we define $x_p(t)$ as the periodic repetition of $x(t)$ with period T:

$$x_p(t) = \sum_{n=-\infty}^{\infty} x(t-nT) \tag{B-36}$$

This $x_p(t)$ has a FS expansion of the form Eqs. (B-32) and (B-33). The Poisson formula asserts that the FS coefficients $\{a_k\}$ are obtained by sampling $X(\omega)$ at the harmonic frequencies $k\omega_0$ and scaling by $1/T$. Thus $x(t)\leftrightarrow X(\omega)$ implies the Poisson sum formula

$$\sum_{n=-\infty}^{\infty} x(t - nT) = \frac{1}{T} \sum_{k=-\infty}^{\infty} X(k\omega_0) e^{jk\omega_0 t} \qquad (\text{B-37})$$

where
$$\omega_0 T = 2\pi$$

The progenitor signal is any aperiodic signal possessing a Fourier transform; it need not be time-limited.

Tables B.1 and B.2 and Eq. (B-37) can be used to establish properties of Fourier series and a table of FS coefficients. For example, the shifting property

$$x(t - t_0) \leftrightarrow e^{-j\omega t_0} X(\omega)$$

suggests the FS pair

$$x_p(t - t_0) \leftrightarrow e^{-jk\omega_0 t_0} \frac{X(k\omega_0)}{T} \qquad (\text{B-38})$$

For another example, the transform pair $\delta(t) \leftrightarrow 1$ and the Poisson sum formula lead directly to

$$\sum_{n=-\infty}^{\infty} \delta(t - nT) = \frac{1}{T} \sum_{n=-\infty}^{\infty} e^{jk\omega_0 t} \qquad (\text{B-39})$$

In turn, the Fourier transform of both sides of Eq. (B-39) gives entry 7 in Table B.2:

$$\mathfrak{F}\{\sum_n \delta(t - nT)\} = \frac{1}{T} \sum_k \mathfrak{F}\{e^{jk\omega_0 t}\} = \frac{2\pi}{T} \sum_k \delta(\omega - k\omega_0)$$

Finally, the symmetry theorem allows us to interchange time and frequency in Eq. (B-37) to obtain the frequency-domain version of the Poisson sum Formula, namely that

$$x(t) \leftrightarrow X(\omega)$$

implies
$$\frac{1}{T_s} \sum_{k=-\infty}^{\infty} X(\omega - k\Omega_s) = \sum_{n=-\infty}^{\infty} x(nT_s) e^{-jnT_s \omega} \qquad (\text{B-40})$$

where
$$\Omega_s T_s = 2\pi$$

Both forms Eqs. (B-37) and (B-40) of the Poisson sum formula demonstrate the adage that sampling in one domain corresponds to periodic repetition in the other.

In the next section, we show that Eq. (B-40) is the discrete-time Fourier transform, that is, the Fourier transform of the sampled signal $\{x(nT_s)\}$.

B.8 THE DISCRETE-TIME FOURIER TRANSFORM (DTFT)

The DTFT is the Fourier transform of an aperiodic sequence $\{x(n)\} \triangleq \{x_a(nT_s)\}$, defined by

$$X(e^{j\theta}) = \sum_{n=-\infty}^{\infty} x(n)e^{-jn\theta} \tag{B-41}$$

where $\theta = \omega T_s$ is the normalized frequency. From Eqs. (B-1) and (B-3), we recognize this transform as the (two-sided) Z transform of $x(n)$ evaluated on the unit circle $z = e^{j\theta}$. Therefore, the properties of the DTFT are special cases of corresponding properties of the Z transform, provided only that $X(z)$ converge on $|z| = 1$.

With $z = e^{j\theta}$ in Eq. (B-7), we obtain the inversion formula

$$x(n) = \frac{1}{2\pi} \int_{-\pi}^{\pi} X(e^{j\theta})e^{jn\theta}\, d\theta \tag{B-42}$$

Equations (B-41) and (B-42) are the discrete-time counterparts to the purely analog transform pair of Eqs. (B-29) and (B-30). For sampled data, the time function is discrete, while the frequency function depends on the continuous variable $\theta = \omega T_s$.

Some properties of the Discrete-Time Fourier Transform.

1. **Periodicity.** We formed the sequence $\{x(n)\}$ by time-sampling the analog signal $x_a(t) \leftrightarrow X_a(\omega)$. But by the Poisson sum formula Eq. (B-40), the Fourier transform of the sampled signal is the periodic repetition of the transform of the analog signal, that is,

$$X(e^{j\theta}) = \sum_{k=-\infty}^{\infty} x(k)e^{-jk\theta} = \frac{1}{T_s}\sum_{k=-\infty}^{\infty} X_a(\omega - k\Omega_s) = X(z)\big|_{z=e^{j\theta}} \tag{B-43}$$

where $\Omega_s = 2\pi/T_s$ is the radian sampling frequency and $\theta = \Omega T_s$ is the normalized frequency. Hence the transform is periodic in ω with period Ω_s or periodic in θ with period 2π.

2. **Even-odd symmetry.** If $x(n)$ is real, the magnitude and phase of $X(e^{j\theta})$ are respectively even- and odd-symmetric about $\theta = 0$ and $\theta = \pi$.

3. **Convolution.** If $x(n) \leftrightarrow X(e^{j\theta})$ and $h(n) \leftrightarrow H(e^{j\theta})$, then

$$x(n) * h(n) \leftrightarrow X(e^{j\theta})H(e^{j\theta})$$

$$x(n) \cdot h(n) \leftrightarrow \frac{1}{2\pi}X(e^{j\theta}) * H(e^{j\theta})$$

$$= \frac{1}{2}\int_{-\pi}^{\pi} X(e^{j(\theta-\lambda)})H(e^{j\lambda})\, d\lambda$$

4. Parseval's theorem.

$$\sum_{n=-\infty}^{\infty} |x(n)|^2 = \frac{1}{2\pi} \int_{-\pi}^{\pi} |X(e^{j\theta})|^2 \, d\theta \tag{B-44}$$

5. Shifting theorems. If $x(n) \leftrightarrow X(e^{j\theta})$, then

$$x(n+m) \leftrightarrow e^{jm\theta} X(e^{j\theta}) \tag{B-45}$$

$$e^{j\alpha n} x(n) \leftrightarrow X(e^{j(\theta-\alpha)})$$

6. Decimation in time (undersampling). Let

$$y(n) = \begin{cases} x(n) & n = 0, \pm M, \pm 2M, \ldots \\ 0 & \text{otherwise} \end{cases}$$

then $$Y(e^{j\theta}) = \frac{1}{M} \sum_{m=0}^{M-1} X(e^{j(\theta-2\pi m/M)}) \tag{B-46}$$

These properties–convolution, Parseval, and shifting–follow directly from the corresponding Z-transform theorems Eqs. (B-9) through (B-13) and (B-16).

B.9 PERIODIC SEQUENCES AND DISCRETE FOURIER SERIES

We can obtain the discrete-time counterpart to Eq. (B-37) by sampling the analog signals $x(t)$ and $x_p(t)$ at intervals $T_s = T/N$ apart [that is, N equally-spaced samples of $x_p(t)$ are taken every period] to generate the aperiodic and periodic sequences $\{x(nT_s)\}$ and $\{x_p(nT_s)\}$, respectively. (See also Fig. B-1.) For notational simplicity, we suppress the T_s in the arguments of the sampled signals. The periodic sequence is therefore

$$x_p(n) = \sum_{\ell=-\infty}^{\infty} x(n - \ell N) \tag{B-47}$$

By analogy with the continuous-time case, we expand $x_p(n)$ in a set of functions $\{\phi_k(n)\}$ that are periodic in x with period N and orthogonal on $[0, N-1]$. Thus we require

$$\phi_k(n) = \phi_k(n + N)$$

$$\sum_{n=0}^{N-1} \phi_k(n) \bar{\phi}_\ell(n) = C\delta_{n-\ell} \tag{B-48}$$

In the following section on the DFT, we show that a sampling of $e^{jk\omega_0 t}$, $\omega_0 = 2\pi/T$, generates the required orthogonal set; that is,

$$\phi_k(n) = e^{jk\theta_0 n} \qquad \theta_0 = 2\pi/N \tag{B-49}$$

is periodic and orthogonal:

$$\sum_{n=0}^{N-1} e^{j(k-\ell)n\theta_0} = N\delta_{k-\ell} \tag{B-50}$$

Therefore, the discrete Fourier series expansion of $x_p(n)$ is

$$\sum_{k=0}^{N-1} \alpha_k e^{jk(2\pi/N)n} \tag{B-51}$$

and inversely, $$\alpha_k = \frac{1}{N} \sum_{n=0}^{N-1} x_p(n) e^{-jk(2\pi/N)n} \tag{B-52}$$

Both time function $x_p(n)$ and FS coefficient α_k are periodic in their arguments with period equal to N.

The link between aperiodic and periodic sequences is provided by the discrete Poisson sum formula:

$$x(n) \leftrightarrow X(e^{j\theta})$$

implies $$x_p(n) = \sum_{\ell=-\infty}^{\infty} x(n - \ell n) = \sum_{k=0}^{N-1} \alpha_k e^{jkn\theta_0} \tag{B-53}$$

$$\alpha_k = \frac{1}{N} X(e^{j\theta}) \Big|_{\theta=k\theta_0} \qquad \theta_0 = 2\pi/N$$

The final link between analog and discrete FS coefficients is simply that the FS coefficient for the sampled periodic signal is the periodic repetition of the FS coefficient for the analog periodic signal, or

$$\alpha_k = \sum_{\ell=-\infty}^{\infty} a_{k-\ell n} \tag{B-54}$$

where $$x_p(t) \leftrightarrow a_k$$
$$x_p(n) \leftrightarrow \alpha_k$$

The example shown in Fig. B-1 illustrates these properties and shows the consequences of sampling and periodic repetition in the time and frequency domains. Figure B-2 is a graphic table of discrete Fourier series pairs.

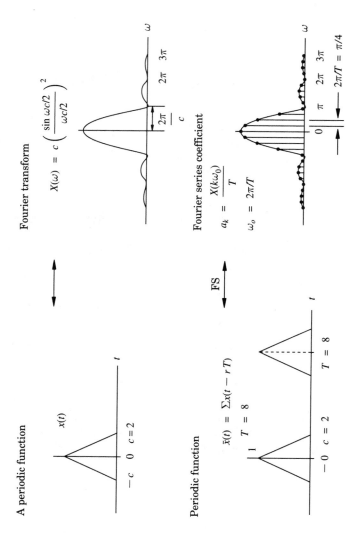

Figure B-1 *Illustrating relations among samples and periodic repetitions in time and frequency domain.*

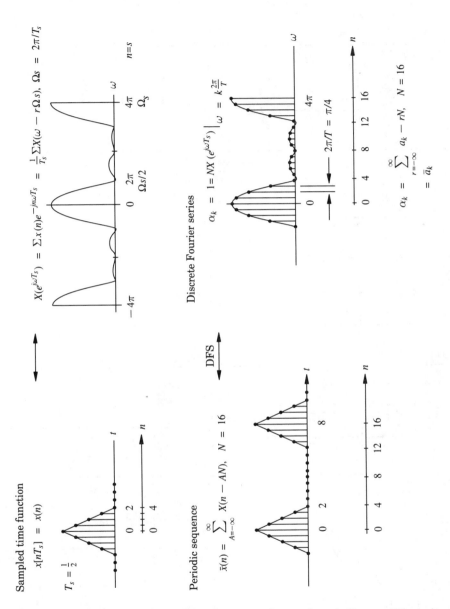

Figure B-1 *Illustrating relations among samples and periodic repetitions in time and frequency domain.*

Discrete Fourier Series Pairs

$$x(k) = \sum_{l=0}^{N-1} \alpha_l W^{lk} \longleftrightarrow \alpha_m = \frac{1}{N} \sum_{k=0}^{N-1} x(k) W^{-mk} \quad W = e^{j2\pi=N}$$

$x(k)$	α_l
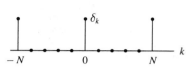	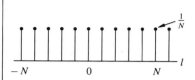
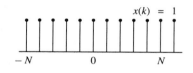	$\alpha_l = \frac{1}{N} e^{-jml2\pi=N}$
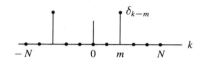 $x(k) = 1$	$\alpha_l = \delta_l$
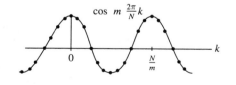 $x(k) = \sum_m \delta_{k - mL}$	$K = N/L$, all integers $1/L$
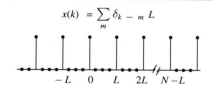 $\cos m \frac{2\pi}{N} k$	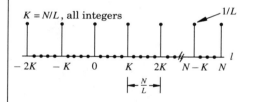
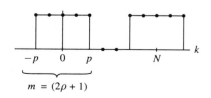 $m = (2\rho + 1)$	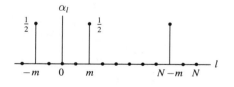 $\dfrac{1}{N} \dfrac{\sin ml\pi=N}{\sin l\pi=N}$ (See Fig. 5.21.)

Figure B-2   *Illustrating discrete Fourier series pairs.*

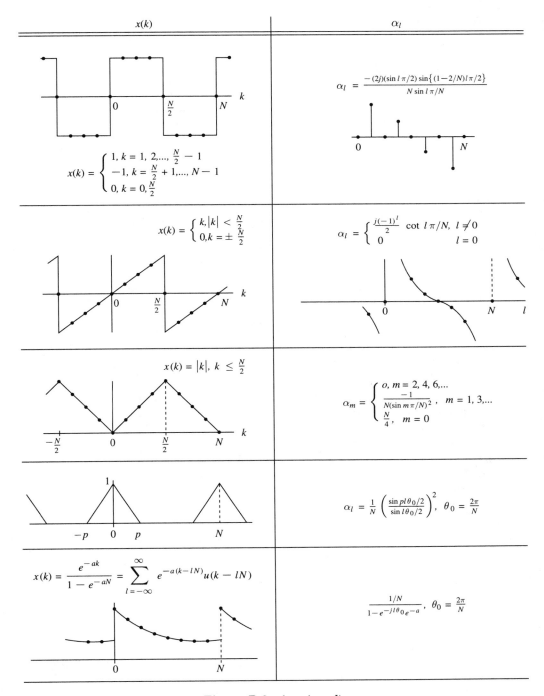

Figure B-2   (continued)

The term, discrete Fourier transform or DFT is now commonly used in lieu of "discrete Fourier series coefficient." Accordingly, the formal properties and manipulations of discrete Fourier series/periodic sequences are expressed in the DFT format in the next section.

B.10 THE DISCRETE FOURIER TRANSFORM (DFT)

The DFT of a sequence of N samples $\{x_0, x_1, ..., x_{N-1}\}$ is defined by

$$X(k) = \sum_{n=0}^{N-1} x(n)e^{-j2\pi kn/N} \qquad k = 0, 1, ..., N-1 \qquad \text{(B-55a)}$$

Because of the ubiquity of the factor $e^{-j2\pi kn/N}$, it is common practice to use the abbreviation

$$W_N = e^{j2\pi/N}$$

Therefore the defining equation may be written

$$X(k) = \sum_{n=0}^{N-1} x(n)W_N^{-nk} \qquad \text{(B-55b)}$$

The subscript N is frequently omitted when it is understood from context. Note that W_N is the principal Nth root of unity.

It is important to note that the DFT is cyclic with period N in both domains. This follows from the sampled nature of $x(n)$ and $X(k)$; see any standard introduction to the DFT, for example Bergland (1969), Bracewell (1965), Cochran et al. (1967), Oppenheim and Schafer (1975), Rabiner and Gold (1975). Hence all subscripts and indices are taken modulo N and, in particular, convolution in these domains is cyclical. This convention will be assumed throughout in what follows.

B.10.1 FOURIER AND Z TRANSFORMS

Consider the Z transform of our sequence $x(n)$, $0 \le n \le N-1$. This is given by

$$X(z) = \sum_{n=0}^{N-1} x(n)z^{-n} \qquad \text{(B-56)}$$

From this we see immediately that $X(k)$ is the Z transform of $x(n)$ evaluated at N equally spaced points on the unit circle. This has the following consequences:

1. The domain of the DFT is sometimes equated with the periphery of the unit circle; as a result of this the frequency variable is written as an angle θ, in the region $(-\pi, \pi)$. The situation is complicated further by the fact that the discrete-time Fourier transform may also be expressed as $X(z)$, $z = e^{j\theta}$, where θ is a continuous variable. It is usually necessary to see from context which transform is intended.

2. On the unit circle, $\theta = \pi$ corresponds to $f = f_s/2$ and to $k = N/2$. Hence we have the mappings

$$\theta = 2\pi f/f_s \tag{B-57a}$$

$$k = N\theta/2\pi \tag{B-57b}$$

and $$k = Nf/f_s \tag{B-57c}$$

3. The lower half of the unit circle corresponds to negative frequencies and to $N/2 < k < N$; hence the negative-frequency components of the transform appear in the upper half of the range $(0, N - 1)$.

B.11 PROPERTIES OF THE DFT

Most of these properties parallel the corresponding properties of the continuous Fourier transform. We will abbreviate the DFT of $x(n)$ as $\mathfrak{F}[x(n)]$.

1. **Orthogonality.** The function W_N^{nk} is orthogonal over the interval $[0, N - 1]$. That is,

$$\sum_{k=0}^{N-1} W_N^{mk} \overline{W}_N^{nk} = N\delta_{m-n} \tag{B-58}$$

The proof requires some ingenuity. First, rewrite Eq. (B-58) as

$$J(m - n) = \sum_{k=0}^{N-1} W_N^{k(m-n)}$$

For $m = n$, the summand is $W_N^0 = 1$, hence the sum is N. For $m \neq n$, let $m - n = p$, $p \neq 0$; then we may write

$$J(p) = \sum_{k=0}^{N-1} W_N^{kp}$$

This is a finite sum of a geometric series; by the well-known formula,

$$J(p) = \frac{1 - W_N^{Np}}{1 - W_N^p} \qquad p \neq 0$$

But W_N is the Nth root of unity; hence the numerator vanishes and $J(p) = 0$.

2. **Linearity.** For any constants a and b,

$$\mathfrak{F}[ax(n) + by(n)] = a\mathfrak{F}[x(n)] + b\mathfrak{F}[y(n)] \qquad \text{(B-59)}$$

3. **Real and imaginary time functions.** If $X(k) = \mathfrak{F}[x(n)]$, then for $\{x(n)\}$ real, $X(N - k) = \overline{X}(k)$, where the overbar indicates the complex conjugate. For imaginary $\{x(n)\}$, $X(N - k) = -\overline{X}(k)$.

4. **Even and odd time functions.** Let $X(k) = \mathfrak{F}[x(n)]$; then if $x(N - n) = x(n)$, $X(N - k) = X(k)$. If $x(N - n) = -x(n)$, then $X(N - k) = -X(k)$. Notice that, in these two properties, because of the circularity of the domains, $N - n$ and $N - k$ take the place of $-n$ and $-k$, respectively.

B.12 THE INVERSE TRANSFORM

Let $X(k) = \mathfrak{F}[x(n)]$. Then the inverse transform is given by

$$x(n) = \frac{1}{N} \sum_{k=0}^{N-1} X(k) W_N^{nk} \qquad \text{(B-60)}$$

Proof:

$$\frac{1}{N} \sum_{k=0}^{N-1} X(k) W_N^{nk} = \frac{1}{N} \sum_{k=0}^{N-1} \sum_{m=0}^{N-1} x(m) W_N^{-mk} W_N^{nk}$$

$$= \frac{1}{N} \sum_{m=0}^{N-1} x(m) \sum_{k=0}^{N-1} W_N^{k(n-m)}$$

By the orthogonality property proved above, the second sum is $N\delta_{n-m}$. Hence

$$\frac{1}{N} \sum_{k=0}^{N-1} X(k) W_N^{nk} = \frac{1}{N} \sum_{m=0}^{N-1} x(m) N\delta_{n-m}$$

$$= x(n)$$

B.13 OTHER PROPERTIES OF THE DFT

1. **Symmetry.** If $X(k) = \mathfrak{F}[x(n)]$, then $X(n) = \mathfrak{F}[Nx(-k)]$. Proof: By the inversion formula, Eq. (B-60),

$$Nx(-n) = \sum_{k=0}^{N-1} X(k)W^{-nk} \qquad \text{(B-61)}$$

The proof follows if we exchange the roles of the subscripts n and k.

2. **Convolution.** Recall from our earlier discussion that the convolution of two sequences of length N is cyclical. We can take this into account by taking all indices and subscripts modulo N. Then cyclical convolution of the time functions corresponds to multiplication of the transforms: if $X(k) = \mathfrak{F}[x(n)]$ and $Y(k) = \mathfrak{F}[y(n)]$, then $\mathfrak{F}[x(n) * y(n)] = X(k)Y(k)$.

3. **Transforms of delta functions.**

$$\mathfrak{F}[\delta(n)] = 1$$

$$\mathfrak{F}[\delta(n - p)] = e^{-j2\pi kp/N} \qquad \text{(B-62)}$$

4. **Transforms of complex exponentials.** Let $x(n) = e^{j2\pi nf/N}$. If f is an integer, then $x(n + N) = x(n)$; hence the whole time function is the same as the periodic extension of the portion in $[0, N)$. *In this case only*, we may apply the symmetry principle to Property 3 to get

$$X(k) = \delta(k - f) \qquad \text{(B-63)}$$

Transforms of sine and cosine functions, subject to the same period-icity condition, may be found by using the usual equivalences; for example, for $x(n) = \cos(2\pi nf/N)$, we get

$$X(k) = \tfrac{1}{2}[\delta(k - f) + \delta(k + f)] \qquad \text{(B-64)}$$

5. **Shifting.** $\mathfrak{F}[x(n - m)] = e^{-j2\pi mk/N}\mathfrak{F}[x(n)]$. The periodicity of the DFT domains means that shifting, like convolution, must be cyclical; again, we must take all indices and subscripts modulo N. Let $y(n) = x(n-m)$; then by the shifting property of $\delta(n)$, $y(n) = x(n) * \delta(n - m)$. The result follows from the convolution property and the transform of $\delta(n - m)$.

6. **Differencing.** Let $\nabla x(n) = x(n) - x(n - 1)$, with the subscripts taken modulo N, as usual. Then $\mathfrak{F}[\nabla x(n)] = (1 - e^{-j2\pi k/N})\mathfrak{F}[x(n)]$. This follows from the shifting property.

7. **Time reversal.** If $\mathfrak{F}[x(n)] = X(k)$, then $\mathfrak{F}[x(N - n)] = X(N - k)$. This follows from the substitution of $x(N - n)$ for $x(n)$ in the defining equa-tion.

8. **Complex conjugate.** If $\mathfrak{F}[x(n)] = X(k)$, then $\mathfrak{F}[\bar{x}(n)] = \bar{X}(N - k)$. This also follows from substitution in the defining equation.

9. Autocorrelation. We define the cyclical autocorrelation of $x(n)$ as

$$r_m = \sum_{n=0}^{N-1} x(n)\bar{x}(n-m) \tag{B-65}$$

with all indices modulo N. Since $r_m = \bar{r}_{-m}$, we may also write

$$r_m = \sum_{n=0}^{N-1} \bar{x}(n)x(n+m) \tag{B-66}$$

But this is $\bar{x}(n) * x(-n)$. From the convolution property and from properties 7 and 8, above

$$R_k = \mathfrak{F}[r_m] = X(k)\bar{X}(k)$$
$$= |X(k)|^2 \tag{B-67}$$

10. Parseval's theorem: If $X(k) = \mathfrak{F}[x(n)]$ then

$$\sum_{n=0}^{N-1} |x(n)|^2 = \frac{1}{N} \sum_{k=0}^{N-1} |X(k)|^2 \tag{B-68}$$

REFERENCES

G. D. Bergland. "A guided tour of the fast Fourier transform." *IEEE Spectrum*, v. 6, pp. 41–62, July 1969.

R. Bracewell. *The Fourier Transform and Its Applications*. New York: McGraw-Hill, 1965.

W. T. Cochran et al. "What is the fast Fourier transform?" *IEEE Trans. Audio and Electroacous.*, v. AU-15, pp. 45–55, June, 1967.

A. V. Oppenheim and R. W. Schafer. *Digital Signal Processing*. Englewood Cliffs, N. J.: Prentice-Hall, 1975.

A. Papoulis. *The Fourier Integral and Its Applications*. New York: McGraw-Hill, 1962.

————. *Signal Analysis*. New York: McGraw-Hill, 1977.

L. R. Rabiner and B. Gold. *Theory and Application of Digital Signal Processing*. Englewood Cliffs, N. J.: Prentice-Hall, 1975.

APPENDIX C

■ ■

PROBABILITY AND STOCHASTIC PROCESSES

The following notes are intended to provide a minimal background in probability theory as it is used in this text.

C.1 EVENTS AND PROBABILITIES

The classic models of probability are the tossing of a coin or a die. Each toss represents an experiment, and we assume that we know what the possible outcomes are, as of course we do in the case of a coin or a die. Let S be the set of all possible outcomes and let the event E be any subset of S. Then the probability of E, written $P\{E\}$, or occasionally Prob $\{E\}$, is a number having the following properties:

1. $P\{E\} \geq 0$ for all E.

2. $P\{S\} = 1$.

3. If E and F are disjoint subsets of S, then[†]

$$P\{E \cup F\} = P\{E\} + P\{F\}$$

[†] $A \cup B$ is the union of sets A and B, that is, the set containing all elements of A or B or both. $A \cap B$ the intersection of sets A and B, that is, the set containing only elements common to A and B. The complement of a set A is the set of all things in S that are not elements of A.

These are the axioms of probability. Notice that they say nothing about how to estimate the probability, only about what probabilities are like, how they behave. The estimates must be obtained independently by reason, intuition, or experiment. Probability is also occasionally defined as the number of favorable outcomes divided by the number of all possible outcomes, assuming all outcomes equally likely; but since the phrase, "equally likely" refers to the probabilities of the individual outcomes, this definition is circular. If we have a good intuitive reason to believe that the outcomes are in fact equally likely, this is frequently a satisfactory basis for estimating probabilites, however.

From these axioms we can deduce immediately,

1. Let E' represent the complement of E. Then $P\{E'\} = 1 - P\{E\}$.

2. If $\emptyset$ is the empty set, then $P\{\emptyset\} = 0$.

3. If E and F are not disjoint, then we can write $E \cup F$ as the union of three disjoint sets,

$$E \cup F = (E \cap F) \cup (E \cap F') \cup (E' \cap F)$$

Since $(E \cap F') \cup (E \cap F) = E$, and similarly for $(E' \cap F)$, we have

$$P\{E \cup F\} = P\{E\} + P\{F\} - P\{E \cap F\} \qquad \text{(C-1)}$$

In the balance of this appendix, we will follow conventional usage and write $A \cap B$ as AB and $A \cup B$ as $A+B$ in order to avoid the clumsiness of set notation.

The conditional probability of an event E given another event F, written $P\{E|F\}$, is defined as

$$P\{E|F\} = P\{EF\}/P\{F\} \qquad \text{(C-2)}$$

E and F are independent if $P\{EF\} = P\{E\}P\{F\}$. In that case, the conditional probabilities are the same as the unconditional probabilities: that is, $P\{E|F\} = P\{E\}$ and $P\{F|E\} = P\{F\}$.

Bayes' theorem provides a way of turning conditional probabilities around. Since

$$P\{AB\} = P\{A|B\}P\{B\} = P\{B|A\}P\{A\}$$

we have $\qquad P\{A|B\} = P\{B|A\}\dfrac{P\{A\}}{P\{B\}} \qquad \text{(C-3)}$

In general, suppose $A_1, A_2, ..., A_n$ are a partitioning of the set S, and let B be a subset of S. Then

$$P\{A_i|B\} = \frac{P\{B|A_i\}P\{A_i\}}{\sum\limits_{k=1}^{n} P\{B|A_k\}P\{A_k\}} \qquad \text{(C-4)}$$

Proof: We have $P\{A_i|B\} = P\{B|A_i\}P\{A_i\}/P\{B\}$. To determine $P\{B\}$, we note that $B = BS = B(A_1 + A_2 + \cdots + A_n) = BA_1 + BA_2 + \cdots + BA_n$, all mutually exclusive. Hence

$$P\{B\} = P\{BA_1 + BA_2 + \cdots + BA_n\}$$

$$= \sum_{k=1}^{n} P\{BA_k\}$$

$$= \sum_{k=1}^{n} P\{B|A_k\}P\{A_k\}$$

This leads to Eq. (C-4).

EXAMPLE A: surplus store has four boxes of integrated circuits.

Box 1 has 2,000 components, 5 percent defective;
Box 2 has 500 components, 40 percent defective;
Box 3 has 1,000 components, 10 percent defective;
Box 4 has 1,000 components, 10 percent defective.

1. If we select a box at random and remove a circuit from the box at random, what is the probability that the device will be defective?
      Let A_i be the ith box; clearly the boxes are mutually exclusive and $S = A_1 + A_2 + A_3 + A_4$. Let D be the subset of defective devices. We assume that selecting a box at random means $P\{A_i\} = 1/4$. Then

$$P\{D\} = P\{D|A_1\}P\{A_1\} + P\{D|A_2\}P\{A_2\} + P\{D|A_3\}P\{A_3\} + P\{D|A_4\}P\{A_4\}$$

$$= (0.25)(0.05 + 0.4 + 0.1 + 0.1) = 0.1625$$

2. Suppose the device is indeed defective. What is the probability that it came from box 2?
      This is $P\{A_2|D\}$. By Bayes's theorem,

$$P\{A_2|D\} = \frac{P\{D|A_2\}P\{A_2\}}{P\{D\}} = \frac{(0.4)(0.25)}{0.1625} = 0.615$$

Note that we knew $P\{D|A_2\}$ in advance; this is termed an *a priori* probability. $P\{A_2|D\}$, the one we found working backward with the aid of Bayes' theorem, is termed an *a posteriori* probability.

C.2 RANDOM VARIABLES

A random variable (rv) is a number whose value depends on the outcome of some experiment. That is, if S is the set of all possible outcomes, then to each element ζ of S we assign a number (usually but not necessarily real) $X(\zeta)$. The probability of any particular $X_i = X(\zeta_i)$ is thus the probability of the outcome ζ_i itself. From the axioms of probability, we know that $P\{X_i\} > 0$. We assume that the outcomes are mutually exclusive; hence we have $\sum P\{X_i\} = 1$.

The variable may be continuous or discrete. We characterize rvs by their distribution functions, their probability densities, and their statistics. The distribution function of a random variable X is defined as

$$F_X(x) = P\{X \leq x\} \tag{C-5}$$

The probability density function (pdf) is the derivative of the distribution:

$$f_X(x) = \frac{d}{dx} F_X(x) \tag{C-6}$$

This may also be written $f_X(x)\,dx = P\{x < X \leq x + dx\}$. For a discrete rv we will use $p(x)$ for the probability that its value will be equal to x. It may also be convenient to write p_i for $p(x_i)$.

For a continuous rv, we have:

1. $0 \leq F(x) \leq 1$.

2. $F(x)$ is monotone nondecreasing.

3. $F(\infty) = 1$.

4. $P\{x_1 < X \leq x_2\} = F(x_2) - F(x_1)$.

5. Since $f(x) = \dfrac{dF}{dx}$,

$$P\{x_1 < X \leq x_2\} = \int_{x_1}^{x_2} f(x)\,dx \tag{C-7}$$

The statistics of greatest interest are the mean, the variance, and the moments. The mean, or expected value, of a continuous random variable X is variously written μ_X, η_X, $\bar{X}$, $E\{X\}$, or $\langle X \rangle$ and is given by

$$\langle X \rangle = \int_{-\infty}^{\infty} x f_X(x)\,dx \tag{C-8}$$

where $f_X(x)$ is the pdf. For a discrete rv,

$$\langle X \rangle = \sum_i x_i p_i \qquad \text{(C-9)}$$

where the sum is taken over all possible values of X. The mean of a set of N observations of a variable is estimated by

$$\hat{\mu}_X = \frac{1}{N} \sum_i x_i \qquad \text{(C-10)}$$

where the sum is taken over all the observations.

The mean of any function $g(X)$ of X is similarly

$$\langle g(X) \rangle = \int_{-\infty}^{\infty} g(x) f(x) \, dx \qquad \text{(C-11)}$$

for continuous random variables and, for discrete random variables,

$$\langle g(X) \rangle = \sum_i g(x_i) p_i \qquad \text{(C-12)}$$

Note that the expectation is a linear operator: for constants a and b,

$$\langle a g(X) + b h(X) \rangle = a \langle g(X) \rangle + b \langle h(X) \rangle$$

The moments of a rv are the expected values of its powers. The nth moment of X, written m_n, is given by

$$m_n = E\{X^n\} = \int_{-\infty}^{\infty} x^n f(x) \, dx \qquad \text{(C-13)}$$

and the integral is replaced by a sum in the case of discrete rvs, as usual. Note that this means that the mean is the first moment.

The variance of a continuous rv X, written σ_X^2 or var $\{X\}$, is defined by

$$\sigma_X^2 = \int_{-\infty}^{\infty} (x - \mu_X)^2 f_X(x) \, dx$$

$$= \int_{-\infty}^{\infty} x^2 f_X(x) dx - \mu_X^2 \qquad \text{(C-14)}$$

$$= m_2 - \mu_X^2$$

The quantity, σ_X, is the standard deviation of X. For a discrete rv,

$$\sigma_X^2 = \sum_x (x - \mu_X)^2 p(x)$$

$$= \sum_x x^2 p(x) - \mu_X^2 \tag{C-15}$$

$$= m_2 - \mu_X^2$$

We can estimate the variance of N observations of a variable by

$$\hat{\sigma}_X^2 = \frac{1}{N} \sum_i (x_i - \mu_X)^2 \tag{C-16}$$

where the sum is taken over all the observations.

Two or More Random Variables.   For two random variables X and Y, we have the distribution function

$$F_{X,Y}(p, q) = P\{X \le p, Y \le q\} \tag{C-17}$$

and the joint probability density

$$f_{X,Y}(p, q) = \frac{\partial^2}{\partial p \partial q} F_{X,Y}(p, q) \tag{C-18}$$

If we let the random vector $\mathbf{x}^T = [X_1, X_2]$, then we may write the mean as

$$\langle \mathbf{x} \rangle = \boldsymbol{\mu}_X = [\langle X_1 \rangle, \langle X_2 \rangle]^T \tag{C-19}$$

and this definition may be extended to any number of random variables.
    The probability that X and Y take on values lying in a given region D is given by

$$P\{(X, Y) \in D\} = \int \int_D f_{XY}(x, y)\, dx\, dy \tag{C-20}$$

For example,

$$P\{x_1 < X \le x_2, y_1 < Y \le y_2\} = \int_{y_1}^{y_2} \int_{x_1}^{x_2} f_{XY}(x, y)\, dx\, dy \tag{C-21}$$

The joint moments of two rvs X and Y are given by

$$m_{pq} = E\{X^p Y^q\} = \int_{-\infty}^{\infty} \int_{-\infty}^{\infty} x^p y^q f_{XY}(x, y)\, dx\, dy \tag{C-22}$$

The covariance of two random variables X and Y is

$$C_{XY} = \langle (X - \mu_X)(Y - \mu_Y) \rangle \tag{C-23}$$
$$= \langle XY \rangle - \mu_X \mu_Y$$
$$= m_{11} - m_{10} m_{01}$$

The correlation coefficient of X and Y is their covariance divided by the product of their standard deviations:

$$r_{XY} = \frac{\langle (X - \mu_X)(Y - \mu_Y) \rangle}{\sigma_X \sigma_Y} \tag{C-24}$$

Conditional Density Functions.   Starting from $P\{A|B\} = P\{AB\}/P\{B\}$, we define

$$F_{X|M}(x|M) = P\{(X \le x)|M\} = \frac{P\{(X \le x), M\}}{P\{M\}} \tag{C-25}$$

and $\qquad f_{X|M}(x|M) = \dfrac{dF_{X|M}(x|M)}{dx}$ $\qquad\qquad$ (C-26)

$F_{X|M}$ and $f_{X|M}$ are bona fide distribution and density functions in all respects. Thus

$$\langle g(X)|M \rangle = \int g(x) f_{X|M}(x|M) \, dx \tag{C-27}$$

In particular, we can show

$$f_{X|Y}(x|y) = \frac{f_{XY}(x,y)}{f_Y(y)} \tag{C-28}$$

and the conditional mean, given by

$$\langle X|y \rangle = \int x f_{X|Y}(x|y) \, dx \tag{C-29}$$

is a function of y.

The correlation matrix of a random vector $\mathbf{x}$ is defined as

$$\mathbf{R} = \langle \mathbf{x}\mathbf{x}^T \rangle \tag{C-30}$$

and the covariance matrix is defined as

$$\mathbf{C} = \langle (\mathbf{x} - \boldsymbol{\mu}_x)(\mathbf{x} - \boldsymbol{\mu}_x)^T \rangle \tag{C-31}$$

The diagonal elements of **C** are the variances of the individual variables X_i, and the off-diagonal elements are the covariances $C_{X_i X_j}$.

There are a number of standard probability density functions, of which the most important for our purposes is the Gaussian density, also called the normal density. For a single random variable, the Gaussian pdf is defined by

$$g(x) = \frac{1}{\sqrt{2\pi}\sigma} \exp\left[\frac{-(x-\mu)^2}{2\sigma^2}\right] \tag{C-32}$$

Note that if a rv is known to be Gaussian, then its pdf is completely determined by its mean and variance.

For two zero-mean random variables with correlation coefficent r, the Gaussian density function is

$$g(x,y) = \frac{1}{(2\pi\sigma_x\sigma_y\sqrt{1-r^2})} \exp\left[-\frac{1}{2(1-r^2)}\left(\frac{x^2}{\sigma_x^2} - 2r\frac{xy}{\sigma_x\sigma_y} + \frac{y^2}{\sigma_y^2}\right)\right] \tag{C-33}$$

The multivariate Gaussian density function is most economically written in vector-matrix form

$$g(\mathbf{x}) = (2\pi)^{-n/2}|\mathbf{C}|^{-1/2} \exp[-\tfrac{1}{2}(\mathbf{x} - \boldsymbol{\mu}_x)\mathbf{C}^{-1}(\mathbf{x} - \boldsymbol{\mu}_x)^T] \tag{C-34}$$

For two zero-mean random variables, this form reduces to Eq. (C-33).

Independent, Uncorrelated, and Orthogonal RVs. Recall that if two events A and B are independent, then $P\{AB\} = P\{A\}P\{B\}$. Similarly, two rvs X and Y are independent if

$$F_{XY}(x,y) = F_X(x)F_Y(y) \tag{C-35a}$$

or
$$f_{XY}(x,y) = f_X(x)f_Y(y) \tag{C-35b}$$

The variances of independent rvs are additive: if X and Y are independent, then

$$\text{var}\{X + Y\} = \text{var}\{X\} + \text{var}\{Y\} \tag{C-36}$$

If $Z = X + Y$, then the pdf for Z is the convolution of the pdfs for X and Y:

$$f_z(z) = \int_{-\infty}^{\infty} f_X(z - v)f_Y(v)\, dv \tag{C-37}$$

Two rvs are uncorrelated if $r_{XY} = 0$, $C_{XY} = 0$. (These are equivalent definitions.) A set of rvs are uncorrelated if the off-diagonal elements of $\mathbf{C}$ are all zero.

Two rvs are orthogonal if $\langle XY \rangle = 0$, and a sequence of rvs is orthogonal if they are pairwise orthogonal, that is, if $\langle X_i X_j \rangle = 0 \; \forall i \neq j$. In this case, their second moments are additive; that is,

$$E\left\{\sum_i X_i^2\right\} = \sum_i E\{X_i^2\} \tag{C-38}$$

Random Variable Transformations. Given $f_X(x)$ and a transformation $y = \phi(\cdot)$, we can obtain $f_Y(y)$ via

$$f_Y(y) = \frac{f_X(x)}{|\phi'(x)|}\bigg|_{x=\phi^{-1}(y)} \qquad \phi'(x) = \frac{d\phi}{dx} \tag{C-39}$$

where ϕ^{-1} is the functional inverse of ϕ.

For two random variables X_1 and X_2, let

$$Y_1 = \phi_1(X_1, X_2) \tag{C-40}$$

and

$$Y_2 = \phi_2(X_1, X_2)$$

Let the transformations ϕ_1 and ϕ_2 have inverses θ_1 and θ_2, respectively:

$$X_1 = \theta_1(Y_1, Y_2)$$
$$X_2 = \theta_2(Y_1, Y_2) \tag{C-41}$$

Then

$$f_{Y_1 Y_2}(y_1, y_2) = \frac{f_{X_1 X_2}(x_1, x_2)}{|J|}\bigg|_{\substack{x_1 = \theta_1(y_1, y_2) \\ x_2 = \theta_2(y_1, y_2)}} \tag{C-42}$$

where J is the Jacobian

$$J = \begin{vmatrix} \dfrac{\partial \phi_1}{\partial x_1} & \dfrac{\partial \phi_1}{\partial x_2} \\[2mm] \dfrac{\partial \phi_2}{\partial x_1} & \dfrac{\partial \phi_2}{\partial x_2} \end{vmatrix} \tag{C-43}$$

C.3 STOCHASTIC PROCESSES

It is customary to model noise and measurement errors as random, or stochastic, processes. We may define a stochastic process as a time-varying rv. As in the case of the rv, we associate a time function $X(\zeta, t)$ with every possible outcome ζ of an experiment. The ensemble of all these time functions (called *sample functions*) constitutes the random process $\{X(t)\}$. Notice that if we sample this process at some particular time t_0, we obtain a random variable.

We may consider the statistics of a stochastic process in two ways. If we fix t and consider the rv $X(t_0)\}$, we can obtain statistics over the ensemble; thus, for example, $\langle X(t_0) \rangle$ is the ensemble average. If we fix ζ, that is, consider a particular sample function, then we have a time function and the statistics we obtain are temporal; thus for example, $\langle X(\zeta_k) \rangle$ is the time average. Roughly, if the time average is equal to the ensemble average, we say the process is ergodic.

The question of ergodicity is important because in practice, we often have access to only one sample function. If we are considering atmospheric disturbances as a stochastic process, then the noise we observe (for example, with an oscilloscope) is a single sample function and the only one with which we can work: we cannot go back in time and rerun the atmosphere to obtain a different sample function. Because of this, we can generally work only with temporal statistics, and it is important to be sure that the temporal statistics we obtain (for example, the measured noise power) are truly representative of the process as a whole.

If we consider the rv $X(t_1)$, then we may describe it by a pdf $f_{X_1}(x; t_1)$ in the usual way. Similarly, $X(t_2)$ may be described by $f_{X_2}(x; t_2)$. The pair $X(t_1)$ and $X(t_2)$ are thus two rvs with joint density $f_{X_1 X_2}(x_1, x_2; t_1, t_2)$ defined as

$$f_{X_1 X_2}(x_1, x_2; t_1, t_2)\, dx_1\, dx_2$$
$$= P\{x_1 < X(t_1) \leq x_1 + dx_1, x_2 < X(t_2) \leq x_2 + dx_2\} \qquad \text{(C-44)}$$

(Where no ambiguity will result, we will omit the $X_1 X_2$ subscript.) Thus,

1. The mean of the process at time t_1 is

$$E\{X(t_1)\} = \int x f_{X_1}(x, t_1)\, dx = \int x f(x, t_1)\, dx = \eta(t_1) \qquad \text{(C-45)}$$

2. The *autocorrelation* is

$$R(t_1, t_2) \triangleq E\{X(t_1)X(t_2)\}$$
$$= \int\int x_1 x_2 f(x_1, x_2; t_1, t_2)\, dx_1\, dx_2 \qquad \text{(C-46)}$$

3. The *autocovariance* (or sometimes just covariance) is

$$C(t_1, t_2) \triangleq E\{[X(t_1) - \eta(t_1)][X(t_2) - \eta(t_2)]\}$$
$$= R(t_1, t_2) - \eta(t_1)\eta(t_2) \qquad \text{(C-47)}$$

4. For two distinct processes $X(t)$ and $Y(t)$, the cross-correlation is

$$R_{XY}(t_1, t_2) \triangleq E\{X(t_1)Y(t_2)\}$$

$$= \int\int xy f_{XY}(x, y; t_1, t_2)\, dx\, dy \qquad \text{(C-48)}$$

[The processes $X(t)$ and $Y(t)$ are assumed to be real. For complex processes, $R_{XY}(t_1, t_2) \triangleq E\{X(t_1)Y^*(t_2)\}$.]

The processes X and Y are

1. Orthogonal if $E\{X(t_1)Y(t_2)\} = 0, \forall t_1, t_2$

2. Uncorrelated if $E\{X(t_1)Y(t_2)\} = E\{X(t_1)\}E\{Y(t_2)\}, \forall t_1, t_2$

3. Independent if $X(t_1), X(t_2), ..., X(t_m)$ are independent of $Y(t_1'), Y(t_2'), ...,$ $Y(t_n'), \forall t_1, ..., t_m, t_1', ..., t_n'$

The process X is wide-sense stationary (WSS) if

$$E\{X(t)\} \text{ is a constant}$$

and
$$R(t_1, t_2) = R(t_1 - t_2)$$

that is, if the autocorrelation depends only on the time difference.

  Ergodicity.   To be ergodic, a process must be stationary. A stationary process is ergodic if all statistics can be obtained from a single sample function. In particular let $\eta = E\{X(t)\}$, an ensemble average. We define

$$\bar{X}_T = \frac{1}{2T}\int_{-T}^{T} x(t)\, dt \qquad \text{a time average of duration } 2T \qquad \text{(C-49)}$$

The mean is ergodic if

$$\lim_{T\to\infty} \bar{X}_T = \eta.$$

Similarly, let $R(\tau) = E\{X(t)X(t + \tau)\}$ and define

$$\bar{R}_T(\tau) = \frac{1}{2T}\int_{-T}^{T} x(t)x(t + \tau)\, dt \qquad \text{(C-50)}$$

[Again, $X(t)$ is assumed real; for X complex, $R(\tau) = E\{X(t)X^*(t + \tau).]$ Then the autocorrelation is ergodic if $\lim_{T\to\infty} \bar{R}_T(\tau) = R(\tau)$.

Correlation Properties. (for WSS functions). Let $\eta = E\{X(t)\}$ and $R(\tau) = E\{X(t)X(t + \tau)\}$. (We assume X real throughout.) Then

1. $R(\tau) = R(-\tau)$

2. $R(0) = E\{X^2(t)\}$

3. $|R(\tau)| \leq R(0)$

4. $\lim_{\tau\to\infty} R(\tau) = \eta^2$

$$\text{(C-51)}$$

For X and Y real processes, $R_{XY}(\tau) = E\{X(t + \tau)Y(t)\}$ is the cross-correlation. We have

1. $|R_{XY}(\tau)|^2 < R_{XX}(0)R_{YY}(0);$ $\qquad\qquad$ (C-52)

2. $|R_{XY}(\tau)| < \frac{1}{2}[R_{XX}(0) + R_{YY}(0)].$

Power Spectral Density. The power spectral density (psd) of a WSS process is the Fourier transform of its autocorrelation.

$$S(\omega) = \int_{-\infty}^{\infty} R(\tau)e^{-j\omega\tau}\, d\tau \qquad\qquad \text{(C-53)}$$

$$R(\tau) = \frac{1}{2\pi} \int_{-\infty}^{\infty} S(\omega)e^{j\omega\tau}\, d\omega \qquad\qquad \text{(C-54)}$$

The psd has the following properties:

1. $S(\omega)$ is real;

2. $S(\omega) \geq 0$

3. If $X(t)$ is real, $S(\omega)$ is an even function of ω;

$$\text{(C-55)}$$

4. $E\{X^2(t)\} = R(0) = \frac{1}{2\pi} \int_{-\infty}^{\infty} S(\omega)\, d\omega$

Linear Systems. Let L be a linear time-invariant system with stable impulse response $h(t)$ and corresponding transfer function $H(\omega)$. If the input $X(t)$ is WSS with mean η_X and autocorrelation $R_{XX}(\tau)$ with corresponding psd $S_{XX}(\omega)$, then the output $Y(t)$ is also WSS and is characterized by

1. $\eta_Y = H(0) \cdot \eta_X$

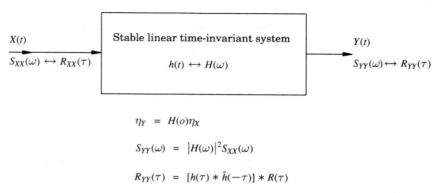

$$\eta_Y = H(o)\eta_X$$

$$S_{YY}(\omega) = |H(\omega)|^2 S_{XX}(\omega)$$

$$R_{YY}(\tau) = [h(\tau) * \bar{h}(-\tau)] * R(\tau)$$

Figure C-1 *Input-output relationships for WSS signals through linear, time-invariant system.*

2. $R_{YY}(\tau) = [h(\tau) * h(-\tau)] * R_{XX}(\tau)$ (C-56)

3. $S_{YY}(\omega) = |H(\omega)|^2 S_{XX}(\omega)$

These properties are illustrated in Fig. C-1.

C.4 RANDOM SEQUENCES

The preceding development of a random process can be readily particularized to the discrete-time case. We simply sample the process $X(t)$ at times $t_1, t_2, \ldots$ to generate a time-ordered sequence of rvs $X(t_1), X(t_2), \ldots$. Hence the properties and definitions for the pair of rvs $X(t_1)$ and $X(t_2)$, given in Eqs. (C-44) through (C-48), carry over intact when these are viewed as members of a sequence. In particular, the discrete autocorrelation is recognized as simply the sampling of the continuous-time autocorrelation function. Also, the definitions of uncorrelated, orthogonal, and independent sequences parallel those given previously.

For convenience, we assume that the samples are taken at uniformly spaced intervals $t_k = kT_s$ and that T_s is normalized to unity. Then

$$X(t_k) = X(kT_s) = X(k) \tag{C-57}$$

and
$$R(m, n) = E\{X(m)X(n)\} \tag{C-58}$$

A WSS sequence is one in which

$$E\{X(k)\} = \text{constant} \tag{C-59}$$

and
$$E\{X(n)X(n+k)\} = R(k)$$

The power spectral density of the sequence $\{X(k)\}$ is the Z transform of the autocorrelation:

$$S_D(z) = \sum_{k=-\infty}^{\infty} R(k)z^{-k} \tag{C-60}$$

and inversely,
$$R(k) = \frac{1}{2\pi j} \oint S_D(z)z^{k-1}\, dz \tag{C-61}$$

where the contour of integration is the unit circle. But on that contour, $z = e^{j\omega T_s} = e^{j\theta}$ (where $\theta \triangleq \omega T_s$), and it follows from Fourier theory (App. B) that

$$S_D(e^{j\theta}) = \sum_{k=-\infty}^{\infty} R(k)e^{-jk\theta} = \mathcal{S}(\theta) \tag{C-62}$$

is the discrete-time Fourier transform of $R(k)$. The inverse is

$$R(k) = \frac{1}{2\pi} \int_{-\pi}^{\pi} \mathcal{S}(\theta)e^{jk\theta}\, d\theta \tag{C-63}$$

Hence $\mathcal{S}(\theta)$ is periodic in the normalized frequency θ with period 2π. Putting $k = 0$ in Eq. (C-62) gives the mean-squared value

$$E\{X^2(t)\} = R(0) = \frac{1}{2\pi} \int_{-\pi}^{\pi} \mathcal{S}(\theta)\, d\theta \tag{C-64}$$

$R(k)$ and $\mathcal{S}(\theta)$ are the discrete counterparts to $R(\tau)$ and $S(\omega)$ for the analog signal. Furthermore from the sampling theorem, we know that if $R(k)$ is a sampling of $R(\tau)$, then its Fourier transform $\mathcal{S}(\theta)$ is the periodic repetition of $S(\omega)$, the Fourier transform of $R(\tau)$. Hence

$$\mathcal{S}(\theta) = \frac{1}{T_s} \sum_{k=-\infty}^{\infty} S(\omega - k\Omega_s) \tag{C-65}$$

where
$$\Omega_s \triangleq 2\pi/T_s \quad \text{and} \quad \theta \triangleq \omega T_s$$

Ergodicity.   The discrete-time counterparts to Eqs. (C-49) and (C-50) for WSS sequences are the time averages

$$\bar{X}_M = \frac{1}{2M+1} \sum_{k=-M}^{M} X(k) \tag{C-66}$$

$$\bar{R}_M(n) = \frac{1}{2M+1} \sum_{k=-M}^{M} X(k)X(k+n) \tag{C-67}$$

As before, the mean and autocorrelation estimates are ergodic if

$$\lim_{M \to \infty} \bar{X}_M = \eta = E\{X(k)\} \tag{C-68}$$

and

$$\lim_{M \to \infty} \bar{R}_M(k) = R(k) \tag{C-69}$$

Linear Systems. The response $Y(n)$ of a linear, time-invariant, stable filter with weighting sequence $h(n)$ and transfer function $H(z)$ to a real WSS input $X(n)$ is also WSS with properties that parallel Eq. (C-56):

$$\eta_Y = H(1)\eta_X$$
$$S_{YY}(z) = H(z)H^*(1/z^*)S_{XX}(z) = Z\{R_{YY}(k)\}$$
$$\mathcal{S}_{YY}(\theta) = |H(e^{j\theta})|^2\mathcal{S}_{XX}(\theta) \tag{C-70}$$
$$R_{YY}(k) = h(k) * h^*(-k) * R_{XX}(k)$$

where the X and XX subscripts refer to the input, the Y and YY subscripts refer to the output, and the asterisk superscript indicates the complex conjugate. These properties are illustrated in Fig. C-2.

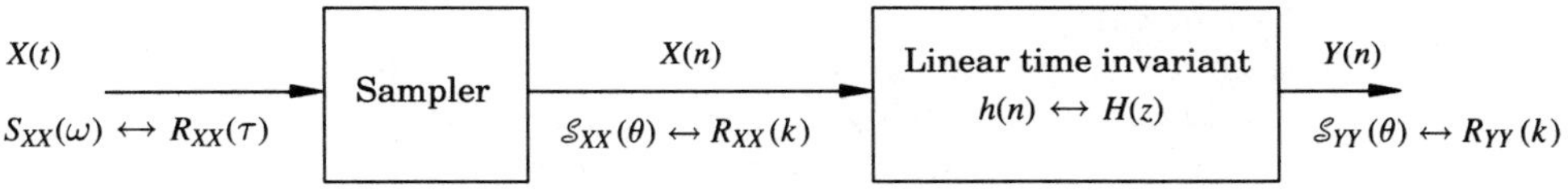

Sampling: $R_{XX}(k) = R_{XX}(kT_s)$ $\theta \triangleq \omega T_s$

$$\mathcal{S}_{XX}(\theta) = \frac{1}{T_s} \sum_{k=-\infty}^{\infty} S_{XX}(\omega - k\Omega_s) \qquad \Omega_s \triangleq 2\pi = T_s$$

Filtering: $\mathcal{S}_{YY}(\theta) = |H(e^{j\theta})|^2\mathcal{S}_{XX}(\theta)$

$$R_{YY}(k) = [h(k) * \bar{h}(-k)] * R_{XX}(k)$$

Figure C-2 *Input-output relationships for WSS signal through sampler followed by linear, time-invariant system.*

APPENDIX D

■ ■

WINDOWING FUNCTIONS

The need to apply time weighting to a function before Fourier transformation arises in the design of FIR filters and in spectral analysis, as described in Chaps. 5 and 7, respectively. Beginning texts in DSP differ in their coverage of this topic; this appendix provides a list of the most generally useful weighting, or windowing, functions and their characteristics. For an exhaustive summary, we refer the reader to Harris (1978).

We will define each window and its transform, and give (1) the width in samples of the main lobe at the 3-dB point, (2) the height of the largest side-lobe, (3) the rate at which the side-lobe amplitudes decrease, and (4) the locations of the zeroes in the transforms. Time functions and transforms will be given in units of samples, since this is usually what is of interest, with peak amplitudes scaled to unity. We will describe the windows as centered about $n = 0$; this removes the linear-phase term from the transform and usually allows us to write the form of the window more economically. In the following, $\text{rect}_N(n)$ is a rectangular pulse of width N, and the tranforms are normalized to $W(0) = 1$.

The *rectangular window*, Fig. D-1(a), is our starting point. We have

$$w(n) = \text{rect}_N(n) = \begin{cases} 1, & |n| \leq (N-1)/2 \\ 0 & \text{otherwise} \end{cases} \tag{D-1}$$

and the normalized transform, Fig. D-1(b) and (c), is

$$W(k) = \frac{\sin(\pi k)}{N \sin(\pi k/N)} \tag{D-2}$$

$$= D_N(k)$$

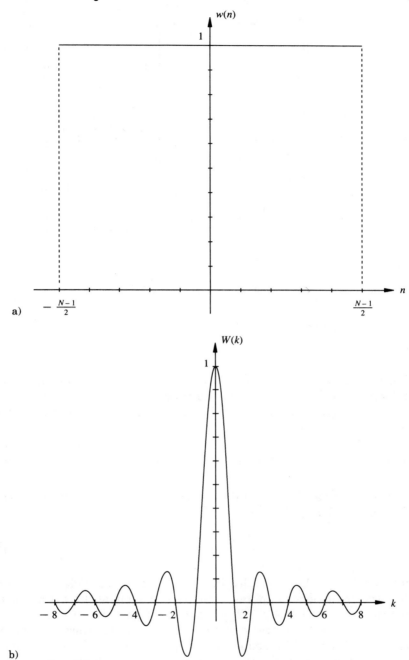

Figure D-1   *(a) The rectangular window; (b) its transform.*

$D_N(k)$ is the Dirichlet kernel (see Chap. 7). The main lobe is 0.89 samples in width; the largest side-lobe is 13 dB below the main lobe, and the

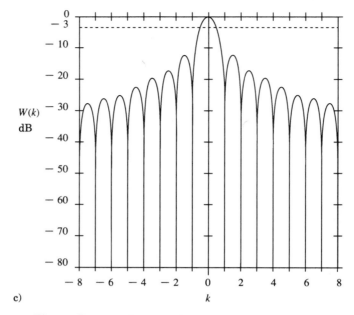

W(k)
dB

c)

Figure D-1   (c) Rectangular window transform.

side-lobes fall off at 6 dB per octave. The first zeroes are at $k = \pm1$; subsequent zeroes are at integer multiples of k.

The triangular window, Fig. D-2(a), is also called the Bartlett window and is given by

$$w(n) = (1 - 2|n|/N)\operatorname{rect}_N(n) \tag{D-3}$$

It can be viewed as the convolution of two rectangles of width $N/2$. Hence its transform, Fig. D-2(b) and (c), is given by

$$W(k) = [D_{N/2}(k/2)]^2 \tag{D-4}$$

The main lobe is 1.28 samples in width; the largest side-lobe is 27 dB below the main lobe, and the side-lobes fall off at 12 dB per octave. The first zeroes fall at $k = \pm2$; subsequent zeroes are at even multiples of k. When power spectral density is found by transforming the auto-correlation function of a sequence, the usual method for estimating the autocorrelation imposes a triangular weighting, as described in Chap. 7; hence this window may arise automatically from this procedure.

The *raised-cosine* window, Fig. D-3(a), is also known as the *Hann* (or *Hanning*) window and is given by

$$w(n) = \tfrac{1}{2}(1 + \cos 2\pi n/N)\operatorname{rect}_N(n) \tag{D-5}$$

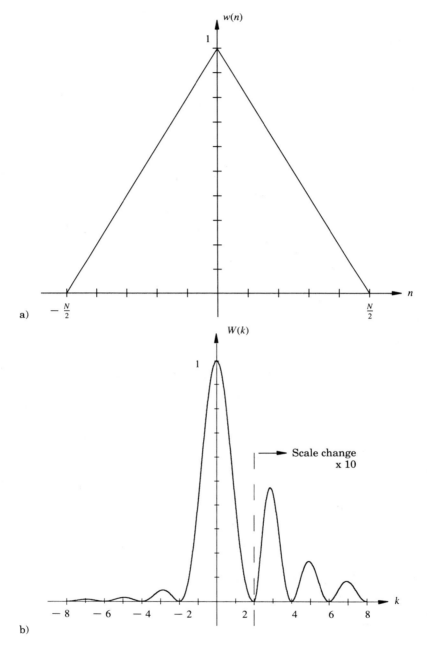

Figure D-2   *(a) The Bartlett window; (b) its transform.*

This window is the sum of a constant and a cosine, both multiplied by a rectangular window; hence by the convolution property, the transform, Fig. D-3(b) and (c), can be written

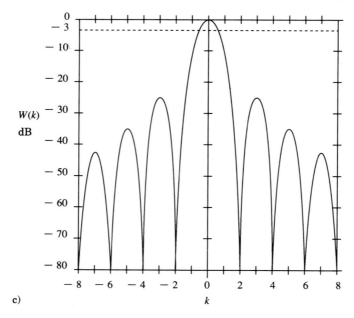

c)

Figure D-2   *(c) Bartlett window transform.*

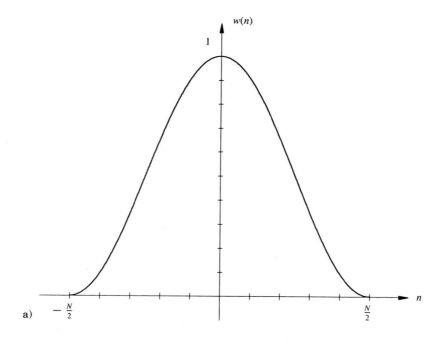

a)

Figure D-3   *(a) The Hanning window.*

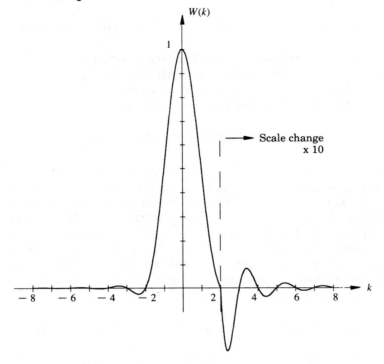

b)

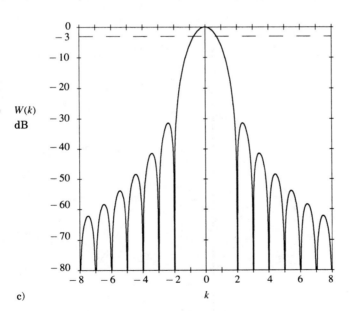

c)

Figure D-3   *(b) and (c) Hanning window transforms.*

$$W(k) = D_N(k) + \tfrac{1}{2}[D_N(k-1) + D_N(k+1)] \qquad \text{for large } N \quad \text{(D-6)}$$

The width of its main lobe is 1.44; the largest side-lobe is 32 dB below the main lobe, and the side-lobes fall off at -18 dB/octave. The first zeroes are at $k = \pm 2$; subsequent zeroes are at integer multiples of k.

The *Hamming* window, Fig. D-4(a), is defined as

$$w(n) = [a + b\cos(2\pi n/N)]\operatorname{rect}_N(n) \qquad\qquad \text{(D-7)}$$

where the constants a and b are 0.5435 and 0.4565, respectively. The Hamming window is the result of an ingenious manipulation of the Dirichlet kernel. Its transform can be analyzed as the sum of Dirichlet kernels, like the Hanning window, but the Hamming window adjusts the magnitudes of these kernels to provide complete cancellation among these three components at $k = \pm 2.5$. This puts a zero in the middle of what used to be the largest side-lobe; the two smaller side-lobes that result are greatly reduced in amplitude. We may write the transform, Fig. D-4(b) and (c), as

$$W(k) = D_N(k) + \tfrac{1}{2}(b/a)[D_N(k-1) + D_N(k+1)] \qquad \text{for large } N \quad \text{(D-8)}$$

The width of the main lobe is 1.30 samples; the largest side-lobe is 43 dB down; and the side-lobes fall off at -6dB/octave. The first zeroes are

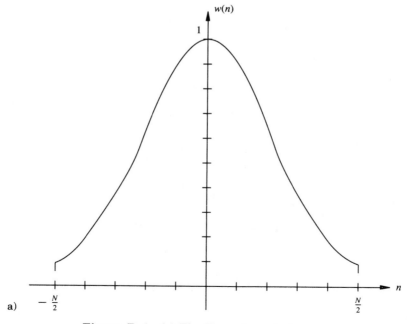

a)

Figure D-4 *(a) The Hamming window.*

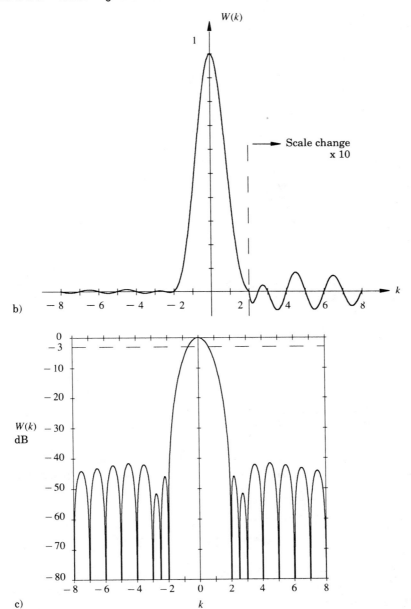

b)

c)

Figure D-4   *(b) and (c) Hamming window transforms.*

at $k = \pm 2$, and subsequent zeroes are at integer multiples of k, except for the extra zeroes at $k = \pm 2.5$.

The Blackman window, Fig. D-5(a), is an extension of Hamming's cancellation scheme to five kernels:

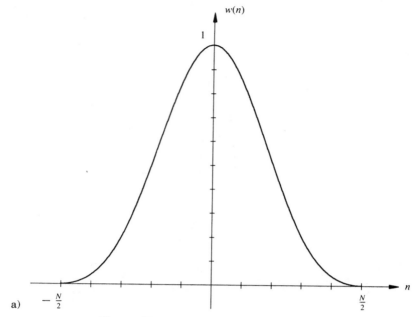

$$w(n)$$

a) $-\dfrac{N}{2}$ $\dfrac{N}{2}$

Figure D-5   *(a) The Blackman window.*

$$w(n) = [a + b\cos(2\pi n/N) + c\cos(4\pi n/N)]\,\mathrm{rect}_N(n) \qquad \text{(D-9)}$$

The coefficients a, b, and c, are commonly taken to be 0.42, 0.50, and 0.08, respectively. These values do not provide exact cancellation, but they have the advantage that $w(-N/2) = w(N/2) = 0$. As a result, the window is continuous at its boundaries and the side-lobes fall off at -18 dB/octave. The transform, Fig. D-5(b) and (c), may be written, rather awkwardly

$$W(k) = D_N(k) + 0.5\{(b/a)[D_N(k-1) + D_N(k+1)]$$
$$+ (c/a)[D_N(k-2) + D_N(k+2)]\} \qquad \text{(D-10)}$$

The highest side-lobe is 58 dB down from the main lobe; this reduction in amplitude is paid for by a rather broad main lobe, whose 3-dB width is 1.68 samples. The first zeroes fall at $k = \pm 3$; subsequent zeroes are at integer multiples of k.

More ambitious windows can by found by various optimization procedures. The resulting weighting functions are generally expensive to compute, but they are occasionally used. Of these, the most important are the *Kaiser* window and the *Dolph-Chebyshev* window.

The optimization problem leading to the Kaiser window is this: find a time-limited function whose transform has the minimum energy outside the main lobe. The solution, found by Slepian, Pollack, and Landau (1961), is based on the prolate spheroidal wave functions. A relatively

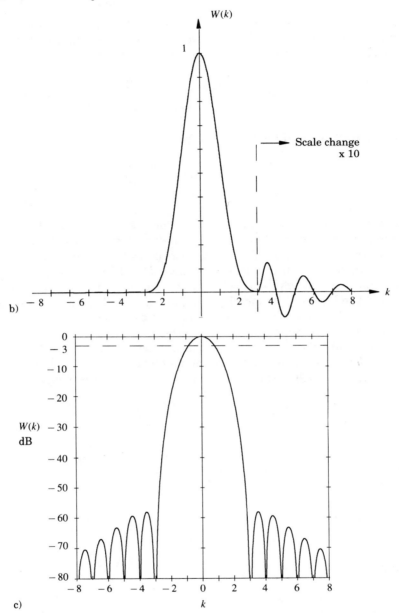

Figure D-5   *(b) and (c) Blackman window transforms.*

tractable approximation to this solution is known as the Kaiser window. This window takes the form

$$w(n) = \frac{I_0[\pi a\sqrt{1 - (2n/N)^2}]}{I_0(\pi a)} \, \text{rect}_N(n) \qquad \text{(D-11)}$$

where I_0 is the zero-order modified Bessel function of the first kind. The constant a controls the tradeoff between the width of the main lobe and the level of the side-lobes. Its transform is given by

$$W(k) = \frac{2\sinh(\pi\sqrt{a^2 - k^2})}{I_0(\pi a)\sqrt{a^2 - k^2}} \tag{D-12}$$

for the main lobe, and for the side-lobes,

$$W(k) = \frac{2\sin(\pi\sqrt{k^2 - a^2})}{I_0(\pi a)\sqrt{k^2 - a^2}} \tag{D-13}$$

This is the non-aliased form; no closed-form expression for the aliased form is known. The window and its transform are shown in Fig. D-6(a), (b), and (c) for $a = 2.0$. The main lobe is 1.43 samples wide and the largest side-lobe is 46 dB down. Figures D-6(d), (e), and (f) show the window and transform for $a = 2.5$; here the main lobe is 1.57 samples wide and the largest side-lobe is 57 dB down. In all cases, the side-lobes fall off at 6 dB/octave. The zeroes are located at $k = \pm\sqrt{i^2 + a^2}, i = 1, 2, ...$

The *Dolph-Chebyshev* window arises from another approach to optimization: find the window with the narrowest main lobe for a given side-lobe level. We said that side-lobe suppression is purchased at the cost of widening the main lobe; hence to minimize the widening of the

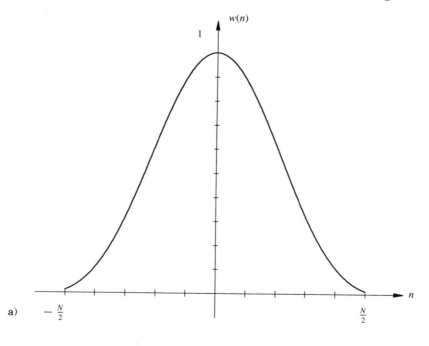

w(n)

1

$-\frac{N}{2}$

$\frac{N}{2}$

n

a)

Figure D-6   (a) The Kaiser window ($a = 2.0$).

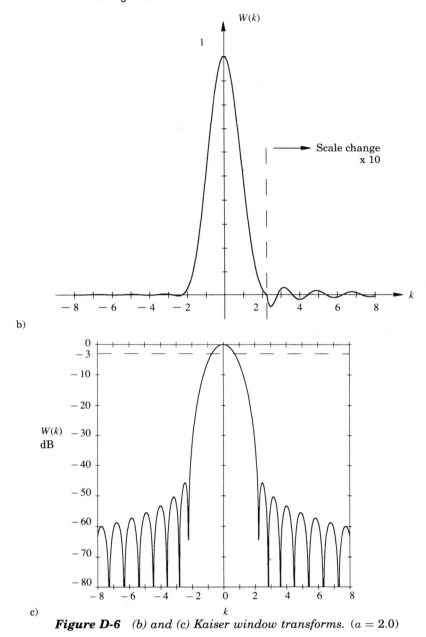

b)

c)

Figure D-6   *(b) and (c) Kaiser window transforms.   (a = 2.0)*

main lobe, we do not make any side-lobe smaller than necessary. In consequence, all the side-lobes have the same magnitude. The transform is
given by

$$W(k) = A \cosh\{N \cosh^{-1}[\alpha \cos(\pi k/N)]\} \qquad \text{(D-14)}$$

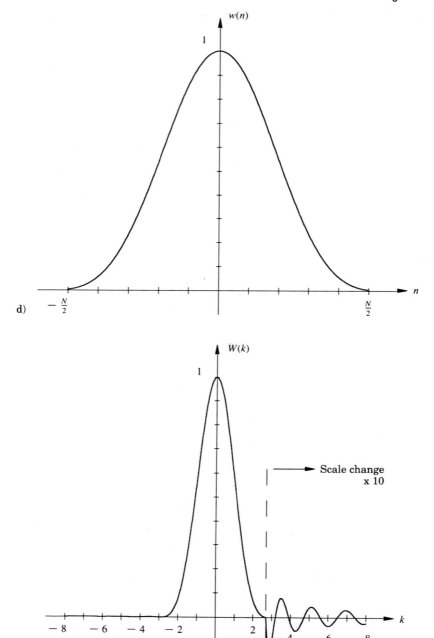

Figure D-6   *(d) the Kaiser window (a = 2.5); (e) its transform.*

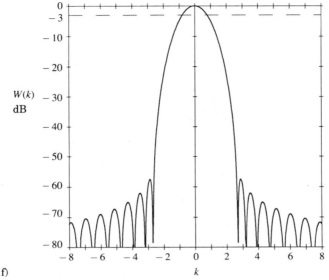

Figure D-6 *(f) Kaiser window transform. (a = 2.5)*

for the main lobe and for the side-lobes,

$$W(k) = A \cos\{N \cos^{-1}[\alpha \cos(\pi k/N)]\} \tag{D-15}$$

where

$$A = 1/\cosh(N \cosh^{-1} \alpha)$$

is a scale factor to make $W(0)$ unity. The parameter α determines the tradeoff between the width of the main lobe and the level of the side-lobes. A Dolph-Chebyshev window is shown in Fig. D-7(a) and (b); in this example, the side-lobe level is -60 dB and the width of the main lobe is 1.44. The zeroes are given by

$$k = (N/\pi) \cos^{-1}\{1/\alpha \cos[(i + 1/2)\pi/N]\} \qquad i = 0, 1, 2, ... \tag{D-16}$$

This would seem to be the ultimate window function, but it is awkward to compute (normally one computes the transform and finds the time window by an inverse DFT); and the constant side-lobe level can be a trap.

Many physical phenomena have spectra with a few large peaks and then a general falloff toward the high frequencies. Unless the side-lobes of the Dolph-Chebyshev window are kept at a very low level, they may show up at high frequencies and obscure components in that range. (This is well illustrated in a figure in Harris's paper.) Therefore, although the Dolph-Chebyshev window represents a theoretical optimum, it may not

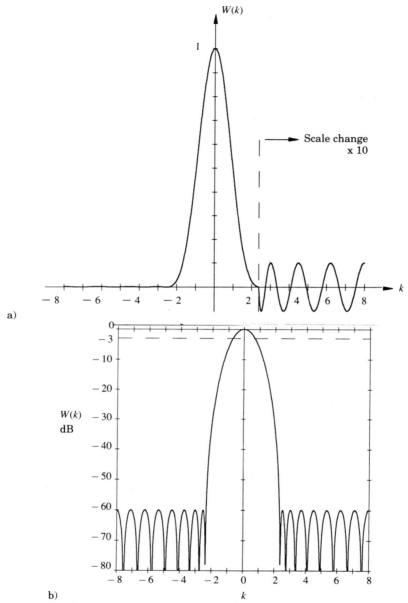

a)

b)

Figure D-7 *Transform of the Dolph-Chebyshev window.*

be practical. Relatively crude windows like the Hanning or Hamming windows provide a desirable rolloff in their side-lobe structure, and these are probably the most frequently used windows; for applications where a peak side-lobe level of -43 dB is still too high, the Kaiser window seems the reasonable choice.

REFERENCES

R. B. Blackman and J. W. Tukey. *The Measurement of Power Spectra*. New York: Dover Books, 1958.

F. J. Harris. "On the use of windows for harmonic analysis with the discrete Fourier transform." *Proc. IEEE*, vol. 66, no. 1, pp. 51–83, Jan. 1978.

H. Landau, H. Pollak. "Prolate-Spheroidal Wave Functions, Fourier Analysis, and Uncertainty—II," *Bell System Technical Journal*, Vol. 40, pp. 65–84, Jan. 1961.

D. Slepian, H. Pollak. "Prolate-Spheroidal Wave Functions, Fourier Analysis, and Uncertainty—I," *Bell System Technical Journal*, Vol. 40, pp. 43–64, Jan. 1961.

APPENDIX E

■ ■

LINEAR PREDICTION AND THE LEVINSON RECURSION

The problem addressed by linear prediction is as follows: given a linear, time-invariant, discrete-time system whose parameters are not known, estimate the forthcoming output sample from a linear combination of input samples, past output samples, or both. This procedure, which might seem at first glance to be restricted to a relatively narrow field of interest, has turned out to have a wide range of applicability in digital signal processing. It has applications in system modelling, speech and image processing, and the problem of maximum-entropy spectral analysis, and it has given rise to a filter structure of special interest, the lattice filter. These topics have mostly been covered in the body of the text; the purpose of this appendix is to provide a general background in linear prediction. Other references that the reader may find useful are Makhoul (1975), Markel and Gray (1976), and Parsons (1986).

E.1 PREDICTION AND MODELLING

If the system inputs and outputs are $\{x(n)\}$ and $\{y(n)\}$, respectively, then the estimate is written

$$\hat{y}(n) = -\sum_{i=1}^{N} a_i y(n-i) + \sum_{j=0}^{M} b_j x(n-j) \tag{E-1}$$

where the hat indicates an estimate. (The first sum is preceded by a negative sign for notational convenience later on.) We observe that Eq. (E-1) is essentially identical in form to the difference equation of a linear, discrete-time, time-invariant system, for example, Eq. (1-7). Hence if the system parameters were known, we could find $y(n)$ exactly by using the coefficients of the system's difference equation for $\{a_i\}$ and $\{b_i\}$. This suggests that in the case of an unknown system, solving Eq. (E-1) would provide a model of the system as well, since if we define $a_0 = 1$, we may write

$$A(z)Y(z) = B(z)X(z)$$

from which
$$\hat{H}(z) = \frac{Y(z)}{X(z)} = \frac{B(z)}{A(z)} \tag{E-2}$$

where the hat again indicates an estimate.

The problem as stated here is the most general form. The parameters obtained from the solution provide estimates of the poles and zeroes of the system model, and Eq. (E-1) is termed the mixed pole-zero model. Two simpler, related models are the all-zero model

$$\hat{y}(n) = \sum_{j=0}^{M} b_j x(n-j) \tag{E-3}$$

and the all-pole model

$$\hat{y}(n) = -\sum_{i=1}^{M} a_i y(n-i) \tag{E-4}$$

The all-zero model is also known as the moving-average (MA) model, the all-pole model as the autoregressive (AR) model, and the mixed pole-zero model as the autoregressive moving-average (ARMA) model.

Of these models, the one of greatest interest, and the one on which we will concentrate here, is the all-pole model. There are two reasons for this. First, it is the simplest to compute; the presence of zeroes results in simultaneous nonlinear equations, which are computationally expensive to solve. Second, in many applications we may have access only to the $\{y\}$ sequence.

The general solution to the all-pole equations Eq. (E-4) proceeds as follows: for any set of parameters $\{a_i\}$, the error in estimating $y(n)$ is given by

$$e(n) = y(n) - \hat{y}(n)$$

$$= \sum_{i=0}^{N} a_i y(n-i) \tag{E-5}$$

where a_0 is taken as 1. We seek to miminize this error in the least-squares sense: that is, we wish to minimize

$$E = \langle e^2(n) \rangle$$

By the orthogonality principle (see for example Papoulis, 1986), this minimum can be found by choosing the $\{a_i\}$ so that the error is orthogonal to the samples $y(n-i)$, $i = 0, 1, ..., N$. Thus we require

$$\langle e(n)y(n-j)\rangle = 0 \qquad j = 1, 2, ..., N$$

where the angle brackets indicate the expected-value operation. Substituting Eq. (E-5) and exchanging the summation and expected-value operations, we have

$$\sum_{i=0}^{N} a_i \langle y(n-i)y(n-j)\rangle = 0 \qquad j = 1, 2, ..., N \qquad \text{(E-6)}$$

If y is a stationary signal, then $\langle y(n-i)y(n-j)\rangle$ is the autocorrelation function

$$r_{i-j} = \langle y(n-i)y(n-j)\rangle$$

and Eq. (E-6) may be written

$$\sum_{i=0}^{N} a_i r_{i-j} = 0 \qquad j = 1, 2, ..., N \qquad \text{(E-7)}$$

The orthogonality principle also tells us that the resulting minimum mean-squared error is given by

$$E = \sum_{i=0}^{N} a_i r_i \qquad \text{(E-8)}$$

E.2 THE LEVINSON RECURSION

The system Eq. (E-7) has a recursive solution that is computationally cheap and of considerable theoretical interest. In what follows, we will assume that the time series is real; the generalization to complex series is not difficult. In that case, $r_{-i} = r_i$, and we can replace r_{n-i} with $r_{|n-i|}$. We will also write $a(i)$ instead of a_i, since this will simplify our notation. With these changes, we can now combine Eqs. (E-7) and (E-8) as follows:

$$\begin{bmatrix} r_0 & r_1 & r_2 & \cdots & r_p \\ r_1 & r_0 & r_1 & \cdots & r_{p-1} \\ r_2 & r_1 & r_0 & \cdots & r_{p-2} \\ \cdots & \cdots & \cdots & \cdots & \cdots \\ r_p & r_{p-1} & r_{p-2} & \cdots & r_0 \end{bmatrix} \begin{bmatrix} a(0) \\ a(1) \\ a(2) \\ \cdots \\ a(p) \end{bmatrix} = \begin{bmatrix} E_p \\ 0 \\ 0 \\ \cdots \\ 0 \end{bmatrix} \qquad \text{(E-9)}$$

There are a number of ways to derive the recursive solution; perhaps the best is due to Burg (1975), and we will use that here. Suppose we have a solution for some value of p, say, n, and let this solution be $\mathbf{a}_n$; then

$$
\begin{bmatrix}
r_0 & r_1 & \cdots & r_n \\
r_1 & r_0 & \cdots & r_{n-1} \\
\cdots & \cdots & \cdots & \cdots \\
r_n & r_{n-1} & \cdots & r_0
\end{bmatrix}
\begin{bmatrix}
a_n(0) \\
a_n(1) \\
\cdots \\
a_n(n)
\end{bmatrix}
=
\begin{bmatrix}
E_n \\
0 \\
\cdots \\
0
\end{bmatrix}
$$

We wish to use this solution as a basis for solving

$$
\begin{bmatrix}
r_0 & r_1 & \cdots & r_n & r_{n+1} \\
r_1 & r_0 & \cdots & r_{n-1} & r_n \\
\cdots & \cdots & \cdots & \cdots & \cdots \\
r_n & r_{n-1} & \cdots & r_0 & r_1 \\
r_{n+1} & r_n & \cdots & r_1 & r_0
\end{bmatrix}
\begin{bmatrix}
a_{n+1}(0) \\
a_{n+1}(1) \\
\cdots \\
a_{n+1}(n) \\
a_{n+1}(n+1)
\end{bmatrix}
=
\begin{bmatrix}
E_{n+1} \\
0 \\
\cdots \\
0 \\
0
\end{bmatrix}
$$

Because the $\mathbf{R}$ matrix is Toeplitz and symmetric, the jth row from the bottom is always the reverse of the jth row from the top; hence any solution will always work with the vectors reversed. In particular for $p = n$,

$$
\begin{bmatrix}
r_0 & r_1 & \cdots & r_n \\
r_1 & r_0 & \cdots & r_{n-1} \\
\cdots & \cdots & \cdots & \cdots \\
r_n & r_{n-1} & \cdots & r_0
\end{bmatrix}
\begin{bmatrix}
a_n(n) \\
a_n(n-1) \\
\cdots \\
a_n(0)
\end{bmatrix}
=
\begin{bmatrix}
0 \\
0 \\
\cdots \\
E_n
\end{bmatrix}
$$

In that case, let us assume a solution of the form

$$
\begin{bmatrix}
a_{n+1}(0) \\
a_{n+1}(1) \\
\cdots \\
a_{n+1}(n) \\
a_{n+1}(n+1)
\end{bmatrix}
=
\begin{bmatrix}
a_n(0) \\
a_n(1) \\
\cdots \\
a_n(n) \\
0
\end{bmatrix}
+ k_{n+1}
\begin{bmatrix}
0 \\
a_n(n) \\
\cdots \\
a_n(1) \\
a_n(0)
\end{bmatrix}
$$

where k_{n+1} is a constant to be determined. Then we have

$$
\begin{bmatrix}
r_0 & r_1 & \cdots & r_{n+1} \\
r_1 & r_0 & \cdots & r_n \\
\cdots & \cdots & \cdots & \cdots \\
r_n & r_{n-1} & \cdots & r_1 \\
r_{n+1} & r_n & \cdots & r_0
\end{bmatrix}
\left\{
\begin{bmatrix}
a_n(0) \\
a_n(1) \\
\cdots \\
a_n(n) \\
0
\end{bmatrix}
+ k_{n+1}
\begin{bmatrix}
0 \\
a_n(n) \\
\cdots \\
a_n(1) \\
a_n(0)
\end{bmatrix}
\right\}
$$

$$
=
\begin{bmatrix}
E_n \\
0 \\
\cdots \\
0 \\
q
\end{bmatrix}
+ k_{n+1}
\begin{bmatrix}
q \\
0 \\
\cdots \\
0 \\
E_n
\end{bmatrix}
$$

where
$$q = \sum_{i=0}^{n} a_n(i)r_{n+1-i}$$

The appeal of this assumed solution is that it gives a right-hand vector which is entirely zero except for the top and bottom elements. We expect the top element to be nonzero, since it will be the new prediction error; hence we need only find a way of getting rid of the bottom element. We can do this by setting

$$k_{n+1} = \frac{-q}{E_n} = -\frac{1}{E_n}\sum_{i=0}^{n} a_n(i)r_{n+1-i} \tag{E-10}$$

This sets the bottom element to zero and results in a new prediction error of

$$E_{n+1} = E_n + k_{n+1}q$$

or
$$E_{n+1} = E_n(1 - k_{n+1}^2) \tag{E-11}$$

Hence we have a way of finding the order-$(n+1)$ solution from the order-n solution; this can clearly be applied to any order. Specifically,

$$a_{n+1}(j) = a_n(j) + k_{n+1}a_n(n+1-j) \tag{E-12}$$

where k_{n+1} is given by Eq. (E-10). This means that we can start with order 0 and work up to any desired order by repeating this recursive step as required.

Since E_n is the mean-squared prediction error of an order-n predictor, E_0 must be the error in the absence of any prediction; this is clearly r_0. The recursive solution therefore proceeds as follows:

1. For $n = 0$, $E_0 = r_0$ and $a_0(0) = 1$

2. For $n = 1$ to p do:
   a. $k_n = -1/E_n \sum_{i=0}^{n-1} a_{n-1}(i)r_{n-i}$.   (E-13)
   b. $a_n(n) = k_n$; $a_n(0) = 1$.
   c. For $i = 1$ to $n - 1$ do

   $$a_n(i) = a_{n-1}(i) + k_n a_{n-1}(n-i)$$

   d. $E_n = E_{n-1}(1 - k_n^2)$.

Notice that Eq. (E-11) implies that $|k_i| \leq 1$. Furthermore if $|k_i| = 1$ at any point in the recursion, then the prediction error vanishes and a higher-order predictor is neither necessary nor possible. Note also that if $k_p = 1$, the predictor coefficients are symmetrical; this can be seen by evaluating Eq. (E-12) with $k_p = 1$.

The parameters $\{k_p\}$ are clearly the key to this process. They are known as reflection coefficients and also as partial-correlation (PARCOR) coefficients; the reasons for these names may be found in Makhoul (1975), Markel and Gray (1976), or Parsons (1986). They carry as much information about the predictor as the predictor coefficients themselves, and they throw into relief particulars about the predictor's behavior that is not immediately evident from inspection of the $\{a_i\}$ parameters.

E.3   RECURRENCE RELATIONS

Given the recursive solution to the predictor equations, it may not be surprising that there are a number of recurrence relations among linear predictors of consecutive orders; there are also similar relations among the prediction errors. For economy, it is convenient to define a reverse predictor

$$b_p(j) = a_p(p + 1 - j)$$

Note that as a consequence of the convention that $a_p(0) = 1$ and $a_p(j) = 0$ for all $j < 0$ and $j > p$, we have $b_p(p + 1) = 1$ and $b_p(j) = 0$ for all $j < 1$ and $j > p + 1$.

Since we are dealing with stationary data and $r_i = r_{-i}$, the coefficients $\{b_i\}$ have this property: given any samples $y(n-1)$ through $y(n-p)$, the b coefficients provide a least-squares estimate of $y(n - p - 1)$.

This reverse predictor allows us to write Eq. (E-12) more economically:

$$a_p(j) = a_{p-1}(j) + k_p b_{p-1}(j) \tag{E-14}$$

and similarly $\qquad b_p(j) = b_{p-1}(j - 1) + k_p a_{p-1}(j - 1) \tag{E-15}$

If we take Z transforms of $\{a_i\}$ and $\{b_i\}$ and observe that

$$B_p(z) = z^{-(p+1)} A_p(1/z) \tag{E-16}$$

we may write

$$A_p(z) = A_{p-1}(z) + k_p B_{p-1}(z) \tag{E-17}$$

$$B_p(z) = z^{-1}[B_{p-1}(z) + k_p A_{p-1}(z)]$$

We can form similar recurrence relations among the prediction errors. Let $e_p^+(n)$ be the forward prediction error for an order-p predictor, and let $e_p^-(n)$ be the reverse prediction error:

$$e_p^+(n) = \sum_{i=0}^{p} a_p(i)y(n-i)$$

$$e_p^-(n) = \sum_{i=1}^{p+1} b_p(i)y(n-i)$$

Then the recurrence relations between consecutive prediction errors are isomorphic to the recurrence relations between predictors:

$$e_p^+(n) = e_{p-1}^+(n) + k_p e_{p-1}^-(n) \tag{E-18}$$

$$e_p^-(n) = e_{p-1}^-(n-1) + k_p e_{p-1}^+(n-1)$$

If we take Z transforms of these error sequences, we have

$$E_p^+(z) = E_{p-1}^+(z) + k_p E_{p-1}^-(z) \tag{E-19}$$

$$E_p^-(z) = z^{-1}[E_{p-1}^-(z) + k_p E_{p-1}^+(z)]$$

If we wish to transform a coefficient sequence $\{a_i\}$ into the corresponding $\{k_i\}$, we can divide through by a gain factor so that $a_0 = 1$ and then run Eq. (E-12) backward as follows: From Eq. (E-15), we have

$$-k_p b_p(j+1) = -k_p b_{p-1}(j) - k_p^2 a_{p-1}(j)$$

With Eq. (E-14), this gives us

$$a_{p-1}(j) = \frac{1}{1-k_p^2}[a_p(j) - k_p b_p(j+1)] \tag{E-20}$$

Since $k_p = a_p(p)$, we can find $\mathbf{a}_{p-1}$ from this; knowing $\mathbf{a}_{p-1}$, we can find k_{p-1} and hence get $\mathbf{a}_{p-2}$, and so forth. This method will work as long as all of the $\{k_i\}$ have magnitudes less than 1.

E.4 PROPERTIES

1. If the parameters $\{k_i\}$ are constrained to have magnitudes less than unity, then all of the zeroes of the filter will fall within the unit circle. This is of particular interest if lattice filters are constructed from Eq. (E-19) as described in Chapter 5, since this constraint will guarantee that the inverse filter will be stable.

There is no easy proof of this important property; we proceed by induction on the order of the predictor, following Benedir and Picinbono (1987). It is trivially true for $p = 1$. We will show that if it is true for $p = n - 1$, then it must be true for $p = n$.

Let r be a root of $A_n(z) : A_n(r) = 0$. Then by Eq. (E-17),

$$A_{n-1}(r) + k_n B_{n-1}(r) = 0$$

or
$$\frac{A_{n-1}(r)}{B_{n-1}(r)} = -k_n \qquad \text{(E-21)}$$

If $\{r_i\}$ are the roots of $A_{n-1}(z)$, then we may factor $A_{n-1}(z)$ as follows:

$$A_{n-1}(z) = \prod_{i=1}^{n-1}(1 - r_i z^{-1})$$

and by Eq. (E-16) we also have

$$B_{n-1}(z) = z^{-(n-1)} A_{n-1}(1/z)$$

$$= \prod_{i=1}^{n-1}(z^{-1} - r_i)$$

Hence Eq. (E-21) becomes

$$\frac{A_{n-1}(r)}{B_{n-1}(r)} = \prod_{i=1}^{n-1}\frac{(1 - r_i r^{-1})}{(r^{-1} - r_i)} = -k_n$$

Since $|k_n| < 1$, this means

$$\prod_{i=1}^{n-1}\frac{|1 - r_i r^{-1}|}{|r^{-1} - r_i|} < 1 \qquad \text{(E-22)}$$

We now show that our inductive hypothesis $|r_i| < 1$ taken together with Eq. (E-22), guarantees that $|r| < 1$. We proceed indirectly: Suppose $|r| \geq 1$ and consider the expression,

$$(1 - |r|^{-2})(1 - |r_i|^2)$$

If $|r_i| < 1$ and $|r| \geq 1$, then this product will be nonnegative. But since

$$|1 - r^{-2}||1 - r_i^2| = |1 - r^{-1} r_i|^2 - |r^{-1} - r_i|^2$$

it follows that $|1 - r^{-1} r_i|^2 \geq |r^{-1} - r_i|^2$, or equivalently that

$$|1 - r^{-1} r_i| \geq |r^{-1} - r_i|$$

Hence if $|r|$ were ≥ 1, then every factor in the product Eq. (E-22) would be unity or greater, and so the condition of Eq. (E-22) would not be met. Hence $|r|$ must be < 1 and the inductive step is proved.

As a corollary if $|k_i| \leq 1$, then the roots of $A_p(z)$ will not fall outside the unit circle.

2. If the previous constraint is applied and the final parameter, k_p, is set to unity, then all zeros of the filter will lie exactly on the unit circle. This follows from the previous property and the fact that $k_p = a_p(p)$. This is true because the finite, nonzero roots of

$$1 + a_p(1)z^{-1} + \cdots + a_p(p-1)z^{-(p-1)} + a_p(p)z^{-p}$$

are the same as those for

$$z^p + a_p(1)z^{p-1} + \cdots + a_p(p-1)z + a_p(p)$$

It is well known from the theory of equations that for such a polynomial, the constant is equal to the product of the magnitudes of the roots. But if the roots cannot fall outside the unit circle, then the only way the product of their magnitudes can equal unity is for the magnitude of each root to be exactly unity.

3. For any set of parameters $\{k_i\}$, a set of equivalent parameters $\{a_i\}$ can be found using Eq. (E-12). This correspondence is not bijective, however (Benidir and Picinbono, 1987; and Mazel and Hayes, 1988). For example, consider the order-3 predictor; repeated application of Eq. (E-13) gives

$$\mathbf{a}_3 = \begin{bmatrix} 1 \\ k_1 + k_1k_2 + k_2k_3 \\ k_2 + k_1k_3 + k_1k_2k_3 \\ k_3 \end{bmatrix}$$

It is clear that there is no parameter set $\{k_1, k_2, 1\}$ that will map to any set $\{1, a_3(1), a_3(2), 1\}$, $a_3(1) \neq a_3(2)$, for if $k_3 = 1$, then $a_3(1) = k_1 + k_1k_2 + k_3 = a_3(2)$. Hence the mapping is not onto. Further if $k_2 = -1$, and $k_3 = 1$, then

$$\mathbf{a}_3 = [1, -1, -1, 1]^T$$

for any k_1; hence the mapping is not one-to-one.

E.5 THE CHOLESKY DECOMPOSITION

In all the foregoing, the time series being predicted was assumed stationary. In a nonstationary time series, the autocorrelation is not a function of the lag only, but of specific times. In this case, the auto- correlation matrix is not Toeplitz, and the Levinson recursion cannot be used. Although the purpose of this appendix is to develop the Levinson recursion and some of its properties, the non-stationary case is important enough to require some discussion. The nonstationary case has come to be called the covariance method, while the stationary case is known as the auto-correlation method; this terminology persists in spite of the fact that in most cases the data are zero-mean and the matrix $\mathbf{R}$ is in either case a covariance matrix.

Starting with the normal equations

$$\mathbf{Ra} = \mathbf{r}$$

where the autocorrelation matrix $\mathbf{R}$ is no longer Toeplitz, we can find the predictor coefficients by factoring $\mathbf{R}$ into two triangular matrices:

$$\mathbf{R} = \mathbf{LU} \qquad\qquad \text{(E-23)}$$

$\mathbf{L}$ is a lower-triangular matrix, and $\mathbf{U}$ is an upper-triangular matrix with all 1s on its main diagonal. This factorization is variously known as the Cholesky decomposition, the Crout reduction, or simply as LU decomposition.

Once the factorization has been accomplished, solution takes place in two steps: solve

$$\mathbf{Lx} = \mathbf{r} \qquad\qquad \text{(E-24a)}$$

then solve

$$\mathbf{Ua} = \mathbf{x} \qquad\qquad \text{(E-24b)}$$

Since these are both triangular systems, the solution proceeds simply and efficiently.

This factorization is not greatly different from a solution by means of Gaussian elimination. It offers the advantage that the $\mathbf{L}$ and $\mathbf{U}$ matrices can be placed in the same storage locations as the original $\mathbf{R}$ matrix; on the other hand, the pivoting normally associated with Gaussian elimination is slightly more awkward. The principal interest for us lies in the fact that the intermediate vector $\mathbf{x}$ bears some similarity to the reflection coefficients of the Levinson recursion; indeed if the Cholesky decomposition is applied to a Toeplitz matrix, the elements of $\mathbf{x}$ are the reflection coefficients.

In speech coding, when the data are treated as nonstationary and the Cholesky decomposition is used, the elements of $\mathbf{x}$ are frequently called reflection coefficients or occasionally (and more correctly) generalized reflection coefficients. In such cases, the elements of $\mathbf{x}$ are transmitted as if they were reflection coefficients, and the filter coefficients may even be obtained from **x** by means of the Levinson decomposition. We can get away with this because for large N, the difference between the autocorrelation and covariance methods is generally slight.

REFERENCES

M. Benidir and B. Picinbono. "Extensions of the stability criterion for ARMA filters." *IEEE Trans. Acous. Speech and Sig. Proc.*, vol. ASSP-35, no. 4, pp. 425–432, Apr. 1987.

J. P. Burg. "Maximum Entropy Spectral Analysis." Ph. D. Thesis, Stanford University, 1975.

J. Makhoul. "Linear prediction: a tutorial review." *Proc. IEEE*, vol. 63, no. 4, pp. 561–580, Apr. 1975.

J. D. Markel and A. H. Gray, Jr. *Linear Prediction of Speech*. Berlin: Springer-Verlag, 1976.

D. S. Mazel and M. H. Hayes III. "Reflections on Levinson's Recursion." *ICASSP-88*, vol. 3, pp. 1632–1635, Apr. 1988.

A. Papoulis. *Probability, Random Variables, and Stochastic Processes*. New York: McGraw-Hill, 1986.

T. W. Parsons. *Voice and Speech Processing*. New York: McGraw-Hill, 1986.

APPENDIX F

■ ■

BRIEF REVIEW OF NUMBER THEORY

Number theory can be described briefly as the mathematics of the integers and polynomials with respect to such matters as divisibility, factorization, and primeness. This introduction is designed to provide the background necessary for Chap. 4 and for other places within the body of the text where number-theoretic topics arise. For more thorough coverage, the reader should consult the references listed at the end of this appendix. In the following material, we will use "number" to mean integer unless otherwise specified, but occasionally when speaking formally, we will take care to say "integer."

F.1 DIVISIBILITY AND FACTORIZATION

As children we learn that the process of division consists of finding, for any two numbers a and b, two other numbers q and r for which $b = aq + r$. The number q is of course the quotient of b/a and r is the remainder. We quickly learn that this process can be carried out with any two numbers provided the divisor is not zero. In number theory, we define divisibility more narrowly however.

An integer b is said to be *divisible* by a nonzero integer a if

$$b = ac \qquad \qquad \text{(F-1)}$$

for some integer c. That is, b is divisible by a if division yields a remainder of 0. We write $a|b$, which is read "a divides b," with the unspoken qualification "with zero remainder." If a does not divide b, that is, if the remainder is not zero—we write, $a \nmid b$. Clearly, $1|x$ for all integers x. Further, every integer $a|0$ provided $a \neq 0$. If $a|b$, then a is a divisor of b.

Theorem F-1.   The relation $|$ has the following properties:

1. It is transitive, since if $a|b$ and $b|c$, then $a|c$.

2. $a|b$ implies that either $|a| \leq |b|$ or $b = 0$.

3. It distributes over addition, since if $a|b$ and $a|c$, then $a|(b+c)$.

4. As a special case of property 4, if mb is any integer multiple of b and $a|b$, then $a|mb$.

5. If $a|b$, then for any nonzero multiplier m, $ma|mb$.

6. From properties 4 and 5, it follows that if $a|b$ and $a|c$, then $a|(mb+nc)$ for any integers m and n.

A positive number $p > 1$ is said to be *prime* if it has no positive divisors other than 1 and p. Numbers that are not prime are said to be *composite*. By convention, the number 1 is considered to be neither prime nor composite.

Theorem F-2.   Any number $n > 1$ can be written as a product of a finite number of primes. These primes are the *factors* of n; finding them is known as *factoring* or sometimes as *factorization*.

If the number is prime, then its factorization is simply $n = n$. If it is composite, then there are at least two numbers other than 1 and n that divide n. Suppose q is the least of these numbers; then q is one of our prime factors. We can recover the remaining prime factors recursively as follows: find the smallest factor of n/q; suppose this number is r. Then r is likewise prime and by property 5 above, it is a factor of n. Next find the smallest factor of n/qr; suppose this is s. Then s is a prime factor of n. Next find the smallest factor of n/qrs; continue this way until the new number to be factored is prime; this is the last of the factors of n. This algorithm always terminates, because $n > n/q > n/qr > n/qrs > \cdots$ and since these numbers are all integers, there can be only a finite number of them.

For example, 630 has a factor of 2. 630/2 = 315; the smallest factor of 315 is 3. 315/3 = 105; the smallest factor of 105 is 3. 105/3 = 35; the smallest factor of 35 is 5. 35/5 = 7; 7 is prime and we are done. The factors of 630 are thus 2, 3, 3, 5, and 7. It is customary to write the factors in ascending order and to express repeated factors as powers; thus $630 = 2 \cdot 3^2 \cdot 5 \cdot 7$. This is known as the *standard form*; we write

$$n = \prod p_i^{a_i} \tag{F-2}$$

where the product is over all distinct prime factors of n.

If we are given an unknown number to factor, the process must proceed by trial and error, except for certain special cases like multiples of 2, 3, 5, or 9. If repeated trials of possible factors fail, we can give up and announce primeness the first time $n/q < q$. If q is a factor of n, q and n/q cannot both be greater than the square root of n; hence if n be composite, at least one factor of n must be $\leq \sqrt{n}$.

Theorem F-3.   (The Fundamental Theorem of Arithmetic). Every number has a unique factorization in standard form. (Proof omitted.)

The *greatest common divisor* of two nonzero integers m and n is the largest positive integer k that divides both m and n. We write, $k = (m, n)$. If $k|m$, then every factor of k is a factor of m, and similarly for n. Then if we have factored m and n, we can find k by finding the factors common to both m and n and multiplying them. For example, $1170 = 2 \cdot 3^2 \cdot 5 \cdot 13$ and $1584 = 2^4 \cdot 3^2 \cdot 11$; hence $(1170, 1584)$ can be found by selecting 2 and 3^2 and multiplying them: $(1170, 1584) = 18$.

We owe the following simple algorithm for finding (a, b) to Euclid:

1. Set $m = a$ and $n = b$.

2. Divide m by n and find the remainder; that is, find

$$m/n = q + r/n$$

3. If $r = 0$, then $n = (a, b)$. Otherwise, replace m by n and n by r and repeat step 2.

This algorithm will always terminate, since in each step $r < n$. For example, to find $(1170, 1584)$, we write

$$1584/1170 = 1 + 414/1170$$
$$1170/414 = 2 + 342/414$$
$$414/342 = 1 + 72/342$$
$$342/72 = 4 + 54/72$$
$$72/54 = 1 + 18/54$$
$$54/18 = 3 + 0$$

Hence $(1170, 1584) = 18$.

Two numbers a and b are termed *coprime* if $(a, b) = 1$. Notice that 1 is coprime to every number, including 1. The numbers in a set of more than two numbers $\{a, b, c, d, ...\}$ are said to be coprime if every pair of numbers in the set is coprime. (Some authors reserve the term coprime for a pair of numbers and refer to a larger set as a set of pairwise coprime numbers.)

F.2 PRIME NUMBERS

The series of prime numbers begins 2, 3, 5, 7, 11, 13, 17, 19, 23, ...
All prime numbers up through any finite limit N can be found by the
following simple but laborious technique:

1. List the numbers from 2 to N.

2. Let $i = 2$;

3. Cross out all multiples of i except i itself, since by property 4, above,
   multiples of i are composite.

4. Let i be the smallest number not yet crossed out. If $i \leq \sqrt{N}$, then
   return to step 3.

5. The numbers remaining are the primes $\leq N$.

This algorithm is attributed to Eratosthenes and is commonly known
as the sieve, since the primes are the numbers which slip through the
crossing-out process.

   Theorem F-4. (Euclid's Second Theorem). The number of primes
is infinite. The proof is indirect: suppose the number were finite and let
P be the last prime. Construct the number $Q = (2 \cdot 3 \cdot 5 \cdot 7 \cdot \cdots \cdot P) + 1$. That
is, Q is 1 greater than the product of all the primes from 2 through P.
The remainder of Q when divided by any of these primes is 1. Hence Q
does not have a prime factor $\leq P$. But this contradicts our assumption,
for then either Q itself is prime or else Q has a prime factor $> P$. Hence
P is not the last prime and in fact there is no last prime.
   There is no known formula for generating all primes. That is, there
is no $f(n)$ that takes on all prime values, and only prime values, for all
integer values of n. [This statement must be qualified: it is easy to find
a formula for primes if we are allowed to use special constants which
function as disguised tables of primes: see Hardy and Wright (1960),
Dudley (1983), or Schroeder (1984).]
   There is likewise no finite exact formula for $\pi(n)$, the number of
primes less than or equal to n, although there are some good approximations. The simplest of these is due to Euler:

$$\pi(n) \approx n / \ln n \qquad\qquad (\text{F-3})$$

This approximate formula can be used to obtain an estimate of the approximate value of the kth prime:

$$p_k \approx k \ln k \qquad\qquad (\text{F-4})$$

This estimate gives us a general idea of where to look (within a rather wide margin) for p_k. To give an idea of how good these approximations are, we note that there are 78,498 primes less than 1,000,000 (Hardy and Wright, 1960); the $x/\ln x$ approximation gives 72,382. The 664,999th prime is 10,006,721; the $x \ln x$ approximation is 8,916,001. It is characteristic of these estimates that they tend to be a few percent too low.

The prime-counting function $\pi(n)$ must not be confused with Euler's *totient function* $\phi(n)$ which gives the number of integers less than n which are *coprime* with n. If p is prime, then for any integer power of p,

$$\phi(p^n) = p^{n-1}(p-1) \tag{F-5}$$

This can be seen as follows: there are $p^n - 1$ numbers less than p^n. Of these, $p^{n-1} - 1$ are multiples of p and hence not coprime. Hence $\phi(p^n) = p^n - p^{n-1}$ and this gives us Eq. (F-5).

For example, for $p^n = 9$, $\phi(9) = 6$; the numbers which are coprime with 9 are 1, 2, 4, 5, 7, 8. We state without proof that if a and b are coprime, then $\phi(ab) = \phi(a)\phi(b)$. It follows from this that $\phi(n)$ can be computed for any n; this is because if n has the factorization

$$n = \prod_i p_i^a$$

then $\phi(n)$ is given by

$$\phi(n) = n \prod_i (p_i - 1)/p_i \tag{F-6}$$

For example, $225 = 5^2 \cdot 3^2$ and $\phi(225) = (225 \cdot 4 \cdot 2)/(5 \cdot 3) = 120$. Note that for n prime, $\phi(n)$ reduces to $n - 1$, as we would expect.

The search for a prime-number formula has produced some unsuccessful tries of considerable theoretical interest. Fermat conjectured that numbers of the form

$$F_n = 2^{2^n} - 1 \tag{F-7}$$

were prime. For $n = 0, 1, 2, 3$, and 4, these numbers are indeed prime, but F_5 is composite, and no other Fermat primes are currently known.

Mersenne speculated that numbers of the form

$$M_p = 2^p - 1 \tag{F-8}$$

where p is prime, were themselves prime numbers. We know many more Mersenne primes than Fermat primes, but this rule also fails: M_{11} is not prime.

F.3 CONGRUENCES AND RESIDUES

We say that an integer a is congruent to b modulo some positive integer m if $m|(a - b)$; this is written

$$a \equiv b \,(\text{mod } m) \tag{F-9}$$

In this case, b is a *residue* of a modulo m. If $0 \le b < m$, then b is the *least positive residue* of a modulo m. The qualifier modulo m is usually abbreviated as (mod m).

Notes on the use of the words "modulo" and "mod": The number m is the *modulus*; a few writers say "to the modulus m." Most writers find it less verbose to use the Latin dative; hence the term "modulo." Careless use of the words "modulo" and "mod" gives rise to a certain amount of confusion, which we will try to sort out here. First, we have the use of modulo and mod given in the definition above. Second, many programming languages provide a function or operator named mod which returns the remainder of a/m. If a is positive, then a mod m returns the least positive residue of a modulo $|m|$; but if a is negative, then a mod m returns the residue b such that $-m < b \le 0$.

Third, in some signal-processing applications, one speaks of "doing arithmetic modulo m." This usually means replacing the result of any arithmetic operation by its least positive residue modulo m. For example, in a cyclical convolution of length 17, all subscripts will be computed modulo 17. Hence if we add a subscript 11 and a lag of 13, the sum will be 24, which is replaced by 7. We use expressions such as "$(a + b)$ mod 17," which appears at first glance to be the same as the mod function of programming languages, but is not, because if $(a + b)$ is negative, for example -20, we usually wish to have its least positive residue, 14, rather than -3, which is what the computer mod function would give us.

The general practice seems to be to use "a mod b" to mean the least residue of a modulo b and to use the computer mod function with great care. We distinguish between congruences and modular arithmetic by writing $a \equiv b$ (mod m) for the congruences and $a = b$ mod m when a is the least positive residue of b (mod m).

A *residue class* (mod m) is the set of all integers congruent to a given residue (mod m). There are clearly m residue classes, and as we run through the integers we repeatedly cycle through the m residue classes. A *complete residue system* modulo m is a set of m numbers containing one number from each of the residue classes. The simplest example of such a system, and the one we usually have in mind, is the set $\{0, 1, ..., m-1\}$; this is known as the *least complete residue system* (mod m).

We use residue classes and modular arithmetic frequently in everyday life. The set of all Mondays is a residue class (mod 7), and the days of the week are a complete residue system (mod 7). Similarly, the set of all Januaries is a residue class (mod 12). Residues can be computed using noninteger values as well; in the circle, angles can frequently be reckoned modulo 2π. The frequency domain in sampled-data systems is a complete residue system modulo the sampling rate f_s, and the phenomenon of aliasing may be more understandable if it is appreciated that $f_s + p$ belongs to the same residue class (modulo f_s) as p.

Theorem F-5. Congruences have the following properties. (All the congruences in this list are modulo m.)

1. They are symmetric: if $a \equiv b$, then $b \equiv a$.

2. They are reflexive: $a \equiv a$.

3. They are transitive: if $a \equiv b$ and $b \equiv c$, then $a \equiv c$.

4. They are additive: if $a \equiv b$ and $c \equiv d$, then

$$(a + c) \equiv (b + d)$$

5. As a special case of property 4, if $a \equiv b$, then for any multiplier k, $ka \equiv kb$.

6. From properties 4 and 5, if $a_1 \equiv b_1$, $a_2 \equiv b_2$, ..., then for any multipliers $k_1, k_2, ...,$

$$k_1 a_1 + k_2 a_2 + ... \equiv k_1 b_1 + k_2 b_2 + \cdots$$

7. If $a \equiv b$, then so are all integer powers of a and b: $a^k \equiv b^k$.

Since congruence is symmetric, reflexive, and transitive, it is an equivalence relation; it thus partitions the integers into equivalence classes, which are the residue classes mentioned above.

It is not necessarily true that if $ka \equiv kb \,(\text{mod } m)$, then $a \equiv b \,(\text{mod } m)$. The equivalent congruence depends on the value of (k, m). Suppose $(k, m) = q$. Let us remove the factor q from k and m: $k' = k/q$ and $m' = m/q$. The first congruence tells us that $m|(ka - kb)$, that is, that $k(a - b) = mz$. Cancelling the factor q from both sides gives us $k'(a - b) = m'z$, or, equivalently, $a \equiv b \,(\text{mod } m')$. Hence we have proved

Theorem F-6. If $ka \equiv kb \,(\text{mod } m)$ and $(k, m) = q$, then $a \equiv b(\text{mod } m/q)$. In particular, $ka \equiv kb \,(\text{mod } m) => a \equiv b \,(\text{mod } m)$ only if m and k are coprime.

This theorem gives us a further result: if (a) runs through a complete residue system modulo m, then so does $(ka + b)$ provided $(k, m) = 1$. For example, suppose $m = 7$ and let (a) be (Sunday, Monday, Tuesday, Wednesday, Thursday, Friday, Saturday). Since $(2, 7) = 1$, if we take every other weekday, all the days still appear: (Monday, Wednesday, Friday, Sunday, Tuesday, Thursday, Saturday). Further experimentation shows that if $b = 3$, we still get all days of the week. Hence if we order the residue classes, as we have done here, multiplying by k and adding b merely changes their sequence. Finally, if $\{a\}$ runs through a complete system of residues coprime to m, then so does $\{ka\}$ (but not necessarily $\{ka + b\}$).

A *linear congruence* has the form

$$ax \equiv b \ (\text{mod} \ m) \qquad\qquad (\text{F-10})$$

Whether such a congruence can be solved for x depends on a, m, and b. For example, $2x \equiv 3(\text{mod} \ 6)$ has no solution. To see when and how such a congruence may be solved, suppose first that $(a, m) = 1$. Then as x runs through the complete residue system $0, 1, ..., m - 1$, ax will likewise run through the same residue system, as we have just seen, although not necessarily in the same order. Hence it is only a matter of time before we hit a value x_0 for which ax_0 belongs to the same residue class as b, and we will have a solution. Once we have found this solution, we can stop, since the only other values of x which will satify the congruence will belong to the same residue class as x_0.

Next, suppose $(a, m) = q$. If $q|b$, then using Theorem F-6, we can write $x \equiv b/q(\text{mod} \ m/q)$. Then as x runs through the system $0, 1, ..., m-1$ once, it will also run through the system $0, 1, ..., m' - 1$ a total of q times. On each pass through this shorter residue system, it will hit a value belonging to the same residue class as b/q and we will have a solution; hence in this case we have a total of $q = (a, m)$ solutions. Suppose, however, that $q \nmid b$: then b/q is not an integer and there is no integer x that will satisfy the congruence. This proves

Theorem F-7. If $(a, m)|b$, then the congruence $ax \equiv b(\text{mod} \ m)$ has exactly (a, m) distinct (that is, non-congruent) solutions.

Theorem F-8. (Euler's Theorem). For a and b coprime,

$$a^{\phi(b)} \equiv 1 \ (\text{mod} \ b) \qquad\qquad (\text{F-11})$$

Consider a complete system of residues modulo b. Of this system, there must be $\phi(b)$ residue classes coprime to b. If these coprime residues are $\{r_i\}$, then $\{ar_i\}$ runs through the same set of classes, since $(a, b) = 1$. Then the product

$$\prod_i ar_i \equiv \prod_i r_i \ (\text{mod} \ b)$$

We can factor $a^{\phi(b)}$ out of the left-hand product:

$$a^{\phi(b)} \prod_i r_i \equiv \prod_i r_i \ (\text{mod} \ b)$$

The two products cancel and give us Eq. (F-11). For example, we have $\phi(9) = 6$, $(4, 9) = 1$, and

$$4^6 = 4,096 \equiv 1 \ (\text{mod } 9),$$

and indeed $4,096 = 9 \cdot 455 + 1$. A special case of Euler's theorem is

Theorem F-9.   (Fermat's theorem).

$$a^{p-1} \equiv 1 \ (\text{mod } p) \tag{F-12}$$

where p is a prime. We can use the same example:

$$4^6 = 4,096 \equiv 1 \ (\text{mod } 7)$$

and indeed $4,096 = 7 \cdot 285 + 1$.

Powers of n modulo a prime p show an interesting behavior. Let $a^k \equiv n \ (\text{mod } p)$. For $a = 2$ and $p = 7$, we have

$$k : 0 \quad 1 \quad 2 \quad 3 \quad 4 \quad 5 \quad 6 \quad 7 \quad 8 \quad 9 \quad \cdots$$
$$n : 1 \quad 2 \quad 4 \quad 1 \quad 2 \quad 4 \quad 1 \quad 2 \quad 4 \quad 1 \quad \cdots$$

For $a = 3$ and $p = 8$, we have

$$k : 0 \quad 1 \quad 2 \quad 3 \quad 4 \quad 5 \quad 6 \quad 7 \quad 8 \quad 9 \quad \cdots$$
$$n : 1 \quad 3 \quad 2 \quad 6 \quad 4 \quad 5 \quad 1 \quad 3 \quad 2 \quad 6 \quad \cdots$$

For $a = 4$ and $p = 7$, we have

$$k : 0 \quad 1 \quad 2 \quad 3 \quad 4 \quad 5 \quad 6 \quad 7 \quad 8 \quad 9 \quad \cdots$$
$$n : 1 \quad 4 \quad 2 \quad 1 \quad 4 \quad 2 \quad 1 \quad 4 \quad 2 \quad 1 \quad \cdots$$

In every case, the sixth power of a is congruent to 1, as we would expect from Fermat's theorem, Eq. (F-12). Furthermore, higher powers repeat the cycle, since

$$a^{6+k} = a^k a^6 \equiv a^k \ (\text{mod } 7)$$

as we would expect from Theorem F, property 5. Hence in every case the cycle repeats after six steps. In some cases, however, the cycle length is shorter. The cycle length is called the *order* of a modulo p, written $\text{ord}_p a$. Since the sequence must repeat every $p - 1$ steps, it is clear that

Theorem F-10.

$$\text{ord}_p a | p - 1 \tag{F-13}$$

Since $2^3 \equiv 1 \pmod 7$, we could consider 2 as the cube root of 1 modulo 7. Indeed, we say that a number a is the *primitive kth root of unity* (mod p) if k is the smallest positive exponent for which $a^k \equiv 1 \pmod p$. When $ord_p a = p - 1$, however, we can make a stronger statement, and we say then that a is a *primitive root* of p. In this case, the powers of a from 0 through $p - 1$ are a complete set of residues (mod p), excluding 0.

Theorem F-11.   (The Chinese Remainder Theorem). Let N have m coprime factors:

$$N = N_1 N_2 \cdots N_m$$

Then if $0 \le k < N$, the congruences

$$k_j \equiv k \pmod{N_j} \qquad j = 1, 2, ..., m \tag{F-14}$$

can be solved for k as follows: Let $Q_i = N/N_i$; then Q_i is prime to N_i and there exists exactly one root a_i to

$$Q_i a_i \equiv 1 \pmod{N_i} \tag{F-15}$$

Then
$$k = \sum_{i=1}^{m} Q_i a_i k_i \tag{F-16}$$

Proof: We must show that Eq. (F-16) satisfies Eq. (F-14). For conciseness, let $p \bmod q$ denote the least positive residue of p (mod q). Since congruences are additive (Theorem F-5, property 4), we may write

$$k \bmod N_j = [\sum_{i=1}^{m} Q_i a_i k_i] \bmod N_j$$

$$= \sum_{i=1}^{m} [Q_i a_i k_i \bmod N_j]$$

Next note that if $i \ne j$, then $Q_i a_i \bmod N_j = 0$, since N_j is a factor of Q_i. Hence all terms for which $i \ne j$ drop out; the remaining term is simply k_j, since $Q_i a_i \bmod N_j = 1$.

EXAMPLE : Let $N = 20$, $N_1 = 4$, and $N_2 = 5$. Then $Q_1 = 5$ and $a_1 = 1$, since $Q_1 a_1 = 5$, $5 \equiv 1 \pmod 4$, and $5 \equiv 0 \pmod 5$; and $Q_2 = 4$ and $a_2 = 4$, since $Q_2 a_2 = 16$, $16 \equiv 0 \pmod 4$, and $16 \equiv 1 \pmod 5$. Hence

$$k = 5k_1 + 16k_2$$

Checks:

1. $k_1 = 3$ and $k_2 = 2$. $k = 15 + 32 = 47 \equiv 7 \pmod{20}$. $7 \equiv 3 \pmod 4$ and $7 \equiv 2 \pmod 5$.

2. $k_1 = 2$ and $k_2 = 3$. $k = 10 + 48 = 58 \equiv 18 \pmod{20}$. $18 \equiv 2 \pmod 4$ and $18 \equiv 3 \pmod 5$.

F.4 POLYNOMIALS

There is a closer relationship between integers and polynomials than might be thought upon casual consideration. Our chief interest in polynomials is centered about division, factorization, congruences, and the polynomial version of the Chinese remainder theorem.

For our purposes, a polynomial in x is an expression of the form

$$p(x) = a_n x^n + a_{n-1} x^{n-1} + \cdots + a_1 x + a_0$$

Polynomials are added and multiplied in the conventional manner; two polynomials are equal if and only if the coefficients of like powers of the argument are equal. In digital signal-processing applications, we are most likely to be interested in polynomials with integer coefficients, but we will not impose that requirement here; life is simpler if we require that the coefficients (and x) be elements of a *field*, for example, the real or complex numbers.

The *degree* of a polynomial is the number n. A *monic* polynomial is one whose leading coefficient a_n is 1.

In arithmetic when we divide two integers a and b, the process yields two integers q and r such that

$$b = aq + r$$

where $0 \le r < a$. Similarly, in algebra, when we divide two polynomials $a(x)$ and $b(x)$, the process gives us two polynomials $q(x)$ and $r(x)$ such that

$$b(x) = a(x)q(x) + r(x)$$

where the degree of $r(x)$ is less than the degree of $a(x)$.

We can define divisibility in a manner parallel to the case of integers. A polynomial $b(x)$ is divisible by a polynomial $a(x)$ if

$$b(x) = a(x)c(x)$$

for some polynomial $c(x)$. Equivalently we can say that $b(x)$ is divisible by $a(x)$ if the division yields a remainder $r(x) = 0$ and write $a(x)|b(x)$. If $a(x)$ does not divide $b(x)$, we write, $a(x) \nmid b(x)$. A polynomial is *irreducible* if it is divisible only by a constant or by a polynomial of degree n. This is roughly equivalent to primeness, but we reserve the term *prime polynomial* for an irreducible monic polynomial. It should be noted that whether a polynomial is reducible depends on the field. For example, $x^2 + 1$ is reducible over the complex field: $x^2 + 1 = (x + j)(x - j)$, but not over the reals, since there is no real number whose square is -1.

Just as integers can be factored into a product of primes, so polynomials can be factored into a product of a constant and one or more prime polynomials. For example, $3x^3 + 6x^2 - 3x - 6$ has the factors 3, $(x + 1)$, $(x - 1)$, and $(x + 2)$. If we are in the field of complex numbers, $3x^3 + 6x^2 + 3x + 6$ has the factors 3, $(x + j)$, $(x - j)$, and $(x + 2)$, but in the field of reals, the factors are 3, $(x^2 + 1)$, and $(x + 2)$ since here $x^2 + 1$ is irreducible. If the field permits factorization into polynomials of degree 1, then by the fundamental theorem of algebra, there will be n such factors.

The greatest common divisor of two polynomials $p(x)$ and $q(x)$ is the highest-degree polynomial $d(x)$ for which $d(x)|p(x)$ and $d(x)|q(x)$. If we allowed the leading coefficient of $d(x)$ to take on any value, then $d(x)$ would not be unique; hence we require that $d(x)$ be monic. Two polynomials whose greatest common divisor is 1 are coprime, and as in the case of integers, we can extend this definition to sets of more than two polynomials.

Two polynomials $a(x)$ and $b(x)$ are congruent modulo a third polynomial $m(x)$ if $m(x)|a(x) - b(x)$, and the notation is similar to the case of integers:

$$b(x) \equiv a(x) \ [\text{mod} \ m(x)]$$

Theorem F-12. (The Chinese Remainder Theorem). Let $P(z)$ have m irreducible polynomial factors:

$$P(z) = P_1(z)P_2(z) \cdots P_m(z)$$

Then the congruences

$$Y_j(z) \equiv Y(z) \ [\text{mod} \ P_j(z)] \qquad j = 1, 2, ..., m \qquad \text{(F-24)}$$

can be solved for $Y(z)$ as follows: Let $Q_i(z) = P(z)/P_i(z)$; then there exists exactly one $A_i(z)$ for which

$$Q_i(z)A_i(z) \equiv 1 \ [\text{mod} \ P_i(z)] \qquad \text{(F-25)}$$

Then
$$Y(z) \equiv \sum_{i=1}^{m} Q_i(z)A_i(z)Y_i(z) \ [\text{mod} \ P(z)] \qquad \text{(F-26)}$$

Proof: The proof parallels the proof of the scalar form exactly. We must show that Eq. (F-26) satisfies Eq. (F-24). As in that proof, let p mod q denote the least positive residue of p (mod q). Since congruences are additive, we may write

$$Y(z) \bmod P_j(z) = [\sum_{i=1}^{m} Q_i(z)A_i(z)Y_i(z)] \bmod P_j(z)$$

$$= \sum_{i=1}^{m} [Q_i(z)A_i Y_i(z) \bmod P_j(z)]$$

Next note that if $i \neq j$, then $Q_i(z)A_i(z) \bmod P_j(z) = 0$, since $P_j(z)$ is a factor of $Q_i(z)$. Hence all terms for which $i \neq j$ drop out; the remaining term is simply $Y_i(z)$ since $Q_i(z)A_i(z) \bmod P_j(z) = 1$.

REFERENCES

G. H. Hardy and E. M. Wright. *An Introduction to the Theory of Numbers*. Oxford: Oxford University Press, 1960.

L. K. Hua. *Introduction to Number Theory*. Berlin: Springer-Verlag, 1982.

I. Niven and H. S. Zuckerman. *An Introduction to the Theory of Numbers*. New York: John Wiley, 1980.

M. R. Schroeder. *Number Theory in Science and Communication*. Berlin: Springer-Verlag, 1984.

INDEX

DATE DUE FOR RETURN

Astronomy and Astrophysics Series Volume 9

The Science of Space-Time

Pachart

Astronomy and Astrophysics Series

General Editor: A.G. Pacholczyk

ii

Pachart Publishing House
1130 San Lucas Circle, Tucson, Arizona 85704

iii